Advances in Science, Technology & Innovation

IEREK Interdisciplinary Series for Sustainable Development

Advances in Science, Technology & Innovation (ASTI) is a series of peer-reviewed books based on important emerging research that redefines the current disciplinary boundaries in science, technology and innovation (STI) in order to develop integrated concepts for sustainable development. It not only discusses the progress made towards securing more resources, allocating smarter solutions, and rebalancing the relationship between nature and people, but also provides in-depth insights from comprehensive research that addresses the **17 sustainable development goals (SDGs)** as set out by the UN for 2030.

The series draws on the best research papers from various IEREK and other international conferences to promote the creation and development of viable solutions for a **sustainable future and a positive societal** transformation with the help of integrated and innovative science-based approaches. Including interdisciplinary contributions, it presents innovative approaches and highlights how they can best support both economic and sustainable development, through better use of data, more effective institutions, and global, local and individual action, for the welfare of all societies.

The series particularly features conceptual and empirical contributions from various interrelated fields of science, technology and innovation, with an emphasis on digital transformation, that focus on providing practical solutions to **ensure food, water and energy security to achieve the SDGs.** It also presents new case studies offering concrete examples of how to resolve sustainable urbanization and environmental issues in different regions of the world.

The series is intended for professionals in research and teaching, consultancies and industry, and government and international organizations. Published in collaboration with IEREK, the Springer ASTI series will acquaint readers with essential new studies in STI for sustainable development.

ASTI series has now been accepted for Scopus (September 2020). All content published in this series will start appearing on the Scopus site in early 2021.

Baojie He · Cristina Piselli · Hirushie Karunathilake · Ali Cheshmehzangi · Shady Attia · Amos Darko
Editors

Towards the Framework of Livable and Resilient Cities

Springer

Editors
Baojie He
School of Architecture and Urban Planning
Chongqing University
Chongqing, China

Hirushie Karunathilake
Division of Physical and Computational Sciences
University of Pittsburgh
Pittsburgh, PA, USA

Shady Attia
Head of the Sustainable Building Design Lab
Liège
Sustainable Architecture and Building
Technology
Belgique, Belgium

Cristina Piselli
Department of Architecture (DIDA)
University of Florence
Florence, Italy

Ali Cheshmehzangi
School of Architecture, Design and Planning
University of Queensland
Brisbane, QLD, Australia

Amos Darko
Department of Construction Management
University of Washington
Seattle, WA, USA

ISSN 2522-8714 ISSN 2522-8722 (electronic)
Advances in Science, Technology & Innovation
ISBN 978-3-031-97848-7 ISBN 978-3-031-97849-4 (eBook)
https://doi.org/10.1007/978-3-031-97849-4

This Springer imprint is published by the registered company Springer Nature Switzerland AG
The registered company address is: Gewerbestrasse 11, 6330 Cham, Switzerland

Preface

The goal of climate action was underscored and raised in the Sustainable Development Goals (Goal 13) to call for urgent actions to constrain and adapt to climate change. After a decade, the action capacity has been significantly improved across different nations, regions, and small island countries. However, the milestones of climate action are far slower than the speed of climate change. Despite an average temperature increase of 1.1–1.2 °C above pre-industrial levels globally, many areas have already experienced a temperature increase of 1.5 °C. Not only the extensive environmental, economic, social impacts, and irreversible situation, but also many areas have already been locked in climate change. Under extreme heat, flooding, bushfires, and droughts, such areas are no longer suitable for human survival, especially among indigenous groups.

Cities are the main human settlements. The science of climate change shows that cities are drivers of climate change and environmental deterioration, while a wide range of facts demonstrate that cities and their citizens are the victims of changing climate and deteriorating environments. Therefore, cities must play a leading role in combating climate change. The COP 29 held in Baku, Azerbaijan, in November 2024 kept stimulating the actions for positive climate mitigation and adaptation by clarifying the mechanisms, tools, and signals of decarbonization and resilience and delivering the announcement that cities must secure the future of urban climate action. To enable tangible and actionable practices of urban planning, design, construction, and management, not only mitigation and adaptation strategies and technologies are required, but also governmental and financial institutions must be activated and mobilized. Different stakeholders should be involved in climate talks to exchange ideas, share knowledge, explore solutions, and mobilize resources.

We have been engaged in promoting positive climate mitigation and adaptation by leading the conferences of Climate Change and Environmental Sustainability (CCES) and Health & Environmental Resilience and Livability in Cities (HERL). The CCES conference contributed to the understanding of the impact of climate change on both urban and rural environments and the role of renewable energy in mitigating the effects of climate change. The HERL conference contributed to innovations in climate change mitigation and adaptation strategies, including environmental analysis and sensing methodologies, incorporating smart systems & the Internet of things (IoT), as well as big data analytics, further developing environmental stability and human liveability. The combination of CCES and HERL offered an important opportunity for scientists, researchers, policymakers, practitioners, and others to conduct interdisciplinary discussions and collaborations on mitigating climate change, enhancing urban climate resilience, and promoting urban sustainable development.

These conferences are a critical platform enabling decision-makers and relevant stakeholders to gain environmental, governmental, financial, social, and technological frontiers upon which mankind has to put in place to combat the harmful effects of climate change on the Earth. The scope is the undertaking and shared commitment to advancing knowledge and solutions in the realm of urban health and environmental resilience. We are also proud to persist in holding the conferences to promote national and international advocacy and actions, such as the United Nations Sustainable Development Goals, the Paris Agreement, and Carbon Neutrality. It is important to magnify discussions from diverse perspectives and take steps toward more resilient, liveable, and sustainable urban environments worldwide.

Baojie He
Cristina Piselli

Acknowledgement

The Book **Towards the Framework of Livable and Resilient Cities** includes the proceedings from the conferences Climate Change and Environmental Sustainability (CCES, 4th Edition) and Health & Environmental Resilience and Livability in Cities (HERL, 3rd Edition).

The chairs and co-chairs of the conferences and the editors dedicate this book to all conference organizers and participants from the industry, governmental organizations, and academia who have been actively engaged with conference preparation, keynote speech moderation, and presentations.

The conference chair of CCES-4 and editor of this book, Baojie He, expresses thanks to the keynote speakers and moderators:

Prof. Dr. Muhammad Azam Khan
Abdul Wali Khan University Mardan, Pakistan
Dr. Joni Jupesta
Bogor Agricultural University, Indonesia
Dr. Phanos Hadjikyriakou
2050 Materials, Italy
Associate Professor Beta Paramita
Universitas Pendidikan Indonesia, Indonesia

Baojie He also expresses the support from the Ministry of Education of China and the National Natural Science Foundation of China (Grant No. W2523040, and W2422003).

The conference chair of HERL-3 and editor of this book, Cristina Piselli, expresses thanks to the other chair of the conference, Prof. Anna Laura Pisello, and the two co-chairs, Dr. Ilaria Pigliautile and Dr. Eliana Martinelli from the University of Perugia, Italy. Moreover, she acknowledges the keynote speakers and moderators:

Dr. Antonella Valentini,
Department of Architecture (DIDA), University of Florence, Italy
Dr. Jorge S. Dolado,
Head of the Ceramic and Cement-Based Materials (C2M) group, Centre for Materials Physics (CFM, CSIC-UPV/EHU)
Dr. Bertug Ozarisoy,
School of the Built Environment and Architecture, London South Bank University, UK
Dr. Claudia Fabiani,
Department of Engineering, University of Perugia, Italy

Organization

Scientific Committee

Elhadi Matallah	Biskra University, Algeria
Mohammad Taleghani	Leeds Beckett University, UK
Reza Roshan	Golestan University, Iran
Simone Forastiere	Department of Architecture (DIDA), University of Florence, Italy
Roberta Jacoby Cureau	Department of Engineering, University of Perugia, Italy
Christos Grapas	Department of Engineering, University of Perugia, Italy
Giulio Hasanaj	Department of Architecture (DIDA), University of Florence, Italy
Elena Tarpani	Department of Engineering, University of Perugia, Italy
Ilaria Pigliautile	Department of Engineering, University of Perugia, Italy
Carolina Santini	Department of Architecture (DIDA), University of Florence, Italy
Mattia Manni	Norwegian University of Science and Technology-NTNU, Trondheim, Norway
Francesco Marchini	Department of Engineering, University of Perugia, Italy
Ioannis Kousis	University of New South Wales-UNSW, Sydney, Australia
Andrea Silei	Department of Architecture (DIDA), University of Florence, Italy
Silvia Cavagnoli	Department of Engineering, University of Perugia, Italy
Veronica Martins Gnecco	Department of Engineering, University of Perugia, Italy
Mateus Bavaresco	Federal University of Santa Catarina-UFSC, Florianópolis, Brazil
Hirushie Karunathilake	University of Pittsburgh, USA
Sanjeeva Witharana	University of Moratuwa, Sri Lanka
Yang Liu	Wuhan University, China
Xiao Liu	South China University of Technology, China
Yu Luo	Kitakyushu University, Japan
Chunping Miao	Chang'an University, China
Hanxiang Zhan	University of Malaya, Malaysia
Chenxi Li	Xi'an University of Architecture and Technology, China
Guojin Qin	Southwest Petroleum University, China
Lingye Yao	Chinese University of Hong Kong, Hong Kong, China
Idris Temitope Bello	The Hong Kong Polytecnic University, Hong Kong, China
Ayyoob Sharifi	Hiroshima University, Japan
Li Xie	Industry Rep, China
Zaheer Allam	Curtin University, Australia
Hicham Arrar	University of Liege, Belgium
Mohamed Elhadi Matallah	Biskra University, Algeria
Mark K. Ansah	University of West England, UK
Wenque Liu	Hong Kong Polytechnic University, Hong Kong, China
Kevin M. Muiruri	University of Washington, USA
Caleb Debrah	Hong Kong Polytechnic University, Hong Kong, China
Al-Obaidy Muheeb	University of Liege, Belgium

The editors warmly thank all the reviewers who have contributed their authority to the double-blind review process to ensure the quality of this publication.

Contents

Renewable Energy and Sustainable Building Technologies

Climate Adaptation Strategies and Environmental Quality

Introduction

Charting Livable and Resilient Cities

Baojie He, Cristina Piselli, Hirushie Pramuditha Karunathilake, Ali Cheshmehzangi, Shady Attia, and Amos Darko

Abstract

Cities are facing severe challenges of climate change and environmental deterioration. This leads to diverse reflection and action on urban values, where sustainable development goals are the consensus of future development direction. Positive climate mitigation and adaptation are needed, while creating healthy, resilient, and livable built environments are vital to ensure citizen's health, safety, and prosperity. However, understandings of theories, methods, frameworks, and strategies for creating climate-resilient and livable environments remain unclear, particularly due to diverse geographic, cultural, social, climatic, economic, technological, and political contexts. The 4th Climate Change and Environmental Sustainability (CCES) and the 3rd Health & Environmental Resilience and Livability in Cities (HERL) international conferences were held in China and Italy, respectively, to address the research, science, and practice challenges to deepening the pathways, actions, plans, and policies. This book collects a total of 32 papers from these two conferences accepted for publication. These papers characterize four key themes, including (i) Urban Design, Public Spaces, and Resilience, (ii) Renewable Energy and Sustainable Building Technologies, (iii) Climate Adaptation Strategies and Environmental Quality, and (iv) Behavioral and Policy Approaches to Climate Change Mitigation. Typically, these four themes form a closed loop to support the creation of livable and resilient cities, by defining urban contexts, mitigation solutions, adaptation solutions, and socio-economic-political participation. Moreover, these papers present the experiences, lessons learnt, findings, and practices from various nations worldwide, such as Australia, Bosnia and Herzegovina, China, Denmark, Dominican Republic, Egypt, India, Italy, Mexico, Norway, Saudi Arabia, Slovenia, Spain, Thailand, the Netherlands, Togo, United Arab Emirates, and the United States. The contents in this book will provide comprehensive, tangible, and implementable frameworks, solutions, pathways, and paradigms to improve urban livability and resilience.

Keywords

Climate resilience · Urban livability · Climate change mitigation and adaptation · Behavioral change · Social approach

1 Climate Change, Urbanization and Solutions

The world is now facing several challenges. First, climate change is an indisputable fact with increasingly frequent, intense, and severe climate-related disasters. The redline of 1.5 °C increase in average surface temperature above the

B. He (✉)
Centre for Climate–Resilient and Low–Carbon Cities, School of Architecture and Urban Planning, Key Laboratory of New Technology for Construction of Cities in Mountain Area, Ministry of Education, Chongqing University, Chongqing 400045, China
e-mail: baojie.unsw@gmail.com

B. He · A. Cheshmehzangi
School of Architecture, Design and Planning, The University of Queensland, Brisbane 4072, Australia

C. Piselli
Department of Architecture (DIDA), University of Florence, 50121 Florence, Italy

H. P. Karunathilake
Division of Physical and Computational Sciences, University of Pittsburgh, Bradford Campus, Bradford, PA 16701, USA

S. Attia
Sustainable Building Design Lab, Department of UEE, Faculty of Applied Sciences, University of Liege, 4000 Liege, Belgium

A. Darko
Department of Construction Management, University of Washington, Seattle, WA, USA

B. He et al. (eds.), *Towards the Framework of Livable and Resilient Cities* Advances in Science, Technology & Innovation, https://doi.org/10.1007/978-3-031-97849-4_1

pre-industrial era has been recorded in many regions of the world (WMO, 2025). This is evidenced by a variety of disasters such as strong precipitation, drought, extreme heat events, and hurricanes. These disasters have caused significantly environmental, social, economic, and health consequences. Second, urbanization is a great trend of human history. The migration of rural residents towards cities or the on-site transformation of their citizenship has really changed the living environments they settle. Despite elevated living quality in various aspects, many city-specific problems and challenges, broadly in environmental, social, cultural, governmental, and economic aspects emerge (Masson et al., 2020; Ma et al., 2025). For instance, many cities are under pressure of land, food, production, energy, water, income, equality, inclusivity, health, safety, and so on. Overall, after long-term research efforts, global nations, organizations, and stakeholders have made several consensuses: (i) Global climate is changing due to greenhouse gas emissions, (ii) Climate change is approaching to redline, exceeding which climate-related impacts will be devastating and irreversible, (iii) Climate mitigation and adaptation should be implemented in parallel to not only decelerate climate change but also to reduce climate-related disasters and protect people and cities from impacts, and (iv) Actions for climate change mitigation and adaptation such as Sustainable Development Goals, Paris Agreement, and Net Zero Action should be implemented across different scales. To move forward, there is a need to open a new chapter to deepen understandings of climate change mitigation and adaptation, as well as urban livability and resilience. A comprehensive framework for urban resilience and livability is highly needed (Sharifi, 2023).

Cities are complex and dynamic. As a result, relevant understandings can be diverse due to the difference in geographic, cultural, social, climatic, economic, technological, and political contexts. Moreover, it is not easy to capture actual experience and lessons through only assumed urban spaces and models. There is a need to conduct piloting projects and experimentations in actual contexts of cities. For instance, extreme heat events can cause significant drought events, thereby leading to wildfire and blackout events in many nations (Brown et al., 2023). The wildfire events have been widely reported in Australia, Europe, and the United States, causing the death of animals and vegetation. Local populations can be affected, mainly in terms of land destruction, property damage, and air pollution (Xu et al., 2024). Many tropical cities are frequently attacked by hurricanes, and small island nations are threatened by the rising sea level. These examples demonstrate the differences in climate-related disasters, which clearly shape the climate mitigation and adaptation priorities of these cities. City-specific urban climate change presents diverse characteristics as well. For instance, heat island effects, or urban thermal environments in a broader scope, are characterized by temperature, humidity, solar radiation, and wind speed (Orkomi and Ameri, 2024). Therefore, the need to regulate these microclimatic variables varies. Mitigation and adaptation performance of various strategies can also define city context (Qi et al., 2023), because the performance is dependent on climatic background, urban morphology, local culture, population demographic structure, and citizens' preference (He and Chen, 2025). Overall, understandings and actions towards climate resilience and urban livability should be built upon specific urban contexts.

Decarbonization is the mitigative solution to climate change. Many studies have clearly defined the decarbonization pathways in terms of electrification, energy efficiency, carbon capture and storage, green fuels, recycling (Sinha et al., 2024). Such pathways are interlinked with many sectors such as building and construction, traffic, agriculture, forestry, manufacturing, and so on, and the whole life cycle carbon solutions have been implemented well. In cities, extensive studies have framed how to achieve a net zero carbon built environment, based on which the net zero guides have been developed by the Global Alliance for Buildings and Construction (GABC), World Green Building Council (WGBC), Low Energy Transformation Initiative (LETI), Royal Institute of British Architects (RIBA), American Institute of Architects (AIA), and International Energy Agency (IEA). The framework of energy efficiency (e.g., HVAC, lighting, equipment, shading and glazing, and fabric and size) and renewable energy uptake (e.g., on-site renewables, energy storage, off-site renewables) has been widely accepted (Prasad et al., 2023). However, the ways to achieve energy efficiency and renewable energies should be piloted, tailoring to different urban contexts, to generate successful formulas. Typically, active design and renewable uptake means a large amount of initial investment, the availability of energy resources, and the usage of advanced and implementable technologies. For instance, the construction of biomass production industry is firstly dependent on the availability of agricultural and forestry waste, domestic garbage, livestock and poultry manure, etc. Solar electricity production through photovoltaic panels has been mostly promoted through building integrated photovoltaic (BIPV) techniques, while the availability of usable rooftops and façade remains a key concern (Ni et al., 2023). Overall, strategies for climate change mitigation should be systematically explored within different contexts for references.

Adaptation to climate change is also crucial to reduce climate-related disasters, improve urban living quality, and protect health and safety. Adaptation should generally address flooding, overheating, and air pollution, as well as many other challenges such as hunger, noise, and equality. Adaptation intervention may directly alleviate the impacts of these challenges. For instance, water-sensitive urban design through adding nature-based landscapes (e.g., parks, green spaces, wetlands, strips, and permeable pavements) into cities to enhance water penetration, absorption, and retaining (Song, 2022). Including cooling interventions such as green-blue infrastructures and innovative materials (e.g., reflective materials, permeable materials) in open spaces and building envelopes can help reduce urban ambient temperatures and improve outdoor thermal comfort (He et al., 2023). Even though, the adaptation solutions may take effects through adding sheltering structures and facilities. This can reduce citizens' exposure to extreme heat and protect their health and safety (Cureau et al, 2024). Accordingly, the adaptation solutions should be highly tailored to citizens' daily activities. In addition, each city is the only field for adding interventions for climate adaptation and environmental quality, while more than one disaster and challenge may occur over time. This suggests the importance of coping with several challenges simultaneously through adaptation. For instance, many cities, especially those in tropical and subtropical climates, may suffer the alternating effects of urban heat and flooding in summer seasons. Co-benefits approach should be stressed to leverage the multiple benefits of urban design solutions to more than one disaster and challenge (Dinh et al., 2024).

Actions for mitigation and adaptation should be implemented by different stakeholders across diverse scales. International organizations have played a leading role in improving global awareness and knowledge, formulating guidelines and frameworks, and promoting governmental communication and collaborations. Only if such actions are implemented at national, provincial, and city scales, they can address climate change and environmental deterioration. There is a need to transfer international guidelines and frameworks into local regulations and specifications from the policy perspective (Caratù et al., 2023). On stakeholders, apart from urban managers, planners, and designers, community and industry stakeholders also play a critical role in implementing mitigation and adaptation actions. For instance, community engagement promotes the understanding of local needs, advanced technologies, and critical challenges, thereby social and economic support for societal changes. On the one hand, industry stakeholders should take the role to decarbonize through upgrading technologies for energy efficiency and renewable energy uptake. On the other hand, they should implement adaptation solutions by providing more livable environments to protect workers' health and safety. The roles and responsibilities in the coming years in creating livable and resilient cities should be clarified through deep communication and collaboration (Cui et al., 2024). A framework of Industry-University-Research-Application should also be developed to enhance communication among different stakeholders to address economic, technological, social, and cultural challenges in practical implementation. In particular, the university and research organizations play critical roles in disseminating new findings and advanced technologies by knowledge transfer. Overall, leveraging behavioral and policy approaches is important to enhance social participation and avoid economic, technological, social, and governmental isolation.

2 Pathways to Livable and Resilient Cities

To deepen understandings and actions to positive climate mitigation and adaptation, it is essential to transfer international and national guidelines and frameworks to city-specific policies, regulations, plans, solutions, and practices. In particular, due to the differences in geographic, cultural, social, climatic, economic, technological, and political contexts, each city should develop action plans that can fit into local characteristics, based on the experience and lessons of others.

The authors of this introductory chapter actively chaired the 4th International Conference on Climate Change and Environmental Sustainability (CCES) with the support from Chongqing University (China), University of Queensland (Australia), Liège University (Belgium), and University of Washington (United States), and the 3rd Health & Environmental Resilience and Livability in Cities (HERL) with the support from University of Perugia (Italy) and University of Florence (Italy). These two conferences were aimed at providing an opportunity for scientists, researchers, policymakers, practitioners, and others to conduct interdisciplinary discussions and collaborations on mitigating urban climate change, enhancing urban climate resilience, and promoting urban sustainable development and livability in cities. After critically screening abstract submissions and blindly reviewing full submissions, a total of 32 conference papers were accepted and edited into the book titled "**Towards the Framework of Livable and Resilient Cities**". These papers are divided into four parts, including "Urban Design, Public Spaces, and Resilience" (Part 1), "Renewable Energy and Sustainable Building Technologies" (Part 2), "Climate Adaptation Strategies and Environmental Quality" (Part 3), and "Behavioral and Policy Approaches to Climate Change Mitigation" (Part 4).

2.1 Urban Design, Public Spaces, and Resilience

The first part about ***"Urban Design, Public Spaces, and Resilience"*** consists of nine chapters. This part explores how to improve urban resilience to environmental, social, and climatic challenges and disasters, with the focus on urban social needs, walkability, livability, agroecology, habitability, safety, sustainability, resilience, and stormwater management, through innovative planning and design of urban public spaces.

Chapter "A conceptual framework designed to fulfill the necessities of urban development" contributed to the identification of the critical challenge of neglecting social needs while advancing urban intelligence and promoting ecological sustainability. A holistic methodological framework that can act as a fundamental basis for future cities was proposed to connect social needs and urban construction. Human needs were structured by a set of indicators including subsistence (existence), health and safety, identity and belonging, creativity and innovation, leisure, knowledge, freedom, flexibility and renewal, self-esteem, power, relationships and participation, and justice, with their requirements demonstrated in four areas including smart infrastructure, buildings & land uses, smart transportation, and open spaces & landscape. This framework enhances city planners and designers' capability of building humane cities and communities.

With the motivation of improving urban walkability, Chapter "Type of public open space as a condition for walkability in the town centre of a small town" investigated the relationship between public open spaces and walkability in Slovenia based on questionnaire survey and statistical analysis. Smaller towns are more likely to limit traffic and create pedestrian-friendly environments, therefore fostering a pleasant walking experience and broadening social and environmental well-being. Well-paved or structured roads were also important to ensure walkability, while there was no clear association between parks and walkability in Slovenian towns, which was different from the findings in existing studies. The size, planar morphological characteristics, and geographical location of towns played critical roles in enhancing walkability.

Chapter "Children's city architecture is not only playing grounds" assumed that contemporary architecture mostly ignored children's needs. Such an assumption was verified though the analysis of city structures and architectural characteristics of Uruk, Rome, Cairo, and Paris, which discouraged the inclusiveness in creating sustainable cities and communities. The analysis of livability of outdoor design cases for children in Oslo, Norway, and Riyadh, Saudi Arabia showed the progress in improving the built environment for children, while the primary architectural solution in such two cities was playground. More studies are needed to shape the theories, framework, and practices of architectural design for children.

Chapter "Urban agroecology and the well-being of older adults: Exploring the impacts of residential-scale pesticide-free vegetable production in Chiang Mai's urban landscape" points out that integrating urban agriculture and sustainable food systems is important to promote health and environmental resilience. However, whether vulnerable populations have enjoyed the benefits of urban agroecology is an open question. The in-depth interview among older adults in Chiang Mai, Thailand, verified multiple benefits including recreation, direct health benefits, physical exercises, and clean production. Socioeconomic benefits of urban farming lie in emerging economic opportunities and the therapy of personal grief and loss associated with old age. Urban agricultural initiatives demonstrate a blueprint for fostering health and environmental resilience in urbanizing cities globally.

Chapter "Revitalizing urban street dynamics via strategies for climate control: A tool for design and evaluation" shows the motivation of urban space reconfirmation and adaptation to maintain urban habitability in the increasingly hot era. The authors developed the Urbankea tool to improve urban designers' capacity of modifying urban spaces and positioning urban elements to enhance human thermal comfort outdoors. The case study in Seville, Spain, demonstrated the applicability of this tool. The analysis showed that the combination of low-transmissive textiles, cooling radiant methods, and water irrigation generated prominent cooling benefits, where there could be an enhancement in the comfort index. This overcame the inadequate performance of only using vegetation or new materials.

Chapter "Effect of natural-based cool pavements in urban open spaces liveability" aims to address urban heat challenges due to heat waves and heat islands and improve urban livability with the cooling solution of natural-based cool pavements. Focusing on urban open spaces, experimental and simulation analysis Utrera, Seville, Spain, showed that compared with conventional pavements, there was a 20 °C reduction in the surface temperature with the application of natural-based cool pavements. A 40% improvement in the comfort range was identified, based on the assessment indicator of physiological equivalent temperature. On the one hand, this approach is novel to avoid water evaporation while improving thermal comfort. On the other hand, this approach ensures entertainability and increases livability and safety under climate change.

Chapter "Traditional stormwater management for urban resilience: An analysis of "Ahar and Pyne" system in sone floodplains in the suburbs of Patna, Bihar, India" presents an Indian version of water-sensitive urban design method, the Ahar and Pyne system, to cope with urban flooding and enhance water conservation. The study provided a thorough analysis of the Ahar and Pyne system by onsite observation, based on which sustainable construction suggestions were proposed. The effectiveness of this system can be affected by heavy rainfall, river conditions, topography, sand mining, and silt accumulation, while deepening riverbeds and constructing permeable screens were modern engineering methods to overcome such challenges. Overall, the Ahar and Pyne system is an ancient water management system that should be reshaped and improved to enhance water management, urban resilience, and sustainability.

Chapter "Designing for disaster: A holistic approach to architectural resilience" investigated how to enhance resilience and adaptability to counterbalance existing and emerging crises such as natural disasters and climate change through infrastructural design. Based on literature review, case study analysis, and analytical framework, this chapter showed the limitation of current architecture designs in terms of interdisciplinary collaboration and flexible regulations. There is a need to develop a proactive and holistic design paradigm that can comprehensively and deeply integrate sustainability, aesthetics, and user-centered principles. The real-life studies of Dutch mentality Amsterdam housing development and the Brooklyn Waterfront Greenway demonstrated how to achieve architectural resilience systematically and how infrastructure design can better enhance disaster recovery and mitigation. In conclusion, architectural resilience should be integrated into modern design for resilience and adaptability.

Chapter "Sustainability in campus building design from micro to macro scale" explores how to design campus buildings sustainably across different scales. This chapter showed sustainable campus indicators in five dimensions such as campus physical operations, institutional characteristics and goals, innovation & design process for green transformations, sustainability engagements for a sustainable society, and policy governance and administration. Based on sustainability assessment of 16 universities around the world, the analysis indicated that investment-relevant and cost-driven indicators and sustainability goals had not been implemented well. This resulted in three criteria in terms of objectives of sustainable design criteria, technical criteria, and process criteria to overcome the challenges in the future. Furthermore, suggestions for limiting environmental impacts were proposed in five dimensions including planning sustainable areas, water conservation, material and resource conservation, energy efficiency, and indoor environmental quality improvement. Overall, this part is of value to provide innovative frameworks, solutions, pathways, and paradigms to improve urban sustainability and resilience.

2.2 Renewable Energy and Sustainable Building Technologies

The second part about ***"Renewable Energy and Sustainable Building Technologies"*** includes eight chapters. The topics in this part cover the uptake of renewable energy into buildings, the production of renewables (e.g., solar energy, hydrogen energy, and biomass energy), the improvement of energy efficiency driven by smart technologies, the regulation of indoor thermal environments, and the cost analysis of passive design.

Chapter "Building Integrated Photo Voltaic Panels for Achieving Net Zero Energy in Tropical Climatic Conditions" stresses the energy usage of buildings plays a critical role in sustaining future cities, where adoption of building integrated photovoltaics (BIPV) is a promising approach to net zero energy buildings. This is proved by the results of one-year-round onsite experiment in Chennai, India. Basically, the analyses showed the achievability of net zero energy in the first two weeks of February 2024, when the sun is in the southern hemisphere. In the remaining period of the year, when the solar resources are more available and accessible, the target of net zero energy can be easily achieved in cities like Chennai.

Chapter "Effect of degradation and size selection for the photovoltaic system installed in a nearly zero energy building in Tropical Climate" is motivated by the uptake of PV systems into buildings for energy transitions and climate change mitigation. However, the electricity production performance of PV modules decreases over years, and the performance can be affected by climate change. Based on dynamic simulations, this chapter demonstrated the PV performance in a residential nearly zero energy building in tropical Santo Domingo, Dominican Republic. The results verified that PV aging resulted in production performance reduction. However, the targets of net zero energy building could be always met when the PV installation capacity was over 3.60 kWp, regardless of climate change and PV degradation.

Chapter "Increasing renewable energy share in urban areas by valorizing local wastes: coupling anaerobic digestion and small gasification units" presents an alternative renewable energy method based on the exploitation of local wastes. This solution was experimented by the conversion of a conventional wastewater treatment plants into biorefineries for processing urban wastes (e.g., sludge, pruning, and garden waste) in León, Spain. In this chapter, the energy recovery potential of various technologies was compared. The drying process of sewage sludge cost excessive energy when using thermal technologies, while such a method showed more pronounced potential than single gasification. With extra pruning and garden waste, the electricity production could significantly increase, up to 80% even when the poplar productivity was the lowest.

To decouple health problems with air pollution caused by the burning of fossil fuels, Chapter "Green Hydrogen production potential from agricultural biomass in Togo's coastal region" focuses on clean energy (SDG 7) by exploring the methods to produce green hydrogen through leftover maize and rice. The estimation based on the national statistics showed that great potential of using agricultural residue (e.g., maize straw, maize cob, rice straw, and rice husk) for hydrogen production in the study area of Maritime region of Togo. The results indicate that maize straw showed the largest potential with the amount of 24.18 GT, followed by maize cob (6.37 GT), while rice straw and rice husk only produced 0.50 GT and 0.30 GT. Overall, rice and maize residue open a door for the Togo's renewable energy uptake and energy structure transition, while maize residue plays a dominant role. Overall, this part provides a comprehensive reference for energy structure transition, particularly tailoring to the building sector, through renewable energy, smart management, passive design, energy use reduction, and energy efficiency improvement.

Chapter "Intelligent energy management systems: A systematic review of technologies, challenges, and opportunities" underlines the importance of smart technologies in improving building energy use efficiency. Therefore, the authors are dedicated to promoting the integration of architecture design and emerging smart technologies (e.g., Artificial Intelligence, the Internet of Things, and machine learning) by a comprehensive review of the technologies, challenges, and opportunities of intelligent energy management systems. Particular attention was paid to the smart grid technologies and renewable uptake in the literature review. This literature highlights the advantages of smart technologies, i.e. the Internet of Things devices for data collecting, artificial intelligence for decision-making and analysis, and intuitive user interfaces for resident participation. Meanwhile, the uptake of renewable energy sources is a promise for energy efficiency improvement. However, the barriers of high initial costs, privacy and security concerns, and interoperability issues should be addressed in promoting Intelligent energy management systems.

Chapter "Constructions of green façade and effects on thermal environment: Simulations with CFD" is motivated to leverage the synergistic benefits of heat island mitigation, traffic noise reduction, and air pollution reduction, through vertical greenery systems, especially the green façade. The numerical simulations were performed to compare the shading and thermal regulation effects of four types of green façade configurations in Guangzhou, China. The results reported that the different configurations (e.g., two horizontal planting patterns and two vertical planting patterns) resulted in a reduction of 0.67–3.24 °C in average temperature of buffer space, while indoor temperatures exhibited a reduction of 1.27–3.45 °C on average. Wall temperature reductions were more marked, by 7.68 °C on the south wall and 13.24 °C on the west wall in two horizontal configurations. This study could prioritize the adoption of horizontal planting patterns in cities like subtropical Guangzhou, China.

Chapter "Cost analysis for passive design buildings in cold climate regions in China: A preliminary review" points out that passive design for creating sustainable building is critical to address climate change, but the adoption of passive design is generally associated with the high initial cost. However, there is a need to determine the adoption of passive design or active design in practice. In the context of Zhangjikou, China, the authors estimated the cost of passive design and active design of ten detached houses (information: 2 floors, 6 m in height, 150 m^2 in floor area, and 300 m^2 in cumulative living area) over the building life cycle. The average initial cost of passive design was about \$500,000, higher than \$420,000 for active buildings, while the operational cost of active buildings (\$13,000 on average) was higher than that of passive buildings (\$7,000 on average). Overall, passive buildings are more energy-efficient and less costly in the long run. With an assumption of energy price increase, the analysis showed that passive buildings can better address economic variabilities.

Chapter "TCAD modeling of bifacial silicon solar cells" presents the technology computer-aided design (TCAD) simulations to explore how to improve the efficiency of silicon bifacial solar cell through changing the physical parameters of the base and isotype layer of the solar cells. The results show that the efficiency of solar cells with the n^+-p-p^+ structure was better than that of p^+-n-n^+ structure, about 5% in value. The illumination on the back and front side of bifacial sensitive solar cells could generate different efficiency: on the back side, the short-circuit current could reach the highest value with a doping concentration of 10^{19} cm^{-3} for the p^+ layer in the n^+-p-p^+ structure and 10^{19} cm^{-3} for the n^+ layer in the p^+-n-n^+ structure; on the front side, the maximum solar cell efficiency could be achieved when the doping concentrations were 10^{18} cm^{-3} for the p^+ layer in the n^+-p-p^+ structure and 10^{19} cm^{-3} for the n^+ layer in the p^+-n-n^+ structure. When illuminating on both sides, the short-circuit photocurrent density increased by 1.57 times and the output power increased by 1.56 times.

2.3 Climate Adaptation Strategies and Environmental Quality

The third part about ***Climate Adaptation Strategies and Environmental Quality*** includes eight chapters, focusing on exploring strategies for climate change adaptation and environmental quality improvement. This part covers topics of urban heat island mitigation, urban flooding alleviation, neighborhood resilience enhancement, urban food sufficiency, and indoor environmental quality improvement. The potential strategies include circular economy, nature-based solutions, community engagement, information and communication technologies, urban geometry modification, hydrophilic architecture, wind shelter, mechanical ventilation, and urban digital modeling.

To better adapt to climate change, Chapter "Climate change adaption projects: Three case studies" demonstrates three international cases such as the "Rainproof" project in

Amsterdam, the "Climate Resilient Neighborhood" in Copenhagen, and the "Urban Forest Strategy" in Melbourne. These three cases show innovative experience of addressing urban flooding based on nature-based solutions, enhancing neighborhood resilience based on community engagement, and mitigating climate change by integrating information and communication technologies. Overall, it is important to synergize nature-based solutions, community engagement, and information and communication technologies to generate a holistic and multicriteria approach to effective climate change adaptation, by ensuring ecological, territorial, environmental, technological, climatic, social, and economic coordination.

Chapter "The influence of outdoor space geometry and greenery configuration on students' thermal perception during moderate seasons in an arid climate" aims to improve the outdoor thermal comfort with the modification of space geometry and space greenery. The case study was conducted in two university spaces at arid El-Giza, Egypt, based on field measurement of microclimatic variables and questionnaire survey of subjective thermal sensation. The results indicate that a space with good geometric configuration, despite no greenery, could provide thermally comfortable environments for pedestrians and users in spring and fall, while most participants felt slightly warm and direct solar radiation in the space with good ground greenery. The results can provide urban planners and architects with references to creating comfortable environments through designing geometric shades and avoiding reflected and emitted heats from surrounding walls in arid cities like El-Giza.

Chapter "Revitalizing food deserts through hydrophilic architecture: A case study of proposing community gardens in university of Arizona dormitories" highlights the importance of hydrophilic design in addressing the water scarcity and food shortages in arid areas. Based on the empirical case of hydrophilic units on building facades, at the University of Arizona in Tucson, US, authors adopted desiccant materials like Calcium Chloride to extract water from low-humidity environments. Such a method presented pronounced capacity of water collection. It is potential to generate 47 L of water per day if the units could be well installed on the southern façade of the Maricopa Hall, which can meet the water needs of planting leafy greens, legumes, herbs, fruits and vegetables, thereby enhancing food security. Overall, hydrophilic architecture is promising to support Sustainable Development Goals in Goal 2 (Zero Hunger), Goal 6 (Clean Water and Sanitation), Goal 11 (Sustainable Cities and Communities), Goal 12 (Responsible Consumption and Production), and Goal 13 (Climate Actions).

Chapter "Wind sheltering techniques for reducing high temperature air penetration in spaces with outdoor acclimatization strategies" suggests that urban ventilation is generally meaningful to enhance heat dissipation, thereby mitigating heat island effects and improving outdoor thermal comfort, while urban ventilation may elevate zoning temperature if the wind carries heat sources. There proposed a hypothesis that properly designing wind shelters is critical to preventing hot air penetration, at which time adopting active cooling strategies can regulate the thermal environments of targeted zones. This hypothesis was verified based on the CFD simulations of urban canyon-like designs and windbreaking structures. Overall, this chapter offers urban planners and designers with alternative understandings of regulating wind for urban thermal quality improvement.

Chapter "Importance of design of air emitters in the healthcare facilities: Computational fluid dynamic maps for air distributions under Mediterranean climate" concentrates on the improvement of indoor environmental quality of healthcare buildings. The numerical study explored to what extent mechanical ventilation could reduce infections and improve thermal quality in the context of a patient's room of "A. Cardarelli" hospital, Campobasso, Italy. Compared with natural ventilation, the adoption of mechanical ventilation improved thermal environment uniformity and reduced the air residence time by 70% and 80% with the central diffuser and vent, respectively. This case study showed that a wall air intake was the best solution, while future studies should refine result accuracy with the consideration of internal environmental factors, uncertainty in pathogen behavior, human preference, compliance with protocols, and insufficient or obsolete data.

Chapter "Potential heat island phenomenon in Florence: Microclimatic assessment through digital urban modelling" developed an urban heat island assessment model to accurately assess microclimatic conditions at local scales, in comparison to Satellite-based coarse results. This model was applied to two areas-a road intersection area and a continuous road segment-in Florence, Italy. The results showed that N-S streets exhibited stronger heat stress than W-E streets. Typically, unshaded areas exhibited significantly strong heat stress, and urban ventilation enhancement was not capable of mitigating heat stress. Overall, the construction of refined heat island maps enables decision makers to identify areas with discomfort alleviation needs and add interventions for mitigation.

Different from residential and commercial buildings, Chapter "Evaluation of the air quality inside a Hindu temple: A case of Mahakaleshwar temple Ujjain" concentrates on religious buildings with the purpose of investigating air quality in Hindu temple. Based on a case study of Mahakaleshwar Temple, this chapter presented indoor air quality and infection risk. This temple experienced insufficient natural ventilation and poor air circulation, and thermal discomfort was a critical issue with the symptoms of suffocation, irritability, and nasal congestion. The simulations further showed that elevated CO_2 levels, reduced oxygen levels, and increased temperature and humidity had worsened indoor environments. Implement mechanical ventilation units should be prioritized, and natural ventilation are also important to enhance thermal comfort and mitigate indoor thermal challenges.

Chapter "Study on circular building design criteria of human settlements" reflects integration of global strategies of circular economy and decarbonization into the building sector. To enable stakeholders to achieve so, this chapter is designated to develop a multi-criteria decision-making evaluation framework for residential circular building design. Based on mind mapping and expert interview, a framework consisting of four dimensions and 24 indicators was constructed; and based on analytical hierarchy process, and Eigenvector methods, the weight of each dimension and indicator was determined. Overall, this framework provides a standardized reference for stakeholders to perform residential circular building projects, especially in the pre-project planning phase. Overall, this part offers technological, management, and methodological strategies for improving climate change adaptation capacity and environmental quality.

2.4 Behavioral and Policy Approaches to Climate Change Mitigation

The fourth and last part about ***"Behavioral and Policy Approaches to Climate Change Mitigation"*** mainly explores how to improve the participation of different stakeholders and decision makers (e.g., individuals, corporates, local government, community residents, university students and staff) in climate change mitigation and environmental quality improvement. This part consists of seven chapters covering sustainability science, climate finance, heat mitigation and adaptation strategies, decarbonization, urban regeneration, healthy built environments, and residential living quality.

Chapter "Sustainability Science: A guide to dealing with problems in climate change and solutions for policy practice" aims to advance science and solutions for sustainable development with an extensive review of definitions, principles, theories, and practices of sustainability science. Basically, the critical question of how sustainability science acts as an education driver to assist in addressing climate change was delineated. The sustainability science in coping with climate change has evolved into a wide-scope, multidisciplinary scientific discipline, while there remain gaps in knowledge transfer and scientific practice to implement solutions or policies. A science, technology and innovation (STI) model was suggested as a transformative model to address such gaps. Moreover, the integration of STI model and policy science was expected to build a more equitable society, through better collaboration, communication, and knowledge transfer to address more global challenges.

Chapter "Climate finance for resilience future: Pioneering steps of the United Arab Emirates (UAE) in the Gulf region" is motivated to integrate Islamic finance into environmental sustainability in the Gulf region. This region is dominant by oil and gas production, playing a crucial role in global Climate Actions (Goal 13). This chapter explored environmental sustainability and its impacts on climate mitigation practices, based on the interrelationships among climate finance, Islamic finance, and environmental sustainability, in UAE which is a pioneering nation in Climate Actions. The results in UAE showed that Shariah-compliant financial instrument, particularly green sukuk, was an effective tool to mobilize capital for climate mitigation and adaptation projects. Based on this, it is suggested that the Shariah-compliant instruments should be developed, while addressing socioeconomic challenges by fostering awareness, building capacity, and promoting strategic partnerships between relevant government entities and private sectors, to escalate Islamic finance's role in climate finance, thereby promoting environmental sustainability goals.

Increasingly frequent, intense, and severe heat waves associated with climate change have significantly exacerbated heat stress of pedestrians while traveling, given which the Chapter "Strategies for enhancing travel choice adaptation in hot weather conditions: Insights from stakeholder consultation" aims to provide effective mitigative and adaptive strategies for combating heat stress and reducing heat vulnerability during commuting. This chapter investigated the acceptance of different strategies among stakeholders at individual, community, and governmental levels. The authors developed a system of mitigation and adaptation strategies and delivered opinion collection among 13 stakeholders during a workshop on "Gendered approaches for addressing adaptation capacity to hot weather" on 5th August 2023 in Delhi, India. The investigation identified the prioritized solutions among individual, community, and governmental representatives. For instance, individuals generally accept the solution of wearing weather-appropriate clothing and staying hydrated. Community-led strategies like capacity building and awareness campaigns are widely supported, and government-led strategies like developing green spaces and implementing climate-sensitive urban design are appreciated. The results provide urban decision makers with empirical references for urban heat mitigation and adaptation implementation.

Chapter "Mitigating climate change: A literature review based on individual and corporate carbon reduction behaviors" highlights individuals and corporations' significant role in advancing decarbonization activities, followed by a literature review of individuals and corporations' decarbonization behaviors. This chapter concluded factors affecting individuals' behaviors included personal norms, individual attitudes and perceptions, green commitment, green experience, social networks, green technology, cost, policy and education. Factors affecting corporations' behavior include corporate financial performance, board diversity, corporate social responsibility, corporate technology, information technology, green finance, carbon prices, and policy environment. This study is of significance to close existing gaps and barriers in individuals and corporations' participation by providing suggestions on sharing data, publicity and education, knowledge sharing, incentives, and carbon performance evaluation and supervision system.

Chapter "The potential civic role of universities in the urban regeneration of local communities: A mutual learning process" stresses the important role of universities (i.e. expanding their research strategies beyond universities and encompassing social and civic engagement within communities) in urban regeneration. Based on the exemplar of San Siro neighborhood for mutual learning, this chapter demonstrated that connections with university activities could advance residents' understanding of their daily life demands, while off-campus communities could be living labs to enable students to understand and perform innovative skills and shape their critical thinkings. Overall, developing mutual learning between university and community is a transformative model to enhance research and community engagement, thereby making the impact of research and science advancement more significant.

Chapter "The impact of the built environment from the local, opportunities and challenges: Affordable regional instrument for healthy housing in Queretaro, Mexico" stresses the importance of social technology as an inclusive regional scale process, so that the authors explored how to use social technology to address the challenges in promoting healthy built environments in the Metropolitan Area of Queretaro, Mexico. A local measurement instrument was developed, covering seven aspects including indoor air quality, hydrothermal range, acoustic range, humidity, lighting range, electromagnetic field, and security and accessibility. Overall, this methodological instrument could narrow the gap of dispersed knowledge in various media and disciplines, providing an implementable and affordable way for all local inhabitants in creating healthy built environments.

Chapter "Building typologies and the effects of spatial configurations on the quality of living case study: The city of Sarajevo" presents a motivation of improving residential living quality (i.e. improving occupant's health and reducing air pollution) with the focus on the modification of urban structure (e.g., form, dimension). This chapter investigated air quality, sunshine quantity, and noise volume in a designated area with diverse building types such as administrative buildings, a school, kindergarten, commercial buildings and residential buildings in Novo Sarajevo, Bosnia and Herzegovina. This chapter presented original findings on living quality. For instance, the designated area suffered more severe air pollution than surrounding areas, despite lower traffic flow in this area. The wintertime heating showed a stronger impact than traffic. Regarding the building form, the closed building layouts were more favorable than linear building layouts in improving air quality, in the absence of strong air flow. Overall, this part presents valuable methods and implications for adopting behavioral and policy approaches to climate change mitigation and environmental quality improvement.

3 Concluding Remarks

The Paris Agreements, Sustainable Development Goals, and Net Zero Actions have been formulated by international organizations and many nations have widely accepted them for promotion. The COP 29 in November 2024 stresses cities' roles in counterbalancing climate change. To achieve positive climate mitigation and adaptation and environmental quality, cities must transfer existing high-level policies, regulations, plans, solutions, and practices to local ones for implementation. The formula for successful implementation is not uniform given cities' complex and dynamic nature. Each city must explore their own ways with collecting experience and lessons from other cities and actual piloting projects, tailoring to geographic, cultural, social, climatic, economic, technological, and political contexts. This book, based on the highlights of two conferences of CCES-4 and HERL-3, offers comprehensive, tangible, and implementable frameworks, solutions, pathways, and paradigms to improve urban livability and resilience.

The understandings of climate change mitigation and adaptation and environmental quality improvement were discussed and communicated among conference participants. This book collects the main outcomes to provide valuable references for experts from Global North and Global South, as well as Small Island Developing Nations. This book covers four key themes: (i) Urban Design, Public Spaces, and Resilience, (ii) Renewable Energy and Sustainable Building Technologies, (iii) Climate Adaptation Strategies and Environmental Quality, and (iv) Behavioral and Policy Approaches to Climate Change Mitigation. These themes correspond to the needs of diverse urban contexts, mitigation solutions, adaptation solutions, and socio-economic-political participation to support the creation of livable and resilient cities. In particular, this book reports the practices and actions of various cities from nations worldwide such as Australia, Bosnia and Herzegovina, China, Denmark, Dominican Republic, Egypt, India, Italy, Mexico, Norway, Saudi Arabia, Slovenia, Spain, Thailand, the Netherlands, Togo, United Arab Emirates, and the United States. The book prioritizes the integration of climate mitigation and adaptation, not only to decelerate climate change through decarbonization, but also to control climate-related disasters and reduce climate-related impacts on urban lives and production. Beyond scientists and researchers, this book values the roles of different stakeholders in building livable and resilient cities, which is demonstrated by community engagement and communications and corporate economic investment. Overall, this book charts a contextual-technological-social-economic-policy framework for references.

References

Brown, P.T., et al.: Climate warming increases extreme daily wildfire growth risk in California. Nature **621**(7980), 760–766 (2023). https://doi.org/10.1038/s41586-023-06444-3

Caratù, M., Pigliautile, I., Piselli, C., Fabiani, C.: A perspective on managing cities and citizens' well-being through smart sensing data. Environ Sci Policy **147**, 169–176 (2023). https://doi.org/10.1016/j.envsci.2023.06.012

Cui, Y., Yin, M., Cheng, X., Tang, J., He, B.J.: Towards cool cities and communities: preparing for an increasingly hot future by the development of heat-resilient infrastructure and urban heat management plan. Environ. Technol. Innov. **34**, 103568 (2024). https://doi.org/10.1016/j.eti.2024.103568

Cureau, R.J., et al.: On urban microclimate spatial-temporal dynamics: evidence from the integration of fixed and wearable sensing and mapping techniques. Environ. Res. **262**, 119795 (2024). https://doi.org/10.1016/j.envres.2024.119795

Dinh, N.T.T., Tran, J., Hensher, M.: Measuring and valuing the health co-benefits of climate change mitigation: a scoping review. Lancet Planet. Health **8**(6), e402–e409 (2024). https://doi.org/10.1016/S2542-5196(24)00095-0

He, B.J., Chen, X.: Synergizing low-carbon transportation and heat adaptation: identifying suitable routes and integrating mitigation and adaptation infrastructure. Build. Environ. **269**, 112448 (2025). https://doi.org/10.1016/j.buildenv.2024.112448

He, B.J., Wang, W., Sharifi, A., Liu, X.: Progress, knowledge gap and future directions of urban heat mitigation and adaptation research through a bibliometric review of history and evolution. Energy Build. **287**, 112976 (2023). https://doi.org/10.1016/j.enbuild.2023.112976

Ma, F., et al.: The disparities and development trajectories of nations in achieving the sustainable development goals. Nat. Commun. **16**(1), 1107 (2025). https://doi.org/10.1038/s41467-025-56076-6

Masson, V., Lemonsu, A., Hidalgo, J., Voogt, J.: Urban climates and climate change. Annu. Rev. Environ. Resour. **45**(1), 411–444 (2020). https://doi.org/10.1146/annurev-environ-012320-083623

Ni, P., Yan, Z., Yue, Y., Xian, L., Lei, F., Yan, X.: Simulation of solar radiation on metropolitan building surfaces: a novel and flexible research framework. Sustain. Cities Soc. **93**, 104469 (2023). https://doi.org/10.1016/j.scs.2023.104469

Orkomi, A.A., Ameri, M.: Investigating the intensity of urban heat island and the impacts of local climate using verified WRF data: a case study of Rasht. Northern Iran. Sustain. Cities Soc. **106**, 105405 (2024). https://doi.org/10.1016/j.scs.2024.105405

Prasad, D., et al.: Delivering on the Climate Emergency: Towards a Net Zero Carbon Built Environment. Springer (2023). https://doi.org/10.1007/978-981-19-6371-1

Qi, J., Ding, L., Lim, S.: Application of a decision-making framework for multi-objective optimisation of urban heat mitigation strategies. Urban Clim. **47**, 101372 (2023). https://doi.org/10.1016/j.uclim.2022.101372

Sharifi, A.: Resilience of urban social-ecological-technological systems (SETS): a review. Sustain. Cities Soc. **99**, 104910 (2023). https://doi.org/10.1016/j.scs.2023.104910

Sinha, A., et al.: Diverse decarbonization pathways under near cost-optimal futures. Nat. Commun. **15**(1), 8165 (2024). https://doi.org/10.1038/s41467-024-49868-9

Song, C.: Application of nature-based measures in China's sponge city initiative: current trends and perspectives. Nat.-Based Solutions **2**, 100010 (2022). https://doi.org/10.1016/j.nbsj.2022.100010

WMO: WMO confirms 2024 as warmest year on record at about 1.55 °C above pre-industrial level. World Meteorological Organization (2025). https://wmo.int/news/media-centre/wmo-confirms-2024-warmest-year-record-about-155degc-above-pre-industrial-level

Xu, R., et al.: Global, regional, and national mortality burden attributable to air pollution from landscape fires: a health impact assessment study. Lancet **404**(10470), 2447–2459 (2024). https://doi.org/10.1016/S0140-6736(24)02251-7

Urban Design, Public Spaces, and Resilience

A Conceptual Framework Designed to Fulfill the Necessities of Urban Development

Shaimaa S. Alsayed

Abstract

Creating humane and enduring communities presents a meaningful endeavor, given planners' vast enthusiasm for forging intelligent cities or solely emphasizing the ecological aspect of sustainability. The study aims to establish a comprehensive framework that connects human needs to the process of urbanization to meet the needs of urban dwellers. Furthermore, the study seeks to identify indicators that accurately gauge the degree to which various aspects and elements of urbanization within a city fulfill these needs. The research method utilized in this study involved conducting a comprehensive analysis of existing literature to investigate the correlation between human urban needs and the field of urban construction. Consequently, a set of universal guidelines and metrics was devised to assist in fulfilling these needs. The result was the creation of a comprehensive framework that serves as a fundamental basis for the establishment of future cities. This framework outlines approaches for realizing human requirements through urban development, thereby facilitating the assessment of cities and the conceptualization of human needs in the future. Consequently, it indicates the prosperity and extent of human values that cities will embody in the future.

Keywords

Human Needs · Urban Components · Future City · Urban Cities · Urban Human Needs (UHN)

S. S. Alsayed (✉)
Department of Architecture Engineering, College of Engineering and Information Technology, Buraydah Private Colleges, Buraydah 51418, Saudi Arabia
e-mail: myshaaaarch@gmail.com

1 Introduction

Cities live up to their understanding of human needs and requirements (Hassan & Megahed, 2021). Cities exist to serve and meet these human requirements (Hallsmith, 2003). Thus, the relationship between the city and humans mutually affects each other through the built environment. Urbanization's built environment must integrate with nature, meet sustainability standards, and meet human needs (Aboria et al. 2024; Hassan, 2023; Hassan & Megahed, 2022). However, there is a conflict between cities and humans. People perceive cities as living entities that necessitate proper development to meet their needs, evolving their shape and working methods. However, they fail to recognize that cities are merely vessels carrying human activities, meeting their needs and desires, not their own. Modern trends overlook real human needs, such as environmental sustainability and smart cities. While these trends may benefit cities in the future, they should prioritize the needs of users, who are humans (Gibbs et al. 2013). The emphasis on modern trends ignores the main goal of cities, which is to meet the needs of people through urbanization.

The significance of human needs is crucial in enhancing individuals' capacity as catalysts for sustainable development. This concept posits that by establishing living conditions that facilitate the fulfillment of residents' needs, encompassing social well-being and empowerment, it has the potential to elevate overall life satisfaction levels. Consequently, this could foster stronger connections between individuals and their place of habitation, as well as their inclination to engage in actions that promote the general living environment (Rodzoś, 2019). These trends focused on a specific aspect of city planning, failing to address the emergence of diseases and epidemics like COVID-19 (N. A. Megahed & Hassan, 2022; N. Megahed & Ghoneim, 2020). It showed the flaw in governments' control over the built environment and did not provide health safety for citizens. Indeed, there has been an outbreak in many cities as climate change continues to this day, with

B. He et al. (eds.), *Towards the Framework of Livable and Resilient Cities* Advances in Science, Technology & Innovation, https://doi.org/10.1007/978-3-031-97849-4_2

the presence of global warming problems and the emergence of many social and economic problems such as unemployment, the disparity in the classes of society, increased poverty in many countries, the occurrence of social isolation, and a decrease in community interaction and participation among residents and decision-makers as well.

As a result, those needs must be identified to determine which of them may help build good-quality architecture (Mutică, 2013) and transform these trends into methods and tools that serve human needs while maintaining a cycle of operation in the cities where they live. The efficiency of such cities is evident in their development within the framework of achieving human needs and creating an appropriate environment to meet both physiological and psychological needs while preserving the natural environment and facilitating human life without canceling or erasing their role in supporting existence. As a result, those needs must be identified to determine which of them may help build good-quality architecture (Mutică, 2013) and transform these trends into methods and tools that serve human needs while maintaining a cycle of operation in the cities where they live. Therefore, this article poses the question of identifying the human needs that dictate the construction of cities in the future, and exploring the potential to fulfill all these needs through the various pillars of city development.

The concept of the modern city and its definition as an economic and social system have sparked criticism. Bunse (Bunkse, 1979) argues that this definition is narrow and fails to address basic needs. Ignoring local factors and individual needs results in a social system devoid of any emotional connections with its members. Community development initiatives must prioritize the fulfillment of human needs (Maclean, 1983). The incapacity of inadequately planned urban areas to satisfy human needs necessitates a thorough examination of various challenges encountered by cities, including transportation, energy, water, housing, amenities, and employment, to enhance the humaneness of cities (Lowe, 1991). As a result, there exists a research void in empirical investigations specifically focusing on the requirements and choices of individuals for sustainable urban amenities (Ji et al. 2021). There appears to be a deficiency in the ability to intricately address human needs and principles (Luckan, 2021).

Hence, scientific research began to focus on city planning from a different perspective; some see modernist ideas in city building as less than perfect. Planning to create a new vision requires a dialogue with sociologists (Mutică, 2013), psychologists, and anthropologists, starting with an individual's needs study. Thus, a study was submitted on the most important human needs that architecture must meet. Human needs encompass a range of individual, economic, social, and political necessities that individuals need to realize their objectives, lead a contented existence, and contribute to societal progress (Cardoso et al. 2022; Mutică, 2013; Tay & Diener, 2011). Various perspectives within communities have approached the identification of human needs, drawing from psychological and sociological frameworks (Hallsmith G., 2003). Consequently, research has shifted its focus towards integrating the concept of meeting human needs into urban development plans, a concept that modern urban trends have largely overlooked. (Seliverstov et al. 2020) have established that cities generate new needs (Neufeld and Wohl, 2022) and confirm that, in the face of technological and cultural changes, standard housing models have exceeded their capacity to keep up. Housing used to provide basic needs but now requires diversity and meeting individual desires.

Many researchers have addressed the concept of human needs and critiqued their relationship to urbanization (Greene, 1992). The concept of community design was introduced to meet basic human needs within broad concepts that encompass cognitive terms, architectural arts, visual arts, and urban design principles. Through Aliyeva's vision (Aliyeva et al. 2019), the rapid development of cities has imposed a new and diverse set of needs such as housing, work, social interaction, entertainment, transportation of goods and people, the need for nature, and green spaces for recreation and relaxation. Dietrich (2021) believes that the city focuses not only on its residents but also on other services and facilities. This means that the city has a strong interaction with its surroundings. Therefore, it studies, from a planning perspective, the demand and resources to meet the city's needs to make it livable internally and with its surroundings to meet its needs. Hallsmith G. (2003) discussed the importance of meeting human needs for sustainable cities and proposed a new approach to urban planning that focuses on developing social, economic, and environmental health systems and governance. Caprotti (2018) emphasizes the significance of giving human needs top priority in future plans for new cities. According to Kochtitzky et al. (2006), urban planning should include assessing and planning for the community's needs in various areas, such as transportation, housing, and others. From here, the needs are connected to different urban sectors or components, and urban planning and public health aim to improve human well-being by emphasizing the evaluation of needs and the provision of services.

Previous research has advocated for the inclusion of human needs in urban planning, particularly in light of technological advancements and external data interference. From this concept, the research continues the researchers' path to bridge the research gap (Foroughi & Daneshgar, 2024). The study outlines an integrated system of urban components and explores how to meet human needs through the latest theories that consider human needs as key to their application in cities, as well as an indicator to measure how well human needs are met in cities.

The study seeks to identify indicators that accurately gauge the degree to which various aspects and elements of urbanization within a city fulfill these needs. It also aims to create a conceptual framework that considers the keys that link human needs and urban components. This paper organizes the rest of its content as follows: The study's method is presented in Sect. 2, which identifies the elements of the conceptual framework and methodology used to determine the relationship between human needs and urban components, as well as the outcomes of this relationship. Section 3 then focuses on the Framework for Realizing Urban Human Needs (UHN) through Urban Cities Components, while Sect. 4 delves into the research findings. Lastly, Sect. 5 offers the conclusions. The paper's results serve as a guide for achieving and measuring human needs through the various elements of urban cities.

2 Methods

The study developed a research method to explore the existing literature on human needs. This study divides its procedures into two sections: first, it identifies the elements of the conceptual framework, then it employs a methodology to establish the connection between human needs and urban components. The outcome of this relationship, a general framework, serves as a set of indicators that urban components must meet to meet human needs.

2.1 Conceptual Framework Determinants

The research is based on an examination of the relationship between human needs and elements of the composition of the city, which will be illustrated in the following points:

2.1.1 Human Needs

Prior research endeavors within this domain have focused on examining the correlation between urbanization and human needs, primarily rooted in two prominent theoretical frameworks: Maslow's hierarchy of needs and Max Neef's conceptualization of human needs. According to recent studies Alsayed (2024), Max Neef's theory del Valle Barrera et al. (2022), and Max-Neef (1992) necessitate certain enhancements to effectively address the limitations associated with the consideration of social and cultural influences, as well as the incorporation of individual and collective needs and the inherently dynamic and evolving nature of these requirements, thereby enabling a more profound comprehension of human needs and the potential for their realization. Urban human needs (UHN) (Alsayed, 2024) is a systematic methodology that aims to clarify the fundamental needs of individuals and establish a connection between these needs and the diverse constituents in urban contexts. Ultimately, this framework aims to promote the evolution of an interconnected urban environment adept at addressing the aspirations and requirements of the citizenry. The construction of this framework was characterized by a thorough process of delineating the fundamental UHN elements: subsistence (existence), health and safety, identity and belonging, creativity and innovation, leisure, knowledge, freedom, flexibility and renewal, self-esteem, power, relationships and participation, and justice. Table 1 shows the concept of each Urban Human Needs element.

The utilization of the UHN theory in the current research was motivated by several key factors that warrant exploration and discussion. Firstly, the fundamental principle of interdependence and complementarity of needs forms the foundation of this theoretical framework, offering a strong and reliable basis for understanding the complex dynamics of cities and societies as interconnected systems. In addition, it is important to note that the UHN theory is one of the newest academic studies to look at humanitarian needs in cities. This shows how relevant and timely it is for solving problems in the present. Lastly, it is imperative to note that the UHN theory draws upon the foundational work of Max Neef, a prominent scholar whose theory emphasizes the integration of needs and has garnered significant acclaim and recognition in various academic research endeavors.

2.1.2 Urban Components

Cities represent intricate and multifaceted systems that emerge from a delicate equilibrium. According to Fan et al. (2023), cities heavily rely on substantial inputs to ensure their sustainability (Hassan et al. 2020; Hassan & Megahed, 2022). They operate as open systems engaged in a continuous interchange of resources, goods, services, individuals, and concepts. The genuine effectiveness of urban centers becomes apparent through their progression within the context of fulfilling human needs and establishing a conducive milieu for addressing both the physiological and psychological requirements of their inhabitants. Consequently, Fig. 1 delineates the 8 fundamental elements that contribute to the formation of cities. These elements can be divided into a set of physical elements and a set of motivational elements to build a civilization and identity in the city that will improve the efficiency of the city, as shown in Fig. 2. This research will focus on examining the relationship of human needs to the four physical elements of Smart Infrastructure, Building and Landuses, Smart Transportation, and Open Spaces & Landscape.

2.2 Generation of Database

The research methodology focuses on examining the components within the conceptual framework and analyzing the correlation between these components and aspects of urban areas. Urban centers can progress in the long term by utilizing indicators that facilitate the fulfillment of human

Table 1. Definitions of the Concept of Human Urban Needs. (Source: By author).

Needs	Definitions of the Concept of Human Urban Needs
Subsistence (Existence)	• Providing decent living and a healthy environment through adaptability • Food supply • Provision of Shelters and clean water sources • Providing fresh air • The ability to move from one place to another
Health and Safety	• It is related to Disease Resistance and Disease Control • Providing a safe environment capable of responding to natural hazards • Safety from crime and safe crossing of the road
Identity and belonging	• Connecting the user to the place and creating a relationship with the space • Control of community behaviors • Preserving the identity and beliefs of the individual and the place • Creating societies that allow their members to live in dignity and harmony with each other
Creativity and Innovation	• Creating capabilities for creativity and producing material requirements creatively and innovatively that parallels continuous developments
Leisure	• It is a sense of excitement, astonishment, relaxation, and getting fun experiences • Getting rid of psychological stress
Knowledge	• Ability to acquire knowledge, information, abilities, and skills required and communicate with others
Freedom	• Freedom to take decisions and actions • Control over life, property, and living in the conditions of its society
Flexibility and Renewal	• Flexibility in changing the shape of society and the city according to modern developments, using technology, and keeping abreast of the times
Self-Esteem	• Including autonomy, privacy, interaction, respect, community standing
Power	• Including creating a generating economic force capable of creating renewable jobs • Raising the Individual's economic level within cities
Relationships and Participation	• Creating a relationship with human surroundings, society, and the environment • Participating in society and public life • Creating a social environment, solidarity, and friendship
Justice	• Equity in access to various services within the city • A sense of justice among members of society

requirements. These indicators aid in the conceptualization of future human needs, providing insights into the potential success and humanitarian aspects of cities. Through a comprehensive review of existing literature about each aspect of urban development and the creation and enhancement of the necessary provisions for humanitarian requirements within the Urban Humanitarian Needs (UHN) components, these objectives are pursued.

The development of a database presents an important step that directly influences the characteristics of the results. This study focused on Scopus, which is widely regarded as the most accessible bibliometric source. Regarding the publication inclusion criteria, a literature search on Scopus was conducted by combining relevant keywords like "human needs," "urban," and "cities" with each element of urban separately ("Smart Infrastructure," "Buildings & Land Uses," "Smart Transportation," and "Open Spaces & Landscape"). The inclusion criteria of this paper focused primarily on human needs and urban environments. When considering the publication filtering criteria, the search process was conducted using the titles, abstracts, and keywords of publication materials.

A total of 337,154,342 and 842 publications were identified on 1 April 2024 for each element of urban separately ("Smart Infrastructure," "Buildings & Land Uses," "Smart Transportation," and "Open Spaces & Landscape"). Furthermore, this study used three filters—year of publication, subject area, and language—to reduce the number of pertinent papers that fit the criteria. This study focused on published research until the end of 2023. This study focused on "Social Science," "Environmental Science," "Engineering," and "Arts and Humanities" as subject s. Publications that were authored in languages other than English were not included. A total of 203, 88, 163, and 541 documents were thus discovered that satisfied the basic filters of each element of urban studies, respectively ("Smart Infrastructure," "Buildings & Land Uses," "Smart Transportation," and "Open Spaces & Landscape"). After the manual screening of the publication titles and abstracts to ensure that the research fits the goals and parameters and a careful examination of the research paper to examine the strategies used to achieve human needs, 40, 36, 60,

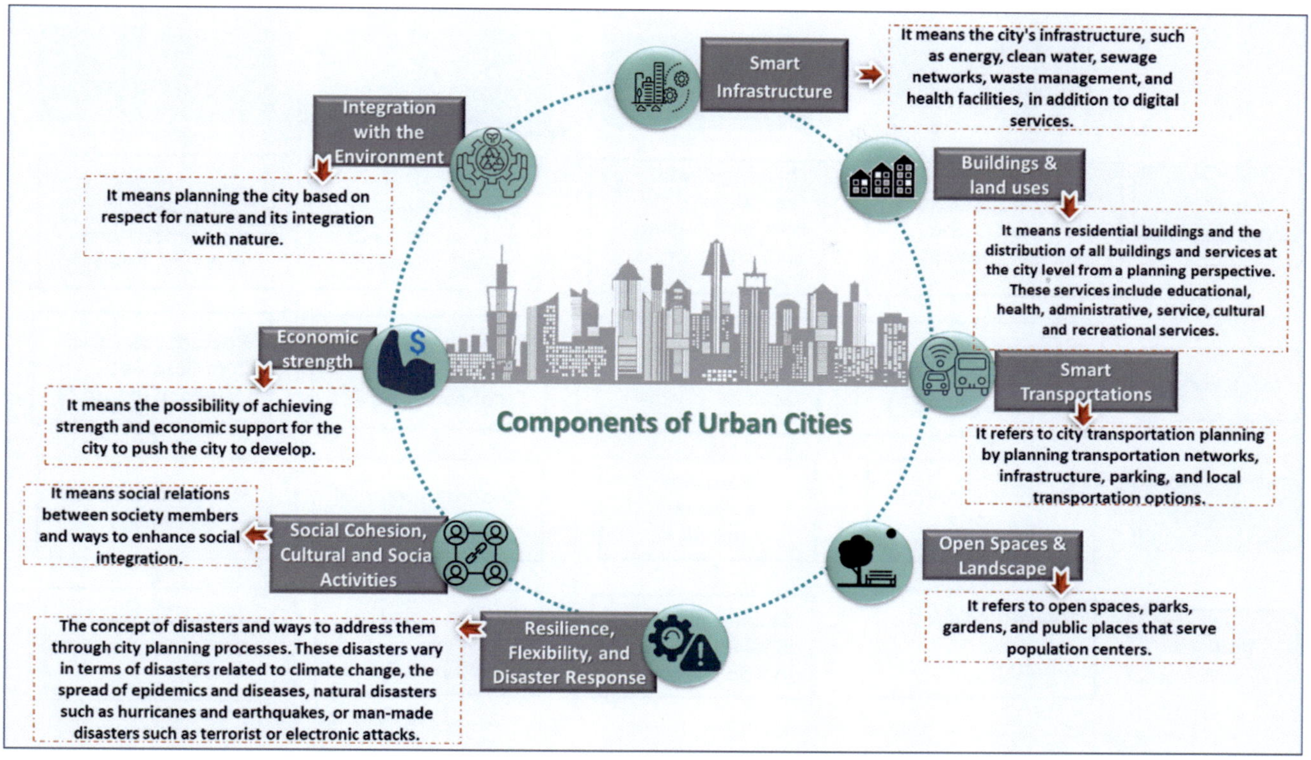

Fig. 1. Concept of urban cities' components. (Source: by Author)

Fig. 2. Classification of urban components. (Source: by Author)

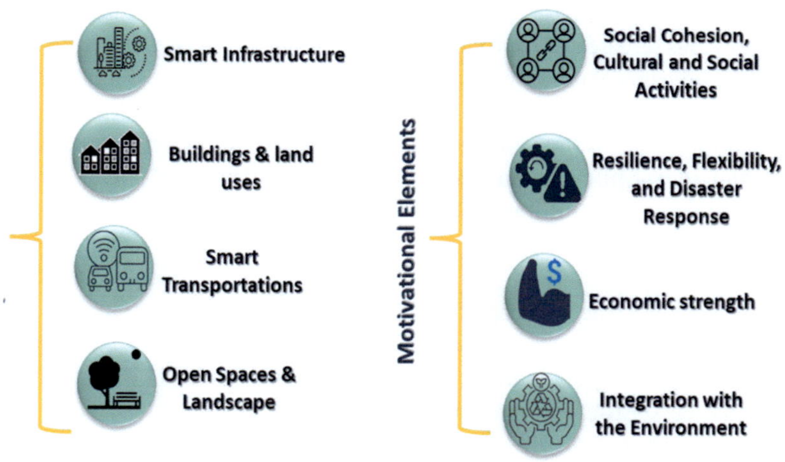

and 90 were selected for each element of urban, respectively ("Smart Infrastructure," "Buildings & Land Uses," "Smart Transportation," and "Open Spaces & Landscape"), as shown in Fig. 3.

3 Framework for Realizing Urban Human Needs (UHN) Through Urban Cities Components

The following framework examines whether the future urbanization of cities and their components can meet human needs. Table 2 (in the Appendix) presents this framework, which reviews human needs and links them to the urban components. The framework also illustrates the potential to fulfill these needs by utilizing the urban components, their standards, and sub-indicators, which gauge urban human needs and assist in gauging the degree of fulfillment of these needs in cities.

To attain this framework necessary for fulfilling urban human requirements, a methodological approach was devised through a series of systematic steps, which include:

- Establishing a comprehensive database for each of the four identified components of urbanism (namely, intelligent infrastructure, smart transportation, buildings & land uses, and open spaces & landscape).

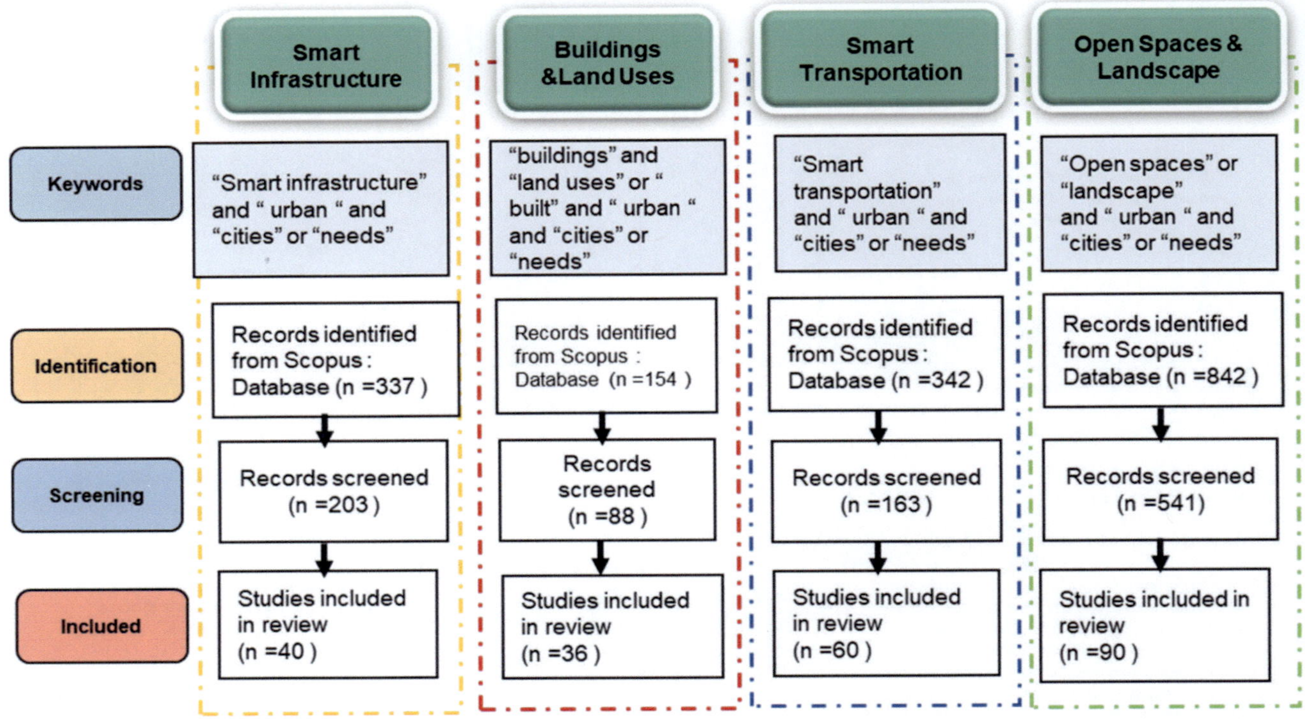

Fig. 3. Workflow of the selection procedure based on the PRISMA method. (Source: by Author)

- Table 1 illustrates the systematic categorization of previous research on each component of urbanism, linking it to elements of urban human needs (UHN) and grounding it in its conceptual framework.
- Through the examination and critical analysis of prior studies, specific criteria and indicators were developed to facilitate the fulfillment of human needs via urbanism components.

In essence, the conceptual framework of UHN, as delineated in the research conducted by Alsayed (2024), concerning urban components, was systematically classified by deconstructing and analyzing existing studies to derive indicators that allow for the assessment of the degree to which urban human needs are satisfied within urban environments.

3.1 Smart Infrastructure

Infrastructure serves as the backbone of the city's cohesion and stability (Wu et al. 2020). Infrastructure supports a range of daily activities in urban infrastructure including the use of energy, water, solid waste disposal, wastewater, transportation, and health services. According to Abdullah et al. (2022a) and Wu et al. (2020), infrastructure plays a crucial role in supporting various daily activities in urban areas. Presently, intelligent urban centers are implementing governance mechanisms aimed at tackling infrastructure challenges and enhancing overall system efficacy by leveraging technology-driven solutions, with Information and Communication Technology (ICT) serving as the cornerstone of such initiatives (Berglund et al. 2020). This enables swift real-time data communication and processing to enhance decision-making processes. Given the aforementioned points, the city's infrastructure emerges as a pivotal component that demands attention from the general populace in the course of their daily tasks. Consequently, it becomes imperative to fulfill these infrastructure requirements to cater to humanitarian necessities (Yang et al. 2021).

As a result, criteria and indicators have been outlined to fulfill these needs (see Fig. 4).

Subsistence: Achieving subsistence necessitates equal access to basic facilities regardless of age or ability (Elhachmi & Kobbane, 2022; Nakashima et al. 2022), measured by city-wide service availability and equitable distribution while preserving air quality (Mete, 2023). It is also crucial to offer digital services to all residents (Vipond et al. 2023), especially low-income families, and to boost adoption rates through specific programs.

Healthy & Safety: Improving safety measures and the dependability of critical infrastructure, such as water and energy systems, assesses the importance of enhancing public safety (Lindley, 1997). Effective emergency and crisis management involves monitoring and enhancing response

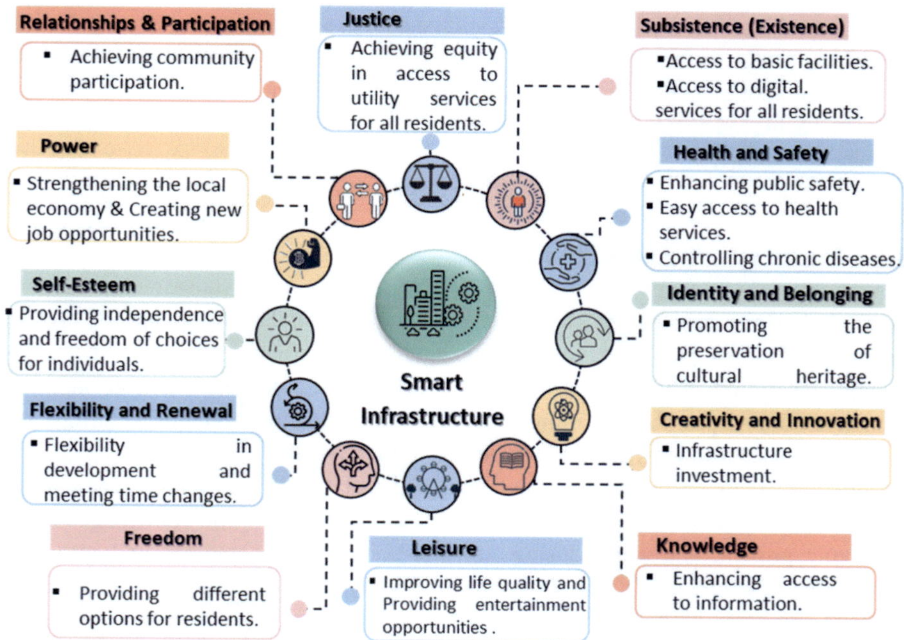

Fig. 4. Framework for achieving HUN through smart infrastructure. (Source: by Author)

times, managing chronic illnesses, and ensuring epidemiological safety (Pacheco et al. 2017; Venkatasubramanian et al. 2014).

Identify & belonging: are attained through preserving cultural heritage and offering various platforms for cultural expression (Ibrahim, 2021). Communication can be enhanced by utilizing digital technology. Social relations can be upheld, tourism can be promoted, and community identity can be reinforced.

Creativity & innovation: Investment in infrastructure encompasses various aspects like resource accessibility (Patvardhan & Santoki, 2019), support for startups, and growth of incubators. Funding availability and entrepreneurship programs are crucial, as are monitoring progress through patent numbers. The integration of IoT power management devices, smart parking systems, and waste management solutions is also important (Abdullah et al. 2022b; Mohamed et al. 2023; Vipond et al. 2023). The applicability of public-private partnerships in smart infrastructure development (Jayasena et al. 2023; Selim and ElGohary, 2020).

Leisure: Achieving well-being is possible with infrastructure that offers leisure opportunities, such as smart parks, sports facilities, and cultural initiatives that meet people's cultural needs (Aramouny, 2021).

Knowledge: According to Ibrahim (2021) It is recognized through improved information access via digital platforms and it fosters learning, awareness, and understanding of various topics to engage with citizens and promote development.

Freedom: According to Kanthavel et al. (2021), freedom refers to the population's ability to choose from a variety of facilities that meet their needs and are affordable.

Flexibility and Renewal: Infrastructure requires resilience to climate change and flexibility for future expansions (Kanthavel et al. 2021). Disaster prediction, response, and monitoring rely on emergency response plans, ample backup energy sources, critical infrastructure that functions effectively during extreme events (Nakashima et al. 2022), intelligent water management systems, adaptability to changing needs and conditions, and scalable infrastructure (Hingmire and Bhaladhare, 2023).

Self-esteem: The need to provide autonomy and freedom of choice for individuals, increase social integration, provide a safe and diverse infrastructure, and achieve privacy (Kumar et al. 2018).

Power: The strength of the infrastructure reflects in its ability to boost the domestic economy through job creation, GDP growth, reduced reliance on fossil fuels, and enhanced energy efficiency. To minimize waste, encourage renewable energy utilization, enhance water management, and decrease traffic emissions (Abdullahi and Sieng, 2023; Costa et al. 2023; Fedajev et al. 2023).

Relationships and Participation: Achieving community participation is enabling citizens' participation in decision-making processes and providing access to information and opportunities for the population to participate in decision-making processes (Li et al. 2023), open data availability, use of digital tools, availability of open data, and use of digital tools to facilitate participation in local government processes and implementation of community participation events (Vipond et al. 2023).

Justice: Equity in access to facilities can be achieved and measured by ensuring access for all residents (Li et al. 2023), regardless of income. All individuals should be able to spend a fair proportion of their income on facilities. The benefits of infrastructure must be distributed fairly. Social realization should be achieved along with ensuring the availability of data privacy and security.

3.2 Buildings and Land Uses

Land use in cities is shaped by economic, social, and cultural factors, and urban planning can manage it effectively for sustainable growth. Building practices can address various human needs and promote flexibility (Abdullah et al. 2022a; Kramer et al. 2014), equity, and innovation. Innovative and sustainable practices in building and land use can create resilient and equitable societies. Affordable housing, health, economic growth, and social justice are just some of the needs these practices can fulfill while also preserving cultural heritage and promoting innovation. As a result, criteria and indicators have been outlined to fulfill these needs (see Fig. 5).

Subsistence: Subsistence ensures the population's access to affordable housing and services (Ghosh, 2019), achieved by diversifying land uses and increasing building density. It also connects residents to green spaces for better mental health, stress reduction, and access to basic services (Hatami et al. 2023).

Health and Safety: Enhancing public safety in residential areas involves boosting walking rates and linking services to promote physical health (Khaddar et al. 2023). Land-use planning that prioritizes crime prevention can enhance safety by protecting individuals from harm, crime, and natural disasters.

Identity and Belonging: This need is achieved by promoting local identity and the adaptive reuse of historical buildings, preserving the community's character by providing new uses for ancient structures (Yusuf et al. 2023).

Creativity and Innovation: Through the integration of smart technology with the design of buildings, we should adopt smart technologies to enhance the functionality and efficiency of buildings and land uses and provide training and educational programs (De Arruda et al. 2023).

Leisure: By providing opportunities for recreational activities, parks, sports facilities, and cultural venues, integrating them with residential buildings, and ensuring simple access to them, buildings and land uses can enhance public welfare and people's quality of life (Huang et al. 2023).

Knowledge: Supporting buildings to access information and knowledge through spaces for libraries, educational institutions, and cultural centers promotes lifelong learning and intellectual development.

Freedom: Achieving social justice lies in the availability and affordability of housing options for different income groups.

Flexibility and Renewal: Utilizing disaster-resistant materials and adaptable, flexible structures with climate changes is crucial for achieving flexibility in building design (Betti et al. 2024). Integrating architectural designs, flexible floor plans, and multipurpose spaces is essential for accommodating changing needs. Supporting building codes requiring flexible design and contingency preparedness plans confirms future changes (Mariano & Rossi, 2023; Sánchez-Silva et al. 2024).

Privacy: individual autonomy, public spaces, and diverse services can foster self-esteem by promoting autonomy and

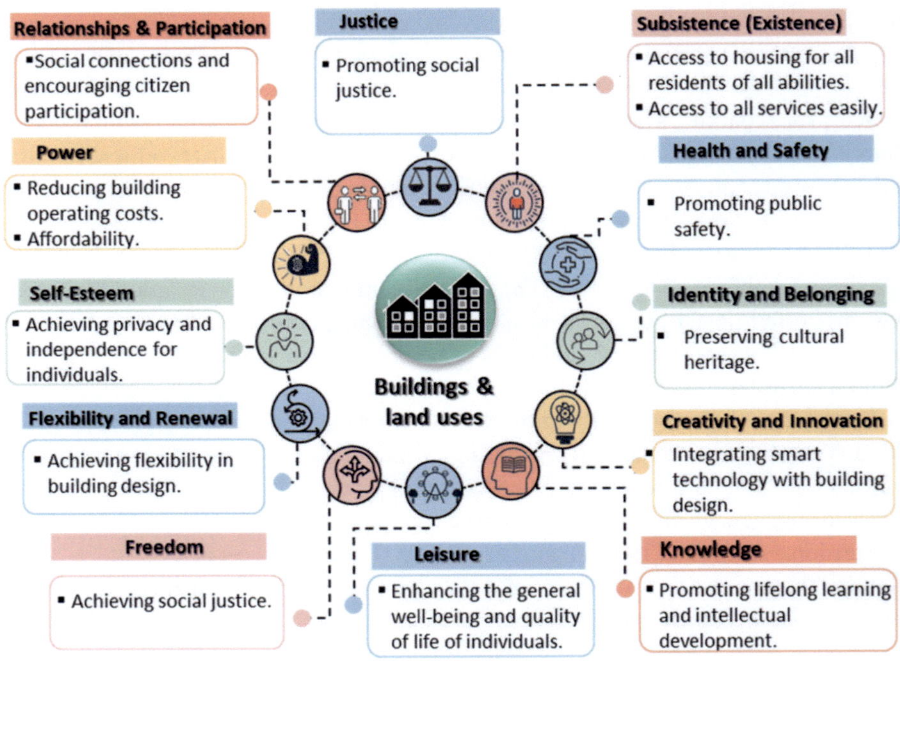

Fig. 5. Framework for achieving HUN through buildings & land uses. (Source: by Author)

self-expression, ultimately enhancing data privacy and security.

Power: The economic strength of buildings and land uses lies in their ability to operate through the design of energy-efficient and affordable buildings.

Relationships & Participation: Increasing social connections and encouraging citizens' participation, including public spaces, parks, assembly areas, and multi-use developments that promote social cohesion (Ghosh, 2019), involving residents in planning and design processes to ensure that buildings and land uses meet their needs, as well as workshops on participatory design, community meetings, and feedback sessions (Dash, 2023).

Justice: Justice is achieved through the promotion of social justice, which can be realized by creating mixed-income societies and reducing social and economic segregation (Fox, 2020; Moser & Avni, 2023).

3.3 Smart Transporation

Urban cities face challenges in moving people and goods efficiently and sustainably. The future of urban transport may involve public transport, electric vehicles, bike sharing, and pedestrian-friendly infrastructure (Aly et al. 2023). Smart Transportation utilizes data and technology to enhance the efficiency, sustainability, and safety of transportation systems. It has the potential to improve lives globally by making transportation more efficient, sustainable, and safe. Smart transportation can address future needs through improved mobility, emission reduction, safety enhancement, security through facial recognition technology, sustainability, and efficiency (David Schrank, 2021; Pinar Caglin, 2020; Seliverstov et al. 2020). As a result, criteria and indicators have been outlined to fulfill these needs (see Fig. 6).

Subsistence: Subsistence involves enhancing accessibility through affordable transport options, including buses, trains, bicycles, and ride-sharing services. It also entails creating infrastructure for walkers and individuals with disabilities, as well as ensuring access to basic services and sanitation. Furthermore, subsistence planning involves considering distances to stations and transit points (Hatami et al. 2023; Porter et al. 2005).

Health and Safety: Improving safety and security is achieved by offering secure transportation choices to reduce accidents and crime, along with accessible emergency services (Cong et al. 2021; Jagatheesaperumal et al. 2024). Decreasing crime and accidents linked to transportation and traffic control contributes to enhancing safety infrastructure (Aldegheishem et al. 2018). Public health is also improved by reducing air pollution from transportation (Taufik et al. 2023).

Identify and Belonging: Enhancing identity and a sense of belonging is achieved through the use of smart transportation systems and various transportation options that allow the population to express their identity and culture and promote cultural diversity.

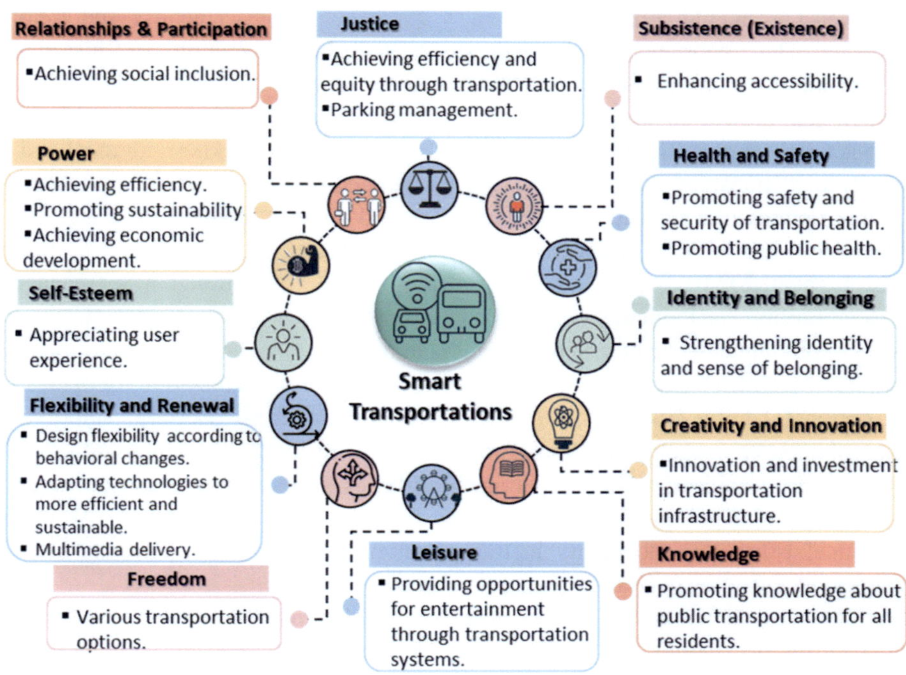

Fig. 6. Framework for achieving HUN through smart transportation. (Source: by Author)

Creativity and Innovation: There are various indicators to measure innovation and infrastructure investment, including tracking traffic, accessibility of transportation options, and transportation-related patents. Transportation efficiency can be improved by controlling travel time, measuring road congestion, and providing traffic information (Slamet et al. 2017).

Leisure: The enhancement of transportation welfare includes bike-sharing programs and walking tours (Li et al. 2021), information on serene parks and picturesque routes, relaxation options during travel, and the presence of shaded trees along transport routes.

Knowledge: Improving public transport knowledge for everyone by sharing transport options on digital platforms, educating about transport safety and sustainability, and enhancing digital connectivity in various communities (Awan et al. 2020).

Freedom: The availability of various transportation options gives individuals freedom of choice based on their wishes and possibilities, providing easy access.

Flexibility and Renewal: Transport flexibility is needed for behavioral changes (Taghipour et al. 2023), promoting public transport use and reducing traffic congestion. Technologies can help reduce congestion, improve fuel efficiency, and lower emissions. Flexibility is attainable through multimodal connectivity, offering various transportation options and integrated services at transportation hubs.

Self-esteem: User service estimation relies on the availability of real-time transportation information (Porter et al. 2005), infrastructure quality, services, and customer satisfaction levels (Awan et al. 2020).

Power: The importance lies in endorsing sustainability and economic progress (Xie & Gu, 2024). It can be gauged through job growth in transport industries, transportation options to job centers (Danilina & Slepnev, 2018), economic activity near transport infrastructure, and prioritizing eco-friendly options like electric vehicles, bicycles, and renewable energy integration to lower carbon emissions (Bakri et al. 2022; Lu et al. 2024).

Relationships and Participation: Social inclusion is attained through community involvement in transport planning, infrastructure investment based on local needs, and access to transport options (Danilina & Slepnev, 2018). Social isolation varies among demographic groups, and transportation availability for people with disabilities and civic participation levels differ among population groups (Wen et al. 2020).

Justice: Efficiency and equity in transportation can be achieved by engaging citizens in planning, improving access for various groups (Eom et al. 2022), ensuring affordability and security, distributing services equitably, and managing parking while considering costs and benefits (Taufik et al. 2023).

3.4 Open Spaces and Land Uses

The importance of open space lies in its recognition of the necessity for cities to offer green areas to residents (Thompson, 2002). Open spaces bring various advantages to individuals and society. Prioritizing open spaces can enhance residents' quality of life and the sustainability of communities. These spaces enhance people's well-being in multiple ways and foster a sense of community and relaxation (Çelikyay, 2020; Ibrahim & Nizarudin, 2019). They provide leisure and entertainment, promote social interaction, and support mental and physical health. Open spaces can also encourage physical activity and support biodiversity by offering wildlife habitats, thus contributing to ecosystem health and sustainability (Jia et al. 2020). As a result, criteria and indicators have been outlined to fulfill these needs (see Fig. 7).

Subsistence: Planning their distribution and connecting them to residential areas is crucial for enhancing access to open spaces and walking (Kurniawan & Roychansyah, 2023). Ensuring sufficiency in open spaces to meet population needs and accommodate recreational facilities is essential for subsistence. The size, quantity, and suitability of open spaces must be considered for providing recreational and green spaces (Eltarabily, 2022; Wang et al. 2022).

Health and Safety: The presence of lighting, surveillance, and security personnel achieves green space security. The low rate of open crime in green spaces serves as a measure of security. The availability of open spaces for physical activities like jogging and cycling, which promote relaxation, stress reduction, and mental regeneration, also contributes to health (Goryacheva et al. 2023).

Identify and Belonging: An opportunity to display the cultural and historical heritage of the city (Khorasgani & Haghighatbin, 2023). Evaluating the presence of cultural monuments or heritage sites in open spaces promotes acknowledgment and conservation of cultural and historical importance in urban open spaces (Goryacheva et al. 2023; Heneidy et al. 2022).

Creativity & innovation: Investment in sustainability through water management, wastewater systems, green cover maintenance (Como et al. 2023), and open-space environmental monitoring systems, along with data and Internet utilization for open space control (Bondar et al. 2022).

Leisure: Improving interrelationships in open spaces: Evaluating features like vegetation and amenities. Offering leisure and relaxation areas. Enhancing air quality and decreasing noise pollution during recreational use (Dogan, 2023; Yu & Lam, 2022).

Knowledge: Assessing educational programs, signs, natural corridors, and interactive establishments to improve environmental education and sustainability awareness (Cheng et al. 2022). Utilizing data and technology to evaluate green space

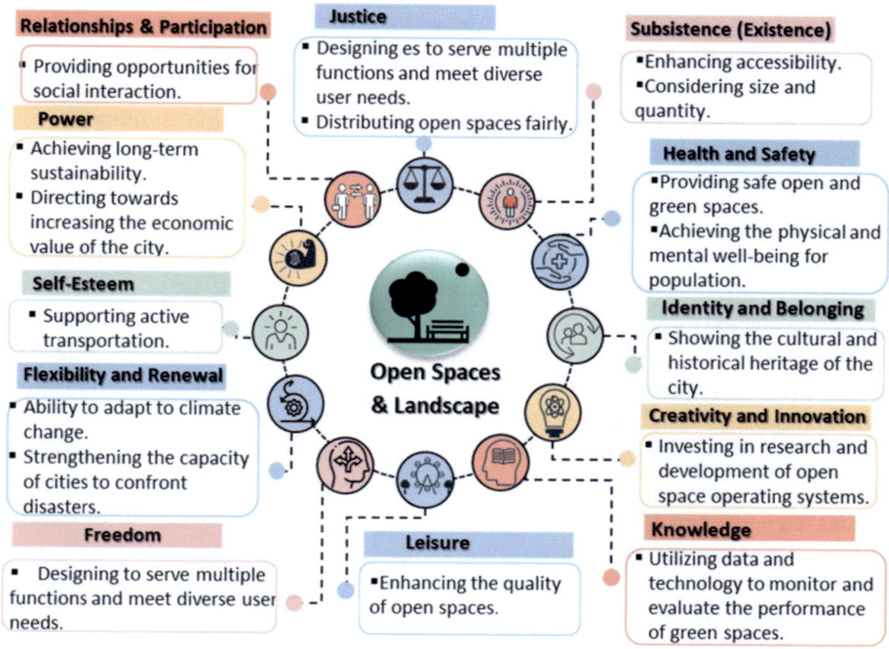

Fig. 7. Framework for achieving HUN open spaces & landscape. (Source: by Author)

performance by measuring user satisfaction through surveys or digital feedback mechanisms (Jim & Hui, 2022). Knowing user behavior in open spaces (Zhao et al. 2022).

Freedom: Designing open spaces for various functions by evaluating available amenities (Haq et al. 2023). It is important to consider the diverse needs of users, taking into account factors such as age, culture, and abilities. Evaluating open spaces' diversity in activities supported and connections between spaces, such as green lanes and transport links.

Flexibility and Renewal: Enhancing adaptability to climate change through climate adaptation and mitigation options (Lin et al. 2022). The evaluation of urban green features and water management systems in open areas is crucial. Encouraging incorporation of climate resilience strategies and green space growth to boost cities' disaster resilience (Mohy & Rasheed, 2022).

Self-esteem: By measuring the extent of pedestrian-friendly infrastructure, dedicated cycling lanes, and bike-sharing systems within open or related spaces, one can achieve self-esteem by supporting active transportation (Nahdatunnisa et al. 2023).

Power: Achieving long-term sustainability by measuring the use of environmentally friendly materials, achieving energy efficiency measures, long-term maintenance plans, and steering towards raising the economic value of the city. This includes measuring the economic benefits of open spaces, such as increasing real estate value and tourism revenue (Gurung & Gowreesunkar, 2023), creating jobs through maintenance, programming, and supporting open-space investment.

Relationships and participation: It involves measuring the frequency and diversity of social activities in open spaces and the extent of public participation (Schrammeijer et al. 2022), participating in the planning and management of open spaces, and strengthening community participation strategies.

Justice: Designing open spaces for various functions to cater to different user needs, such as playgrounds, sports facilities, seating areas, walking trails, and cultural spaces, aims for inclusivity for all age groups, cultures, and abilities (Haq et al. 2023). Ensuring a fair distribution of open spaces across neighborhoods is essential.

4 Discussion

Based on the previous analysis of previous studies, the research presented a general framework to guide planners in planning cities based on human needs, or cities for humans, within the context of future trends. This framework was designed to accommodate the changes that occur over time as a result of fulfilling human needs. The methodology employed led to the development of the necessary framework to meet urban human needs. This was achieved through a series of steps, including the establishment of a comprehensive database for each of the four identified urbanism components (Smart Infrastructure, Buildings & Land Uses, Smart Transportation, and Open Spaces and Landscape). Additionally, we organized prior studies for each urbanism component by associating them with UHN elements and according to their concepts. Finally, through the examination and analysis of past research, criteria and indicators were established to fulfill human needs via urbanism components. This framework was created by taking the UHN idea from Alsayed's research on urbanism components and putting them into logical groups. These groups were made by breaking down and analyzing previous studies to find indicators that show how well human needs are met in cities.

Human needs are interconnected and integrated; one need cannot be prioritized over another to achieve satisfaction among the population.

Human needs and their fulfillment are interconnected with each other, and the integration of cities is achieved by meeting all human needs.

The relative weights of these needs differ from one component to another in urbanization, and their importance varies according to geographical location and cultural changes. Therefore, this framework is flexible and can be developed to suit the changes in different societies, cultures, and times.

The results showed how complexly human needs and smart infrastructure are linked. They emphasized how important smart infrastructure is for human development and meeting a wide range of needs, from basic livelihoods to overall well-being and economic progress. This is done by making sure that everyone has equal access to infrastructure services, regardless of their social or economic status or other factors. It also talked about how important technology is for making infrastructure more efficient, long-lasting, and resilient. It suggested that planning and building infrastructure that works for everyone requires a deep understanding of people's needs and a strong commitment to equality and sustainability. The active participation and engagement of communities in the decision-making process are of paramount importance in formulating a holistic approach to infrastructure planning, which must take into account both technical and social dimensions.

As a discussion by (Y. A. Abdullah et al. 2022b; Huang et al. 2023; Kramer et al. 2014; Zhou et al. 2022). Meeting the diverse needs of humanity through the implementation of effective and strategic land use planning serves as an exceptionally potent mechanism for tackling a variety of pressing social challenges, including, but not limited to, the availability of affordable housing, the enhancement of public health outcomes, the facilitation of sustained economic growth, and the promotion of social justice. Furthermore, this approach is equally efficacious as a formidable instrument for advancing the principles of social justice and ensuring the long-term viability of sustainability initiatives within various communities.

Establishing transportation systems that meet human needs can enhance efficiency. First, smart transportation systems can improve efficiency by optimizing traffic flow, reducing congestion, and improving the reliability of public transportation. This can lead to shorter travel times, reduced fuel consumption, and lower emissions, contributing to a more sustainable and livable city. Second, smart transportation can enhance safety by leveraging technologies such as advanced traffic management systems, intelligent transportation systems (ITS), and connected vehicles. These technologies can help prevent accidents, improve traffic flow, and enhance the safety of pedestrians and cyclists. Additionally, smart transportation can improve security through the use of facial recognition technology and other surveillance systems to deter crime and enhance public safety, as proven in David Schrank, 2021; Hatami et al. 2023; Pinar Caglin, 2020; Porter et al. 2005; Seliverstov et al. 2020.

The interdependence of public spaces, human needs, and urbanization is critical to examine. By comprehensively analyzing the correlation between public spaces and elements such as sustenance, well-being, safety, community belonging, creativity, recreation, knowledge acquisition, autonomy, self-worth, authority, interpersonal relationships, and equitable justice, policymakers and urban planners may formulate more efficacious strategies for establishing sustainable, fair, and resilient urban environments. Moreover, it underscores the crucial role of community engagement in the decision-making processes pertinent to public spaces. By involving citizens in the planning and management phases, it becomes feasible to ascertain that public spaces adequately satisfy the community's requirements and foster social unity. This involvement can yield an enhanced sense of ownership and civic pride, thereby cultivating a more dynamic and sustainable urban habitat.

This framework ultimately aims to encompass all human needs across various aspects of urban areas, with the ultimate goal of enhancing urban life quality and fulfilling the population's satisfaction. People view it as a strategic concept that aims to achieve future sustainability in the context of smart cities, transforming them to benefit humanity instead of the other way around. The primary focus of this framework is on catering to individuals and realizing their aspirations. It is envisioned as a forthcoming avenue for imbuing future smart cities with a more human-centric approach.

Regarding the limitations, numerous potential restrictions must be taken into account. Primarily, the development of the database, a pivotal phase, laid the groundwork for this particular examination. On April 1, 2024, an exploration was carried out utilizing a restricted set of keywords within narrowly defined groupings, yielding specialized outcomes that lacked generalizability. A lot of literature was also left out of this study by using the Scopus database and its three criteria: categories, years (until the end of 2023), publication types, and language; or by carefully reading each article by hand to make sure it fit with the study's goals and scope, as well as its relevance to the topic.

This work strengthened the possibility of achieving human needs by formulating a framework that aids in assessing cities and understanding human needs for the future. The framework for achieving human needs serves as an initial step towards humanizing cities. Researchers can develop this tool through applied studies on cities, providing guidance to urban planners and designers in city planning. With the involvement of programs like Space Syntax or GIS, it serves as an effective assessment tool for specialized studies on urban spaces, enabling the study and fulfillment of human needs in cities.

5 Conclusion

The success of building communities relies on meeting people's needs in biological, psychological, and social environments for security and well-being. Failure diminishes social potential and hinders community building. The creation of enduring and humane communities is a significant task due to planners' focus on intelligent cities and ecological sustainability. A research project aims to link human needs with urbanization processes to cater to urban dwellers. The study yielded the following findings: The study identified indicators to measure the fulfillment of these needs in various urban aspects, evaluated their possibility through urban standards, and analyzed human needs in urban settings. It focused on meeting urban human needs for future changes that were adaptable to various societies.

In the end, the study contributes to achieving human needs and making cities more humane. This framework ultimately seeks to address all human needs across various aspects of urban areas, enhance the quality of urban life, and ensure satisfaction among the populace. Furthermore, the study emphasizes the significance of considering people's needs and requirements in city planning, and advocates for the development of more integrated and humane cities. People perceive it as a methodical approach to achieving long-term sustainability in smart city contexts, reshaping them to prioritize human welfare over other factors. The main focus of this framework lies in catering to individuals and fulfilling their aspirations, positioning it as a prospective avenue for imbuing forthcoming smart cities with human-centric qualities.The research led to a comprehensive framework for future cities, guiding urban development to meet human requirements. This framework aids in assessing cities, understanding human needs for the future, and indicating the values that towns will embody. The findings of this research serve as a new key for researchers to apply their knowledge and conduct more specialized studies on how to achieve human needs through urban design and city planning.

Appendix

Table 2. Framework for Realizing Urban Human Needs (UHN) through Urban Cities Components. (Source: By author)

	Urban human needs	Criteria	Indicators
1. Subsistence (Existence)	**Smart Infrastructure**	**Access to basic facilities**	– The extent to which access to basic facilities such as energy, clean water, waste management, and food is guaranteed for all residents regardless of age or ability – The extent of access to health care services
		Access to digital services for all residents	– The extent of access to digital services for all residents – The effectiveness of implementing programs that assist in providing facilities to low-income families – The rate of adoption of digital services such as government services – Number of public WI FI hotspots
	Buildings & land use	**Access to housing for all residents of all abilities**	– Availability of affordable housing – The extent of diversity of land uses – The percentage of increasing building densities to increase land efficiency
		Access to all services easily	– The extent to which housing is linked to access to green spaces to improve mental health and reduce stress – The extent to which paths are provided linking residential buildings to basic services
	Smart Transportations	**Enhancing accessibility**	– The availability of all means of transportation for all members of society at reasonable prices – Integrating different modes of transportation such as buses, trains, bicycles, and ride-sharing services – The extent to which infrastructure is provided that supports pedestrians and people with disabilities – The extent of providing access to basic services and health facilities through various transportation options – Availability and affordability of transportation options – There is a planning of the distance to stations and transit stations – The quality of transportation infrastructure
	Open Spaces & Landscape	**Enhancing accessibility**	– Providing easy access to open spaces for all residents and easy walking to them – Planning the distribution of open areas and integrating them with residential areas and transportation lines – The extent to which people benefit from the recreational and social activities it provides
		Considering size and quantity	– Providing a sufficient amount of open space to meet the needs of residents – Providing different opportunities for various activities – The size of the open spaces is appropriate to accommodate recreational facilities, green spaces, and facilities
2. Health & Safety	**Smart Infrastructure**	**Enhancing public safety**	– Strengthening safety and security measures and increasing public safety – Reducing high crime rates – Providing reliable infrastructure such as energy and water networks – Improving emergency response times and monitoring and managing traffic
		Easy access to health services	– Ease of access to health services
		Controlling chronic diseases	– Reducing and controlling the incidence of chronic diseases – Increased life expectancy

(continued)

Table 2. (continued)

	Urban human needs	Criteria	Indicators
	Buildings & land use	**Promoting public safety**	– Linking services to residential areas to increase walking rates and enhance the physical health of residents – The effectiveness of crime prevention strategies when planning land use – Providing a safe and secure infrastructure for individuals that protects them from physical harm, crime, and natural disasters
	Smart Transportations	**Promoting the safety and security of transportation**	– Providing safe and secure transportation options that reduce risks of accidents and crime – The quality and effectiveness of the availability of emergency services – The extent to which the level of crime and transportation-related accidents has decreased – Quality of infrastructure in terms of safety characteristics in transportation
		Promoting public health	– The level of air pollution resulting from transportation – Availability of health care services near infrastructure and transportation services
	Open Spaces & Landscape	**Providing safe open and green spaces**	– Measuring factors such as lighting, surveillance, and the presence of security personnel – The extent to which crime rates decrease in open spaces
		Achieving the physical and mental well-being of the population	– The availability of open spaces for physical activities such as jogging and cycling – The presence of elements that promote relaxation, stress reduction, and mental rejuvenation
3. Identity & Belonging	**Smart Infrastructure**	**Promoting the preservation of cultural heritage**	– Providing opportunities for cultural expression and communication using digital technology – Promoting tourism and strengthening community identity
	Buildings & land use	**Preserving cultural heritage**	– Strengthening local identity through urban design and buildings – Adaptive reuse of historic buildings in preserving community character – Providing new uses for old structures
	Smart Transportations	**Strengthening identity and sense of belonging**	– Using intelligent transportation systems to help people feel a sense of belonging and self-esteem – Providing different transportation options that allow residents to express their identity and culture – Availability and accessibility of transportation options that promote cultural diversity and expression
	Open Spaces & Landscape	**Showing the cultural and historical heritage of the city**	– Evaluating the presence of cultural monuments or heritage sites within open spaces – Promoting recognition and preservation of the cultural and historical significance of urban open spaces
4. Creativity & Innovation	**Smart Infrastructure**	**Infrastructure investment**	– Providing access to resources and support for startups and small businesses – Availability of financing for startup companies – Number of patents filed – Strengthening entrepreneurship programs to implement policies that encourage innovation and entrepreneurship – Integrating LOT energy management devices, smart parking systems, and smart waste management solutions
	Buildings & land use	**Integrating smart technology with building design**	– Access to creative spaces and resources – Availability of training and educational programs – Adopting smart technologies to enhance the functions and efficiency of buildings and land uses

(continued)

Table 2. (continued)

	Urban human needs	Criteria	Indicators
	Smart Transportations	**Innovation and investment in transportation infrastructure**	– Promoting innovation in transportation technology, such as providing new ways to track traffic – Availability and accessibility of transportation options that allow people to engage in creative activities – Level of access to transportation options that allow residents to pursue creative activities – The level of investment in research and development in the field of transportation – Number of patents related to transportation technologies – The rate of adoption of new transportation technologies and services – The efficiency of the average travel time to work or school – Level of congestion on roads and highways – Availability of real-time traffic information
	Open Spaces & Landscape	**Investing in research and development of open-space operating systems**	– Investing in ways to achieve sustainability through water management, waste systems, and green cover maintenance – Developing environmental monitoring systems in open spaces – Investing in exploiting data and the Internet and controlling open spaces through them
		Supporting the circular economy	– Supporting the concept of the circular economy, which aims to reduce waste, maximize resource efficiency, and encourage the reuse and recycling of materials – Measuring the adoption of circular economy practices and reducing waste generation – Promoting sustainable consumption and production patterns
		Sustainable entrepreneurship linked to environmental goals **Promoting and supporting creativity and innovation**	– Enhancing cooperation between various stakeholders, including governments, companies and academia – Encouraging partnerships between the public and private sectors in sustainable development projects – Embracing technological progress and innovation to support environmental integration – Exploring smart city technologies, data analysis, and Internet of Things (LOT) devices – Improving resource management, energy efficiency, and transportation systems – Encouraging research and development in the field of sustainable technologies – Supporting startups and companies that focus on environmental solutions
5. Leisure	**Smart Infrastructure**	**Improving life quality**	– Providing entertainment opportunities through smart parks, sports facilities, and cultural initiatives
	Buildings & land use	**Enhancing the general well-being and quality of life of individuals**	– The extent to which buildings and land use provide opportunities for recreational activities – Providing parks, sports facilities, cultural places, and green spaces
	Smart Transportations	**Providing opportunities for entertainment through transportation systems**	– Providing bicycle sharing or walking tour programs – Providing information about quiet parks or scenic routes – Availability of opportunities for relaxation during transportation and the level of coverage of shaded trees along the transportation corridors
	Open Spaces & Landscape	**Enhancing the quality of open spaces**	– The extent to which open spaces are connected – Evaluating the physical characteristics of open spaces, such as the size of vegetation and amenities – Providing places for entertainment and relaxation – The extent of improving air quality and reducing noise pollution – The extent of participation in recreational activities

(continued)

Table 2. (continued)

	Urban human needs	Criteria	Indicators
6. Knowledge	**Smart Infrastructure**	**Enhancing access to information**	– Providing access to information and knowledge through digital platforms – Promoting education, awareness, and understanding of various topics
	Buildings & land use	**Promoting lifelong learning and intellectual development**	– Supporting buildings for access to information and knowledge by providing spaces for libraries, educational institutions, and cultural centers – Availability of public facilities that promote learning and intellectual development
	Smart Transportations	**Promoting knowledge about public transportation for all residents**	– Providing information about transportation options through digital platforms – Educating the population about transportation safety and sustainability – The level of digital communication in different societies
	Open Spaces & Landscape	**Promoting environmental education and sustainability awareness**	– Evaluating the availability of educational programs, interpretive signs, and natural trails in the interactive facility
		Utilizing data and technology to monitor and evaluate the performance of green spaces	– Measuring user satisfaction through surveys or digital feedback mechanisms – Evaluating patterns of use of open spaces using sensors or mobile phone applications
7. Freedom	**Smart Infrastructure**	**Providing different options for residents**	– Providing different options for residents to use the facilities depending on their needs and ability to afford them
	Buildings & land use	**Achieving social justice**	– Availability and affordability of housing options for different income groups
	Smart Transportations	**Various transportation options**	– Availability of various transportation options and giving individuals the freedom to choose based on their desires and capabilities – Providing easy access to it
	Open Spaces & Landscape	**Designing open spaces to serve multiple functions and meet diverse user needs**	– Measuring the presence of amenities such as playgrounds, sports facilities, seating areas, walking paths, and cultural spaces – Taking into account the needs of different age groups, cultural backgrounds, and abilities – Evaluating the diversity of open spaces by measuring the range of activities they support
		Connecting open spaces and integrating them into urban areas	– Evaluating the connection and connections between open spaces, such as the presence of green corridors, pedestrian and cycling paths, and connectivity to public transport – Encouraging the creation of a cohesive and accessible open space system
8. Flexibility and Renewal	**Smart Infrastructure**	**Flexibility in development and meeting time changes**	– Building infrastructure capable of withstanding climate change – The extent of development of real-time monitoring and analysis systems – The availability of emergency response plans – Availability of several backup power sources – Reliability of vital infrastructure during severe weather events – Using smart water management systems – Flexibility and adaptability to changing needs and circumstances – Designing the infrastructure so that it is scalable and capable of facing disasters and emergency changes

(continued)

Table 2. (continued)

	Urban human needs	Criteria	Indicators
	Buildings & land use	**Achieving flexibility in building design**	– Innovation in building design and designing buildings with materials resistant to flood and earthquake disasters – Designing buildings to be adaptable and flexible to changing needs – Integrating architectural designs, flexible floor plans, and multi-purpose spaces to suit different purposes and accommodate future changes – Resilience through planning and adaptation to the effects of climate change – Supporting building codes that require flexible design and emergency preparedness plans
	Smart Transportations	**Design flexibility according to behavioral changes**	– Taking into account encouraging behavioral changes in the population, such as using public transportation, cars, or riding bicycles – The availability of awareness campaigns to reduce traffic congestion
		Adapting smart transportation technologies to make it more efficient and sustainable	– Using technologies to reduce traffic congestion, improve fuel efficiency and reduce emissions
			– Level of preparedness for natural disasters – The durability of transportation infrastructure and services during natural disasters – The level of community resilience after natural disasters
		Multimedia delivery	– Availability of multiple transportation options in transportation centers – Number of bike-sharing stations and parking facilities – The level of integration between different transportation services
	Open Spaces & Landscape	**Ability to adapt to climate change**	– Providing opportunities to adapt to the climate and mitigate its effects – Measuring the presence of features such as urban forests, green roofs, permeable surfaces, and water management systems within open spaces – Promoting the integration of climate resilience strategies and developing green spaces
		Strengthening the capacity of cities to confront disasters	– Evaluating the presence of floodplains – The presence of green infrastructure for rainwater management
9. Self-Esteem	**Smart Infrastructure**	**Providing independence and freedom of choice for individuals**	– Control and effectiveness of providing levels of social integration – The extent to which a safe and diverse infrastructure is provided
	Buildings & land use	**Achieving privacy and independence for individuals**	– Creating public spaces and various services to encourage independence and self-expression – Enhancing data privacy and security
	Smart Transportations	**Appreciating user experience**	– Availability of real-time transportation information – Quality of infrastructure and services – The level of customer satisfaction with transportation options
	Open Spaces & Landscape	**Supporting active transportation**	– Measuring the extent of pedestrian-friendly infrastructure and dedicated cycling lanes – Availability of bicycle-sharing systems within or connected to open spaces
10. Power	**Smart Infrastructure**	**Strengthening the local economy**	– Creating new job opportunities – The extent of Gross domestic product (GDP) growth – Reducing dependence on fossil fuels and improving energy efficiency – Reducing waste by promoting the use of renewable energy sources – Improving water management practices – The ability to withstand the effects of climate change

(continued)

Table 2. (continued)

	Urban human needs	Criteria	Indicators
	Buildings & land use	**Reducing building operating costs**	– Designing energy-efficient buildings that can withstand extreme temperatures – The extent to which sustainability is achieved by constructing buildings using sustainable materials and energy-saving systems – Evaluating the integration of renewable energy sources such as solar panels and wind turbines
		Affordability	– The extent to which residents are guaranteed access to safe, decent, and affordable housing that meets their needs
	Smart Transportations	**Achieving efficiency**	– Achieving efficiency by reducing congestion and travel times while maximizing the capacity of transportation infrastructure
		Promoting sustainability	– Prioritizing environmentally friendly options such as electric vehicles, bicycles, and pedestrian-friendly infrastructure – Reducing traffic emissions – Integrating renewable energy sources into transportation systems to reduce carbon emissions
		Achieving economic development	– The level of job growth in transportation-related industries – Availability of transportation options to employment centers – Level of economic activity near transport infrastructure and services
	Open Spaces & Landscape	**Achieving long-term sustainability**	– Measuring the extent of use of environmentally friendly materials – Achieving energy efficiency measures – Developing long-term maintenance plans
		Directing towards increasing the economic value of the city	– Measuring the economic benefits resulting from open spaces, such as increasing real estate values and tourism revenues – Creating job opportunities through maintenance and programming – Supporting investment in open spaces
11. Relationships & Participation	**Smart Infrastructure**	**Achieving community participation**	– Effectively enabling citizen participation in decision-making processes – Providing access to information and opportunities for residents to participate in decision-making processes – Availability of open data and the use of digital tools to facilitate participation in local government operations
	Buildings & land use	**Social connections and encouraging citizen participation**	– Including public spaces, parks, gathering areas, and mixed-use developments that encourage social cohesion – Involving residents in planning and design processes to ensure that buildings and land uses meet their needs – The effectiveness and existence of workshops on participatory design, community meetings, and feedback mechanisms
	Smart Transportations	**Achieving social inclusion**	– Level of community participation in transportation planning and decision-making – Investing in infrastructure based on studying the local needs and priorities of the population – Level of access to transportation options that are managed in response to community needs – The level of social isolation among different demographic groups – Availability of transportation options for people with disabilities
	Open Spaces & Landscape	**Providing opportunities for social interaction**	– Measuring the frequency and diversity of social activities that take place in open spaces – The extent of public participation and participation in planning and management of open spaces – Strengthening community participation strategies

(continued)

Table 2. (continued)

	Urban human needs	Criteria	Indicators
12. Justice	**Smart Infrastructure**	**Achieving equity in access to utility services for all residents**	– Providing facilities that are accessible to everyone, regardless of their income – The percentage of income that is spent on utilities – Ensuring that the benefits of infrastructure. Are distributed fairly to achieve social justice – Ensuring that data privacy and security are provided
	Buildings & land use	**Promoting social justice**	– Creating mixed-income communities – Reducing social and economic segregation
	Smart Transportations	**Achieving efficiency and equity through transportation**	– Involving citizens in the planning and decision-making process – Providing better access to transportation for people with disabilities, low incomes, and people living in underserved communities – Ensuring that all community members have access to safe and affordable transportation options – Distribution of transportation and infrastructure services across different neighborhoods and residential groups
		Parking management	– Availability and cost of parking facilities – Level of enforcement of parking regulations – Percentage of availability of parking spaces designated for sustainable transportation options
	Open Spaces & Landscape	**Designing open spaces to serve multiple functions and meet diverse user needs**	– The extent of amenities such as playgrounds, sports facilities, seating areas, walking paths, and cultural spaces – The availability of inclusive and accessible spaces for people of all different age groups, cultural backgrounds, and abilities
		Distributing open spaces fairly between neighborhoods and residential communities	– Measuring the extent of accessibility and availability of open spaces in low-income areas – All residents have equal opportunities to benefit from it
		Social and environmental justice in resource distribution	– Measuring the extent of equality in access to clean air, water, and other natural resources – Distribution of environmental benefits and burdens across societal groups

References

Abdullah, N., et al.: IoT-based waste management system in formal and informal public areas in Mecca. Int. J. Environ. Res. Public Health **19**(20) (2022a). https://doi.org/10.3390/ijerph192013066

Abdullah, Y.A., et al.: An insight into domestic solid waste recycling facilities and urban households' behaviour in Shah Alam, Malaysia. IOP Conf. Ser. Earth Environ. Sci. **1019**(1) (2022b). https://doi.org/10.1088/1755-1315/1019/1/012028

Abdullahi, A., Sieng, L.W.: The effect of infrastructure development on economic growth: the case of Sub-Saharan Africa. J. Infrastruct. Policy Dev. **7**(2) (2023). https://doi.org/10.24294/jipd.v7i2.1994

Aboria, S.G., Eleinen, O.M.A., El-Mowafy, B.N., Hassan, A.M.: How urban morphology affects energy consumption and building energy loads? Strategies based on urban ventilation. In: Negm, A.M., Rizk, R.Y., Abdel-Kader, R.F., Ahmed, A. (eds.) Engineering Solutions Toward Sustainable Development, pp. 375–395. Springer Nature, Switzerland (2024)

Aldegheishem, A., et al.: Smart road traffic accidents reduction strategy based on intelligent transportation systems (TARS). Sensors (Switzerland) **18**(7) (2018). https://doi.org/10.3390/s18071983

Aliyeva, Z., Assipova, Z., Bazarbekova, M., Mussagaliyeva, A., Sansyzbayeva, A.: Urban area planning and environment in Kazakhstan's cities: case of Almaty. In: International Multidisciplinary Scientific Geo Conference Surveying Geology and Mining Ecology Management, SGEM, vol. 19, no. 5.2, pp. 775–782 (2019). https://doi.org/10.5593/sgem2019/5.2/S20.097

Alsayed, S.S.: Urban human needs: conceptual framework to promoting urban city fulfills human desires. Front. Sustain. Cities **6** (2024). https://doi.org/10.3389/frsc.2024.1395980

Aly, T.F., Ehab, S., Lotfi, Y.A.: Identifying key urban design attributes for enhanced sense of safety – the case of El-Sherouk city in Cairo. Int. J. Architectural Res. Archnet-IJAR (2023). https://doi.org/10.1108/ARCH-02-2023-0041

Aramouny, C.: Ecologies of leisure: reimagined architectures and landscapes of leisure and infrastructure. Plan J. **6**(1), 197–218 (2021). https://doi.org/10.15274/tpj.2021.06.01.14

Awan, J.H., Memon, S., Shah, A.A., Pathan, K.T.: Proposed framework of smart transportation in Pakistan: issues, challenges, vulnerabilities, and solutions. Int. J. Cyber Warfare Terrorism **10**(4), 48–63 (2020). https://doi.org/10.4018/IJCWT.2020100104

Bakri, R., El-Sayad, Z., Bakr, A.F.: Big data in smart transportation planning: towards an applicable smart socio-economic transportation platform in historical mega cities. WIT Trans. Built Environ. **212**, 13–24 (2022). https://doi.org/10.2495/UMT220021

Berglund, E.Z., et al.: Smart infrastructure: a vision for the role of the civil engineering profession in smart cities. J. Infrastruct. Syst. **26**(2) (2020). https://doi.org/10.1061/(ASCE)IS.1943-555X.0000549

Betti, G., Tartarini, F., Nguyen, C., Schiavon, S.: CBE clima tool: a free and open-source web application for climate analysis tailored to sustainable building design. Build. Simul. **17**(3), 493–508 (2024). https://doi.org/10.1007/s12273-023-1090-5

Bondar, I., Penchuk, O., Lavrenyuk, O., Malaniuk, V., Kulyk, A., Shvets, I.: The latest innovations in landscape design. J. High. Educ. Theor. Pract. **22**(11), 65–78 (2022). https://doi.org/10.33423/jhetp.v22i11.5413

Bunkse, E.V.: The role of a humane environment in Soviet urban planning. Geogr. Rev. **69**(4), 379–394 (1979). https://doi.org/10.2307/214802

Caprotti, F.: Future cities: moving from technical to human needs. Palgrave Commun. **4**(1) (2018). https://doi.org/10.1057/s41599-018-0089-5

Cardoso, R., Sobhani, A., Meijers, E.: The cities we need: towards an urbanism guided by human needs satisfaction. Urban Stud. **59**(13), 2638–2659 (2022). https://doi.org/10.1177/00420980211045571

Çelikyay, H.S.: Empathic design approach to public space organization in the urban environment. In: Planning, Design, Applications Theories, Techniques, Strategies: For Spatial Planners & Designers (2020)

Cheng, S., Zhai, Z., Sun, W., Wang, Y., Yu, R., Ge, X.: Research on the satisfaction of Beijing waterfront green space landscape based on social media data. Land **11**(10) (2022). https://doi.org/10.3390/land11101849

Como, A., Cuomo, A., Perrotta, L.S.: An integrated approach to subsurface water pathways for the sustainable development of the architectural landscape of agro-urban areas. Sustain. (Switzerland) **15**(12) (2023). https://doi.org/10.3390/su15129208

Cong, P., Zhang, Y., Wang, W., Zhang, N.: DND: driver node detection for control message diffusion in smart transportations. IEEE Trans. Netw. Serv. Manage. **18**(3), 3583–3594 (2021). https://doi.org/10.1109/TNSM.2021.3059696

Costa, Á., Cruz, C.O., Sarmento, J.M., Sousa, V.F.: Bundling road and railway infrastructure operators: analysis of the impact on efficiency. J. Infrastruct. Syst. **29**(1) (2023). https://doi.org/10.1061/JITSE4.ISENG-2106

Danilina, N., Slepnev, M.: Managing smart-city transportation planning of "park-and-ride" system: case of Moscow metropolitan. IOP Conf. Ser. Mater. Sci. Eng. **365**(2) (2018). https://doi.org/10.1088/1757-899X/365/2/022002

Dash, A.: Does citizens' participation moderate the relationship between the built environment and their quality of life in Indian smart cities? Transforming Gov. People Process Policy **17**(4), 673–687 (2023). https://doi.org/10.1108/TG-06-2023-0084

Schrank, D., Eisele, B.: 2021 Urban Mobility Report. The Texas A&M Transportation Institute (2021)

De Arruda, R.N., Figueiredo, K., Vasco, D.A., Haddad, A., Najjar, M.K.: Cost-benefit analysis of solar energy integration in buildings: a case study of affordable housing in Brazil. Front. Built Environ. **9** (2023). https://doi.org/10.3389/fbuil.2023.1255845

del Valle Barrera, M., Belloy, P., Mougenot, B., Doussoulin, J.P.: The international impact of Manfred Max-Neef's scholarship: a bibliometric approach. Int. J. Sustain. Dev. **25**(1–2), 1–29 (2022). https://doi.org/10.1504/ijsd.2022.126458

Dietrich, U.: Space needed to make a city sustainable and necessary changes to reach it: the case of Germany. WIT Trans. Ecol. Environ. **253**, 229–240 (2021). https://doi.org/10.2495/SC210201

Dogan, Y.I.: Determining the recreation potential of urban forests in terms of open space sports recreation activities (Denizli Camlik city forest example). J. Environ. Prot. Ecol. **24**(1), 35–45 (2023)

Elhachmi, J., Kobbane, A.: A federated learning approach for water distribution networks monitoring. In: Proceedings - 2022 9th International Conference on Wireless Networks and Mobile Communications, WINCOM 2022 (2022). https://doi.org/10.1109/WINCOM55661.2022.9966455

Eltarabily, S.: Toward a conceptual framework for evaluating the quality of urban open spaces. J. Sustain. Archit. Civil Eng. **31**(2), 58–84 (2022). https://doi.org/10.5755/j01.sace.31.2.31311

Foroughi, R., Daneshgar, F.: Environmental protection through sustainable land management. Environ. Sci. Sustain. Dev., 73–82 (2024)

Eom, J.-K., Lee, K.-S., Ko, S., Lee, J.: Exploring travel mode preference of external trips for smart city transportation planning: Sejong, Korea. Sustain. (Switzerland) **14**(2) (2022). https://doi.org/10.3390/su14020630

Fan, C., Xu, J., Natarajan, B.Y., Mostafavi, A.: Interpretable machine learning learns complex interactions of urban features to understand socio-economic inequality. Comput.-Aided Civil Infrastruct. Eng. **38**(14), 2013–2029 (2023). https://doi.org/10.1111/mice.12972

Fedajev, A., Mitić, P., Kojić, M., Radulescu, M.: Driving industrial and economic growth in central and eastern Europe: the role of electricity infrastructure and renewable energy. Utilities Policy **85** (2023). https://doi.org/10.1016/j.jup.2023.101683.

Fox, N.H.: More than a slogan: or, how we built a social justice program that made our campus more just. Radic. Teach. **116**, 66–75 (2020). https://doi.org/10.5195/rt.2020.714
Ghosh, G.: Landuse planning for social inclusion: an analysis of Bangalore, India. WIT Trans. Built Environ. **182**, 141–150 (2019). https://doi.org/10.2495/UT180141
Gibbs, D., Krueger, R., MacLeod, G.: Grappling with smart city politics in an era of market triumphalism. Urban Stud. **50**(11), 2151–2157 (2013). https://doi.org/10.1177/0042098013491165
Goryacheva, E., Vlasova, I., Dudnik, M.: Creating the architectural-landscape space in the major cities as a tool of harmonious personality development in multinational townspeople. In: E3S Web of Conferences, vol. 371 (2023). https://doi.org/10.1051/e3sconf/202337101059
Greene, S.: City shape: communicating and evaluating community design. J. Am. Plann. Assoc. **58**(2), 177–189 (1992). https://doi.org/10.1080/01944369208975792
Gurung, D.J., Gowreesunkar, V.: Mapping the landscape of tourism cities research: a bibliometric analysis. Int. J. Tourism Cities (2023). https://doi.org/10.1108/IJTC-10-2023-0207
Hallsmith, G.: The key to sustainable cities. In: Meeting Human Needs, Transforming Community Systems. New Society Publishers, Gabriola Island (2003)
Haq, I., Radja, A.M., Syam, S.: Analysis of comfort level in public open space facilities at Anjungan Sungai Mata Allo Enrekang: visitors' perspective. IOP Conf. Ser. Earth Environ. Sci. **1272**(1) (2023). https://doi.org/10.1088/1755-1315/1272/1/012010
Hassan, A.M.: UMC-based models: an integrating UMC performance analysis and numerical methods. Renew. Sustain. Energy Rev. **181**, 113307 (2023). https://doi.org/10.1016/j.rser.2023.113307
Hassan, A.M., ELMokadem, A.A., Megahed, N.A., Abo Eleinen, O.M.: Urban morphology as a passive strategy in promoting outdoor air quality. J. Build. Eng. **29**, 101204 (2020). https://doi.org/10.1016/j.jobe.2020.101204
Hassan, A.M., Megahed, N.A.: COVID-19 and urban spaces: a new integrated CFD approach for public health opportunities. Build. Environ. **204**, 108131 (2021). https://doi.org/10.1016/J.BUILDENV.2021.108131
Hassan, A.M., Megahed, N.A.: Urban planning and development improving urban energy resilience with an integrative. Archit. Eng. **2017** (2022). https://doi.org/10.23968/2500-0055-2022-7-4-17-35
Hatami, F., Rahman, M.M., Nikparvar, B., Thill, J.-C.: Non-linear associations between the urban built environment and commuting modal split: a random forest approach and SHAP evaluation. IEEE Access **11**, 12648–12661 (2023). https://doi.org/10.1109/ACCESS.2023.3241627
Heneidy, S.Z., et al.: Archeological sites and relict landscapes as refuge for biodiversity: case study of Alexandria city, Egypt. Sustain. (Switzerland) **14**(4) (2022). https://doi.org/10.3390/su14042416
Hingmire, A.M., Bhaladhare, P.R.: Advance urban flood control system using fuzzy logic and Internet of Things (IoT) for smart city. Int. J. Recent Innov. Trends Comput. Commun. **11**(6S), 447–456 (2023). https://doi.org/10.17762/ijritcc.v11i6s.6952
Huang, D., Kyttä, M., Kajosaari, A., Xie, X., Zhang, J.: One approach does not fit all settings: exploring the effects of natural and built environments on running pleasantness across places. Build. Environ. **245** (2023). https://doi.org/10.1016/j.buildenv.2023.110961
Ibrahim, M.: Smart infrastructure and integrated platform of smart and sustainable cities. In: Lecture Notes in Networks and Systems, vol. 183 (2021). https://doi.org/10.1007/978-3-030-66840-2_35
Ibrahim, F.L., Nizarudin, N.D.: Human interaction theory for sustainable city. IOP Conf. Ser. Earth Environ. Sci. **268**(1) (2019). https://doi.org/10.1088/1755-1315/268/1/012046
Jagatheesaperumal, S.K., Bibri, S.E., Huang, J., Rajapandian, J., Parthiban, B.: Artificial intelligence of things for smart cities: advanced solutions for enhancing transportation safety. Comput. Urban Sci. **4**(1) (2024). https://doi.org/10.1007/s43762-024-00120-6
Jayasena, N.S., Chan, D.W.M., Kumaraswamy, M.M., Saka, A.B.: Applicability of public-private partnerships in smart infrastructure development: the case of Hong Kong. Int. J. Constr. Manag. **23**(11), 1932–1944 (2023). https://doi.org/10.1080/15623599.2022.2027076
Ji, T., Chen, J.-H., Wei, H.-H., Su, Y.-C.: Towards people-centric smart city development: investigating the citizens' preferences and perceptions about smart-city services in Taiwan. Sustain. Cities Soc. **67** (2021). https://doi.org/10.1016/j.scs.2020.102691
Jia, J., Zlatanova, S., Zhang, K.: Exploring spatial parameters to evaluate human walking accessibility of urban green space. Int. Arch. Photogrammetry Remote Sens. Spat. Inf. Sci. SPRS Arch. **54**(3/W1), 73–80 (2020). https://doi.org/10.5194/isprs-archives-XLIV-3-W1-2020-73-2020
Jim, C.Y., Hui, L.C.: Offering green roofs in a compact city: benefits and landscape preferences of socio-demographic cohorts. Appl. Geogr. **145** (2022). https://doi.org/10.1016/j.apgeog.2022.102733
Kanthavel, D., Sangeetha, S.K.B., Keerthana, K.P.: An empirical study of vehicle to infrastructure communications - an intense learning of smart infrastructure for safety and mobility. Int. J. Intell. Netw. **2**, 77–82 (2021). https://doi.org/10.1016/j.ijin.2021.06.003
Khaddar, S., Rahman Fatmi, M., Winters, M.: How daily activities and built environment affect health? A latent segmentation-based random parameter logit modeling approach. Travel Behav. Soc. **33** (2023). https://doi.org/10.1016/j.tbs.2023.100624
Khorasgani, A.M., Haghighatbin, M.: Regeneration of historic cities: reflections of its evolution towards a landscape approach. ISVS E-J. **10**(5), 60–78 (2023)
Kochtitzky, C.S., et al.: Urban planning and public health at CDC. MMWR Morb. Mortal. Wkly Rep. **55**(Suppl 2), 34–38 (2006)
Kramer, A., Lassar, T.J., Federman, M., Hammerschmidt, S.: Urban land institute. In: Building for Wellness: The Business Case (2014)
Kumar, S., Mahapatra, B., Kumar, R., Turuk, A.K.: Security and privacy solution for I-RFID based smart infrastructure health monitoring. In: International Conference on Technologies for Smart City Energy Security and Power: Smart Solutions for Smart Cities, ICSESP 2018 - Proceedings, 1–5 January 2018 (2018). https://doi.org/10.1109/ICSESP.2018.8376733
Kurniawan, H.B., Roychansyah, M.S.: The social equity of public green open space accessibility: the case of south Tangerang, Indonesia. Geogr. Environ. Sustain. **16**(1), 45–54 (2023). https://doi.org/10.24057/2071-9388-2022-124
Li, D., Wang, W., Huang, G., Zhou, S., Zhu, S., Feng, H.: How to enhance citizens' sense of gain in smart cities? A SWOT-AHP-TOWA approach. Soc. Indic. Res. **165**(3), 787–820 (2023). https://doi.org/10.1007/s11205-022-03047-9
Li, L., Park, P., Yang, S.-B.: The role of public-private partnership in constructing the smart transportation city: a case of the bike sharing platform. Asia Pac. J. Tourism Res. **26**(4), 428–439 (2021). https://doi.org/10.1080/10941665.2018.1544913
Lin, J., Li, D., Brown, R.D.: Microclimatic landscape architecture: from theory to application. Urban Sci. **6**(1) (2022). https://doi.org/10.3390/urbansci6010009
Lindley, J.: Building smart infrastructure to serve travelers and system managers. Public Roads **61**(2), 30–35 (1997)
Liu, Y., Ali, A., Chen, Y., She, X.: The effect of transport infrastructure (road, rail, and air) investments on economic growth and environmental pollution and testing the validity of EKC in China, India, Japan, and Russia. Environ. Sci. Pollut. Res. **30**(12), 32585–32599 (2023). https://doi.org/10.1007/s11356-022-24448-w
Lowe, M.D.: Shaping cities: the environmental and human dimensions. Worldwatch Paper **105** (1991)
Lu, H., Xiao, C., Jiao, L., Du, X., Huang, A.: Spatial-temporal evolution analysis of the impact of smart transportation policies on urban

carbon emissions. Sustain. Cities Soc. **101** (2024). https://doi.org/10.1016/j.scs.2024.105177

Luckan, Y.: Defining the layers of urban complexity: an epistemological shift towards inclusive cities. In: Urban Book Series (2021). https://doi.org/10.1007/978-3-030-81511-0_11

Maclean, A.: Human need and community initiatives in an urban setting. In: Canterbury at the Crossroads (1983)

Mariano, C., Rossi, F.: RivEr/Generation_LAB-Linking resilience with inclusiveness in the urban-built environment of Rome. Sustain. (Switzerland) **15**(6) (2023). https://doi.org/10.3390/su15064774

Max-Neef, M.: Development and human needs. In: Real-Life Economics: Understanding Wealth Creation, pp. 197–213 (1992)

Megahed, N.A., Hassan, A.M.: Evolution of BIM to DTs: a paradigm shift for the post-pandemic AECO industry. Urban Sci. **6**(4) (2022). https://doi.org/10.3390/urbansci6040067

Megahed, N., Ghoneim, E.: Antivirus-built environment: lessons learned from Covid-19 pandemic. Sustain. Cities Soc. **61**, 102350 (2020). https://doi.org/10.1016/j.scs.2020.102350

Mete, M.O.: Geospatial big data analytics for sustainable smart cities. Int. Arch. Photogr. Remote Sens. Spat. Inf. Sci. ISPRS Arch. **48**(4/W7–2023), 141–146 (2023). https://doi.org/10.5194/isprs-archives-XLVIII-4-W7-2023-141-2023

Mohamed, N., Upadhyay, R., Jakka, G., Rambabu, P.V., Alfurhood, B.S., Singh, D.P.: Framework for the deployment of intelligent smart cities (ISC) using artificial intelligence and software networking technologies. In: 2023 3rd International Conference on Advance Computing and Innovative Technologies in Engineering, ICACITE 2023, pp. 667–671 (2023). https://doi.org/10.1109/ICACITE57410.2023.10182730

Mohy, H.A.A., Rasheed, K.G.: Morphological perspective of urban resilience through eco urban landscape: Iraq-Basra as a case study. Int. J. Des. Nat. Ecodyn. **17**(6), 807–821 (2022). https://doi.org/10.18280/ijdne.170601

Moser, S., Avni, N.: Analysing a private city being built from scratch through a social and environmental justice framework: a research agenda. Urban Stud. (2023). https://doi.org/10.1177/00420980231211814

Mutică, A.P.: Architecture and human needs. Acta Technica Napocensis Civil Eng. Archit. **55**(4) (2013). http://constructii.utcluj.ro/ActaCivilEng

Nahdatunnisa, N., Wahyudi, S.I., Adi, H.P., Arzal Tahir, M.: Pedestrian path infrastructure in urban public green open space (case study: green open space religious monument Kendari City, Indonesia). Civil Eng. Archit. **11**(5), 2631–2643 (2023). https://doi.org/10.13189/cea.2023.110529

Nakashima, H., Nakamura, T., Hosoi, Y., Konno, K., Kawamura, H.: Smart infrastructure asset management system on metropolitan expressway in Japan. Sustain. Civil Infrastruct. (2022). https://doi.org/10.1007/978-3-030-79801-7_41

Neufeld, C., Wohl, S.: Experiments in living together. In: Future Urban Habitation: Transdisciplinary Perspectives, Conceptions, and Designs (2022). https://doi.org/10.1002/9781119734895.ch14

Pacheco, J., Zhu, X., Badr, Y., Hariri, S.: Enabling risk management for smart infrastructures with an anomaly behavior analysis intrusion detection system. In: Proceedings - 2017 IEEE 2nd International Workshops on Foundations and Applications of Self* Systems, FAS*W 2017, pp. 324–328 (2017). https://doi.org/10.1109/FAS-W.2017.167

Patvardhan, N., Santoki, S.: Connecting governments' accessibility policy to its visually challenged stakeholders for an all-inclusive smart infrastructure. Int. J. Innov. Technol. Exploring Eng. **8**(11 Special), 280–285 (2019). https://doi.org/10.35940/ijitee.K1054.09811S19

Pinar Caglin, B.E.: Analysis of the global future cities components as governance tools for urban projects. UN-Habitat (2020)

Porter, C.D., Ten Siethoff, B., Smith, V.: Impacts of comprehensive planning and smart growth initiatives on transportation: case studies of five regions. Transp. Res. Rec. **1902**, 91–98 (2005). https://doi.org/10.3141/1902-11

Qi, X., Mei, G., Piccialli, F.: Resilience evaluation of urban bus-subway traffic networks for potential applications in IoT-based smart transportation. IEEE Sens. J. **21**(22), 25061–25074 (2021). https://doi.org/10.1109/JSEN.2020.3046270

Rodzoś, J.: The concept of human needs in sustainable development of cities Teoria potrzeb ludzkich w rozwoju zrównoważonym miast, vol. 14 (2019)

Sánchez-Silva, M., et al.: Moving toward resilience and sustainability in the built environment. Struct. Saf. (2024). https://doi.org/10.1016/j.strusafe.2024.102449

Schrammeijer, E.A., van Zanten, B.T., Davis, J., Verburg, P.H.: The advantage of mobile technologies in crowdsourcing landscape preferences: testing a mobile app to inform planning decisions. Urban Forestry Urban Green. **73** (2022). https://doi.org/10.1016/j.ufug.2022.127610

Selim, A.M., ElGohary, A.S.: Public–private partnerships (PPPs) in smart infrastructure projects: the role of stakeholders. HBRC J. **16**(1), 317–333 (2020). https://doi.org/10.1080/16874048.2020.1825038

Seliverstov, S., Seliverstov, Y., Gavkalyk, B., Fahmi, S.: Development of transport infrastructure organization model for modern cities with growing effectiveness. Transp. Res. Procedia **50**, 614–625 (2020). https://doi.org/10.1016/j.trpro.2020.10.073

Slamet, W., Made, K., Tubagus, P., Agus, S., Sri Wiwoho, M.: Internet of Things (IoT) as green city economic development smart transportation system. MATEC Web Conf. **138** (2017). https://doi.org/10.1051/matecconf/201713807015

Taghipour, A., Ramezani, M., Khazaei, M., Roohparvar, V., Hassannayebi, E.: Smart transportation behavior through the COVID-19 pandemic: a ride-hailing system in Iran. Sustain. (Switzerland) **15**(5) (2023). https://doi.org/10.3390/su15054178

Taufik, S.N.B.M., Ishak, S.Z., Mohd Yusoff, Z., Mohd Jaafar, M.N.: Enabling intelligent transportation system (ITS) application for smart parking monitoring system (SPMS) at urban environment. Plann. Malaysia **21**(4), 176–189 (2023). https://doi.org/10.21837/pm.v21i28.1325

Tay, L., Diener, E.: Needs and subjective well-being around the world. J. Pers. Soc. Psychol. **101**(2), 354–365 (2011). https://doi.org/10.1037/a0023779

Thompson, C.W.: Urban open space in the 21st century. Landsc. Urban Plan. **60**(2), 59–72 (2002). https://doi.org/10.1016/S0169-2046(02)00059-2

Venkatasubramanian, K.K., Mukherjee, T., Gupta, S.K.S.: CAAC-an adaptive and proactive access control approach for emergencies in smart infrastructures. ACM Trans. Auton. Adapt. Syst. **8**(4) (2014). https://doi.org/10.1145/2555614

Vipond, N., Kumar, A., James, J., Paige, F., Sarlo, R., Xie, Z.: Real-time processing and visualization for smart infrastructure data. Autom. Constr. **154** (2023). https://doi.org/10.1016/j.autcon.2023.104998

Wan, D., et al.: Spatial distribution and accessibility measurements for elderly day care centers in china's urban built-up area: the case of Tianjin Nankai District. Buildings **12**(9) (2022). https://doi.org/10.3390/buildings12091413

Wen, R., et al.: A new approach of generating travel demands for smart transportation systems modeling. In: SAE Technical Papers, April 2020 (2020). https://doi.org/10.4271/2020-01-1047

Wu, S., et al.: Evaluation of smart infrastructure systems and novel UV-oriented solution for integration, resilience, inclusiveness, and sustainability. In: 5th International Conference on Universal Village, UV 2020 (2020). https://doi.org/10.1109/UV50937.2020.9426194

Xie, X., Gu, L.: A comparative study on smart transportation policies of Beijing and Shanghai based on three-dimensional framework. J. Urban Aff. (2024). https://doi.org/10.1080/07352166.2024.2311160

Yang, F., Wen, X., Aziz, A., Luhach, A.K.: The need for local adaptation of smart infrastructure for sustainable economic management. Environ. Impact Assess. Rev. **88** (2021). https://doi.org/10.1016/j.eiar.2021.106565

Yu, A., Lam, S.K.N.: Assessing open space in Scotland: reliability and construct validity of the open space scale. Sustain. (Switzerland) **14**(22) (2022). https://doi.org/10.3390/su142215203

Yusuf, D.A., et al.: Quest for an innovative methodology for retrofitting urban built heritage: an assessment of some historic buildings in Kano Metropolis, Nigeria. Buildings **13**(8) (2023). https://doi.org/10.3390/buildings13081899

Zhao, J., Wang, L., Ye, Q., Zhao, Q., Wei, S.: Association of environmental elements with respondents' behaviors in open spaces using the direct gradient analysis method: a case study of Jining, China. Int. J. Environ. Res. Public Health **19**(14) (2022). https://doi.org/10.3390/ijerph19148494

Zhou, Y., Wang, M., Lin, S., Qian, C.: Relationship between Children's independent activities and the built environment of outdoor activity space in residential neighborhoods: a case study of Nanjing. Int. J. Environ. Res. Public Health **19**(16) (2022). https://doi.org/10.3390/ijerph19169860

Type of Public Open Space as a Condition for Walkability in the Town Centre of a Small Town

Mateja Volgemut

Abstract

The appeal of a built environment influences its walkability, and efforts to enhance this appeal vary based on the size of the city and its geographical context. In small towns, features like paved surfaces and promenades improve walkability, while reduced traffic contributes to tranquillity, especially in urban centres. The presented study conducted in Slovenia explores the correlation between public open spaces and walkability. Using spatial analysis, a survey questionnaire, and statistical analysis, the study answers the research question: Is there a correlation between different types of public open space and walkability? This dual approach, combining objective spatial analysis with subjective user feedback, is crucial for understanding how public open spaces contribute to walkability in small urban environments, and also ensures the statistical validation of the results. Residents of 33 towns answered the questionnaire, and the spatial analysis results are shown for two towns where walkability was rated below average (Litija = 3.05) and above average (Slovenske Konjice = 3.62; average = 3.27). The research findings clearly show a connection between walkability and the prevalence of promenades or paved surfaces in smaller towns. However, the results from Slovenian towns differ from previous studies, as it was impossible to establish a clear connection between parks and walkability. This study highlights the significance when analysing walkability of considering the size of the town, the morphological characteristics of its ground plan type, and its geographical location. These factors are essential for understanding how urban design impacts pedestrian accessibility and movement within different environments.

Keywords

Livability · Public open space · Small towns · Walkability · Town centre

1 Introduction

Many studies focus on walkability in large cities, but it is also important to examine walkability in small towns. Small towns are at human scale, with shorter distances and unique sociological ties. Insights from studies of large cities cannot be fully transferred to smaller settlements, highlighting the need for more focused research. It is also essential to understand and study walkability in the context of a particular geographic area, which may be more influential in small towns than in larger urban areas (Stefánsdóttir et al., 2024).

Despite a growing body of research, significant gaps remain in walkability studies. Venerandi et al. (2024), in their synthesis of 45 studies published between 2005 and 2023, highlight that many existing walkability metrics fail to incorporate detailed urban morphological analyses. These studies also often neglect the subjective experiences of people, which can offer crucial insights into how urban spaces are perceived and used by pedestrians. This lack of consideration for the user's perspective can lead to incomplete assessments of walkability and urban livability.

Both perceived walkability, as assessed by residents, and audited walkability, based on objective environmental evaluations, are essential for understanding how walkability affects urban living. Research indicates that combining subjective assessments with objective data provides a comprehensive view of walkability, as each offers unique insights into the pedestrian experience (Jensen et al., 2017). This dual approach enables planners to address both the physical infrastructure and the personal experiences that shape walkability in urban environments.

M. Volgemut (✉)
Faculty of Architecture, University of Ljubljana, Ljubljana, Slovenia
e-mail: mateja.volgemut@fa.uni-lj.si

B. He et al. (eds.), *Towards the Framework of Livable and Resilient Cities* Advances in Science, Technology & Innovation, https://doi.org/10.1007/978-3-031-97849-4_3

Recent studies reveal that walkability in small towns has significant impacts on public health, urban sustainability, and social equity. Key findings emphasise the importance of functional amenities—like shops, schools, and health-care facilities—alongside safe, accessible pedestrian pathways (Doescher et al., 2014). These amenities help reduce car dependency and encourage active mobility, fostering healthier and more environmentally-friendly communities.

Green spaces contribute to quality of life by offering recreational areas, improving air quality, and fostering community interaction. However, the impact of green spaces on walkability in small towns remains complex, often depending on the specific urban context. Research has found that while green spaces can promote walking, their influence varies based on factors such as accessibility, design, and integration into the surrounding urban environment. In some cases, green spaces improve walkability, while in others their effects are less clear, suggesting that the wider environment and other factors (size) also play an important role (Liu et al., 2023).

2 Walkability and Public Open Space

Various authors from different scientific and professional fields have defined public open space (hereinafter: POS) in different ways, focusing on different areas such as use and function (Carr et al, 1992; Low, 2000), management (Carmona et al, 2008), control (Scruton, 1984), ownership (Brown, 2006; Kohn, 2004; Staeheli & Mitchell, 2008), law (Lofland, 1998), physical appearance (Tibbalds, 1992; Zukin, 1995; Madanipour, 2003), and sociological function (Gehl, 1996; Gehl & Gemzǿe, 2000; Mitchell, 2003). For the purpose of this paper, POS is defined as unbuilt open space that is accessible to everyone under the same conditions, regardless of ownership. Jackson (1984) noted that the POS is increasingly becoming a space for various uses other than a space for people's well-being. Other phenomena have also impoverished the image, functions, and physical appearance of POS. The introduction of the car into towns in modern times has caused the space to become interrupted (Lefebvre, 1991) and taken over by motorised traffic (Gehl & Gemzøe, 2000), which occupied it in place of pedestrians. At the same time, buildings were retreating to the middle of the building plot in modernist cities and a large area of the plot remained without a function; an undesigned open space (Sennett, 1977). And so, POS has become internalized in the buildings in which public life takes place, but has disappeared from outside (Madanipour, 2003). But the main and most important function of POS is to support public life, which is connected with walkability.

Only POS that fulfils the universal positive qualities (Carmona et al, 2008) – maintained, accessible, aesthetic, comfortable, durable, functional, recognisable, safe, robust, green, non-polluting, and fulfilling – provides support and opportunity for public life. Good connections between services that attract people are also important. The connections between them need to be easy to use, attractive, comfortable, and well-marked (Llewelyn Davies, 2000). These are the conditions for POS to be attractive to pedestrians, and to be consistent with the concepts of urban space design (Carmona et al, 2008). These concepts also determine the degrees to which POS enables freedom of movement (Tibbalds, 1992) and to which it is designed for pedestrians (New urbanism, 1993) and so becomes lively (Jacobs & Appleyard, 1987) as a result of social interactions between people.

The quality of the built environment in terms of walkability has become an essential element of urban planning and design during the last decade. The attractiveness of walking comes from the balanced development of urban areas and public services, which increases quality of life and raises the level of people's satisfaction (Lee et al., 2016). Areas that are better equipped with POS and public services, such as urban centres, are more walkable (Badland et al, 2010). Walkability comes from the assets of the built environment, especially POS, that encourage residents to walk to work, recreation, and services.

Encouraging walking is a key measure in the transition of towns to sustainable mobility through measures such as closing town centres to motorised traffic and the transformation of areas for motorised traffic into POS for pedestrians. Sustainable mobility is also achieved indirectly by strengthening public transport. The sustainable development of towns is also ensured (Agenda for Sustainable Development until 2030, 2015) in part through sustainable mobility (European Green Deal, 2019).

A study that focuses on the attractiveness for walking of town centres in Slovenia is presented in this paper. In particular, the relationship between POS and walkability; more precisely the relationship between different types of POS and walkability. The research question is: Is there a correlation between different types of POS and walkability?

3 Methods for Researching the Interaction Between Public Open Space and Walkability in Small Towns

The goal of this paper is to provide a detailed analysis of two towns, examining how different types of POS are linked to walkability. Additionally, it aims to investigate the factors contributing to the lower walkability assessment in one town compared to the higher walkability rating in the other.

Slovenia is committed to polycentric urban development, with 82 urban areas meeting the criteria for small towns, 10 for medium-sized towns, and one for a large town (Zavodnik Lamovšek et al, 2008). On the basis of the functional criteria, 33 towns were selected (Volgemut et al., 2021) for the study from the large number of small towns. Based on morphological and formal criteria, the selection of towns was narrowed down to two towns. Three research methods (Vertelj Nared & Zavodnik Lamovšek, 2015) were applied, and a spatial analysis was conducted across two towns. Additionally, residents from 33 towns participated in a survey questionnaire, and the results were analysed using statistical methods.

Two towns were selected from the group of small towns due to their differing walkability scores. They were selected also because they have a similar number of inhabitants (SURS, 2024) in their municipalities (Litija 15.834, Slovenske Konjice 15.485) and in their central settlements (Litija 6.710, Slovenske Konjice 5.233). Building development in Slovenian towns was very modest until the mid-20th century, so there are few development periods: medieval, classical, and modernist periods. The towns can be defined according to the type of ground plan as different combinations of (in)complete, medieval, classical, and modernist ground plans (Drozg, 1998). Both selected towns feature a combination of medieval and modernist ground plan, with Litija having an incomplete ground plan and Slovenske Konjice a complete ground plan.

This paper presents the results of the spatial analysis and the survey for the two selected towns, along with comparisons to survey results from a broader group of 33 towns.

3.1 Spatial Analysis

The spatial analysis was carried out in three steps that defined 1. The location of public services at the local, primary level (education, district court, health, public administration) and the type of buildings in the town, 2. The location, function, and type of POS (types of POS are presented on Fig. 1 and Fig. 2 and are adapted from Carmona, 2010), and 2.3. The location and area of the town centre made as a synthesis of the first two steps.

The town centre is the area where there is an interaction between public services and POS. Public services are services of general interest (hereinafter: SGI). Services in this paper refer to SGI (Nared et al, 2016) and services of general economic interest (other services) all together (ESPON SeGI 2013). This paper shows the presence and the area of the different types of POS within each town centre, as shown in

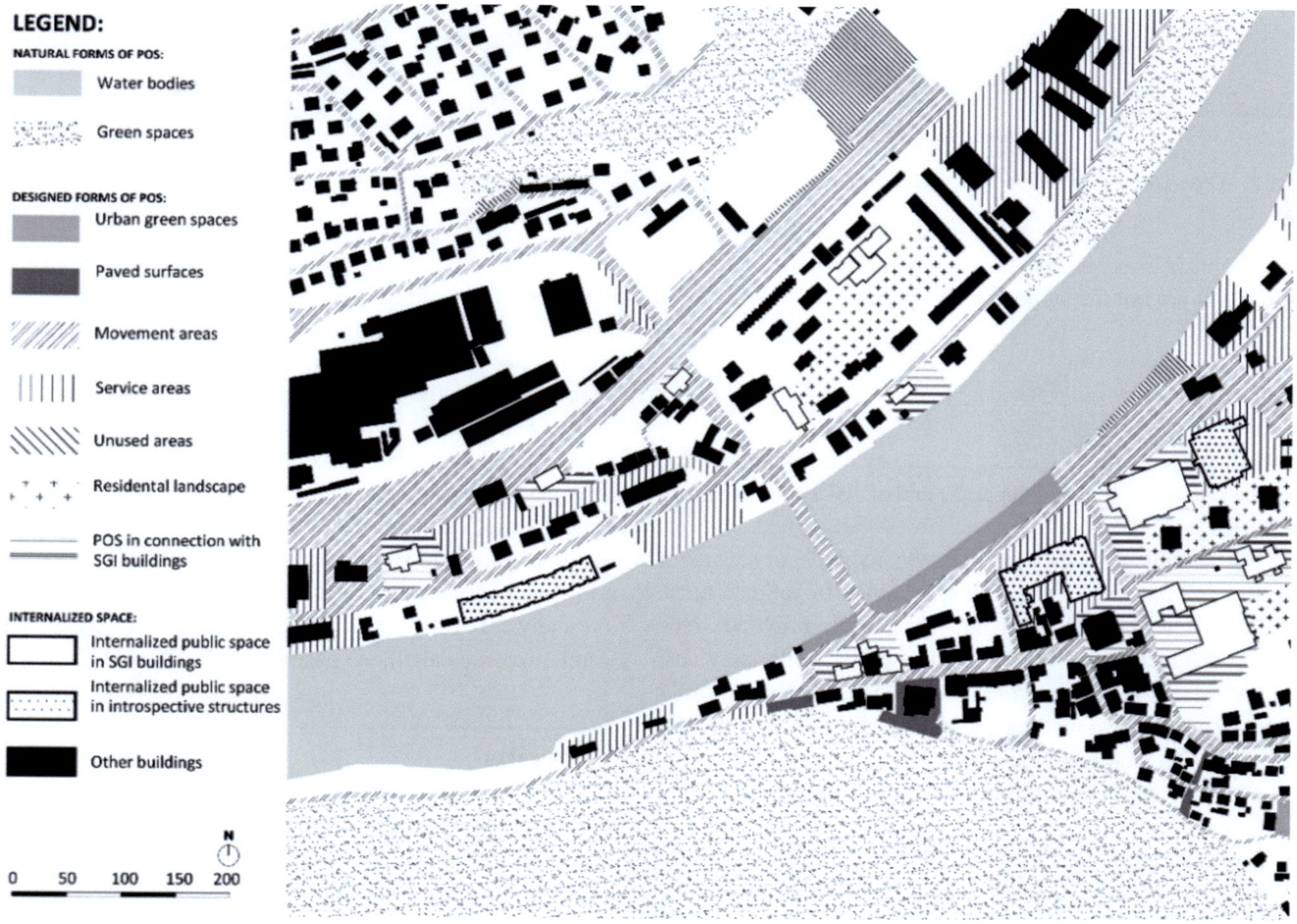

Fig. 1. Different types of public open spaces in Litija. Data source: GURS (2016) and author, visualization by the author.

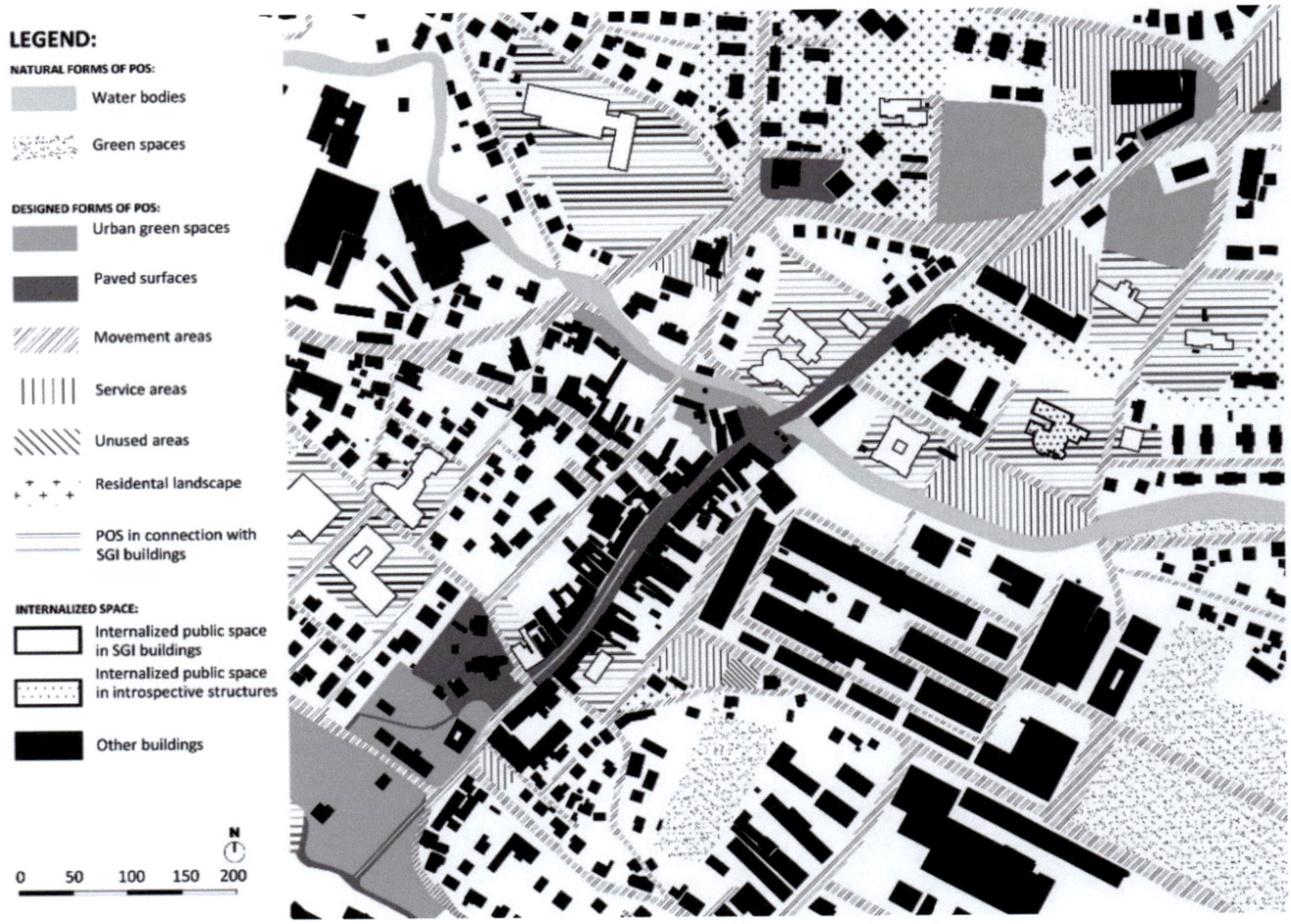

Fig. 2. Different types of public open spaces in Slovenske Konjice. Data source: GURS (2016) and author, visualization by the author.

Fig. 1 and 2. There are types of internalized space on Figs. 1 and 2 that are not the subject of this paper.

3.2 Survey

Table 1 presents the content of the survey questionnaire. The online survey was carried out at the end of 2018 and beginning of 2019.

The spatial analysis identified the location of the town centre, and the survey questionnaire aimed to validate these findings (Question 6). The analysis also highlighted areas where residents most frequently use POS, public services, and other services. Respondents identified areas they would recommend to town visitors and marked their own residential locations. To facilitate Question 6, the town centre map was divided into distinct morphological zones labelled A-J (Fig. 5 and Fig. 6).

Respondents shared their perspectives on the availability of POS in their respective town centres and the overall pedestrian appeal of the corresponding urban centre. Since the survey focused on the town centre, residents also discussed the frequency of their visits to various POS types. Additionally, the survey gathered information on preferred modes of travel within the town centre, the typical walking distances residents are willing to cover, overall satisfaction with life in the town, and the importance of specific town assets for quality of life.

The spatial analysis results are displayed solely as a map in this paper. Numerical data on the area of each POS type was also collected, but is not presented here. The survey questionnaire results are summarised using descriptive statistics, and inferential analyses (specifically correlation and regression analysis) are applied to address the research question. Correlations are calculated based on variables (1–4).

4 Results

This chapter presents the spatial analysis results, survey questionnaire results, and statistical analyses results. First, the results of the spatial analysis comparing the two cities are presented.

Table 1. Survey questionnaire details for research on town centre walkability (NT – the number of survey respondents across 33 towns: Ajdovščina, Brežice, Cerknica, Črnomelj, Domžale, Gornja Radgona, Grosuplje, Idrija, Ilirska Bistrica, Kamnik, Kočevje, Krško, Lenart, Lendava, Litija, Ljutomer, Ormož, Postojna, Radovljica, Sevnica, Sežana, Škofja Loka, Slovenj Gradec, Slovenska Bistrica, Slovenske Konjice, Šentjur, Šmarje, Tolmin, Trbovlje, Trebnje, Velenje, Vrhnika, Žalec); NL – the number of survey questionnaire respondents for Litija; NSK – the number of survey questionnaire respondents for Slovenske Konjice). Data source and visualization by the author.

QUESTION AND QUESTION TYPE	TOPIC	QUESTION AND RESPONDENT COUNT	RESPONSE OPTIONS
1. QUESTION (Closed questions)	LIVING STANDARDS	What factors contribute most to improving the living environment? (NT = 7110, NL = 378, NSK = 105)	1.) Residential environment and POS 2.) Transport and communal infrastructure 3.) Availability of jobs 4.) Accessibility to services 5.) A healthy environment 6.) A sense of community
2. QUESTION (Likert ten-point scale)		Share your satisfaction with living in your town. (NT = 2488, NL = 123, NSK = 40)	1.) Extremely dissatisfied 2.) … 9.) 10.) Extremely satisfied
3. QUESTION (Closed questions)	MODE OF TRAVEL AND DISTANCE	How do you most often travel to run everyday errands in your town? (NT = 2583, NL = 134, NSK = 41)	1.) By car 2.) By public transport, by taxi, etc 3.) By bike, by scooter, by rollerblades, etc 4.) On foot
4. QUESTION (Closed questions)		What is the length of the route you would still walk? (NT = 2553, NL = 132, NSK = 41)	1.) Up to 250 m (3 min' walk) 2.) Up to 500 m (6 min' walk) 3.) Up to 750 m (9 min' walk) 4.) Up to 1000 m (12 min' walk)
5. QUESTION (Closed questions)	POS	How frequently do you use? 5a.) Promenade (variable 1) (NT = 2387, NL = 124, NSK = 39) 5b.) Central square (variable 2) (NT = 2291, NL = 112, NSK = 34) 5c.) Central park (variable 3) (NT = 2242, NL = 112, NSK = 34) 5d.) POS in relation to public services (NT = 2236, NL = 112, NSK = 33)	1.) It's not located in the centre 2.) Never 3.) Rarely (Annually) 4.) Occasionally (Monthly) 5.) Often (Weekly) 6.) Very often (Daily)
6. QUESTION (Closed questions)		This is a map of the urban centre, divided into areas from A - J. Choose one area where: 6a.) is urban centre (NT = 1888, NL = 99, NSK = 27) 6b.) you use public services most often (kindergarten, court, post office) (NT = 1838, NL = 92, NSK = 25) 6c.) you most often use other services (shop, bar) (NT = 1798, NL = 91, NSK = 23) 6d.) you most often use POS (square, park), (NT = 1799, NL = 88, NSK = 23) 6e.) is the area that people should visit (NT = 1827, NL = 89, NSK = 24) 6f.) your home location (NT = 1831, NL = 90, NSK = 22)	1.) Area is not on the map and 2.) areas from A – J
7. QUESTION (Likert five-point scale)		Evaluate these statements about urban centre 7a.) There is enough POS (promenade, parks) (NT = 1909, NL = 94, NSK = 26) 7b.) It is pedestrian-friendly (variable 4) (NT = 1934, NL = 95, NSK = 26)	1.) Extremely Disagree 2.) Do not agree 3.) Neither agree nor disagree 4.) Agree 5.) Extremely Agree

Table 2. Living environment factors (Question 1, Table 1). Data source and visualization by the author.

TOWN	Residential environment and POS	Transport and communal infrastructure	Availability of jobs	Accessibility to services	Healthy environment	Sense of community
Litija	25.4%	21.4%	9.0%	12.2%	22.8%	9.3%
Slovenske Konjice	21.9%	20.0%	10.5%	11.4%	27.6%	8.6%
33 towns	23.0%	20.3%	12.4%	14.1%	22.3%	7.9%

Litija is neither an old nor a mining town, nor was it built according to a unified urban plan. It has expanded mainly to the east and north, as it is bounded by a hill to the south.

A large part of the town centre is covered by the river. There is a small section of urban green area along the river. Other green spaces in the centre are limited to strips along the road and forest on the hill. Figure 1 shows the analysis of the POS in Litija.

In Litija, there are a few paved surfaces, including an extended sidewalk with a monument, a square next to the church, and another square in front of a commercial building. The road passing through the older part of the town is designed as a movement area, similar to the roads in the newer part of the town. Thus, the road visually functions as a key element in the urban centre. It has two petrol stations, and another is located along a secondary road.

The historic town centre also features POS in connection with SGI buildings: a park and a small square. Several public services are housed in a large building complex, surrounded by a service area, a large car park, and a service yard. This complex includes an expansive internal POS, where most activities are concentrated indoors. Besides the large building complex, there are also low-rise structures with large footprints present in the town centre, along with POS areas designed for vehicles.

In Litija's residential areas, the POS consists of a large park with a children's playground. The residential landscape and POS connected to social and government institutions (such as the square in front of the school, the park next to the pharmacy, and children's playgrounds) play a significant role in the overall provision of POS. Similarly, POS linked to other services, such as the square in front of the commercial building, also contribute notably to the town's POS network.

The urban centre of Slovenske Konjice features a wide, pedestrian-only street serving as a promenade, enhanced by a canal that adds a water element (Fig. 2). The old town centre is bordered to the north by a river, where landscaped urban green spaces are located. At the promenade's southern end, paved areas extend near the church, leading into expansive urban green spaces. At the promenade's northern end lies the marketplace, where buildings, including a library and courthouse, are surrounded by a large square with a statue.

Some types of POS (POS in connection with SGI buildings) next to other public buildings include parking lots, urban green spaces, parks, and squares. Near the elementary school and kindergarten, fenced children's playgrounds and parking lots are provided. Smaller areas of paved surfaces and urban green spaces around public buildings, and POS in the residential landscape enhance the diversity of POS available in the urban centre. The clear delineation between private and public lands is another feature that highlights the urban character of the urban centre.

4.1 Quality of Life

Descriptions of the physical condition of the living environment (Question 1; Table 2) rank at first, third, and fourth place among quality-of-life factors, with the general value of a healthy environment coming in at second place. A healthy environment as a general value is very important to people and this is also reflected in other responses. The lowest on the scale is the 'sense of community' and just above that is the 'availability of jobs'. A similar result is found in Litija where, according to residents, the availability of jobs contributes the least to the quality of living.

The results are also similar in Slovenske Konjice, where inhabitants rank a healthy environment as most important as well-organized residential environment and POS.

People rated their satisfaction with living in towns on a ten-point scale (Question 2; Table 3). The most dissatisfied with living in their town among residents of the 33 towns are residents of Litija (5.22). Satisfaction with living in the town is higher in Slovenske Konjice (6.88), and also higher than the average (6.55).

Table 3. Satisfaction with living in the town (Question 2, Table 1). Data source and visualization by the author.

TOWN	Satisfaction with living in the town
Litija	5.22
Slovenske Konjice	6.88
33 towns	6.55

Table 4. Travel mode for everyday errands (Question 3, Table 1). Data source and visualization by the author.

TOWN	By car	By public transport, by taxi, etc	By bike, by scooter, by rollerblades, etc	On foot
Litija	70.1%	0.0%	0.0%	29.9%
Slovenske Konjice	56.1%	0.0%	12.2%	31.7%
33 towns	69.6%	0.9%	6.7%	22.8%

Table 5. Walking distances in the town centre (Question 4, Table 1). Data source and visualization by the author.

TOWN	Up to 250 m (3 min' walk)	Up to 500 m (6 min' walk)	Up to 750 m (9 min' walk)	Up to 1000 m (12 min' walk)
Litija	3.8%	22.7%	11.4%	62.1%
Slovenske Konjice	2.4%	12.2%	9.8%	75.6%
33 towns	4.2%	15.2%	13.7%	67.0%

4.2 Mode of Travel and Distance

In general, it can be said (Question 3; Table 4) that the residents of the 33 towns use a car to get around (69.6%). More people use a car for this purpose in Litija (70.1%) and less in Slovenske Konjice (56.1%).

In Litija, 29.9% of people walk to conduct their everyday errands, while in Slovenske Konjice, a slightly higher proportion of people (31.7%) walk, both exceeding the average of 22.8%. Neither town has anyone who uses public transport for everyday errands. In terms of alternative transportation for everyday errands, bicycles and similar modes are not used in Litija, while the response rate for their use in Slovenske Konjice is higher at 12.2%, which is also above the average of 6.7%.

A spatial analysis showed that the town centres of the two towns are compact in form. Respondents stated (Question 4; Table 5) that they are willing to walk longer distances (1000 m) in town centres to carry out their everyday errands (67.0%). Similar results were shown in both selected towns. The results significantly deviate from the finding that the majority of residents travel by car. This is likely due to the fact that many residents do not live in the central part of the municipality and therefore do not rely on active forms of transportation for their errands.

It was identified through spatial analysis that the factors influencing the way people travel around the town include the structure of the town, the presence of POS in the town, the location of services, and transport infrastructure (including parking garages, parking spaces, etc.).

4.3 Public Open Space in the Two Selected Towns

In the 33 towns, an average of 26% of residents think that there is no promenade in the town centre (Question 5a; Table 6). In terms of frequency, they use it monthly (19.4%), annually (18.5%), or weekly (18.1%). Some use the promenade every day (10.6%), and some people don't use it at all (7.4%).

The percentage of respondents in Litija who believe there is no promenade is alarmingly high at 55.6%, whereas in Slovenske Konjice, the situation is the opposite, with only 7.7% of respondents sharing this view. The percentage of people who never use the promenade is low in Litija, with 12.1% saying they never use the promenade following from the fact that there is no promenade in the centre. In contrast, in Slovenske Konjice, the promenade is used more frequently, with 28.2% of respondents using it weekly. This reflects the physical condition of the space and highlights the attractiveness for walking in the urban centre.

The results are similar for the frequency of use of the central square. A higher proportion of residents in Litija (29.5%) than the average (12.7%) believes there is no central square in the town (Question 5b; Table 7). Additionally, people in Litija report using the central square mostly on an annual basis (37.5%), which is similar to the average across the 33 towns (31.3%). No one in Litija uses the central square daily, while in Slovenske Konjice, the daily usage rate is higher (11.8%), with more residents stating that they use the central square on a monthly (32.4%) and weekly basis (29.4%). The results align with the findings of the spatial

Table 6. Frequency of use of the promenade (Question 5a, Table 1). Data source and visualization by the author.

TOWN	It's not located in the centre	Never	Rarely (Annually)	Occasionally (Monthly)	Often (Weekly)	Very often (Daily)
	PROMENADE					
Litija	55.6%	12.1%	8.9%	8.9%	10.5%	4.0%
Slovenske Konjice	7.7%	5.1%	15.4%	20.5%	28.2%	23.1%
33 towns	26.0%	7.4%	18.5%	19.4%	18.1%	10.6%

Table 7. Frequency of use of the central square (Question 5b, Table 1). Data source and visualization by the author.

TOWN	It's not located in the centre	Never	Rarely (Annually)	Occasionally (Monthly)	Often (Weekly)	Very often (Daily)
	CENTRAL SQUARE					
Litija	29.5%	9.8%	37.5%	17.0%	6.3%	0.0%
Slovenske Konjice	5.9%	5.9%	14.7%	32.4%	29.4%	11.8%
33 towns	12.7%	7.2%	31.3%	26.8%	16.6%	5.4%

Table 8. Frequency of use of the central park (Question 5c, Table 1). Data source and visualization by the author.

TOWN	It's not located in the centre	Never	Rarely (Annually)	Occasionally (Monthly)	Often (Weekly)	Very often (Daily)
	CENTRAL PARK					
Litija	56.3%	17.9%	14.3%	4.5%	5.4%	1.8%
Slovenske Konjice	8.8%	2.9%	29.4%	26.5%	23.5%	8.8%
33 towns	17.4%	12.8%	32.7%	21.9%	11.2%	4.0%

analysis, which reveal that Litija lacks a central square, with the vehicular road being the dominant element in the urban centre. In contrast, Slovenske Konjice features many paved surfaces, including a designated central square.

The central park is one of the key types of POS in a town, yet people in Litija perceive (Question 5c; Table 8) that there is no park, with more than half of residents (56.3%) stating that this is the case. In contrast, only 17.4% of people across the 33 towns perceive the absence of a park, and even fewer in Slovenske Konjice (8.8%), where residents use the park annually (29.4%) and monthly (26.5%). That aligns with the average usage rates across the 33 towns.

The alignment of the spatial analysis and the survey questionnaire results is once again confirmed, as there is no park in Litija, except for the POS in the residential landscape and the urban green spaces along the river. In contrast, Slovenske Konjice features a central park on the south side of the urban centre.

The analysis reveals that residents most frequently use POS related to public services (4.43), followed by the central square (3.44), the promenade (3.32), and the central park (3.13), based on the calculated averages for the 33 towns. When it comes to using POS in connection with public services (SGI), the results are similar in both towns. People use these spaces on a monthly basis, with slightly more frequent use on a weekly basis in Slovenske Konjice (Question 5d; Table 9).

Figure 3 (Question 7a; Table 1) shows that the presence of POS in Litija is much poorer than in Slovenske Konjice, and than in the group of 33 towns on average. In Litija, a higher proportion of people disagree with the statement that the town is well-equipped with POS (75.5%) compared to those who agree (8.5%). The proportion of undecided respondents is 16.0%. In contrast, residents of Slovenske Konjice are generally very satisfied with the presence of POS (57.8%), with only 23% expressing dissatisfaction and 19.2% undecided. Satisfaction with POS in Slovenske Konjice is even higher than the average for the 33 towns (38.8% satisfied, 41.7% dissatisfied, 19.5% undecided). The residents of Litija rated the presence of POS lower (1.99) than those in Slovenske Konjice (3.35), which rates above the average of 2.91 for all 33 towns.

Next, the attractiveness of the urban centre for walking is assessed (Fig. 4; Question 7b; Table 1). The proportion of dissatisfied residents is highest in Litija (32.6%), while the share of undecided residents is also the highest there (22.1%). In contrast, the highest percentage of residents satisfied regarding the attractiveness of the urban centre for pedestrians is in Slovenske Konjice (73.1%). The proportion of satisfied residents is also relatively high across the 33 towns (55.4%). These differences are reflected in the average ratings: Litija (3.05), Slovenske Konjice (3.62), and the average for 33 towns (3.27).

Table 9. Frequency of POS use in relation to public services (Question 5d, Table 1). Data source and visualization by the author.

TOWN	It's not located in the centre	Never	Rarely (Annually)	Occasionally (Monthly)	Often (Weekly)	Very often (Daily)
	POS IN RELATION TO PUBLIC SERVICES					
Litija	0.9%	2.7%	25.0%	43.8%	18.8%	8.9%
Slovenske Konjice	3.0%	3.0%	12.1%	24.2%	45.5%	12.1%
33 towns	1.3%	3.3%	17.9%	29.9%	26.3%	21.3%

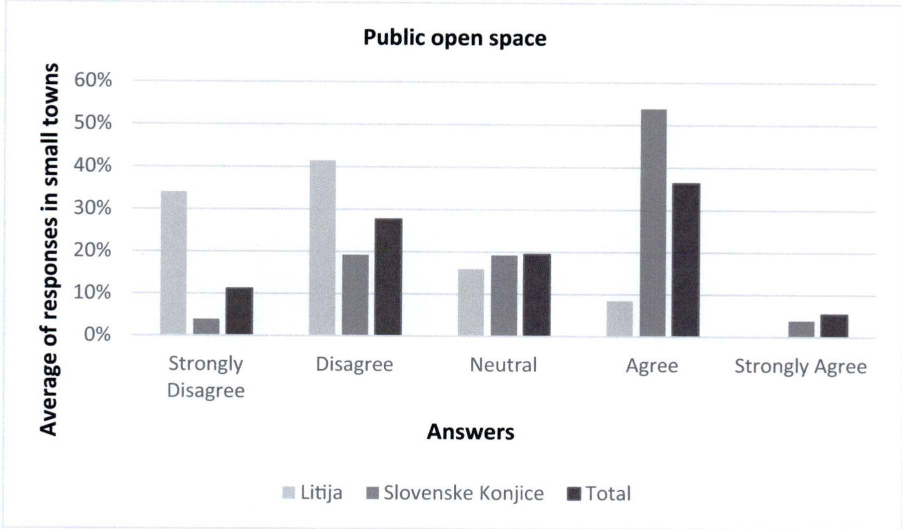

Fig. 3. Public open space presence in the two towns compared to the group of 33 towns (Question 7a, Table 1). Data source and visualization by the author.

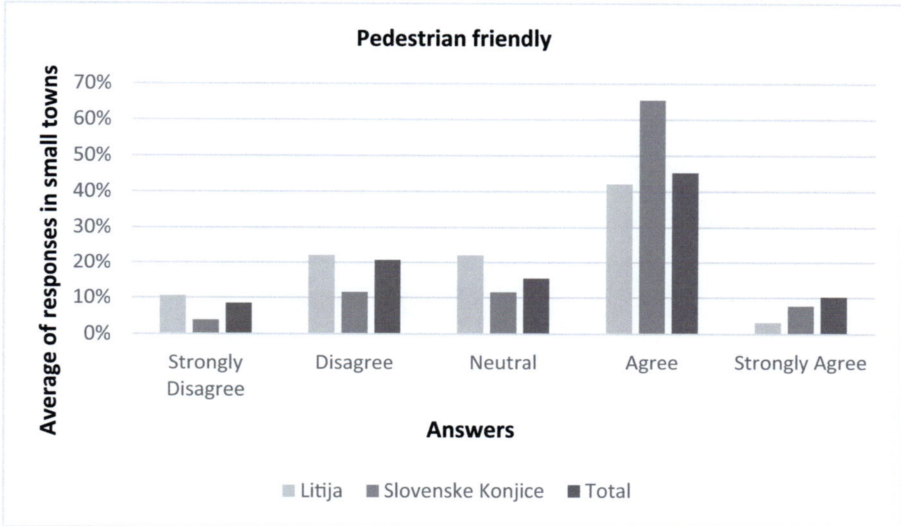

Fig. 4. Attractiveness for walking in the two towns compared to the group of 33 towns (Question 7b, Table 1). Data source and visualization by the author.

4.4 Urban Centre and Other Areas in Towns

The people surveyed defined the roles of individual morphological areas in their towns' urban centres (Question 6).

In Litija, respondents chose area B as the town centre (Question 6a; 50.5%, Fig. 5). The buildings in the area were built before and after World War II, and people would also recommend the area to visitors (Question 6e; 42.7%). The main motif in area B, which was created before the Second World War, is the road and the paved surfaces: the square next to the church and the widened pavement with the monument. Along the road there is a pharmacy with a park, and a bank with a small square. There is also a large service area next to a large building complex where there are public services. In area C, public services are most frequently used by residents (Question 6b; 34.8%). The public services in the area are a kindergarten and a primary school, and there is also a sports hall. A large building complex was built after the war in area C. The POS next to the kindergarten and primary school is a fenced playground, and next to the sports hall is a small square. Residents use other services in area E (Question 6c; 37.4%), where there are buildings (with a service space next to them) that were built after 1998 and that have large built-up areas and low heights with commercial and other services. Respondents consider that the area where they most frequently use POS is not shown on the map (Question 6d; 35.2%). Most respondents live in an area (Question 6f) that is not shown on the map (45.6%), and 22.2% of respondents live in area C.

Residents of Slovenske Konjice identified area E (40.7%) as the urban centre (Question 6a; Fig. 6). This is the medieval area, where POS are most frequently used (Question 6d; 34.8%) and which 58.3% of residents would recommend to visitors (Question 6e). Public services are most commonly

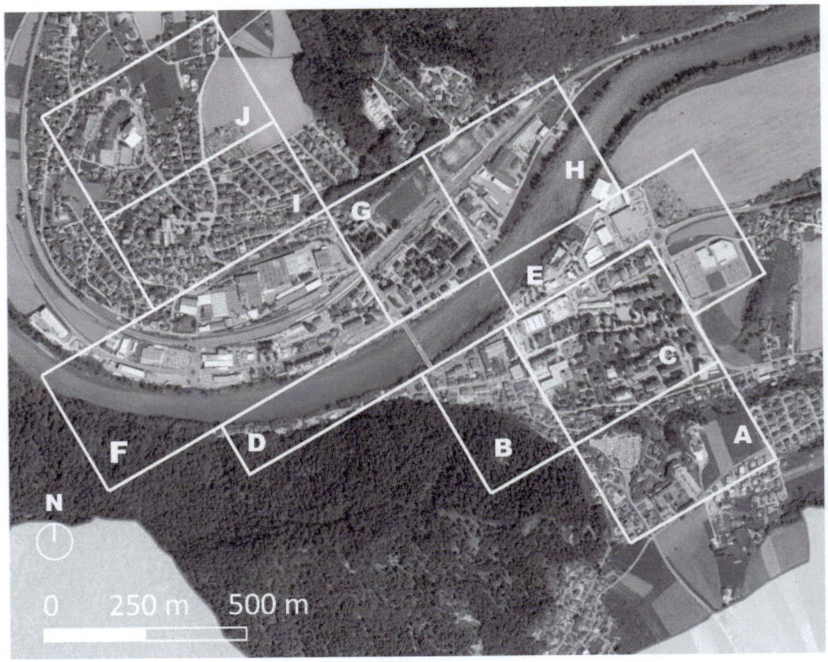

Fig. 5. Morphological areas in Litija (Question 6, Table 1). Data source: GURS (2016) and author; visualization by the author.

Fig. 6. Morphological areas in Slovenske Konjice (Question 6, Table 1). Data source: GURS (2016) and author; visualization by the author.

used in area H (Question 6b; 36%), and include the district court, cultural centre, library, and post office. This area was developed between 1946 and 1997 and features a large square. Area C, developed after 1998, is where people use other services (Question 6c; 34.8%). It consists of low-rise buildings with large built-up areas, mainly for service activities. Extensive urban green spaces are located near the health centre in the southern part, in area B. The inhabitants of Slovenske Konjice also live in an area not shown on the map (72.7%; Question 6f).

4.5 Answer to the Research Question

The research question is further addressed by confirming the results through statistical analysis, which includes data from all 33 towns (variables 1, 2, 3, and 4; Table 1).

The survey revealed that towns with a higher frequency of promenade use (variable 1) also tend to have a better walkability rating (variable 4). The calculated correlation coefficient of 0.71 ($p < 0.05$) indicates a strong correlation between these two variables. Regression analysis of the relationship between promenade use frequency and walkability showed that the regression model is statistically significant ($F(1, 31) = 31.90$, $p < 0.001$). The frequency of promenade use explains 51% of the variation in the walkability rating of the town centre. This suggests that the frequency of promenade use has a significant impact on walkability—the higher the frequency of promenade use, the better the walkability score of the town centre.

There is also a correlation between the frequency of use of the central square (variable 2) ($r = 0.51$, $p < 0.01$) and the central park (variable 3) ($r = 0.36$, $p < 0.01$) and walkability (variable 4). However, the frequency of use of the central square explains 26% of the variation in the walkability score ($F(1, 31) = 10.68$, $p = 0.0001$), while the frequency of use of the central park accounts for only 13% of the variation ($F(1, 31) = 4.73$, $p = 0.0001$). In both cases, it is more challenging to attribute a significant impact of one variable on the other.

We have addressed the research question by demonstrating the correlation and impact between the promenade and walkability. This question has also been indirectly answered through the results of the spatial analysis and the survey questionnaire.

5 Discussion

There are significant differences in POS provision across towns with distinct ground plan types. This article compares two towns with different ground plans, highlighting that the town with an incomplete medieval and modernist ground plan (Litija) lacks a promenade, central park, and central square. In this town, the pedestrian-friendly POS is mostly limited to areas around public institution buildings (POS in connection to SGI buildings) and within residential neighbourhoods (residential landscape). Other types of POS are primarily designed for vehicles, such as roads, parking lots, and gas stations. Furthermore, the potential for developing urban green spaces along the river or transforming the historic part of the town into a more pedestrian-friendly area remains untapped.

In contrast, the town with a complete medieval and modernist ground plan (Slovenske Konjice) is well provided with pedestrian-friendly POS. It features a promenade, central square, and park, with additional pedestrian areas surrounding public institution buildings (POS in connection to SGI buildings) and within residential neighbourhoods (residential landscape), which significantly enhance the overall area of quality POS. Car-oriented spaces, such as roads and parking lots, are situated away from the urban centre, ensuring minimal interference with pedestrian movement.

It was previously confirmed that there are differences in the presence of POS between morphologically distinct groups of towns in Slovenia that differ in terms of ground plan (Volgemut et al., 2023). Towns with a complete medieval and modernist ground plan in Slovenia are better equipped with POS, while towns with other ground plan types are less well equipped. It can also be said that there are differences between towns in the above-mentioned morphological group of towns, which certainly stem from actions taken in the past, such as the closing of town centres and the redesigning of POS. These actions have improved the presence and area of POS for pedestrians and increased the use of active modes of transport.

Although all towns have a continuous network of POS, the types of POS significantly impact walkability. A town with more paved areas and a promenade is generally more appealing to pedestrians than one without a promenade and with more car-dedicated types of POS. An urban centre closed to motorised traffic is more pedestrian-friendly, even though most residents in both towns commute by car due to living in outlying municipal areas far from the centre. Small towns are structured for walking due to the shorter distances within the town, and residents have expressed a willingness to walk longer distances. Within municipalities, sustainable solutions like local public transport, car and bike rentals, and similar measures would be beneficial for bridging the gaps between settlements.

Well-organised residential environment and POS are important assets for residents, which is why a large number of them participated in the survey questionnaire. It was also established that in a city that is poorly equipped with POS, the urban centre is unattractive for walking, and general satisfaction with living in the town is also poor. It was also found that, in general, people most often use POS in connection with SGI buildings but, as the spatial analysis shows, this space must be intended for pedestrians. Other types of POS, such as paved surfaces, promenades, and central squares, however, are more directly related to walkability, while a connection with the central park is not clearly established.

This study makes an important contribution to understanding Slovenian small towns and their characteristics in terms of POS and walkability. When researching walkability, it is essential to consider not only town size but also geographical context. For instance, northern European towns differ significantly from southern towns: in the north, factors like snowdrifts and the maintenance of pedestrian surfaces are key, whereas in the south, providing shade and managing heat stress are more prominent concerns.

This research showed that it is possible to use a survey questionnaire to investigate the characteristics of towns (presence of POS), but only a detailed spatial analysis shows the actual presence and types of POS and the reasons for the (un)attractiveness of urban centres for pedestrians. Despite this, Slovenian towns are relatively well provided with POS, but it turns out that in some towns many improvements are needed. A similar argument can be made for their pedestrian appeal, which is favourable, but a detailed analysis is required to evaluate each town.

This research has provided important insights into the relationship between POS and walkability, as it has been conducted in a way that links the physical characteristics of the built environment with the perceptions of users. The research met all three main objectives: 1.) data was compared between towns, 2.) a heterogenous group of people was included in the survey, so it was possible to generalise the results to the whole population; 3.) indicators (types of POS) that influence walking and the interactions between them were studied.

6 Conclusions

This research into the connection between POS and walkability in small towns in Slovenia acknowledges that town size is a crucial factor to consider, as small towns have their own urban dynamics, sidewalk characteristics, and inherent safety/comfort feelings (Arellana, 2020). Therefore, it is essential to thoroughly examine various types of POS, considering not only their spatial arrangement but also the subjective experiences of users. This dual approach, combining objective spatial analysis with subjective user feedback (Maniei et al., 2024), is crucial for understanding how POS contributes to walkability in small urban environments, and ensures the statistical validation of results.

The importance of studying POS in relation to walkability is emphasised in recent research, which stresses the need to go beyond basic intersectional analysis or a reliance solely on GIS-based measures. While GIS techniques offer valuable spatial data, they often fall short in capturing the aspects of the built environment that influence walkability, such as pedestrian experiences and the human scale of urban spaces.

Therefore, understanding walkability in smaller towns requires focusing on the human experience within specific types of POS, and how these spaces, both in terms of design and use, directly enhance urban mobility (Aparício et al., 2024). Walkability should be studied differently in small settlements to how it is in large cities because these environments have distinct spatial, social, and infrastructural characteristics, so walkability metrics and design principles need to be adapted for city size and context.

As noted by Stefansdottir with colleagues (2024), the context of a town also plays a significant role in shaping its urban characteristics. In the case of this article, the focus is on Slovenia, a country in Central Europe where unique geographical, historical, and climatic factors contribute to the local urban fabric. These contextual aspects, including the climate and the historical development of settlements, are essential when studying walkability. The specific characteristics of the area impact both the design of POS and the lived experiences of residents, influencing how walkability is perceived and achieved.

The significant impact of urban morphology on walkability (Ibrahim et al., 2024) is confirmed again in this study. Towns differ from each other in terms of POS types and attractiveness for walking depending on the ground plan type.

The findings of this paper align with those of other studies, confirming that forms of POS such as promenades and paved surfaces have a more direct impact on walkability in smaller towns than they do in larger urban areas, where broader infrastructural elements often overshadow these spaces. However, the results from Slovenian towns differ in this regard, as no clear connection between parks and walkability could be established. Greening alone does not necessarily enhance walkability (Liu et al., 2023), because other characteristics of parks also matter (Zhang et al., 2020).

Increasing walkability in town centres is closely linked to creating a peaceful atmosphere, which can be achieved by reducing traffic and promoting calm POS. As highlighted by Stefansdottir with colleagues (2024), planning for tranquillity in urban areas is essential, and this goal may be more easily achieved in smaller town centres. In smaller towns, there is often more potential to limit traffic and create pedestrian-friendly environments, therefore enhancing walkability. Such spaces not only foster a pleasant walking experience, but also contribute to broader social and environmental well-being by encouraging people to engage more with their surroundings in a relaxed, safe environment.

This study also provides valuable lessons for similar-sized towns in Central Europe by offering strategies to increase the attractiveness and livability of small towns.

References:

Aparício, J.T., Arsenio, E., Santos, F.C., Henriques, R.: Walkability defined neighborhoods for sustainable cities. Cities **149**, 104944 (2024). https://doi.org/10.1016/j.cities.2024.104944

Arellana, J., Saltarín, M., Larrañaga, A.M., Alvarez, V., Henao, C.A.: Urban walkability considering pedestrians' perceptions of the built environment: a 10-year review and a case study in a medium-sized city in Latin America. Transp. Rev. **40**(2), 183–203 (2020). https://doi.org/10.1080/01441647.2019.1703842

Badland, H.M., Keam, R., Witten, K., Kearns, R.: Examining public open spaces by neighborhood-level walkability and deprivation. J. Phys. Act. Health **7**(6), 818–824 (2010). https://doi.org/10.1123/jpah.7.6.818

Brown, A. (ed.): Contested Space: Street Trading, Public Space, and Livelihoods in Developing Cities. Rugby: ITDG Pub (2006)

Carmona, M., de Magalhães, C., Hammond, L.: Public Space: The Management Dimension, 1st edn. Routledge (2008)

Carmona, M.: Contemporary public space, part two: classification. J. Urban Des. **15**(2), 157–173 (2010). https://doi.org/10.1080/13574801003638111

Carr, S., Francis, M., Rivlin, L.G., Stone, A.M.: Public Space. Cambridge University Press, Cambridge (1992)

Congress for the New Urbanism: Charter for the New Urbanism (1993). http://www.cnu.org/charter. Accessed 3 July 2024

Doescher, M.P., et al.: The built environment and utilitarian walking in small U.S. Towns. Prev. Med. **69**, 80–86 (2014). https://doi.org/10.1016/j.ypmed.2014.08.027

Drozg, V.: Ground plans of Slovenian towns. Geogr. Bull. Bull. Geogr. Relat. Sci. **70**(1998), 75–90 (1998)

European Commission: The European Green Deal (2019). https://commission.europa.eu/strategy-and-policy/priorities-2019-2024/european-green-deal_en. Accessed 10 Jan 2024

Gehl, J.: Life Between Buildings: Using Public Space. Arkitektens Forlarg (1996)

Gehl, J., Gemzφe, L.: New City Spaces. The Danish Architectural Press (2000)

GURS: Surveying and Mapping Authority. Land Parcels, Buildings, Real estate ca-dastre data, Orthophoto and other data. Republic of Slovenia, Ministry of Natural Re-sources and Spatial Planning (2016)

Ibrahim, S., Younes, A., Assem, S.: Impact of neighborhood urban morphologies on walkability using spatial multi-criteria analysis. Urban Sci. **8**(2), 70 (2024). https://doi.org/10.3390/urbansci8020070

Jackson, J.B.: The American Public Space. In: Glazer, N., Lilla, M. (eds.) 1987, pp. 276–291. The Public Face of Architecture, Civic Culture and Public Spaces, The Free Press (1984)

Jacobs, A.B., Appleyard, D.: Toward an urban design manifesto. J. Am. Plann. Assoc. **53**(1), 112–120 (1987). https://doi.org/10.1080/01944368708976642

Jensen, W.A., Brown, B.B., Smith, K.R., Brewer, S.C., Amburgey, J.W., McIff, B.: Active transportation on a complete street: perceived and audited walkability correlates. Int. J. Environ. Res. Public Health **14**(9), 1014 (2017). https://doi.org/10.3390/ijerph14091014

Kohn, M.: Brave New Neighborhoods: The Privatization of Public Space. Routledge (2004)

Lee, S.M., Conway, T.L., Frank, L.D., Saelens, B.E., Cain, K.L., Sallis, J.F.: The relation of perceived and objective environment attributes to neighborhood satisfaction. Environ. Behav. **49**(2), 136–160 (2016). https://doi.org/10.1177/0013916515623823

Lefebvre, H.: The Production of Space. Basil Blackwell (1991)

Liu, Y., Ding, X., Ji, Y.: Enhancing walking accessibility in urban transportation: a comprehensive analysis of influencing factors and mechanisms. Information **14**(11), 595 (2023). https://doi.org/10.3390/info14110595

Davies, L.: Urban Design Compendium. English Partnerships & The Housing Copporation (2000)

Lofland, L.H.: The Public Realm: Exploring the City's Quintessential Social Territory. Routledge, Taylor & Francis (1998)

Low, S.M.: On the Plaza: The Politics of Public Space and Culture. University of Texas Press (2000)

Madanipour, A.: Public and Private Spaces of the City. Routledge (2003)

Maniei, H., Askarizad, R., Pourzakarya, M., Gruehn, D.: The influence of urban design performance on walkability in cultural heritage sites of Isfahan. Iran. Land **13**(9), 1523 (2024). https://doi.org/10.3390/land13091523

Mitchell, D.: The Right to the City: Social Justice and the Fight for Public Space. Guilford Press (2003)

Nared, J., et al.: Polycentric network of centres and accessibility of the population to services of general interest and services of general economic interest. Research report. Research Centre of the Slovenian Academy of Sciences and Arts (2016)

SeGI: Services of General Interest: A European Issue (2013). https://www.espon.eu/topics-policy/publications/evidence-briefs/services-general-interest-european-issue. Accessed 18 May 2023

Sennett, R.: The Fall of Public Man. A. A. Knopf (1977)

Scruton, R.: Public space and the classical vernacular. In: Glazer, N., Lilla, M. (eds) The Public Face of Architecture, Civic Culture and Public Spaces, pp. 13–25. The Free Press (1984)

Staeheli, L.A., Mitchell, D.: The People's Property? Power, Politics, and the Public. Routledge, Taylor & Francis (2008)

Stefánsdóttir, H., Mouratidis, K., Rynning, M.K., Meyer, S.F.: Perceived walkability and daily walking behaviour in a "small city context" – The case of Norway. J. Transp. Geogr. **121**, 104014 (2024). https://doi.org/10.1016/j.jtrangeo.2024.104014

SURS: Statistical Office, Republic of Slovenia. Demographic data (2024). https://pxweb.stat.si/SiStat/sl. Accessed 10 May 2024

Tibbalds, F.: Making People-friendly Towns: Improving the Public Environment in Towns and Cities. Longman (1992)

United Nations, General Assembly: Transforming our world: the 2030 Agenda for Sustainable Development. United Nations (2015). https://sdgs.un.org/2030agenda. Accessed 21 July 2024

Venerandi, A., Mellen, H., Romice, O., Porta, S.: Walkability indices—the state of the art and future directions: a systematic review. Sustainability **16**(16), 6730 (2024). https://doi.org/10.3390/su16166730

Vertelj Nared, P., Zavodnik Lamovšek, A.: Public open space as a contribution to urban development in small Slovenian cities. Urbani Izziv **26**(special issue, supplement), S114–S129 (2015)

Volgemut, M., Fikfak, A., Zavodnik Lamovšek, A.: The importance of open public spaces in small city centres for the provision of services of general interests. Ann. Ann. Istrian Mediterr. Stud. Ser. Historia Et Sociologia **31**(1), 83–98. https://doi.org/10.19233/ASHS.2021.06

Volgemut, M., Fikfak, A., Zavodnik Lamovšek, A.: The impact of public open space on the image of small towns centres in Slovenia. Prostor **31**(2(66)), 140–153 (2023). https://doi.org/10.31522/p.31.2(66).1

Zavodnik Lamovšek, A., Drobne, S., Žaucer, T.: Small and medium-size towns as the basis of polycentric urban development. Geodetic Bull. J. Assoc. Surveyors Slovenia **52**(2), 267–289 (2008)

Zhang, X., Melbourne, S., Sarkar, C., Chiaradia, A., Webster, C:. Effects of green space on walking: does size, shape and density matter? Urban Stud. **57**(16) (2020). https://doi.org/10.1177/0042098020902739

Zukin, S.: The Cultures of Cities. Blackwell Publishers (1995)

Children's City Architecture is Not Only Playing Grounds

Hoshiar Nooraddin

Abstract

This study critically examines contemporary architecture with the hypothesis that it largely overlooks the needs of children. It argues that architecture has evolved since the earliest civilizations to address the needs and visions of adults, often excluding considerations for children. Architectural knowledge is predominantly shaped by adult perspectives, which influence how architecture is studied, applied, and developed to meet architects' visions and design requirements. As a result, children are relegated to mere users of the built environment. The paper specifically focuses on outdoor design to explore these issues. It traces the historical development of outdoor spaces from ancient civilizations to the present, highlighting how cities have traditionally been designed according to adults' spatial understanding and needs, rather than those of children. The study evaluates urban livability for children through case studies in Oslo, Norway, and Riyadh, Saudi Arabia, examining the current state of outdoor spaces in modern cities. The study's contribution lies in emphasizing that architecture is primarily tailored to adults and advocating for the development of architectural knowledge that fully integrates children's needs in all aspects of architectural thinking and design. This approach calls for further research into methods of incorporating children's needs into architectural practice.

Special thanks to my students Majed Binabbas and Faisal Alshammary, and Faisal Alsabeg in Alfaisal university, Riyadh, Saudi Arabia, for assisting with case studies of Riyadh, Saudi Arabia. Special thanks to Architect Jean Nooraddin, Suzan Nooraddin, and Aya Nooraddin at American University of Dubai for assisting with case studies of Oslo, Norway. Many thanks to Professor Einar Lillebye at Cities Architecture for Children organization in Norway for his productive discussions about the topic.

Keywords

Children architecture · Outdoor space · Livable city · Urban design · Urban development · Architecture history · Playing ground

1 Introduction

Children have not been given as much attention in the architectural design as adults, as the design of buildings and cities primarily addresses adult needs. This results in cities dominated by adult-oriented architecture, in terms of function, form, and character. This paper is based on the hypothesis that modern architecture is inherently adult-focused, which is reflected in all aspects of architectural design, including buildings, interiors, urban design, and development strategies. The aim is to highlight this issue in contemporary city architecture and explore how current architectural thinking can be reformed to include children alongside adults. Such a shift could also have significant implications for various fields, including social dynamics, economic development, environmental sustainability, urban psychology, and future urban planning. The study documents and observes architectural solutions for outdoor spaces and their use by children. It includes a historical analysis of major architectural considerations that have influenced the design of outdoor spaces conducive to children's activities. Additionally, two contemporary cities—Oslo and Riyadh—are critically examined to demonstrate how current architectural practices address children's needs in outdoor space design. These cities were chosen for their contrasting local conditions, cultures, social structures, and architectural and urban policies.

H. Nooraddin (✉)
Cities Architecture for Children CAC, Oslo, Norway
e-mail: hoshiarno@gmail.com

B. He et al. (eds.), *Towards the Framework of Livable and Resilient Cities* Advances in Science, Technology & Innovation, https://doi.org/10.1007/978-3-031-97849-4_4

Both the historical and contemporary cases provide evidence that children have often been overlooked in the development of outdoor spaces. While some modern cities show limited consideration for children, others fail entirely to address their needs. This study operates on the understanding that children's architecture in cities involves two intertwined aspects: adults' conceptualizations and definitions of children's spaces, and children's actual expectations, experiences, and interactions with those spaces. The effectiveness of architectural solutions in meeting children's needs depends on how well these two perspectives are integrated.

The study traces how cities inherited this issue from early civilizations and how it has persisted to the present day. Although the topic of children's needs in urban architecture is recognized in contemporary architectural and urban design discourse, the study raises a critical concern: children are still not considered as comprehensively as adults. This research does not aim to provide immediate solutions but seeks to open a new area of inquiry in modern architectural practice, emphasizing the need for further research. References include historical studies of architecture and cities, as well as insights from contemporary architectural movements.

2 Children in the History of Architecture

The architecture of early cities established design principles that evolved over time and influenced present-day urban design. This includes residential areas, roads, commercial districts, ruler complexes, religious buildings, public gathering spaces, and civic buildings (Morris, 1994). These traditional design characteristics are still evident in modern urban planning principles (Adhya, 2022). The architectural characteristics of early cities were primarily shaped by the demands and desires of rulers, with later developments influenced by elite groups designing palaces and common citizens constructing homes with limited resources. The resulting architectural landscape reflected social hierarchies, with rulers and elites building grand, imposing structures to display power, wealth, and dominance (Kostof, 1993; Clark, 2016). In contrast, poorer communities built simple, functional shelters, forming vernacular architecture characterized by practicality and local identity (Oliver, 1997). A study of the historical architecture of these cities reveals a limited focus on children's needs in urban design. Despite architectural advancements, children were rarely considered primary users of outdoor spaces. This pattern has persisted throughout history, from ancient civilizations to the present. The following sections explore examples of cities from different historical periods to illustrate how early architectural practices addressed—or neglected—children's needs in outdoor space design.

2.1 Uruk

Uruk, located in present-day southern Iraq, is recognized as the earliest advanced city in the world with a comprehensive urban structure and distinctive architectural identity. The city emerged as a settlement around 5000 BC and was formally established by King Enmerkar around 4500 BC. Uruk became the birthplace of writing, advanced architecture, mathematics, governance systems, social organization, and many other innovations, making it the most significant city of its time and a cornerstone of human civilization (Van De Mieroop, 1999). These early developments laid the foundations for urban planning and architectural design, including principles that have influenced city structures throughout history. However, while Uruk's monumental buildings, temples, and administrative complexes reflect societal priorities, there is little evidence to suggest that children's needs were directly considered in the city's design.

Uruk developed a well-organized urban structure with consistent architectural solutions for its buildings, structures, and public spaces (Figs. 1 and 2). The architectural advancements were achieved through creative structural designs and the use of durable construction materials, enabling the construction of large, identical buildings and complex structures (Gruseman, 2019). Uruk's social organization followed a clear hierarchy, with the ruler at the top, followed by the wealthy, the wise, military officers, and common citizens at the base. Key architectural landmarks included the ziggurat and royal palaces, which symbolized religious and political power. The city's layout reflected an adult-defined vision of strength and authority. This was evident in the monumental scale of palaces, religious temples, public buildings, and fortified city walls. Decorative wall carvings, spatial dimensions, and functional zoning reinforced this societal structure, prioritizing religious, administrative, and defensive functions.

Fig. 1. Uruk city (Crusemann, 2015).

Fig. 2. The excavation sites (Crusemann, 2015).

2.2 Rome

Rome, established in 509 BC in Italy, became a central hub of power and civilization until the 4th century AD. Roman city planning and architecture were based on military precision and principles of power, reflected in the design, scale, and organization of buildings, streets, and public spaces. This approach created a visually imposing and strategically organized urban environment (Figs. 3 and 4). Ordinary Roman citizens lived in *insulae*, multi-story housing units that combined residential and commercial spaces. As seen in Figs. 5 and 6, *insulae* typically featured shops and workshops on the ground floor, with the shop owners and their families living above or behind the shops. Several *insulae* often formed a city block or courtyard, also referred to as *insulae* (Aldrete, 2009; Dayson, 2010).

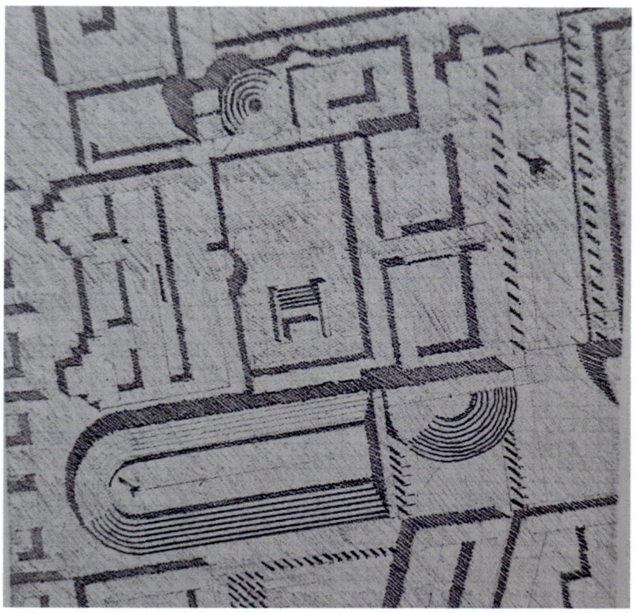

Fig. 3. Rome, Italy Incredible Scale Model of Ancient Rome. Located in EUR by Italo Gismondi (Kostof, 1992)

Fig. 4. Outdoor space, Old Rome (Nardo, 2017)

Fig. 5. Reconstruction of Ostia Insula. https://commons.wikimedia.org/wiki/File:Insula_reconstrustion._Casa_di_Diana._Gismondi.JPG

The social structure in Rome was rigidly hierarchical, dominated by male adults. Children were considered future soldiers and were trained for military service from a young age, typically starting at 14 years (Rawson, 2005). The city's architectural design reflected this societal focus on military readiness, civic pride, and public order, with little emphasis on creating child-friendly spaces. The design of public

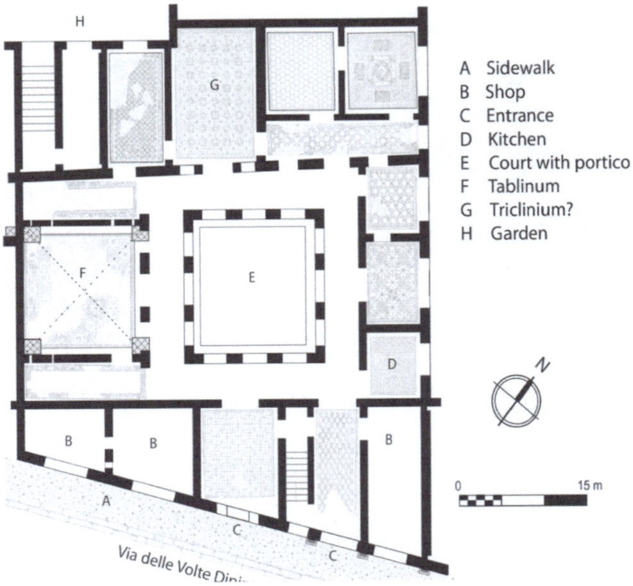

Fig. 6. Plan of Insula (Yegul, 2019)

spaces such as forums, amphitheaters, and bathhouses served adult-oriented purposes, further emphasizing the adult-centric nature of Roman urban planning.

2.3 Cairo

Cairo in Egypt built in 10th century as a planned royal city. It was planned by the leader of the military army Jawhar who intended to reflect the wealth and power of the new rulers of Egypt. The city built to settle the royal family, the army and their clans. The basic design concept was based on clustered buildings with courtyards along narrow streets (Figs. 7 and 8). Design and use of these spaces were guided by local social traditions which divided between private indoor private spaces and public outdoor spaces. This designing and using these spaces organized by social customs and religious rules controlled and applied by male adults.

The children had three main spaces shared with other age groups: the courtyards inside the dwellings, along the streets, and inside the schools' courtyards (Nooraddin, 1996).

2.4 Paris

Paris began as a village and became a town under Roman rule in 52 BC. During the Middle Ages, it developed into the largest city in Europe (Sutcliffe, 1996). The city's structure faced significant urban challenges due to its organic and dense development, resulting in various problems (Fig. 9). What makes Paris unique is its historical role in initiating an architectural revolution in the 19th century through its

Fig. 7. A commercial street in old Cairo, Egypt 18th century (Drawn by Hoshiar Nooraddin)

urban redevelopment based on Haussmann's city plan principles. The new urban development included the creation of wide, straight avenues (Fig. 10), large parks, and monumental structures such as the Eiffel Tower. These efforts aimed to address the city's crowded, dense, and irregular structure, which suffered from severe urban issues, including health, safety, environmental, and social problems. A study of the resulting city architecture reveals that Haussmann's innovative concept sought to meet the needs of the king and the city's upper social while ensuring a safer and healthier urban environment. However, the plan reflected adult-centered thinking, focusing on their needs, reactions, and expectations. It did not include specific principles or considerations for children.

The previous historical cases of Uruk, Rome, Cairo, and Paris affirm scholars' conclusions that ancient cities developed diverse architectural solutions tailored to typical urban functions (Francis, 2017; Kostof, 1992; Rossi, 1984). City plans and architectural features were influenced by social, cultural, economic, and environmental factors (Mumford, 1989; Morris, 1994). However, studying their city structure can show that these cities consistently prioritized adults as primary users in their architectural designs. This adult-centric

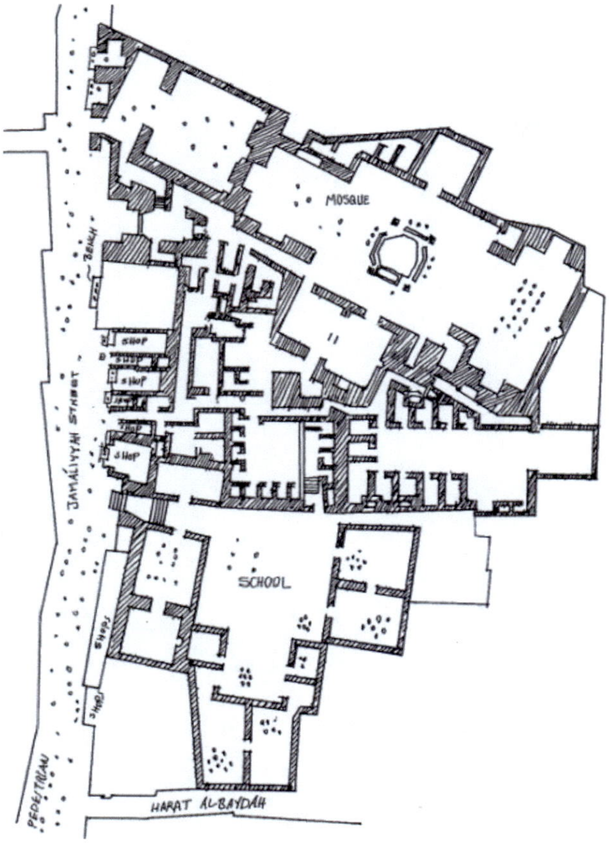

Fig. 8. Plan of clustered of different types (Drawn by Hoshiar Nooraddin)

Fig. 9. Boulevard Montmartre 1897 After building Hausmann Development Plan of Paris. https://en.wikipedia.org/wiki/Paris_in_the_Belle_%C3%89poque

Fig. 10. Street of old Paris 1868. https://www.laphamsquarterly.org/roundtable/old-paris-no-more

focus is evident in palaces, temples, military settlements, theaters, gates, monuments, and residential homes (Crusemann, 2015). In such cities, children had to adapt to spaces designed without their needs in mind, using leftover areas for play and social activities. This exclusion stifled the emergence of dedicated children's architecture, limiting its development from the earliest stages of urban planning.

3 Children in Contemporary Architecture

The French Revolution of 1799 triggered transformative changes in architecture by establishing modern architectural education as source for architectural knowledge (Griffin, 2020). This shift encouraged architects to break free from historical constraints and explore new creative directions. One significant outcome was the rise of functionalism, a movement that became a hallmark of modern architecture. Its core principle was to replace traditional design methods with function-driven architecture (Charles, 2000). Functionalist theory prioritized creating healthy, economical, and utilitarian buildings where form followed function (Aurelia, 2016). An early advocate of functionalist thinking was architect Louis Sullivan, whose 1896 design for the Wainwright Building in St. Louis, Missouri, embodied these ideals. He described his architectural vision as follows:

"It must be tall, every inch of it tall. The force and power of altitude must be in it; the glory and pride of exaltation must be in it. It must be every inch a proud and soaring thing, rising in sheer exultation that from bottom to top it is a unit without a single dissenting line."

Sullivan's approach redefined architectural thought, influencing early 20th-century architects such as Le Corbusier (Jencks, 2000). Le Corbusier advocated for architectural simplicity, emphasizing that buildings should honestly reflect their functions. His theories revolutionized city planning, urban design, and building design, spreading globally and shaping modern architectural traditions.

Despite these advances, functionalist and modernist movements failed to recognize children as distinct users with unique architectural needs. This oversight is symbolized by the use of The Modulor Man (Figs. 11, 12), a measurement system based on adult proportions, rooted in Leonardo da Vinci's Vitruvian Man (Fig. 13). Although intended as a universal design standard, this concept disregarded children entirely.

The early 20th century saw an architectural response to shifting economic, social, and political landscapes following two world wars, the collapse of empires, and the rise of new powers. These turbulent changes gave birth to various architectural movements, but even within architectural education, the architectural knowledge and innovations of different architectural movements, children's needs remained

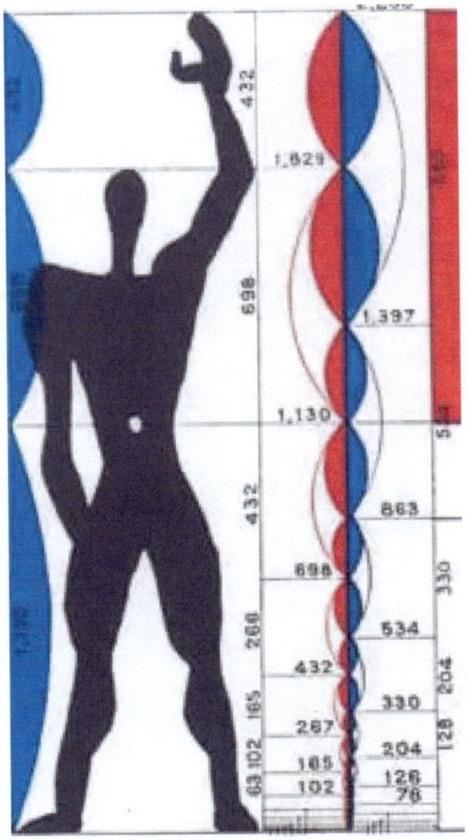

Fig. 11. The golden architectural ratio The Modular Man. https://wiki.dtonline.org/images/thumb/a/ad/CorbusiersModulor.png/600pxCorbusiersModulor.png

overlooked, reinforcing an enduring adult-centric design tradition.

The application of architectural movements from the late 19th century to the 1960s led to significant urban problems due to their limited understanding of the complexities of human living environments. One critical issue was the deterioration of social life in public spaces, as identified in various studies by scholars like Jane Jacobs (1969, 1993). Jacobs criticized urban planning approaches that neglected community life and social interaction. Similarly, Jan Gehl suggested that addressing these issues required a reform of architecture and urban design to create user-friendly environments (Gehl, 2010, 2011). However, these studies and the projects derived from them failed to consider or comprehensively address children's needs in urban spaces. Their proposed reforms still reflected an adult-centric view of the city.

In 1989, the United Nations adopted the *Convention on the Rights of the Child* (UNCRC), establishing children's rights across civil, political, economic, social, and cultural dimensions (UNICEF, UN). This legally binding agreement prompted a new trend in urban design that considered children more thoughtfully, particularly in outdoor play space design. Designers began incorporating innovative outdoor furniture,

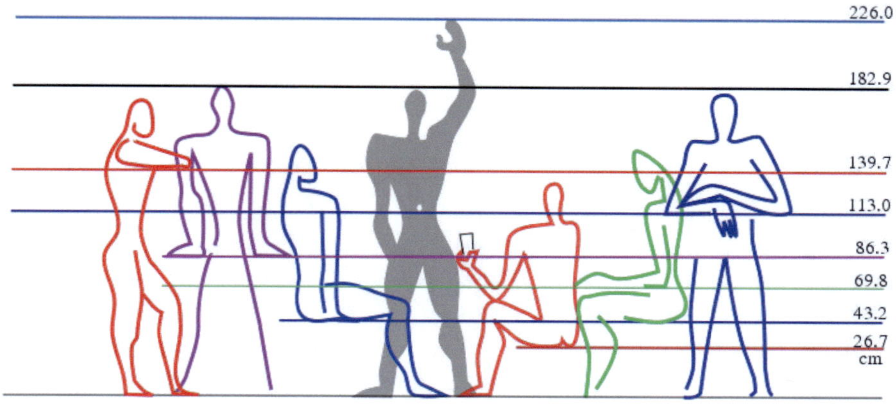

Fig. 12. Using Modular Man design of buildings, interior design and furniture design. https://www.architecturelab.net/wp-content/uploads/2024/06/746px-Modulor_measurements.svg_.jpg

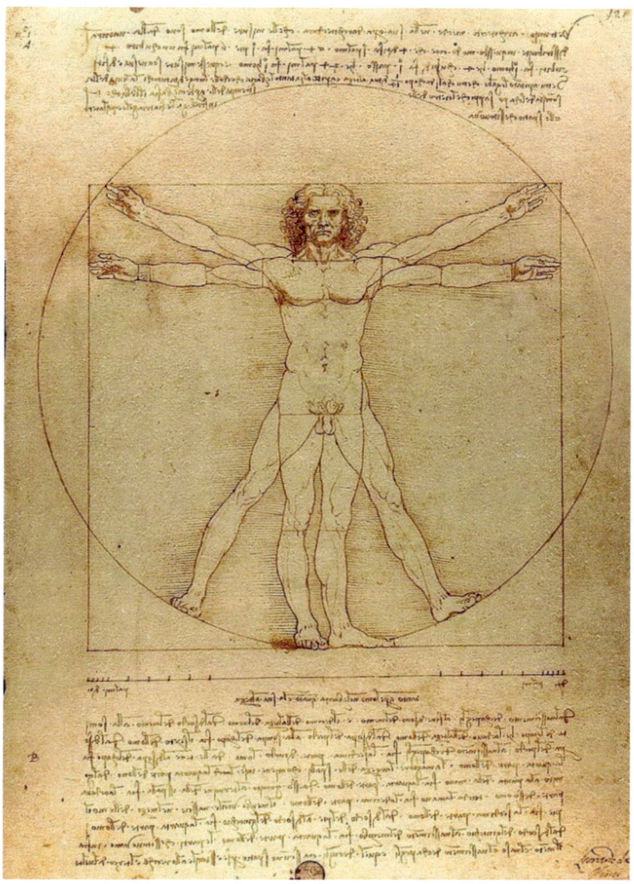

Fig. 13. Leonardo Da Vinci, Vitruvian man. https://commons.wikimedia.org/wiki/File:Da_Vinci_Vitruve_Luc_Viatour.jpg

play structures, and interactive features, reflecting this new awareness (Gutman, 2014). This trend has since influenced architectural and urban design practices aimed at creating more livable, sustainable, and child-friendly cities.

However, the implementation of these principles has varied across countries, leading to diverse design outcomes (Lee, 2015). While some cities have embraced child-friendly design more fully, the approach remains fragmented and limited. These projects still lack a comprehensive architectural framework that addresses all aspects of children's urban experiences (Nooraddin, 2020). Current urban development initiatives, such as creating "happy," "livable," and "sustainable" cities, reflect attempts to align with the goals of the UNCRC (Mohammad, 2016). Nonetheless, these efforts often differ between countries and cities in scale, scope, and effectiveness due to varying interpretations, policy applications, and cultural contexts.

This highlights the need for deeper integration of children's needs into architectural and urban design, moving beyond isolated interventions toward a holistic architectural approach that considers children as essential urban users.

4 Outdoor Design in City's Children Architecture

A growing international trend emphasizes integrating children into urban development strategies to create livable cities (Laura, 2017) (Krysiak, 2020). This shift reflects a rethinking of architectural approaches, particularly after the rise of functionalist architectural movements in the last century, as well as the adoption of national policies prioritizing children's well-being. Several United Nations organizations encourage and support this shift through international commitments and recommendations aimed at improving children's lives in urban environments. Different countries have implemented these recommendations in diverse ways, applying unique architectural and urban design strategies. Various authorities, organizations, and researchers, including

UNICEF, have developed principles for designing child-friendly cities (UNICEF, 2004). These guidelines emphasize age-appropriate designs, suitable facilities, safety measures, and accessibility for all children.

Children and young people have their own imaginations, experiences, needs, and understanding of space in terms of design, function, content, qualities, and characteristics. For example, a child's or young person's bedroom typically has different interior features and contents than an adult's. These differences include elements such as colors, lighting, furniture, and play items. We don't use play and safety only but many more aspects. We design these spaces with accumulated learnings, recognizing that children need indoor environments suited to their age to help them grow in a healthy, stable, enjoyable, and stimulating way while respecting their need for private space. Therefore, although issues such as safety and play are important for children and young people, the architectural design is crucial for children living environment. But do we apply the such principles when designing a house, a building, a structure, a space, and a city?

Realizing such a comprehensive approach to "children's architecture" requires rethinking how cities, buildings, and outdoor spaces are designed. This demands changes in architectural theory and practice, challenging long-established modern international design knowledge.

4.1 Case Study: Oslo, Norway

Oslo, the capital of Norway, lies along the Oslo Fjord and features a green, maritime landscape (Fig. 14, 15). The city has a population of about 700,000, with over 1 million residents in the greater Oslo area. Its population density is approximately 1,400 inhabitants per square kilometer, relatively low compared to other European cities. The climate in Oslo is characterized by cool summers, mild but snowy winters, and long autumns, with spring arriving only by late March. Despite the climate, Oslo has made significant efforts to support children's outdoor life and activities by improving public spaces and facilities (Figs. 16, 17).

The city applies innovative architectural and urban design solutions aimed at enhancing urban livability for all age groups, including children. These efforts extend to the renovation of older urban areas, although the quality and extent of these projects vary depending on the area's social class and available space (Figs. 18, 19). A crucial factor contributing to the success of Oslo's child-friendly urban environment is the country's inclusive social culture. Norwegian society supports equal access to outdoor spaces and public facilities across all social classes and age groups. This cultural aspect ensures that public spaces are actively used and appreciated, making the city's urban design strategies more effective in achieving their goals.

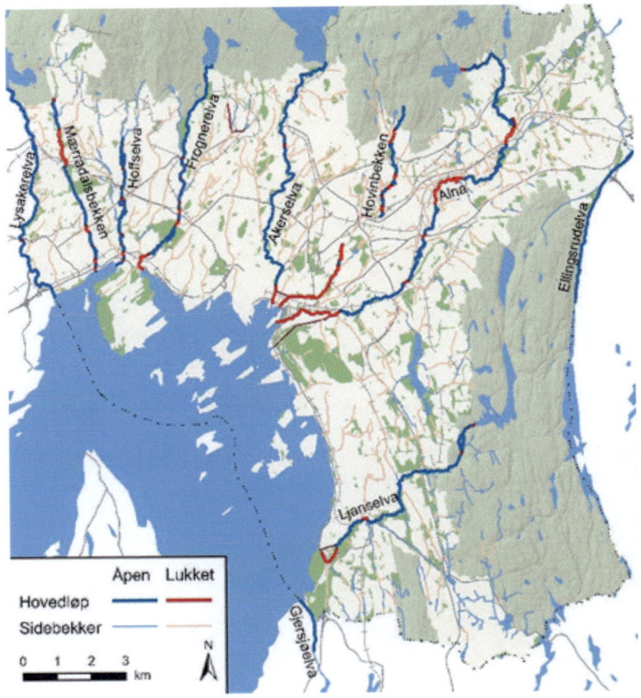

Fig. 14. Plan of Oslo in Norway https://www.osloelveforum.org/kart/

Fig. 15. City Centrum of Oslo. https://www.vartoslo.no/bydel-sentrum-ina-helen-ostby-jernbanetorget/ser-du-hva-som-mangler-her-slik-kjorte-trikken-rundt-i-sentrum/589204

Studying these cases reveals a strong commitment from the authorities to implement development policies aimed at improving outdoor life for children. Observing these areas clearly shows that children actively use these spaces, regardless of the size, location, or quality of the play equipment. The children who use these areas are from all age groups, mostly between 2 and 16 years old, with each group engaging in its own types of play. However, what has been achieved is

Fig. 16. A new dens area with compact design. Bjørvika, Oslo. https://www.visitoslo.com/no/oslo-for-deg/barnas-oslo/lekeplasser/

primarily the creation of play areas within the broader outdoor space of the street or district. Oslo municipality considers playing grounds the basic solution to support children needs in the city, But playing ground is only one aspect of the entire architecture of children outdoor areas in Oslo. It is noticed, that whenever there is a plan to build a new project or improve an existing public outdoor space for children, the primary solution is typically the creation of play areas. However, in

Fig. 18. Playing area under a bridge in a poor district of Oslo. https://www.facebook.com/photo/?fbid=6072770082740796&set=pcb.6072775102740294

Fig. 17. Playing area in middle class area. https://www.visitoslo.com/no/oslo-for-deg/barnas-oslo/lekeplasser/

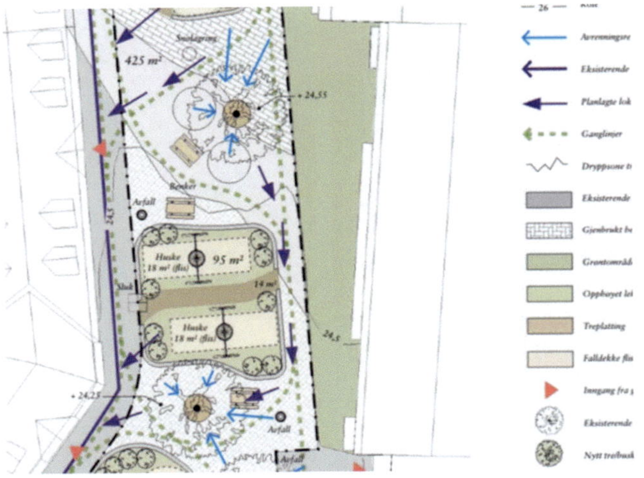

Fig. 19. Renovating playing area, Monradsgate Oslo. https://www.oslo.kommune.no/slik-bygger-vi-oslo/monrads-gate-oppgradering/#gref

Fig. 20. Riyadh traffic network. https://www.google.no/maps/@24.7668288,46.9243814,10.25z?entry=ttu&g_ep=EgoyMDI0MTIxMS4wIKXMDSoASAFQAw%3D%3D

Fig. 21. High density of urban fabric Riyadh. https://milesopedia.com/en/guide/destination/riyadh-saudi-arabia-guide/

reality, these play areas are like islands, isolated within the larger outdoor space of the adults.

The observations show to achieve truly child-friendly outdoor spaces, it is necessary to broaden the authority's perspective on children's needs in public outdoor areas. This means considering all the components of the outdoor environment and their design characteristics, not just one specific element. These components include all components in the city such as street dimensions, pavement types and colors, street lighting, sidewalk design, materials and colors of sidewalks and elevation surfaces, window sizes and locations of buildings, building heights, the relationship between indoor and outdoor spaces, and the design of ground-floor areas within buildings, among others. In essence, it is about how the designer integrates all these elements into the design of outdoor spaces in relation to children. By adopting this comprehensive approach, we can establish a more holistic architectural perspective when designing or upgrading public outdoor spaces. Such thinking must be applied equally across all areas and cities. Therefore, what has been achieved and planned in Oslo cannot yet be considered as a complete "children's outdoor architecture."

4.2 Case of Riyadh, Saudi Arabia

Riyadh, the capital of Saudi Arabia, is located in the heart of the Arabian Peninsula. The city has a population of 7 million, with a population density of approximately 4,300 people per square kilometer. The climate is dominated by dry desert conditions and extremely high temperatures. Riyadh has unique local social traditions that significantly influence the use of outdoor spaces, including how children interact with these environments. The city's urban design policies and architecture reflect these cultural influences. The present urban condition is characterized by high density, a dominant urban traffic network, and limited green areas (Momrah, 2022) (Figs. 20, 21). The city has a mixed population of local Saudis and 40% foreigners from various countries.

The current city plan for Riyadh was developed in 1970 by Doxiadis Associates (DA), and it was later replaced in the 1980s with a new, larger plan due to rapid population growth and urban expansion fueled by the city's economic boom. The authority made several efforts to transform various aspects of

Fig. 22. King Abdullah recreation park with multi outdoor activities for all ages. Exploring King Abdullah Park in Riyadh: A Guide to the City's Magical Green Oasis - Wasalt Blog.

local tradition while preserving the essence of the culture. It is called vision 2030 but the strategy is still in its early stages, and it is too soon to evaluate its full impact.

There are many urban development projects which are done by Ryiadh municipality and the governmental authorities to improve the green and open public spaces in order to achieve livable outdoor spaces. But these projects have various challenges to achieve the targets of urban livability for children mainly because of climate, culture, accessibility, and urban fabric.

The children playing and gathering areas are mainly four different types:

1. **Parks:** The children gathering areas are mainly located in community parks or larger parks which serve the whole district or the city (Fig. 22). These parks are fenced and create outdoor livability within the park boarders not in the neighborhoods. Many of the community parks are not in good condition. Most Saudis don't go to these parks and prefer indoor playing areas.
2. **Indoor Playing and gathering areas in commercial buildings:** They are mostly shopping centers which offer different types of playing areas and fast-food restaurants (Fig. 22), outdoor places with a new design and furniture. They are in open spaces along streets or in parks (Fig. 23).
3. **Playing areas within Residential areas:** The residential areas in general (high, middle, poor economic class) have playing areas but different in different conditions and use. Most of the social pedestrian street life and no children gathering and playing areas in most of the streets in these areas are not designed pedestrian friendly. The playing areas in the residential area are mostly in small parks or open spaces. Many of them not in proper condition and

Fig. 23. Riyadh has large number commercial buildings. Different indoor playing grounds. (Source: By author: Hoshiar Nooraddin)

Fig. 24. Playing area in residential area. (Source; Faisal AlSabeg)

Fig. 25. Playing area in residential area. (Source; Faisal AlSabeg)

need upgrading. The furniture are also not updated and don't cover the needs (Fig. 24, 25).

Planning of the streets of the residential areas are mostly based on prioritizing the car needs, with isolated residential villas and buildings form outdoor spaces, using high fences (can be 5 or 8 m high). The sidewalks are totally occupied and blocked by the owners by planting trees, parking cars, and other needs (Fig. 26). The poor residential districts are totally dominated by clustered houses and streets without sidewalks or any playing areas (Fig. 27).

The observations in Riyadh reveal that local economic, social, and climatic factors strongly influence children's use of outdoor play areas. The extreme heat limits outdoor activities during the day, while conservative social traditions discourage the use of play spaces within neighborhoods. Despite these challenges, the city has not developed alternative urban design solutions that accommodate its unique climatic and cultural context. The lack of comprehensive architectural consideration for children has led to urban environments in Riyadh characterized by design elements that are incompatible with children's well-being. To create a more inclusive urban space, Riyadh's urban design approach must evolve, integrating children's needs into its architectural and urban planning strategies. This shift would require rethinking design principles, policies, and practices to ensure that outdoor spaces become functional, accessible, and supportive environments for children.

Fig. 26. No livability in middle class area. (Source: By author: Hoshiar Nooraddin)

Similar issues were noted in Oslo, where urban design also failed to adopt a holistic perspective that equally prioritizes children and adults. In both cities, architectural and urban development principles remain predominantly adult-oriented, overlooking children's specific spatial and developmental needs.

Fig. 27. Low-class area. (Source; Faisal AlSabeg)

5 Conclusions

Architecture has evolved over thousands of years through accumulated learning in the design and construction of cities, buildings, and spaces. This paper critiques the process, highlighting that it has largely been shaped by adults with the goal of creating environments tailored to adult needs and imagination of what is a city and its architecture. This critique is based on the hypothesis that contemporary city design and architectural knowledge continue to prioritize adults, while children receive only limited consideration.

The cases of Uruk, Rome, Cairo, and Paris affirm that ancient cities developed diverse architectural solutions tailored to typical urban functions. Analyzing the architectural structures of these cities reveal evidence of this long-standing phenomenon, where city structures and architectural characteristics provided no specific design solutions for children.

This adult-centric approach likely stems from the assumption that if a design works well for adults, it will also function adequately for children. This perspective has shaped architectural thinking throughout history, leading to the widespread exclusion of children from architectural design and theoretical discourse. As a result, children's specific needs and experiences have been neglected in the development of cities, buildings, and public spaces since the earliest civilizations. This oversight underscores the need for a more inclusive architectural framework that recognizes children as distinct users requiring thoughtful, age-appropriate design considerations. In recent decades, new movements have emerged aimed at improving livability in cities for all users, including children. However, these movements have led to limited architectural solutions specifically for children, while adult needs still dominate city architecture.

This paper also includes case studies from Oslo, Norway, and Riyadh, Saudi Arabia, to document the persistence of this issue in contemporary cities. The case studies focus on identifying the architecture of outdoor spaces and observing how these spaces are used. The two cities represent distinct countries with different social traditions, cultures, architectural styles, urban design practices, and policies. Both cases show that significant progress has been made in recent decades in improving the built environment for children. In both cities, playgrounds are the primary architectural solution employed to create better environments for children in public outdoor spaces. Both cases of Oslo and Riadh show that they have no other architectural design solutions for children.

In Norway, playgrounds are widely used, with considerations for safety, accessibility, and updates to ensure they are child-friendly. These areas are free to access and are actively used by children and their families. However, in Riyadh, playgrounds face different challenges. Many communities lack proper play spaces, and where they do exist, they are not evenly distributed across the city. This shortage forces families to rely on large public parks or commercial indoor play areas as alternatives.

Both case studies underline the need for a real shift in architectural knowledge and design practices to incorporate a broader understanding of children's needs in outdoor spaces. This requires the development of a comprehensive "children's architecture" for designing public outdoor spaces and implementing urban development strategies. This shift would involve integrating children's needs into architectural education, research, and development strategies. Considering the full range of children's needs in the architecture of outdoor public spaces would transform many aspects of current city architecture, creating a more wholistic built environment that benefits society and generates social and economic improvements.

This reform calls for further studies to cover a broad range of topics and establish a foundation for children's architecture, such as:

- How can architectural theory be modified to include children?
- What are the components of children's architecture?
- How can architecture and urban design address children's functional needs, spatial understanding, behavior, and imagination?
- How can children's needs be used as a source for developing architectural solutions for various areas of the city that address their full range of needs, similar to adults?
- How can existing urban spaces be developed to create a comprehensive living environment for children?
- How can education contribute to the development of children's architectural knowledge?
- How can official urban development strategies incorporate children's architecture?

References

Adhya, A., Plowright, P.: Urban Design Made by Humans: A Handbook of Design Ideas. Routlege (2022)
Aldrete, G.: Daily Life in the Roman City: Rome. Oklahoma University Press, Pompeii and Ostia (2009)
Al-Assad, M., Mehrotra, R.: Shaping Cities. Hatje Cantz Verlag GmbH, Aga Khan Award for Architecture (2016)
Jencks, C.: Le Corbusier and the Continual Revolution in Architecture. The Monacelli Press (2000)
Clark, G.: How cities took over the world: a history of globalization spanning 4,000 years. Guardian (2016)
Ching, F., Jarzombek, M., Vikramaditya, P.: A Global History of Architecture. Wiley (2017)
Crusemann, N., et al.: Uruk: First City of the Ancient World. Getty Trust Publications (2015)
Dayson, S.: Rome A Living Portrait of an Ancient City (Ancient Society and History). The John Hopkins Press (2010)
Dulic, O., Alazic, V.: A note on graphical representations in architecture - diagrams over sketches. In: 4th International Conference Contemporary Achievements in Civil Engineering, Subotica (2016)
Gutman, M.: A City for Children. The University of Chicago Press (2014)
Gehl, J.: Life Between Buildings: Using Public Space. Island Press (2011)
Gehl, J.: Cities for People. Island Press (2010)
Giraldi, L., et al.: Designing for the next generation: children urban design as a strategic method to improve the future in the cities. Des. J. **9**(6) (2017)
Jacobs, J.: The death and life of great American cities. Mod. Libr. (1993)
Jacobs, J.: The Economy of Cities. Random House Inc. (1969)
Jordan, D.: Transforming Paris: the Life and Labors of Baron Haussmann. University of Chicago Press (1995)
Kostof, S.: The City Assembled. Thames & Hudson (1992)
Kostof, S.: The City Shaped. Bulinch Press (1993)
Krsiak, N.: Childe Friendly High-Density Neighborhoods. Cities for Play (2020)
Lee, H., et al.: A metastudy of qualitative research examining determinants of children's independent active free play. Int. J. Behav. Nutr. Phys. Act. **12**(5) (2015)
Morris, A.E.J.: History of Urban Form. Longman Scientific & Technical (1994)
MoMRAH: Kingdom of Saudi Arabia's Report on the Implementation of the New Urban Agenda. Ministry of Municipal, Rural Affairs & Housing. Kingdom of Saudi Arabia (2022)
Mumford, L.: The City in History. A Harvest Book, New York (1989)
Nardo, D.: Cause & Effect: Ancient Rome. Reference Point Press, San Diego (2017)
Nooraddin, H.: Children city architecture. Adv. Soc. Sci. Res. J. **7**(7) (2020)
Nooraddin, H.: Al-Fina: A Study of In between Spaces Along Streets as an Urban Design (1996)
Concept in Islamic Cities of the Middle East with a Case study in Cairo. NTNU, Trondheim, Norway
Oliver, P.: The Encyclopedia of Vernacular Architecture of the World. Oxford Brookes University (1997)
Sutcliffe, A.: Paris: An Architectural History. Yale University Press (1996)
Rossi, A.: The Architecture of the City. The MIT Press (1984)
UNICEF: Building Child Friendly Cities: A Framework for Action. UNICEF Innocent (2004)
Research Centre. https://www.childfriendlycities.org
UNICEF, UN: https://www.unicef.org
Van De Mieroop, M.: The Ancient Mesopotamian City. Oxford University Press (1999)
Yegul, F.: Roman Architecture and Urbanism. Cambridge University Press (2019)

Urban Agroecology and the Well-Being of Older Adults: Exploring the Impacts of Residential-Scale Pesticide-Free Vegetable Production in Chiang Mai's Urban Landscape

Pongsakorn Suppakittpaisarn, Tupthai Norsuwan, Thakoon Panyasai, Nattaya Jaipunya, Rattanaphon Chaima, Kamon Thippachote, and Thanit Saltikulnukarn

Abstract

The emphasis on urban agriculture and sustainable food systems has become imperative for promoting health and environmental resilience. In 2022, the Center of Agricultural Resources System Research, Faculty of Agriculture, Chiang Mai University, initiated the program called 'Development of Pesticide-free Vegetable Production Systems in Chiang Mai's Urban Agroecology'. The program aimed to disseminate knowledge about growing pesticide-free vegetables across general public in Chiang Mai communities. The program is reported to be successful. However, the follow-up well-being benefits of the program, especially for the vulnerable population, might still be needed to confirm the impact and reach of the program.

This research aimed to investigate the well-being benefits of older adults participating in the program one year after the program concluded, focusing on the in-depth interview of five elderly participants from the program in the Ton-Pin community.

While some older adults engaged in pesticide-free vegetable cultivation as a recreational activity, others emphasized its direct health benefits, combining physical exercise and consumption of clean produce. Community knowledge and social networking enhanced the program's reach and impact. Furthermore, some participants viewed urban farming as an economic opportunity, while others use this opportunity navigate personal grief and loss that comes with old age. This study highlights the significance of urban agriculture in enhancing health, fostering environmental resilience, and promoting city livability, especially for the older demographic. The integration of the BCG economic model in Chiang Mai's urban agricultural initiatives offers a sustainable blueprint for other rapidly urbanizing cities globally.

Keywords

Urban agriculture · BCG economic model · older adults · agroecology · environmental psychology

P. Suppakittpaisarn (✉)
Landscape Design and Environmental Management Studio, Department of Plant and Soil Sciences, Faculty of Agriculture, Chiang Mai University, Chiang Mai, Thailand
e-mail: pongsakorn.sup@cmu.ac.th

T. Norsuwan · T. Panyasai · N. Jaipunya · R. Chaima · T. Saltikulnukarn
Agricultural Innovation Research, Integration, Demonstration and Training Center (AIRIDTC), Faculty of Agriculture, Chiang Mai University, Chiang Mai, Thailand
e-mail: tupthai.n@cmu.ac.th

T. Panyasai
e-mail: thakoon.p@cmu.ac.th

N. Jaipunya
e-mail: nattaya.jai@cmu.ac.th

R. Chaima
e-mail: rattanaphon.c@cmu.ac.th

T. Saltikulnukarn
e-mail: thanit.sal@cmu.ac.th

K. Thippachote
Department of Plant and Soil Sciences, Faculty of Agriculture, Chiang Mai University, Chiang Mai, Thailand
e-mail: kamon.thippachote@cmu.ac.th

B. He et al. (eds.), *Towards the Framework of Livable and Resilient Cities* Advances in Science, Technology & Innovation, https://doi.org/10.1007/978-3-031-97849-4_5

1 Introduction and Literature Review

1.1 Nearby Nature and Older Adults' Well-Being

The global demographic is shifting towards an older age range (Rudnicka et al., 2020). Even as medical innovations ensure that individuals age more gracefully and remain active contributors to society, the older adults still face specific challenges that can influence their overall health and quality of life (Rudnicka et al., 2020).

One of the ways we can ensure the health and well-being of people, including older adults, is contact with nature. Contact with nature, like walking among trees, gardening, or viewing natural scenery from the windows, can provide multiple physical and psychological benefits (Sullivan, Frumkin, Jackson, & Chang, 2014). This includes urban agricultural and horticultural activities such as growing vegetables (Shao, Elsadek, & Liu, 2020).

Older adults, in particular, can benefit from these interactions with the natural elements—such as reducing risks of high blood pressure, cardiovascular diseases, depression, and dementia (Matthew H. E. M. Browning, Lee, & Wolf, 2019; Donovan et al., 2013; Mitchell & Popham, 2008; Wichrowski, Whiteson, Haas, Mola, & Rey, 2005). With older adult population expanding and their vulnerabilities (Rudnicka et al., 2020), the study regarding contact with nature, especially urban agriculture, could be beneficial to the design and planning of their daily activities. However, the mechanisms of the benefits should still be explored further (Matthew H. E. M. Browning, Rigolon, & McAnirlin, 2022; Rigolon, Browning, McAnirlin, & Yoon, 2021; Sullivan et al., 2014).

1.2 Urban Agriculture and Co-benefits of Nature

Urban agriculture not only contributes to food security and waste management but also offers multiple environmental and social co-benefits, reinforcing the relationship between humans and nature (Norsuwan et al., 2021). According to the Food and Agriculture Organization (FAO), urban farming can mitigate the Urban Heat Island effect, leading to cooler and more comfortable cities (Saed, 2012). It also provides social engagement. For example, a study by Lovell (2014) emphasized that urban agriculture fosters community engagement. Some studies also suggest that these practices can help in increasing biodiversity within the urban and peri-urban environments (Cabannes & Raposo, 2013; Norsuwan et al., 2021; Zasada, 2011). Further, urban agroecology is an agricultural practice in the urban area where the p.

1.3 Gaps of Knowledge

With this information, researchers, designers, and policy makers start to support the ideas of urban agriculture among citizens, especially older adults. However, there are limitations. While the benefits of urban agriculture are documented, a deeper understanding is lacking, especially in the cultural and geographical contexts of Northern Thailand and towards older adults (Suppakittpaisarn et al., 2023). Most studies focus on urban agriculture in Western or highly urbanized Asian settings (Graefe, Buerkert, & Schlecht, 2019; Weidner, Yang, & Hamm, 2019), with limited insights into regions like Northern Thailand. This is partially because the unique cultural ties between Northern Thai people and their environment are under-represented in current literature (Shrestha, Chaweewan, & Arunyawat, 2017), especially in comparisons to places like Europe, East Asia, and the US (Chen & Feeley, 2014; Finlay, Franke, McKay, & Sims-Gould, 2015; Rosso, Auchincloss, & Michael, 2011). This may influence the outcomes of urban agriculture projects. Finally, there is a lack of long-term studies examining the sustained impact of urban agriculture programs on participants, especially older adults (Taylor & Lovell, 2014).

1.4 This Study

These existing gaps of knowledge guided the research question of this study. That is, what might be an enduring well-being benefits experienced by older adults in Northern Thailand who participated in an urban agriculture program? Thus, this research aimed to investigate the well-being benefits of older adults participating in the program one year after the program concluded, focusing on the in-depth interview of five elderly participants from the program in the Ton-Pin community. The study's findings can provide valuable feedback for future urban agricultural projects targeting senior populations to be more culturally aware and effective in the long run.

Fig. 1. Program activity in a local community. Images by Tupthai Norsuwan.

2 Methods

2.1 Background: The CMU Urban Agroecology Program

In late 2021 to early 2022, the Center of Agricultural Resources System Research[1], Faculty of Agriculture, Chiang Mai University, initiated the program called 'Development of Pesticide-free Vegetable Production Systems in Chiang Mai's Urban Agroecology' (CMU Urban Agroecology Program). The program aimed to disseminate knowledge about growing pesticide-free vegetables across the public in Chiang Mai communities (Fig. 1).

The program achieved these goals by approaching the community leaders and invited interested parties of all ages to attend continuous 1-year support program with events, courses, and on-site consultations. The project collaborated with over 30 community members through the following processes:

- Educating communities on urban agricultural development concepts via lectures and on-site training.
- Organizing online study visits and practical training sessions during the pandemic.
- Providing budget support for agricultural materials.
- Setting up a communication and supporting group through social media.
- Developing community business models, market channels, and certifying Good Agricultural Practices (GAP) and chemical residue standards in produce through AGtrace, a procedure from the Faculty of Agriculture at Chiang Mai University (CMU).

As a results of this initiative, the community was able to expand its vegetable cultivation area by a total of 1,350 square meters, including various planting setups like plots, blocks, pots, shelves, tables, and under-roof structures. Nine members achieved GAP certification.

To expand the results, the project organized a seminar on "Urban Agricultural Insights" involving representatives from municipalities, non-governmental development organizations, and project members to discuss local development strategies through urban agriculture. Community members expressed their desires for municipal support, especially in terms of market development and the inclusion of urban agriculture in local development plans.

2.2 The Follow-Up Interviews

In April 2023, another group of researchers from the Faculty of Agriculture, Chiang Mai University were investigating the relationships between nearby nature and the health of older adults. Adults aged 60 and over were asked to give in-depth interviews regarding their daily life, their environment, and their perceived well-being. Older adults from four Chiang Mai communities participated.

The steps of the interviews were as follows: First, we submitted the protocol to the Chiang Mai University Research Ethics Committee and had the methods approved. Then, we developed a conceptual framework based on environmental design and older adults in Thailand, which can be applied to the urban agro ecological environment (Fig. 2), (Suppakittpaisarn et al., 2023) and trained the researchers to structure the questions.

We recruited the participants via social media and engaging with community informal leaders. Once at the site, we asked a series of questions regarding their surroundings, their daily activities, and their well-being. While we allowed the participants to tell their own stories, if they did not know where to start, we asked them some of the following questions to prompt their responses.

[1] The Center of Agricultural Resources System Research was combined with other organizations to later become Agricultural Innovation Research, Integration, Demonstration and Training Center (AIRIDTC).

Fig. 2. Conceptual framework of environmental design and older adults adapted from Suppakittpaisarn, et al. (2023)

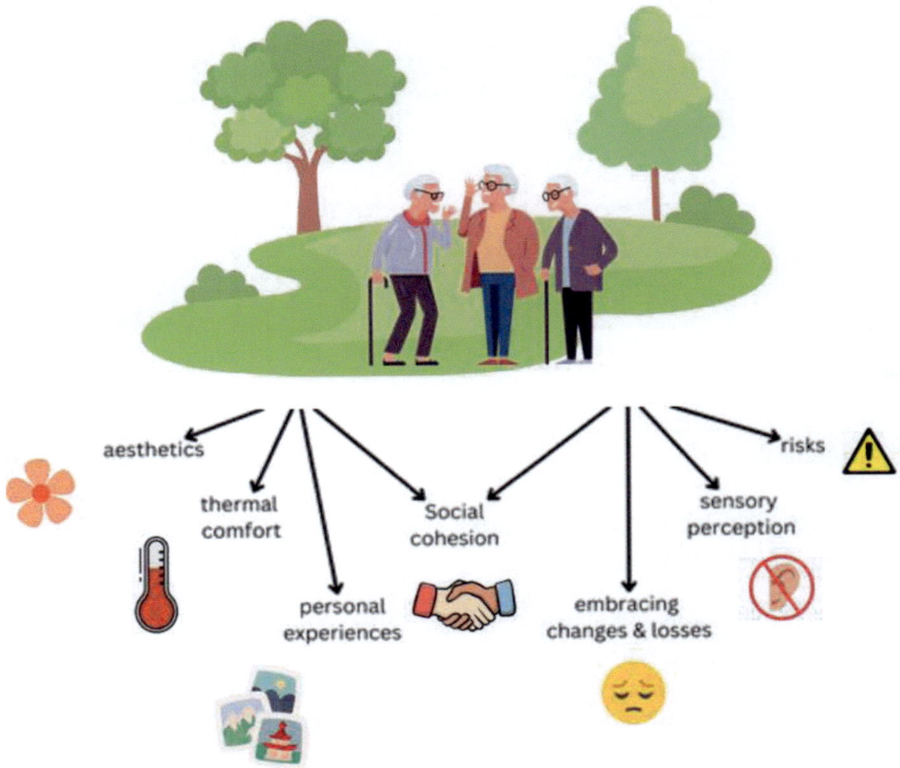

- Please let us know about your daily environment. What is your home like?
- What do you usually do from when you wake up until you go to bed in a typical day?
- Do you feel healthy? Please tell us about your health.

The researchers took reflexive notes as well as audio recordings. The reflexive notes were composed of observed data, such as non-verbal cues, physical environment, or the reflections of the researchers, while the audio recording device was placed openly between the researchers and the interviewees for transparency. The audio was then transcribed into texts by the researchers who conducted the interviewed, along with the tagged reflexive journal.

For analysis, a researcher (PS) conducted a content and thematic analysis of the transcripts and the notes. They also placed the transcripts and notes into a language processing webservice (Chat GPT 4.0) with the prompt: what themes and emotions were detected from this participant's story? The replies were then copied into a word document.

Then the combined analysis of a researcher and the language processing artificial intelligence model were conducted for thematic and sentiment analyses. These steps reflect the guides of recommendations for qualitative interviews (Golafshani, 2003; Jabareen, 2009; Maxwell, 2008). Figure 3 explains the steps of the interview.

2.3 Research Sites

We found in the preliminary investigation that Ton-Pin community via contact of an informal community representative. The community was located south of Chiang Mai Old City (Fig. 4). It had a few older adults who participated in the CMU Urban Agroecology Program. Thus, for these older adults, we requested that they explain deeper about the program and how the program and urban agriculture in general affected their lifestyle and well-being one year after the program has ended.

3 Results

Five participants from Ton Pin community engaged in the CMU Urban Agroecology Program. All five of them still grew fruits and vegetables in varying capacity (Fig. 5). All of them mentioned in the in-depth interview that most initiated the ideas from curiosity and intended it to be a hobby. However, four key links between urban agriculture and health emerged including physical health, economy, sharing, and processing grief. The results could be concluded in the diagram as shown in Fig. 6.

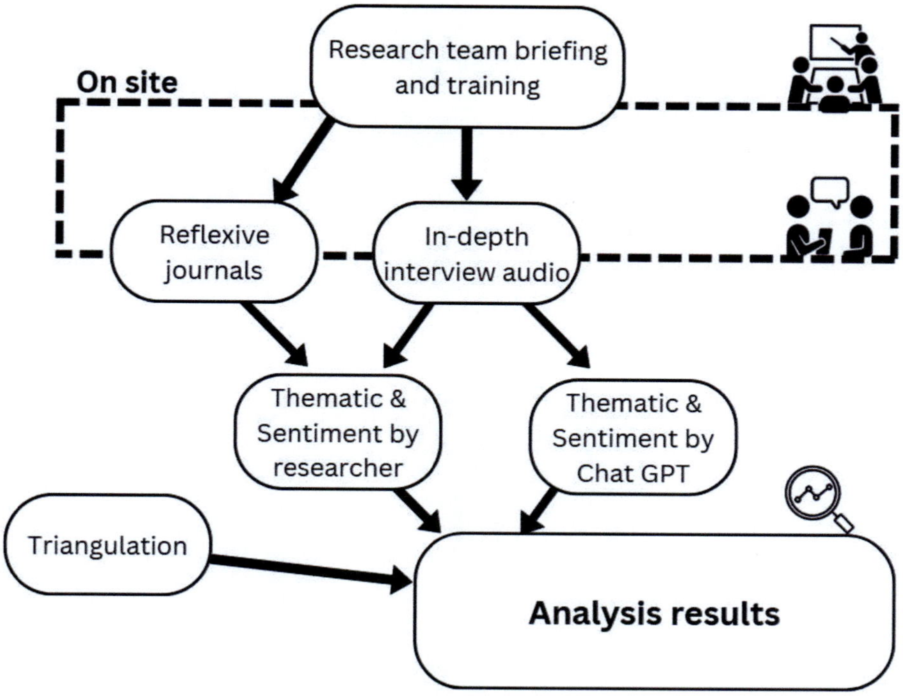

Fig. 3. The interview process of this study. Source: By author.

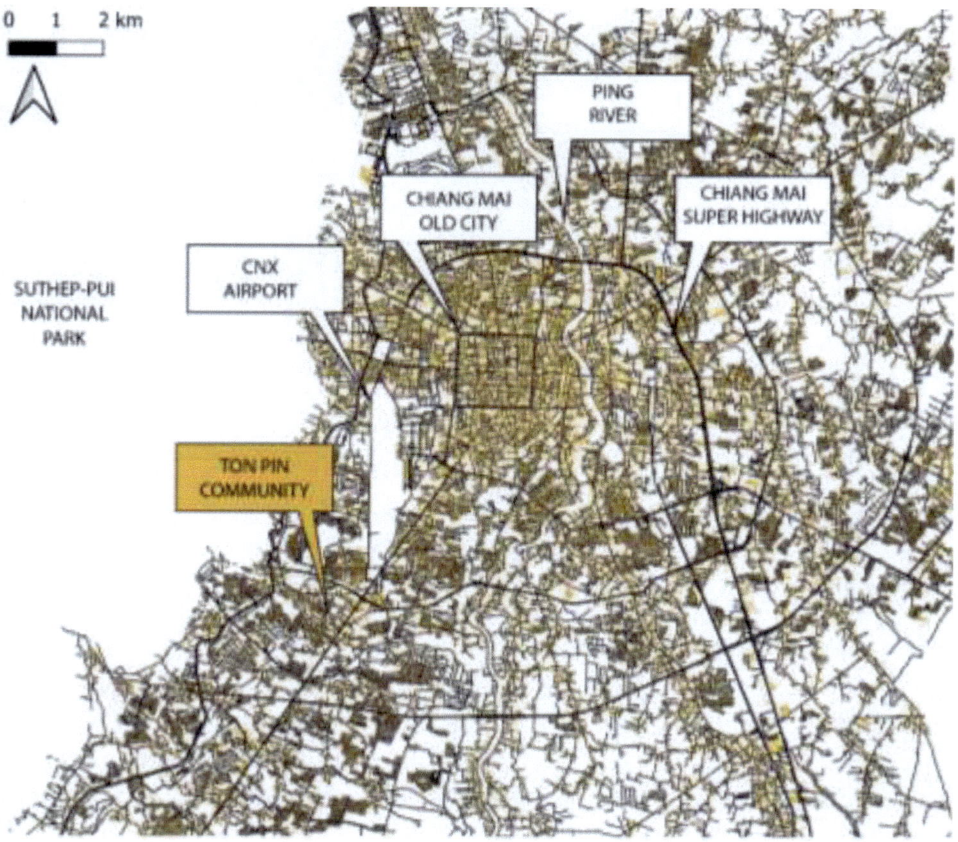

Fig. 4. Ton Pin Community in relations to Chiang Mai geographical landmarks. Source: By author.

Fig. 5. Urban agriculture from Ton Pin Community residences (Source: By authors)

Fig. 6. The diagram concluding the results of the interviews. Source: By author.

Physical Health: Participants reported engaging in urban agriculture helped them get pesticide-free produce. Furthermore, it motivated them and/or their partners to move around and perform more physical activities. Thus, the consistent engagement to the program had kept them physically healthy despite their advancing ages. This physical health gave them a sense of control toward their actions.

> "It's really good for me to move and plant these vegetables. It's a form of exercise for me."
>
> Participant B (Female)

Economy: Participants gained economic benefits from this experience. Some opened a smoothie shop using some of the produce they grew. Others reduced the costs of groceries by the supplies of vegetables and fruits from the urban agriculture program. While the money saved is relatively small, for some participants who recently retired and were living with their family, this contribution gave them a sense of pride.

> "We save money because we no longer have to go buy these vegetables. And they are chemical-free, which are usually pretty expensive."
>
> Participant A (Female)

Sharing: Participants reported sharing activities happening because of urban agriculture. They shared their produce with relatives and neighbors, and they shared their knowledge of pesticide-free urban agriculture with the community members who were interested. One participant invited the researchers to try the longans growing from her front yard. The interesting part of sharing was the positive emotion from giving and teaching, and the social aspects of the urban agriculture program keep the participants feeling that they contribute to the surrounding people in meaningful ways.

> "I got to teach some of this to my friends. It's like this. Older folks like me still have things to teach some folks. It's a good conversation starter, when you know how to grow cabbage and make compost piles."
>
> Participant D (male)

Processing Grief: Like most older adults, the participants experienced losses and changes in many ways, and gardening was used to occupy their minds. One of the participants recently lost her partner, but she told the researcher that the gardening helped her understand that death was a cyclical part of nature. While another had lost their lifestyle due to retirement, and expressed that gardening kept her busy and distracted from the free time she never had before.

> "My husband passed away 2 years ago. But gardening keeps me sane and centered. You learn that plants grow and die. Just a part of life, you know. And without him and with my daughter going to college, growing these [vegetables] helps me focused on something else that wasn't myself."
>
> Participants E (female)

4 Discussion

In this study, the researchers investigated older adults who participated in a CMU Urban Agroecology program one year after their participation. We found that for older adults, engaging in urban agriculture encouraged them to be physically active and eat healthy, provided them with some economic benefits, helped them contribute to the people they cared about, and guided them through grief and losses. These interpretations add to the literature regarding urban agriculture, agroecology, and well-being in older adults.

The findings reflected previous literature outside of the Northern Thailand context while strongly support the existing evidence. As mentioned in the literature review, urban landscapes can provide several ecosystem services, such as stress reduction (Ulrich et al., 1991), reduced health risks (Suppakittpaisarn, Jiang, & Sullivan, 2017), and providing social engagements (Tinsley, Tinsley, & Croskeys, 2002). In addition to these benefits, the results of the interviews suggested that urban agroecology provided special facets towards the health of people because it encourages nutritious food provision, physical activities, opportunities for social interactions, and reduction of some financial burdens. These several factors in combination with the physical and mental health benefits provided by urban nature and landscapes aligned with the health needs for older adults, who are particularly at risks of physical issues based on lower muscle masses and poor diet, lower immune systems, and depression and mental health risks stemming from loneliness and lack of social support (Cahoon, 2012; Chaudhury, Campo, Michael, & Mahmood, 2016; Hong et al., 2018; Wolinsky, Wyrwich, Babu, Kroenke, & Tierney, 2003). The evidence added here, for all four dimensions of older adults' health improvement from urban agroecology, supported the existing potential of growing food, sharing, and having nature-based activities. Growing food at home was related to physical health improvement across ages alongside the straightforward economic benefits (Howarth, Brettle, Hardman, & Maden, 2020; Taylor & Lovell, 2014). Studies showed that sharing and contribution could increase self-esteem in individuals (Lee & Jang, 2010; Neff, 2011), and that spending time gardening, along with artwork and music, could act as a therapeutic aid to those in emotional distress (Ulrich, 1999; Ulrich et al., 1991). Furthermore, the addition of urban agroecology might provide better urban environments with reduced carbon, air pollution, and urban heat island effect, which can provide additional ecosystem services (Abino, Castillo, & Lee, 2014). Combined with the existing literature review, the study showed a cohesive picture that urban agriculture could benefit older adults in several directions.

This study highlights the significance of urban agriculture in enhancing health, fostering environmental resilience, and promoting city livability, especially for the older demographic. The integration of the BCG economic model in Chiang Mai's urban agricultural initiatives offers a sustainable blueprint for other rapidly urbanizing cities globally.

There were limitations to the study. We only studied five older adults in the specific community of Chiang Mai. The participants were all local Chiang Mai-born individuals with a higher middle class economic status and were somewhat educated. Thus, these findings may need to be investigated further towards people with different demographics across the city of Chiang Mai.

5 Conclusion

The program initiated by the Center of Agricultural Resources System Research, Chiang Mai University, emphasizes the crucial roles of urban agriculture in promoting health, environmental resilience, and city livability. This research, which concentrated on older adults from the Ton-Pin community, further emphasized that beyond the obvious benefits of clean produce, urban farming serves as both a recreational activity and a therapeutic means to cope with the challenges of aging. The findings revealed that many elderly individuals not only benefit from the physical exercise involved but also from the psychological comfort and potential economic prospects urban farming provides. Enhanced community knowledge and strengthened social networks played an important role in the program's success. By integrating the BCG economic model into Chiang Mai's urban agricultural practices, a sustainable framework emerges, presenting a viable model for other cities around the world that are also facing rapid urbanization. Overall, urban agriculture has proved its worth as a holistic approach to addressing various facets of well-being, especially for older adults, and offers a promising pathway for the future of urban living.

Acknowledgements This work is partially supported by Chiang Mai University. The extension program is supported by the Thailand Science, Research, and Innovation (TSRI) and Program Management Unit (PMU). Grant number: A15F640032. Recipient: Tupthai Norsuwan. The in-depth interview part is supported by the National Research Council of Thailand (NRCT). Grant number N42A650309. Recipient: Pongsakorn Suppakittpaisarn.

References

Abino, A.C., Castillo, J.A.A., Lee, Y.J.: Assessment of species diversity, biomass and carbon sequestration potential of a natural mangrove stand in Samar, the Philippines. For. Sci. Technol. **10**(1), 2–8 (2014)

Browning, M.H.E.M., Lee, K., Wolf, K.L.: Tree cover shows an inverse relationship with depressive symptoms in elderly residents living in US nursing homes. Urban For. Urban Greening **41**, 23–32 (2019)

Browning, M.H.E.M., Rigolon, A., McAnirlin, O.: Where greenspace matters most: a systematic review of urbanicity, greenspace, and physical health. Landsc. Urban Plan. **217**, 104233 (2022)

Cabannes, Y., Raposo, I.: Peri-urban agriculture, social inclusion of migrant population and Right to the City: practices in Lisbon and London. City **17**(2), 235–250 (2013)

Cahoon, C.G.: Depression in older adults. AJN Am. J. Nurs. **112**(11), 22–30 (2012)

Chaudhury, H., Campo, M., Michael, Y., Mahmood, A.: Neighbourhood environment and physical activity in older adults. Soc. Sci. Med. **149**, 104–113 (2016)

Chen, Y., Feeley, T.H.: Social support, social strain, loneliness, and well-being among older adults: an analysis of the health and retirement study. J. Soc. Pers. Relat. **31**(2), 141–161 (2014)

Donovan, G.H., et al.: The relationship between trees and human health: evidence from the spread of the emerald ash borer. Am. J. Prev. Med. **44**(2), 139–145 (2013). http://ac.els-cdn.com/S0749379712008045/1-s2.0-S0749379712008045-main.pdf?_tid=248591d8-8192-11e5-a630-00000aab0f6c&acdnat=1446490220_f0cab250069a687651887dc06e872756

Finlay, J., Franke, T., McKay, H., Sims-Gould, J.: Therapeutic landscapes and wellbeing in later life: impacts of blue and green spaces for older adults. Health Place **34**, 97–106 (2015)

Golafshani, N.: Understanding reliability and validity in qualitative research. Qual. Rep. **8**(4), 597–606 (2003)

Graefe, S., Buerkert, A., Schlecht, E.: Trends and gaps in scholarly literature on urban and Peri-urban agriculture. Nutr. Cycl. Agroecosyst. **115**, 143–158 (2019)

Hong, A., et al.: Linking green space to neighborhood social capital in older adults: the role of perceived safety. Soc. Sci. Med. **207**, 38–45 (2018)

Howarth, M., Brettle, A., Hardman, M., Maden, M.: What is the evidence for the impact of gardens and gardening on health and well-being: a scoping review and evidence-based logic model to guide healthcare strategy decision making on the use of gardening approaches as a social prescription. BMJ Open **10**(7), e036923 (2020)

Jabareen, Y.: Building a conceptual framework: philosophy, definitions, and procedure. Int. J. Qual. Methods **8**(4), 49–62 (2009)

Lee, E.-J., Jang, J.-W.: Profiling good Samaritans in online knowledge forums: effects of affiliative tendency, self-esteem, and public individuation on knowledge sharing. Comput. Hum. Behav. **26**(6), 1336–1344 (2010)

Maxwell, J.A.: Designing a qualitative study. In: The SAGE Handbook of Applied Social Research Methods, vol. 2 (2008)

Mitchell, R., Popham, F.: Effect of exposure to natural environment on health inequalities: an observational population study. Lancet **372**(9650), 1655–1660 (2008)

Neff, K.D.: Self-compassion, self-esteem, and well-being. Soc. Pers. Psychol. Compass **5**(1), 1–12 (2011)

Norsuwan, T., et al.: Urban agriculture and BCG economic model [in TH]. J. Agric. Res. Extension **38**(3), 100–116 (2021)

Rigolon, A., Browning, M.H.E.M., McAnirlin, O., Yoon, H.V.: Green space and health equity: a systematic review on the potential of green space to reduce health disparities. Int. J. Environ. Res. Public Health **18**(5), 2563 (2021)

Rosso, A.L., Auchincloss, A.H., Michael, Y.L.: The urban built environment and mobility in older adults: a comprehensive review. J. Aging Res. (2011)

Rudnicka, E., Napierała, P., Podfigurna, A., Męczekalski, B., Smolarczyk, R., Grymowicz, M.: The World Health Organization (WHO) approach to healthy ageing. Maturitas **139**, 6–11 (2020)

Saed: Urban farming: the right to what sort of city? Capital. Nat. Social. **23**(4), 1–9 (2012)

Shao, Y., Elsadek, M., Liu, B.: Horticultural activity: its contribution to stress recovery and wellbeing for children. Int. J. Environ. Res. Public Health **17**(4), 1229 (2020)

Shrestha, R.P., Chaweewan, N., Arunyawat, S.: Adaptation to climate change by rural ethnic communities of Northern Thailand. Climate **5**(3), 57 (2017)

Sullivan, W.C., Frumkin, H., Jackson, R.J., Chang, C.-Y.: Gaia meets Asclepius: creating healthy places. Landsc. Urban Plan. **127**, 182–184 (2014)

Suppakittpaisarn, P., Jiang, X., Sullivan, W.C.: Green infrastructure, green stormwater infrastructure, and human health: a review. Current Landscape Ecol. Rep. **2**(4), 96–110 (2017)

Suppakittpaisarn, P., Vienravee, K., Yaipimol, E., Charoenlertthanakit, N., Surinseng, V., Wanitchayapaisit, C.: A conceptual framework development for designing nearby nature for older adults. Landscape Archit. Art **120** (2023)

Taylor, J.R., Lovell, S.T.: Urban home food gardens in the Global North: research traditions and future directions. Agric. Hum. Values **31**, 285–305 (2014)

Tinsley, H.E., Tinsley, D.J., Croskeys, C.E.: Park usage, social milieu, and psychosocial benefits of park use reported by older urban park users from four ethnic groups. Leis. Sci. **24**(2), 199–218 (2002)

Ulrich, R.S.: Effects of gardens on health outcomes: theory and research. In: Marcus, C.C., Barnes, M. (eds.) Healing Gardens: Therapeutic Benefits and Design Recommendations, pp. 25–77. Wiley, New York (1999)

Ulrich, R.S., Simons, R.F., Losito, B.D., Fiorito, E., Miles, M.A., Zelson, M.: Stress recovery during exposure to natural and urban environments. J. Environ. Psychol. **11**(3), 201–230 (1991)

Weidner, T., Yang, A., Hamm, M.W.: Consolidating the current knowledge on urban agriculture in productive urban food systems: learnings, gaps and outlook. J. Clean. Prod. **209**, 1637–1655 (2019)

Wichrowski, M., Whiteson, J., Haas, F., Mola, A., Rey, M.J.: Affects of horticultural therapy on mood and heart rate in patients participating in an inpatient cardiopulmonary rehabilitation program. J. Cardiopulm. Rehabil. Prev. **25**(5), 270–273 (2005)

Wolinsky, F.D., Wyrwich, K.W., Babu, A.N., Kroenke, K., Tierney, W.M.: Age, aging, and the sense of control among older adults: a longitudinal reconsideration. J. Gerontol. B Psychol. Sci. Soc. Sci. **58**(4), S212–S220 (2003)

Zasada, I.: Multifunctional Peri-urban agriculture—a review of societal demands and the provision of goods and services by farming. Land Use Policy **28**(4), 639–648 (2011)

Revitalizing Urban Street Dynamics via Strategies for Climate Control: A Tool for Design and Evaluation

Cristian Romero García, Paulo García-Melgar, Francisco Ruda Sarria, and José Sánchez Ramos

Abstract

The increasing air temperature in urban areas is transforming the habitability of cities. It is crucial to reconfigure and adapt the design of urban spaces to accommodate the growing population and effectively address climate challenges. Comfort should be a priority in urban environments of the future, while also minimizing dependence on active energy systems. The URBANKEA tool was developed with the aim of facilitating a correct choice of strategies to address comfort in urban areas by organizers and architects. The calculation module 'Comfy-Place' allows the design of urban spaces, as well as the correct planning and location of urban elements. This module allows simulations in any location by using interactive maps, incorporating essential variables like climate, occupancy, and necessary technologies to assess thermal comfort in a specific environment. A mesh is employed to simulate outdoor urban space, and users can define this mesh to customize the level of detail as needed. Urban elements such as covers, pavements, and vegetation can be modelled, with all their properties stored in an extensive database. The tool's calculation engine processes the defined input data, providing users with easily interpretable results. Moreover, the tool allows the user to establish the operational range, considering key variables that enable citizens to achieve optimal comfort levels. Results, including surface temperatures and incident radiation on urban elements and individuals, are presented and made accessible to the tool's user. These results allow users to identify the technologies needed to improve outdoor comfort, offering a detailed analysis of the variables that influence their characterization. Its user-friendly interface promotes the embrace of inventive solutions for climate change adaptation, emphasizing improvements like increased vegetation, coverings, or better-defined pavements. All with a focus on enhancing energy efficiency and restoring well-being in public spaces.

Keywords

Calculation tool · Urban planning · Thermal comfort · Urban climate · Urban heat island

1 Introduction

The urban scenery is being impacted by the rise in air temperature, affecting the overall quality of life in cities (IPCC, 2021). Forecasts for future climate trends (Fischer & Schär, 2010) indicate a significant increase in air temperature, leading to more frequent and intense heat waves. This forecast has already come to completion, with 2023 being identified as the warmest year on record characterized by unusually high temperatures and, in some cases, reaching extreme levels (NOAA National Centers for Environmental Information, 2024). By 2050, it is projected that 97% of cities on the Iberian Peninsula will shift to warmer climates, with cold climate zones nearly disappearing (Díaz-López et al., 2021). This implies that current urban planning may not be sufficient to mitigate urban overheating or ensure adaptation to progressively warmer conditions (Nazarian et al., 2022). Even regions traditionally associated with colder climates,

C. Romero García (✉) · P. García-Melgar · F. Ruda Sarria · J. S. Ramos
Department of Energy Engineering, University of Seville, Seville, Spain
e-mail: cromero3@us.es

P. García-Melgar
e-mail: pgarcia21@us.es

F. Ruda Sarria
e-mail: fruda@us.es

J. S. Ramos
e-mail: jsr@us.es

B. He et al. (eds.), *Towards the Framework of Livable and Resilient Cities* Advances in Science, Technology & Innovation, https://doi.org/10.1007/978-3-031-97849-4_6

particularly those at high latitudes are undergoing a shift towards warmer climates (Beck et al., 2018), as evidenced by the observation of extreme temperatures. In the last years, the United Kingdom recorded a temperature exceeding 40 ºC for the first time (Copernicus Climate Change Service (C3S), 2023). As urban environments grapple with the challenges posed by a warmer climate, urban planners are encountering new obstacles related to the risk of overheating in urban settings.

Consequently, traditional urban solutions are insufficient to address this evolving environment, as the concept of thermal comfort needs to expand beyond the physical boundaries of buildings, especially with heat waves becoming more frequent and intense, particularly in summer (Fischer & Schär, 2010). To counter this shift in urban livability, future urban environments should integrate both buildings and their surroundings, improving thermal envelopes and functioning as areas capable of offering natural thermal comfort, thereby reducing the dependence on active energy systems (Santamouris & Kolokotsa, 2016). However, evaluating the impacts of interventions in urban spaces poses a challenge similar to assessing the actual impact of the urban heat island effect in a specific area (Nazarian et al., 2022). For an accurate assessment of this effect, it is essential to understand all the heat fluxes influencing air and surface temperatures (Carmeliet & Derome, 2023) and how they impact individuals' thermal perception (Nazarian et al., 2022).

The Urbankea tool has been created to assist in the design and assessment of urban spaces. Urbankea offers a range of modules intended for creating urban solutions that improve outdoor thermal comfort in different settings. Among these modules is "ComfyPlace," which embodies the goal of this tool by conducting simulations and subsequent analyses of outdoor areas. This module enables simulations in any location worldwide, considering factors like materials, climate, occupancy, and technologies to assess and comprehend thermal comfort in a particular environment.

1.1 Problem Statement

The main problem encountered is the inexistence of a standard simulation procedure of outdoor environments since there is too much casuistry in all the simulations, there are many simulation programs that only take into account a series of specific characteristics for different types of problems. In the last decade, a multitude of simulation tools have appeared, both in terms of mathematical models and 3D simulations (Vurro & Carlucci, 2024). The tools found have several problems, such as high computational costs, many of these tools are based on CFDs, this implies the above and also that specific training is needed, which the end user does not always have to have. Many urban planners do not have information specific about CFDs, in addition to this is the issue associated with the simulation time, since preparing the model to run it, not allowing quick decisions. This raises the need to create a tool with a simple interface that can be used by the general public and not only by experts with specific training but also that can be used by government agents or urban planners.

1.2 Study Aim

The Urbankea tool serves as a valuable resource for designing and evaluating urban areas. It provides a variety of modules designed to develop urban solutions that enhance outdoor thermal comfort across diverse environments. One notable module, "ComfyPlace," epitomizes the tool's objective by facilitating simulations and analyses of outdoor spaces. Through this module, users can conduct simulations in global locations, taking into account variables such as materials, climate conditions, occupancy rates, and technologies to evaluate and understand thermal comfort within a specific setting.

1.3 Review of the Literature

Currently there are many simulation tools all based on various systems of numerical methods and simulation tools (He et al., 2009). The use of CFDs is very common due to the freedom that they represent since the study of thermal comfort in different open spaces has many gas sources and many variables that must be taken into consideration while not (Asawa et al., 2008). Other types of tools are also those that are based on a 3D system in software. This type of software also allows integration with different types of external aerial programs that allow you to make a graphic interpretation of the entire system in an urban space (Asawa et al., 2008). Since there are so many ways to create a simulation tool in outdoor spaces, all the articles that exist must be filtered since in the last decade these tools have increased greatly, all based on the same programs and the same methodology (Vurro & Carlucci, 2024). Specifically, the simulation of open spaces is a challenge due to the large casuistry that exists. To begin with, the heat of the space must be taken into account. This can be done using the geolocation system called GIS systems (Itami et al., 2003).

This system is always based on maps but the complete information is never extracted from them, which is the limitation of these systems, which is why people resort to CFDs or similar tools, however, these tools imply specific training of the person unable to help decision making quick decision (Malczewski, 2004; Vurro & Carlucci, 2024). These tools also need different calculation engines and different comfort models since the needs of the models are different as there is a variation between them, which makes the creation of this type of simulation tools much more complicated (Potchter et al., 2018).

2 Methodology

2.1 General Functionality

The main objective during the development of the "ComfyPlace" tool was to create a user-friendly interface that is easy to understand. The simulation process for an urban area is based on four main variables (Fig. 1): local climate, the layout of the urban space (geometry), how the space is used (usage and occupancy rates), and the technologies used for environmental control.

By utilizing free weather databases like Climate. OneBuilding.Org (Lawrie & K, 2024) or the PVGIS web-based tool (Joint Research Centre. European Commission, 2024) to access Typical Meteorological Year (TMY) data and combining them with future weather generators (*Meteonorm Version 8 - Meteonorm (En)*, n.d.; Rodrigues et al., 2023), urban planners can evaluate the impact of urban retrofits on the livability of outdoor spaces in both current and future climates. A key feature of this tool is its ability to efficiently define an entire avenue or square using a mesh grid, allowing users to specify the spatial resolution, type of use, structures, and materials in each cell. The tool includes a wide range of materials commonly found in urban environments and typical systems used for environmental control in public areas. By considering user-defined usage profiles, the tool evaluates the perceived thermal comfort of individuals and the operational temperature across the area being studied.

The user interface consists of two primary toolbars. The first toolbar enables users to create files, open existing ones, and access the database. Upon creating a new project, a second main toolbar becomes active, offering all the essential functions required to conduct a simulation. These functions involve defining climate data, geometry and spatial arrangement, occupancy and usage settings, and conditioning technologies (Fig. 2a and Fig. 2b).

2.2 Geometry

The geometry module is a key tool that helps users design and analyze real-world scenarios accurately by defining urban elements, surfaces, and vegetation. Users can add urban features to maps, manage properties efficiently, and analyze shading effects and tree coverage for urban environment assessment.

The tool allows users to add various urban elements to an interactive map and accurately determine their real geolocation. Users can assign properties to these elements using the tool's database. The pavement can be defined using a quadrangular mesh, with the option to assign different floors based on design needs. Buildings surrounding the area are highlighted with red lines, indicating potential shade sources at different times of the day. Coverings are shown as red polygons, outlining their shadow zones. Tree shading can be illustrated with red circles labeled from Ai to An.

2.3 Climate and Operational Data

Building thermal simulation tools rely on weather datasets with hourly resolution. To meet this requirement, climate databases are typically structured to align with tools like the Typical Meteorological Years guide (Wilcox & Marion, 2008). These tools commonly support weather file formats such as.epw and.met, streamlining the process for users

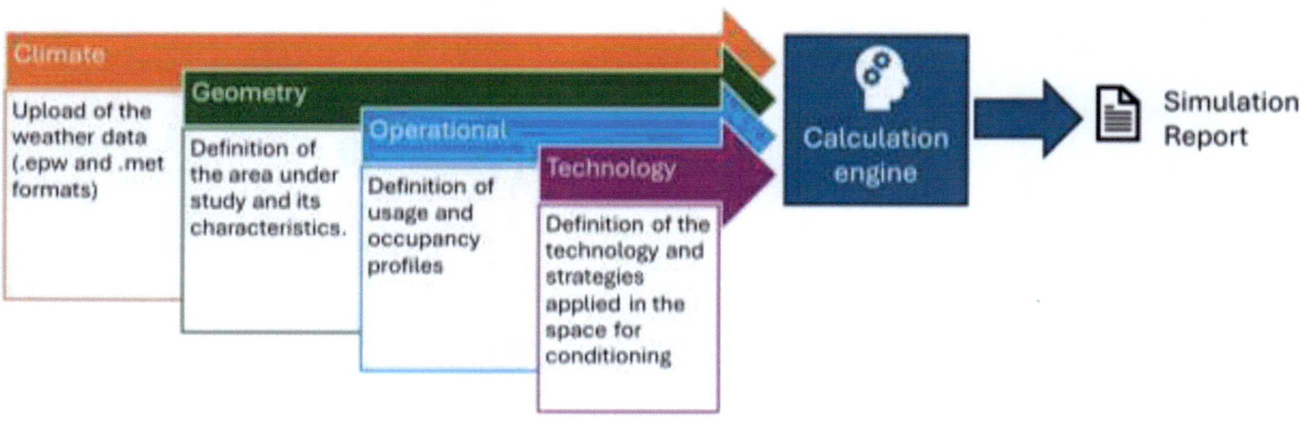

Fig. 1. General operational scheme

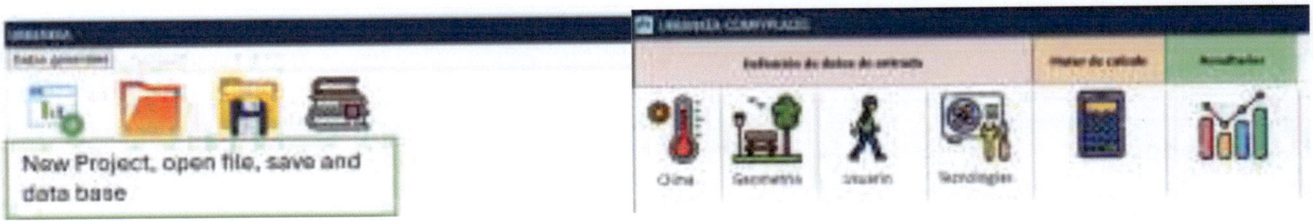

Fig. 2. Main (a) and second main (b) toolbars.

without the need for extensive analysis or data preparation. Once the weather data is uploaded, a statistical report is automatically generated, detailing the average outdoor conditions per month, including temperature profiles, solar irradiation, wind velocity, and direction. Moreover, the climate assistant can suggest design days for simulation during specific periods or entire seasons.

On the other hand, users need to provide detailed operational data by specifying the occupancy rate and profile for each cell within the designated area (Fig. 3). This includes defining a standard person with characteristics like weight, height, metabolic activity, and clothing, which are crucial for accurate calculations. Considering that urban squares often see various physical activities taking place, it is important to consider the level of activity individuals are engaged in. Additionally, determining the population density is necessary, as it will be used later to calculate hourly occupancy as a percentage. Users are advised to provide approximate percentages rather than exact densities for ease of input. These details play a significant role in identifying peak usage times in the urban area, as the occupancy levels in these zones vary throughout the day.

- **Green cells**: Represent different types of pavements.
- **Points**: Represent **trees**.
- **Red surfaces**: Indicate **coverage areas** (covered spaces).
- **Red lines**: Represent obstacles.

2.4 Database and Technologies

This software has an extensive database that can be edited by the user, allowing the recovery of the original database if necessary. Additionally, it allows the user to add elements to all groups of the database consisting of coverings, pavements and floors, facades, vegetation and uses. The database of the tool has five modules:

- Coverings
- Pavements and Floors
- Vegetation
- Façades
- Uses

Each module outlines the functionalities of materials, categorizing them into specific groups. For instance, under the

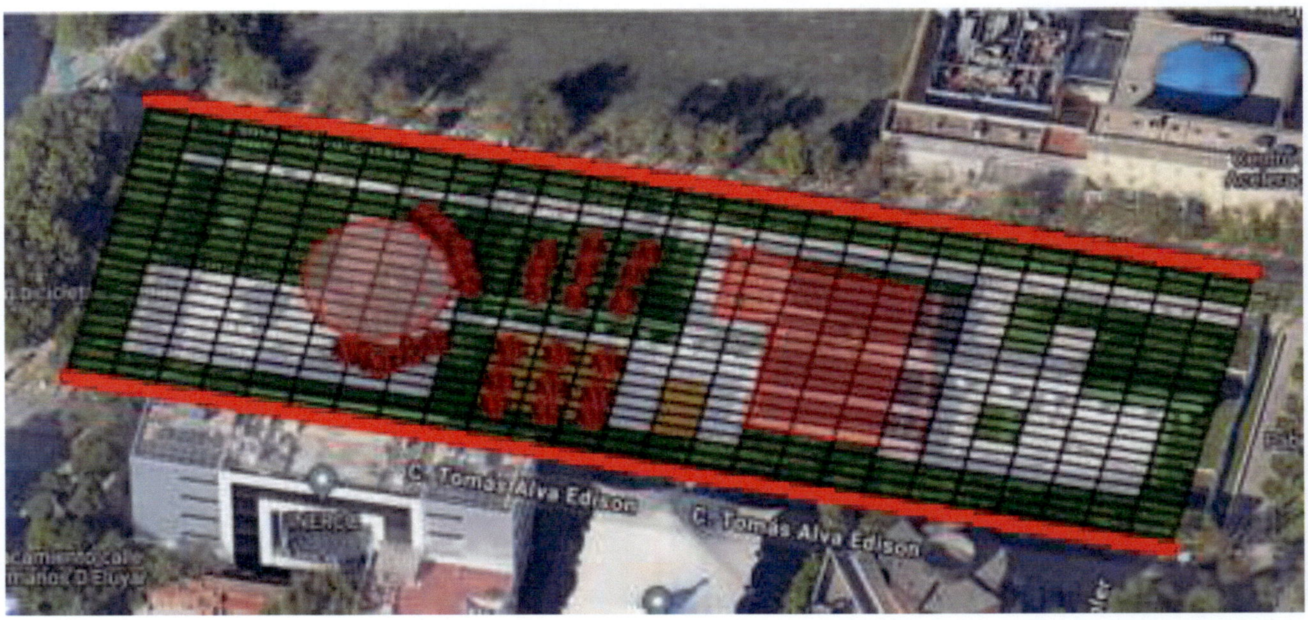

Fig. 3. Geometry definition using the interactive map of the tool

coverings section, materials like textiles, wood, concrete, metal, photovoltaic panels, and vegetation are listed. The pavements and floors section distinguishes between impermeable, permeable, and vegetal materials. In the vegetation category, options include deciduous and perennial leaves, while the façades section differentiates between façades and screens.

The editable properties depend on the accessed material. Users can adjust properties like transmissivity, thickness, reflectivity, permeability, electrical performance (for solar PV panels), density, and specific heat. When it comes to trees, users can modify their transmissivity for winter and summer, adjust tree heights, and dimensions at six levels, and specify tree width and length.

Furthermore, users can add custom materials as needed, which will be integrated into the menus based on their functions. Adding materials is simple with just a single right-click, allowing users to select the desired functionality. Users have full freedom to define materials within the materials section according to their preferences.

In the other hand, the conditioning technologies module is crucial for defining the area of study. Users can utilize active systems within this module to address thermal discomfort caused by high air temperatures. These systems can include outdoor air handling units for sensible cooling or cool surfaces like pavements and covers for radiant cooling. Air handling units supply cold air in large volumes, reducing the indoor temperature well below the outdoor level. They can be complemented by water cooling from a pond, ground cooling, or an evaporative cooler. Alternatively, users can opt for radiant pavements or covers using water from a pond (Fig. 4).

These conditioning technologies can be used in combination or individually. Designers must input design data such as air flows, water flows, and system efficiencies, but the tool also offers predefined values for user assistance. These systems are most effective when integrated into outdoor areas with a certain level of enclosure, achieved through strategies like recessing the floor, incorporating vegetal barriers, or using textile covers. Otherwise, the effectiveness of these systems may be compromised in open spaces.

2.5 Comfot Index

Comfort indices describe how the human body experiences atmospheric conditions, specifically air temperature, humidity, wind, and radiation. Accordingly, various index models exist, including the COMFA and PET models. However, in this case, for the sake of simplicity within the calculation engine, the COMFA model is primarily used, as it offers several distinct advantages.

1. Customization and adaptation to physical activity level (Kenny et al., 2009a)):
   - COMFA: This model considers the individual's level of physical activity, adjusting results according to expenditure.
   - PET: Although these models also incorporate metabolism, they do so in a more generalized way, overall this model need the skin temperature.
2. Adaptation to Microclimates and Urban Environments (Vanos et al., 2012):
   - COMFA: it has a specific focus on microclimatic conditions, accounting for effects such as direct solar radiation and shadows, crucial factors in urban areas with vegetation and structures that cast shadows.
   - PET: Although they evaluate comfort in outdoor spaces, PET is less detailed regarding direct solar radiation.
3. Simplification in calculations and flexibility in Variable Climates (Kenny et al., 2009b):
   - COMFA: Although it incorporates multiple factors, its calculation method is more straightforward, facilitating its application in preliminary environmental studies. This approach allows for greater flexibility and

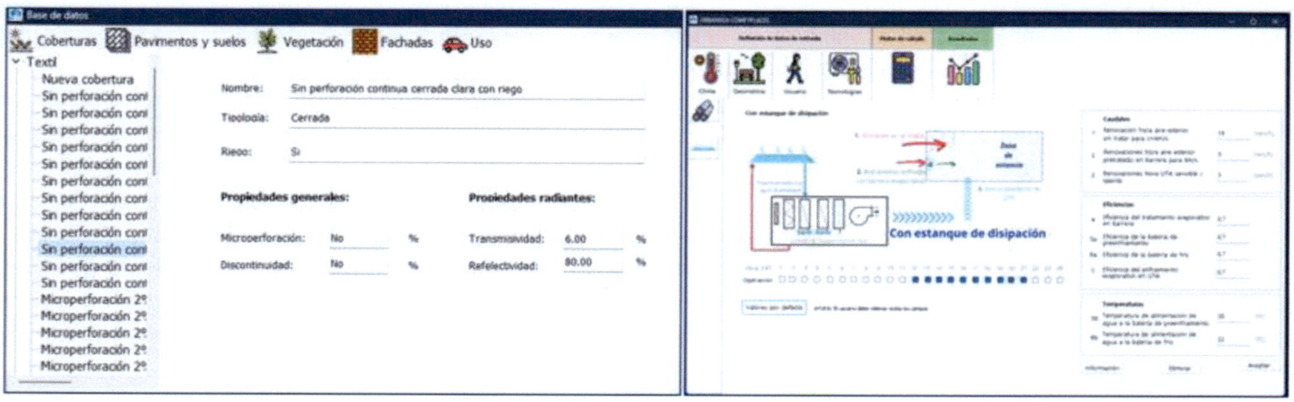

Fig. 4. Interface overview for the database (left) and conditioning technologies (right).

adaptability across different climates, enabling easy configuration.

- PET: They have a more complex calculation process, requiring highly specific data and iterative computations, although they provide more precise results.

For these reasons, the COMFA model is preferred, as it is currently used for simplified engines. Its reduced complexity and more straightforward calculations make it more feasible for programming, with minimal detail required at the user level. COMFA relies directly on climatic conditions, making it accessible and efficient for broader applications without needing extensive user input.

2.6 Calculation Engine

The URBANKEA tool is an open-source platform based on calculation engines developed using real, experimentally derived calculations. For URBANKEA, simplified calculation engines are used, drawn from and based on the computational models established in the experiments conducted during the 1992 EXPO (ÁlvarezDomínguez, 1992). These models included simplified representations of various urban elements, such as pavements, obstacles, facades, vegetation, and coverings. These calculation engines are continually validated through ongoing experimentation and are subject to constant refinement (Amores et al., 2023).

3 Results and Discussion

3.1 Case of Study

The case study centres on an avenue in Seville, located within the Parque Científico y Empresarial de la Isla de la Cartuja (Expo 92 site). This avenue recently underwent a renovation, incorporating bioclimatic strategies to create a pleasant and healthy public space, especially crucial in warmer urban environments. The focus is on the amphitheatre area, featuring seating and an open-air stage. The goal is to assess how different solar canopies enhance comfort and provide ample shading. Various scenarios were evaluated, testing different materials with varying transmissivity and colours, along with the use of water for cooling. Tall cypress trees were planted around the amphitheatre to reduce the influx of hot air from the southwest.

The design day selected to evaluate the effectiveness of the strategies during the avenue's renovation corresponds to Seville's hottest day in the summer of 2023, with a temperature of 44.5 °C. The summer of 2023 saw unusually high temperatures, including multiple heatwaves. The severity of the heatwave is depicted, showing nighttime temperatures above 25 °C and extremely high temperatures during the afternoon (12 h–20 h), making outdoor activities challenging due to the heat. These conditions underscore the importance of adapting urban spaces to ensure livability in cities (Fig. 5).

The strategy's efficacy was evaluated based on its impact on individuals' thermal perception, also known as thermal comfort. This concept refers to how an individual responds to the thermal conditions of their surroundings. It can be quantified using a physiological model representing the human body's energy balance (Departamento de Ingeniería Energética y Mecánica de Fluidos (Universidad de Sevilla), 1994), which is an empirical combination of equations reflecting the body's actual behavior. A widely studied comfort indicator is the Thermal Comfort Index (COMFA) (Brown & Gillespie, 1986), useful for defining comfortable microclimates across various temperature ranges. The eat balance equation at any given moment describes the heat exchange between the body and the external environment.

In all the scenarios examined, vegetation areas were integrated, and additional enhancements were made to the amphitheatre cover (Fig. 5):

- Case 1: Textile cover with 70% transmissivity, translucent (solar absorptivity of 20%)
- Case 2: Textile cover with 40% transmissivity, white color (solar absorptivity of 40%)
- Case 3: Textile cover with 40% transmissivity, white color (solar absorptivity of 40%) with water irrigation (radiant cover).

3.2 Results

In the study, different textile covers were tested to see how they affected thermal comfort in an amphitheatre.

Textile covers with high transmissivity (the first and second types) caused the occupant to experience considerable thermal stress, especially during the central hours of the day. This stress was mainly due to the covers allowing a lot of solar radiation to pass through, leading to COMFA levels exceeding 120 W/m^2 for most of the central hours (Table 2).

On the other hand, when a textile cover with low transmissivity and water irrigation was used, the situation improved significantly. This combination helped lower the surface temperature around the occupant and reduce the overall temperature of the space. Consequently, there was a notable 70% improvement in the comfort index in this setup. The individual felt warmer and more comfortable compared to when high-transmissivity covers were in place (Table 1).

3.3 Discussion

These results has been validated through experimentation by conductin on-site measurements (Amores et al., 2023). Comparing these findings with other programs available on the market, such as ENVI-met, reveals a significant advantage in graphical simplicity. This program is crafted to be accessible for users with less specialized knowledge, enabling realistic and quick approximations based on specific design needs.

However, these tools do have a steep learning curve: their complexity can be overwhelming for new users. For instance, integrating ENVI-met into Grasshopper workflows demands advanced knowledge and can present substantial challenges. Although both interfaces are aimed at non-programmers, users are still expected to have a general understanding of their functionality (Pacifici & Nieto-Tolosa, 2021).

Additionally, it's important to consider that 3D tools have high computational demands, resulting in significant processing costs. Unlike these, the tool we present is GIS-based, allowing users to modify elements within the system, thus supporting broader, less computationally intensive studies.

Table 1. Results for the three cases tested in the tool.

Hour/ COMFA (W/m^2)	Case 1	Case 2	Case 3
12:00	135	127	86
14:00	194	190	111
16:00	220	215	134
18:00	165	140	116
20:00	120	110	81

Table 2. COMFA index valoration scale (Potchter et al., 2018).

Heat load Q (W/m^2)	>150	120-150	85-120	50-85	-50–50
Consideration	Very hot	Hot	Warm	Slightly warm	Comfort

However, this approach does bring certain limitations. The tool is not designed for highly detailed system calculations but rather for general assessments. For users requiring in-depth analysis, more advanced programs, like those mentioned earlier, would be necessary.

4 Conclusion

The Urbankea tool was created to aid in the design and assessment of urban areas. It offers a range of modules tailored for crafting urban solutions that enhance outdoor thermal comfort in diverse settings and climate conditions, enabling the implementation of various strategies in urban environments, including active systems for conditioning outdoor spaces.

The findings from the case study highlight the importance of implementing measures to mitigate the effects of unfavorable climate conditions. The study demonstrated that significant cooling effects were achieved only when combining a low-transmissivity cover with a cooling radiant solution. This resulted in a 70% enhancement in the comfort index, with individuals feeling warm and comfortable. Traditional interventions relying solely on vegetation or new materials were insufficient to address all the challenges posed by a warming climate.

Climate change is causing significant disruptions to global weather patterns. The past summer exemplifies what may become more common in the future. With a warmer climate expected to bring about more intense heat waves, public life

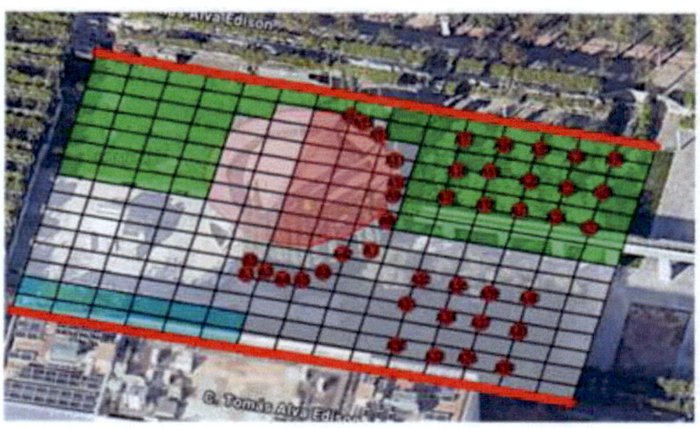

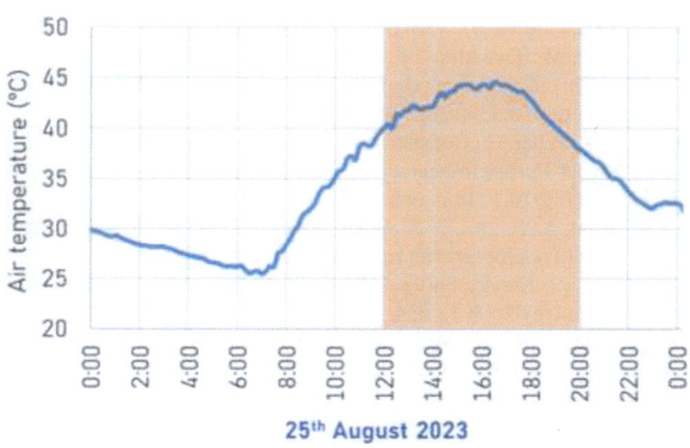

Fig. 5. (left) Study area definition: water ponds, vegetation, trees, conditioned space areas. (right) Outdoor air temperature for the warmest day during the 2023 summer.

in outdoor spaces could become untenable, emphasizing the critical need to adapt urban areas to ensure cities remain livable.

Acknowledgements This study has been funded by the projects "CONSTANCY - Resilient urbanisation methodologies and natural conditioning using imaginative nature-based solutions and cultural heritage to recover the street life" (Grant Agreement PID2020-118972RB-I00) and the project "NATURBEAM - Lighting the way to a greener future to restore urban habitability through nature-based solutions" (Grant Agreement TED2021-130416B-I00) by the Spanish Ministry of Science, Innovation and Universities.

The third author is funded by a public subsidy for the development of the "INVESTIGO Program" (Activity: C23.I1.P03.S01.01 ANDALUCÍA), financed by funds from the "Recovery and Resilience Mechanism".

References

Domínguez, S.A.: Control climático en espacios abiertos : el proyecto Expo'92. In: Álvarez Domínguez, S. (ed.) Centro de Investigaciones Energéticas, Medioambientales y Tecnológicas Ciemat, Secretaría General Técnica (1992)

Amores, T.P., Gutiérrez, Mp.M., Palma, R.M., Delgado, Mc.G., Ramos, J.S., Domíguez, S.Á.: Redefining human thermal perception for the Mediterranean climates. Sci. Talks **7**, 100241 (2023). https://doi.org/10.1016/j.sctalk.2023.100241

Asawa, T., Hoyano, A., Nakaohkubo, K.: Thermal design tool for outdoor spaces based on heat balance simulation using a 3D-CAD system. Build. Environ. **43**(12), 2112–2123 (2008). https://doi.org/10.1016/j.buildenv.2007.12.007

Beck, H.E., Zimmermann, N.E., McVicar, T.R., Vergopolan, N., Berg, A., Wood, E.F.: Present and future Köppen-Geiger climate classification maps at 1-km resolution. Sci. Data **5**, 180214 (2018). https://doi.org/10.1038/sdata.2018.214

Brown, R.D., Gillespie, T.J.: Estimating outdoor thermal comfort using a cylindrical radiation thermometer and an energy budget model. Int. J. Biometeorol. **30**(1), 43–52 (1986). https://doi.org/10.1007/BF02192058

Carmeliet, J., Derome, D.: How to beat the heat in cities through urban climate modelling, Nature Reviews Physics **6**, 2–3 (2023). https://doi.org/10.1038/s42254-023-00673-1

Departamento de Ingeniería Energética y Mecánica de Fluidos (Universidad de Sevilla): Control climático en espacios abiertos: evaluación del proyecto Expo'92. CIEMAT (1994)

Díaz-López, C., Jódar, J., Verichev, K., Rodríguez, M.L., Carpio, M., Zamorano, M.: Dynamics of changes in climate zones and building energy demand. a case study in Spain. Appl. Sci. **11**, 4261 (2021). https://doi.org/10.3390/app11094261

Fischer, E.M., Schär, C., Consistent geographical patterns of changes in high-impact European heatwaves. Nat. Geosci. **3**, 398–403 (2010). https://doi.org/10.1038/ngeo866

He, J., Hoyano, A., Asawa, T.: A numerical simulation tool for predicting the impact of outdoor thermal environment on building energy performance. Appl. Energy **86**(9), 1596–1605 (2009). https://doi.org/10.1016/j.apenergy.2008.12.034

Itami, R., et al.: RBSim 2: simulating the complex interactions between human movement and the outdoor recreation environment. J. Nat. Conserv. **11**(4), 278–286 (2003). https://doi.org/10.1078/1617-1381-00059

IPCC, Climate Change 2021: The Physical Science Basis. Contribution of Working Group I to the Sixth Assessment Report of the Intergovernmental Panel on Climate Change (2021). https://doi.org/10.1017/9781009157896

Joint Research Centre. European Commission: PVGIS Photovoltaic Geographical Information System (2024). https://re.jrc.ec.europa.eu/pvg_tools/en/

Kenny, N.A., Warland, J.S., Brown, R.D., Gillespie, T.G.: Part A: assessing the performance of the COMFA outdoor thermal comfort model on subjects performing physical activity. Int. J. Biometeorol. **53**(5), 415–428 (2009a). https://doi.org/10.1007/s00484-009-0226-3

Kenny, N.A., Warland, J.S., Brown, R.D., Gillespie, T.G.: Part B: revisions to the COMFA outdoor thermal comfort model for application to subjects performing physical activity. Int. J. Biometeorol. **53**(5), 429–441 (2009b). https://doi.org/10.1007/s00484-009-0227-2

Lawrie, L., et al.: Development of Global Typical Meteorological Years (TMYx). Climate.OneBuilding.Org (2024). http://climate.onebuilding.org

Malczewski, J.: GIS-based land-use suitability analysis: a critical overview. Prog. Plan. **62**(1), 3–65 (2004). https://doi.org/10.1016/j.progress.2003.09.002

Meteonorm Version 8 - Meteonorm (en) (n.d.). https://meteonorm.com/en/meteonorm-version-8. Accessed 22 Mar 2024

Nazarian, N., et al.: Integrated assessment of urban overheating impacts on human life. Earths Future **10** (2022). https://doi.org/10.1029/2022EF002682

NOAA National Centers for Environmental Information, Monthly Global Climate Report for Annual 2023 (2024). https://www.ncei.noaa.gov/access/monitoring/monthly-report/global/202313. Accessed 22 Jan 2024

Pacifici, M., Nieto-Tolosa, M.: Comparing ENVI-met and grasshopper modelling strategies to assess local thermal stress and urban heat island effects. In: Urban Microclimate Modelling for Comfort and Energy Studies, pp. 293–316. Springer (2021). https://doi.org/10.1007/978-3-030-65421-4_14

Potchter, O., Cohen, P., Lin, T.-P., Matzarakis, A.: Outdoor human thermal perception in various climates: a comprehensive review of approaches, methods and quantification. Sci. Total. Environ. **631–632**, 390–406 (2018). https://doi.org/10.1016/j.scitotenv.2018.02.276

Rodrigues, E., Fernandes, M.S., Carvalho, D.: Future weather generator for building performance research: an open-source morphing tool and an application. Build. Environ. **233**, 110104 (2023). https://doi.org/10.1016/j.buildenv.2023.110104

Santamouris, M., Kolokotsa, D., (eds)., Urban Climate Mitigation Techniques, Routledge, London (2016). https://doi.org/10.4324/9781315765839

Vanos, J.K., Warland, J.S., Gillespie, T.J., Kenny, N.A.: Improved predictive ability of climate–human–behaviour interactions with modifications to the COMFA outdoor energy budget model. Int. J. Biometeorol. **56**(6), 1065–1074 (2012). https://doi.org/10.1007/s00484-012-0522-1

Vurro, G., Carlucci, S.: Contrasting the features and functionalities of urban microclimate simulation tools. Energy Build. **311**, 114042 (2024). https://doi.org/10.1016/j.enbuild.2024.114042

Wilcox, S., Marion, W.: Users Manual for TMY3 Data Sets (Revised) (2008). https://doi.org/10.2172/928611

Effect of Natural-Based Cool Pavements in Urban Open Spaces Liveability

Paulo García-Melgar, Francisco Ruda Sarria, Cristian Romero García, and MCarmen Guerrero Delgado

Abstract

The prevailing issue of climate change is precipitating a fundamental alteration in weather patterns, resulting in Earth's average temperature rise of up to 1.5 °C. This is causing more frequent, intense, and prolonged heat waves. Another significant climatological phenomenon exacerbated by climate change is the Urban Heat Island effect, which causes cities to experience higher temperature levels than the nearest rural areas. These extreme temperatures have directly affected the liveability of residents and their wellbeing. As the main solution, circular and regenerative economies based on bioclimate and natural resources are proposed. To this purpose, a nature-based solution by cool pavements is chosen. These pavements are cooled by circulating water, which has been collected from rainwater by the urban draining systems and cooled through evaporative cooling using pre-existing water bodies in cities beneath their lower surface. The novelty of this concept lies in the lack of capillarity of the pavements, allowing for effective cooling without evaporation and the associated water usage. This work focuses on studying the impact of these cool pavements on urban areas conceived for pedestrian habitability, which may require additional adjustments, such as shaded terrace areas, to achieve acceptable thermal comfort in an urban environment. By using a digital model validated through an experimental procedure, a multicase simulation is carried out to determine the impact of this solution when integrated into the cities. Through the simulations and the data collected, the PET outdoor comfort model is used to quantify the level of thermal comfort and liveability of citizens. Results show that by primarily using water's sensible heat, the temperature of the pavements can be reduced by up to 20 °C. This approach ensures that leisure and public spaces in cities can offer a comfortable area, thereby enhancing liveability and creating secure spaces to counteract the adverse effects of climate change.

Keywords

Cool Pavements · Natural-Based Solutions · Outdoor comfort · Passive cooling · Urban Heat Island

1 Introduction

The contemporary issue of climate change has instigated a shift in weather patterns, carrying profound implications for life, ecosystems, and the foreseeable future. This shift has resulted in an increase in average temperatures by up to 1.5 °C, leading to more frequent, intense, and prolonged heat waves (International Energy Agency, 2021). An illustrative example is the summer of 2023, which witnessed Spain experiencing its third-highest temperatures on Record (Agencia Estatal de Meteorología, 2023). These extreme temperatures have directly impacted residents' quality of life and well-being, culminating in 2155 fatalities attributed to excessive heat, marking it as the third deadliest summer due to this condition (Escuela Nacional de Sanidad, 2015; Rodríguez, 2023). Despite their relatively small physical footprint on Earth, urban areas exert a disproportionately significant influence on the biosphere due to their dense populations and economic activities (Jansson et al., 2013). The concentration of people,

P. García-Melgar (✉) · F. R. Sarria · C. R. García · M. G. Delgado
Department of Energy Engineering, University of Seville, Seville, Spain
e-mail: paugarmel@us.es

F. R. Sarria
e-mail: fruda@us.es

C. R. García
e-mail: cromero3@us.es

M. G. Delgado
e-mail: mgdelgado@us.es

B. He et al. (eds.), *Towards the Framework of Livable and Resilient Cities* Advances in Science, Technology & Innovation, https://doi.org/10.1007/978-3-031-97849-4_7

assets, and economic pursuits renders cities exceptionally vulnerable, positioning them as focal points for evaluating the effects of climate change (Dawson et al., 2007; Shi et al., 2016). The present goal to tackle these challenges centers on the renaturalization of urban areas to mitigate the impact of these phenomena.

Another significant climatological phenomenon exacerbated by climate change is the Urban Heat Island effect (UHI). The UHI refers to the temperature difference between urban areas and the nearest rural areas and is influenced by two agents: heat storage in the urban envelope and the heat release from anthropogenic activities (Leroy, 2021; Oke et al., 1991).

Pavements have a vital function in the Urban Heat Island (UHI) effect, typically occupying 20–40% of the urban landscape. Their dark hue and low reflectivity play a crucial role in elevating surface temperatures, resulting in the retention and release of substantial heat energy into the surroundings. As a result, there is an increasing focus on the creation and adoption of pavements with reduced surface temperatures, to alleviate the UHI effect and improve thermal comfort. This research aims to present a Nature-Based Solution referred to as 'cool pavements'.

1.1 Problem Statement and Aim of Study

While additional information will be presented in subsequent sections, this research will investigate the feasibility of lowering pavement surface temperature by circulating cold water beneath it. Through literature research, the concept presented in this paper has been found to be an undiscovered cold pavement technology. Contrasting with evaporative pavements, this technique diverges by ensuring the water does not contact the upper surface, negating the necessity for water evaporation. Therefore, cooling occurs as the lower surface of the pavement comes into contact with water. This contact results in a decrease in temperature on the upper surface due to conduction and the sensible heat transfer from the water. This strategy presents a more sustainable alternative, especially given the challenges of climate change and escalating droughts, as it markedly curtails water usage in comparison to evaporative techniques.

The primary goal of this study is to demonstrate the effectiveness of cooling a pavement by using chilled water, thereby assessing its potential to mitigate the Urban Heat Island effect without relying on evaporative cooling and its associated water usage.

To showcase the efficacy of this pioneering concept, the thermal repercussions of implementing this pavement solution in a street in Utrera, Seville are being examined. Given the city's warm climate and the prevailing discomfort during peak solar hours, this study aims to utilize a mathematical model previously experimentally validated to forecast the thermal performance of the pavement. By contrasting a standard pavement with the cool pavement concept, the tangible variance can be ascertained. Subsequently, leveraging the results, outdoor urban comfort indicators can be computed to evaluate the efficacy of the suggested solution.

2 Review of the Literature

In the context of the re-naturalization of urban environments, the implementation of solutions that emulate natural elements within cities, referred to as "Nature-based Solutions" (NBS), is essential for addressing these challenges (European Commission, 2015). There are several categories of NBS, which include green, blue, and grey solutions (Leroy, 2021). Green NBS emphasizes the integration of vegetation, such as trees, parks, and grasslands (Feyisa et al., 2013). Blue NBS concentrates on water features in urban settings, including lakes, rivers, and fountains (Lin et al., 2020). In contrast, grey NBS depends on the morphology and materials utilized in the urban landscape, such as pavements and buildings (Leroy, 2021) Various technologies for cool pavements can be developed, such as reflective pavements, evaporative pavements, or altering their properties for heat accumulation (Anupam et al., 2021).

Reflective pavements are constructed with high reflectivity and emissivity to solar energy to minimize solar radiation absorption and enhance the emission of infrared radiation (Santamouris, 2013). This design prevents radiation absorption within the concrete structure, leading to reduced surface temperatures and heat storage. Although extensively researched and relatively easy to deploy, reflective pavements pose challenges in terms of installation and maintenance costs (Xie et al., 2020), as well as concerns regarding thermal and visual comfort for urban residents (Leroy, 2021).

On the other hand, permeable and water-retentive pavements have more void spaces than traditional pavements, allowing water to permeate through lower layers and the ground (Santamouris, 2013). These pavements retain absorbed water and utilize it for cooling through evaporation, facilitated by their permeability (Wang et al., 2022). Additionally, the filtration capabilities of permeable pavements are believed to enhance groundwater quality (Berland et al., 2017). However, a significant drawback of this pavement type is the water consumption associated with evaporation for cooling purposes.

Another innovative option is the concept of photovoltaic pavement, where a photovoltaic canopy shades the pavement surface. This lowers the surface temperature and generates electricity to cover peak demand (Efthymiou et al., 2016).

There are various experimental proofs of concept for these cold pavement technologies. For instance, Ruas et al. (Ruas et al., 2024) conducted a study comparing various technologies like reflective paving, evaporative paving, and different shading qualities for outdoor paving in Paris, France. While this study assesses comfort indicators based on UTCI, it does not address the visual discomfort associated with reflective pavements.

3 Methodology

3.1 Concept of Solution

The investigated cold pavement concept involves naturally cooling the pavement by utilizing water collected from rainfall and employing sensible cooling methods.

The technology scheme depicted in Fig. 1 illustrates the concept under examination. The central component of this technology revolves around the idea of cooling the pavement by circulating cold water through channels located on the underside of the pavement. Through conduction, the pavement transfers heat to the cold water, effectively lowering the surface temperature of the pavement. The heat transfer primarily occurs through conduction, as the water circulates solely in the lower part of the pavement, lacking the capillarity to ascend to the upper surface and evaporate. This process significantly reduces water usage by eliminating losses through evaporation.

To implement this technology effectively, it is crucial to have access to a water supply that can serve as a coolant for cooling the pavement. This requires meeting two distinct requirements: firstly, having a nearby and sufficient water source, and secondly, being able to cool this water for usage. The first requirement involves the installation of underground cisterns to collect rainwater during heavy rainfall periods. This collected rainwater is then utilized during warmer seasons to cool the pavement when needed. The second requirement is met by using jets that release water droplets large enough to provide cooling without evaporating too quickly. These jets can be seamlessly integrated into urban water features like fountains in public squares, ensuring the complete incorporation of this technology into the cityscape.

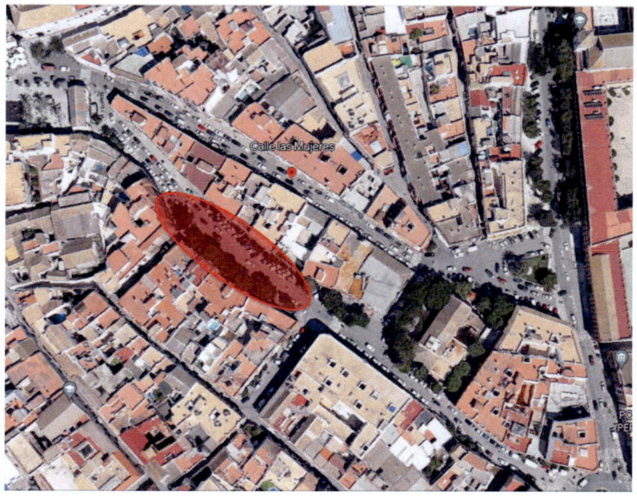

Fig. 2. Location of the intervention zone in Utrera, Seville.

3.2 Intervention Zone

The target location for implementing this solution is situated in Utrera, a city in Seville, Spain. Figure 2 displays a map image of Utrera highlighting the specific intervention area. This area features a boulevard lined with various bar terraces, making it an optimal setting for establishing a naturally cooled space to enhance the outdoor urban comfort of individuals frequenting these terraces.

The intervention area possesses the following characteristics: it comprises two modules totaling 170 m^2 (94 m^2 and 76 m^2 respectively), specifically targeting two terraces associated with different bars on the premises. The available rainwater collection area spans 500 m^2. Moreover, the planned storage volume is 45 m^3. This volume allocation is based on Utrera's water precipitation rate of 450 L/m^2 year. With a collection area of 500 m^2, a potential for water storage of 225 m^3 is estimated. The design strategy adopted a storage capacity to precipitation ratio of 0.2, leading to an accumulation volume of 45 m^3. Figure 3 visually depicts the living area, collection zone, and the rainwater tank's location.

After delineating the intervention and rainwater harvesting areas, the installation schemes for this cold pavement solution need to be established. Two distinct modes are identified: the first involves setting up the infrastructure for rainwater collection, while the second entails the actual installation of the cool

Fig. 1. Concept diagram of cold pavement with sensible cooling.

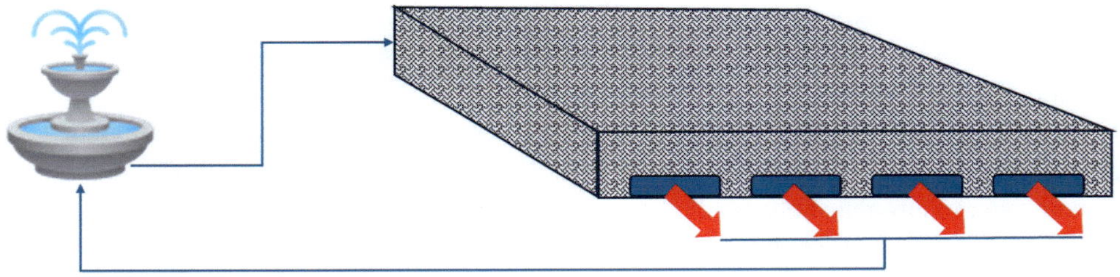

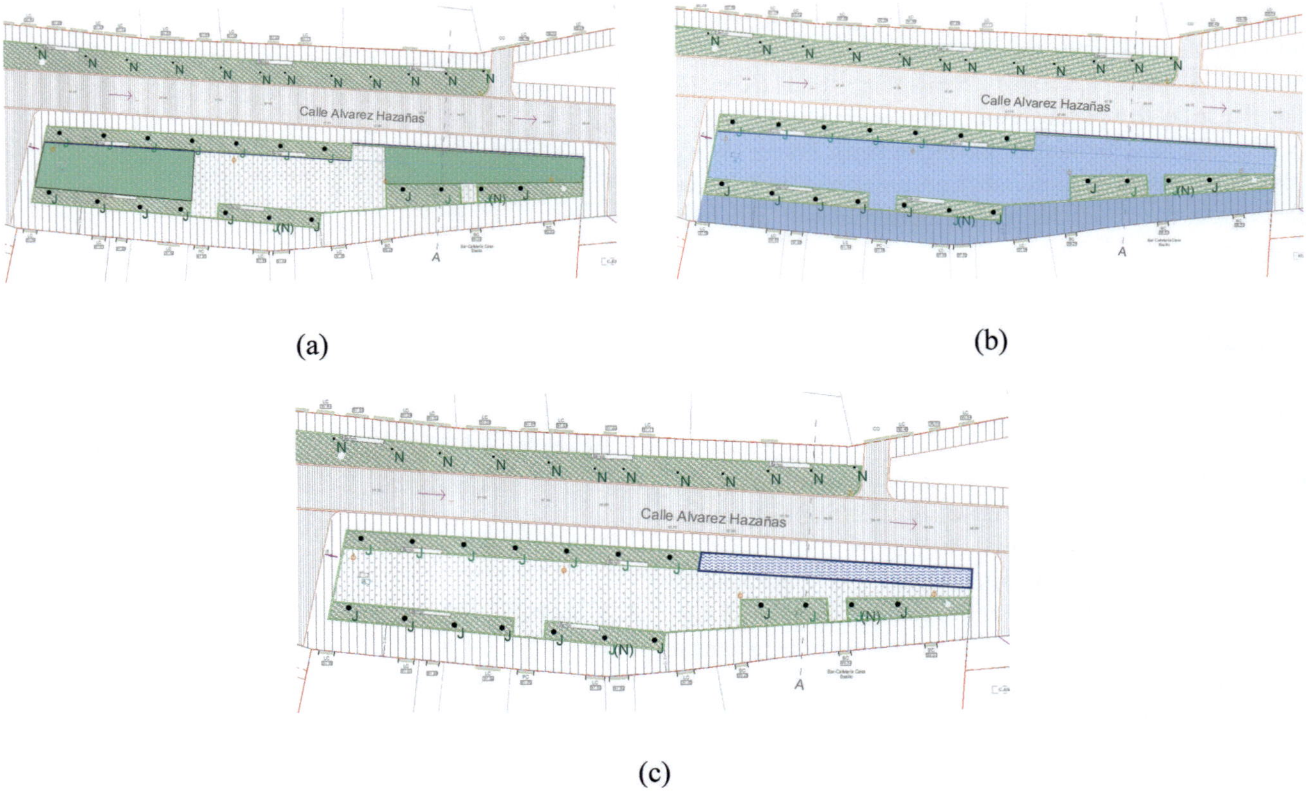

Fig. 3. (a) Living area zone (b) rainfall water collection zone (c) rainwater tank's location.

Fig. 4. Rainfall water collection installation.

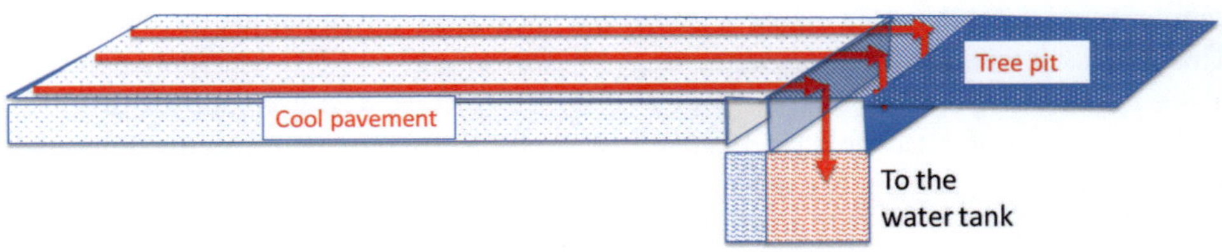

Fig. 5. Cool pavement water circulation installation.

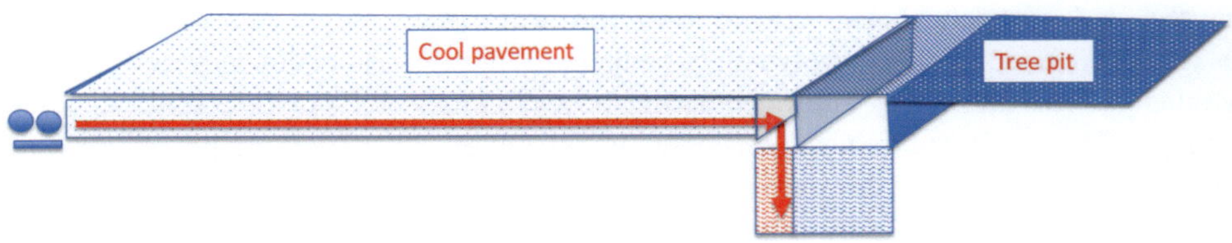

pavements, incorporating the circulation of water beneath the pavement surface.

Figure 4 illustrates the installation and operational process during rainfall. Water descends, guided by gravity due to a slight slope in the terrain, towards a surrounding drain that encircles the pavement, leading to a tree pit. Subsequently, this water flows into a piped system that channels it towards the storage tank.

In terms of implementing cool pavements, Fig. 5 depicts the setup enabling rainwater-collected and evaporatively cooled water to circulate beneath the pavement. This circulation facilitates the cooling of the pavement's surface through conduction. The water is then recirculated to the water tank, establishing a closed and sustainable system with minimal water usage.

3.3 Digital Twin Model

In this section, the mathematical model used as a "Digital Twin" of the thermal behavior of pavements is presented. The numerical model is employed to evaluate the effectiveness of the proposed conceptual solution under the Utrera climate conditions.

$$\begin{bmatrix} 1+\frac{\rho_e\cdot c_e\cdot\Delta x^2}{2\cdot k_e\cdot\Delta t}+\frac{h_e\cdot\Delta x}{k_e} & -1 & 0 & 0 & \cdots & 0 & 0 \\ -k(i) & \frac{C_i\cdot\rho_i\cdot\Delta x^2+(k_{i+1}+k_i)\cdot\Delta t}{\Delta t} & -k(i+1) & 0 & \cdots & 0 & 0 \\ 0 & -k(i) & \frac{C_i\cdot\rho_i\cdot\Delta x^2+(k_{i+1}+k_i)\cdot\Delta t}{\Delta t} & -k(i+1) & \cdots & 0 & \vdots \\ 0 & 0 & -k(i) & \ddots & \cdots & -k(i+1) & 0 \\ \vdots & \vdots & \vdots & 0 & -k(i) & \frac{C_i\cdot\rho_i\cdot\Delta x^2+(k_{i+1}+k_i)\cdot\Delta t}{\Delta t} & -k(i+1) \\ 0 & 0 & 0 & 0 & \cdots & -1 & 1+\frac{\rho_i\cdot c_i\cdot\Delta x^2}{2\cdot k_i\cdot\Delta t}+\frac{h_i\cdot\Delta x}{k_i} \end{bmatrix} * \begin{bmatrix} T_{i,j} \\ T_{i+1,j} \\ T_{i+2,j} \\ T_{i+3,j} \\ \vdots \\ T_{n,j} \end{bmatrix} = \begin{bmatrix} T_{ext}\cdot\frac{h_e\cdot\Delta x}{k_e}+\frac{T_{i,j-1}\cdot\rho_e\cdot c_e\cdot\Delta x^2}{2\cdot k_e\cdot\Delta t} \\ \vdots \\ \frac{T_{i,j-1}\cdot\rho_i\cdot c_i\cdot\Delta x^2}{\Delta t} \\ \vdots \\ T_{int}\cdot\frac{h_i\cdot\Delta x}{k_i}+\frac{T_{i,j-1}\cdot\rho_i\cdot c_i\cdot\Delta x^2}{2\cdot k_i\cdot\Delta t} \end{bmatrix}$$

Fig. 6. Description of the matrix equation terms

To investigate the performance of opaque elements under varying boundary conditions, the analysis commences with the fundamental equation of heat conduction with generation, as outlined in Eq. (1) (ASHRAE, 2017).

$$\frac{\partial}{\partial x}\left(k\frac{\partial T}{\partial x}\right)+\frac{\partial}{\partial y}\left(k\frac{\partial T}{\partial y}\right)+\frac{\partial}{\partial z}\left(k\frac{\partial T}{\partial z}\right)+G=\rho C\frac{\partial T}{\partial t} \tag{1}$$

Particularizing this expression for the unidirectional case without generation, the resulting expression is shown in Eq. (2).

$$\frac{\partial}{\partial x}\left(k\frac{\partial T}{\partial x}\right)=\rho C\frac{\partial T}{\partial t} \tag{2}$$

The equation is determined by the boundary conditions that describe the pavement scenario. The boundary condition on the upper surface pertains to the energy equilibrium at that specific surface. The main factors contributing to this are the absorption of short-wavelength radiation (from the sun), long-wavelength radiant transfer between the pavement surface and the sky, and convective transfer resulting from the pavement surface's interaction with the ambient air.

In the reference (conventional) situation, the bottom surface will have a temperature imposed by the temperature variation of the ground as a function of depth and observation period. Conversely, for a wet pavement scenario, the temperature at this surface corresponds to that of the circulating water.

To solve Eq. (2), numerical modeling is established, discretizing the domain with a sufficiently small mesh so that uniform temperature can be assumed in each cell. For each pavement node, an incremental balance is proposed as follows by Eq. (3).

$$\rho C_p\Delta x_i\left(\frac{\partial T_i}{\partial t}\right)=\left(\frac{T_{i+1}-T_i}{R_i}+\frac{T_{i-1}-T_i}{R_{i-1}}\right) \tag{3}$$

When solving the derivative term concerning time, two potential discretization methods can be employed: either explicitly or implicitly. In this case, the implicit method will be used as it is more complex but will provide more reliable and stable results. The implicit method is shown in Eq. (4)

$$\rho C_p\Delta x_i\left(\frac{T_{i,j}-T_{i,j-1}}{\Delta t}\right)=\left(\frac{T_{i+1,j}-T_{i,j}}{R_i}+\frac{T_{i-1,j}-T_{i,j}}{R_{i-1}}\right) \tag{4}$$

The resulting matrix equation for the implicit method is depicted in Eq. (5) and Fig. 6.

$$[MA]\cdot[T]=[VB] \tag{5}$$

3.4 Outdoor Thermal Comfort (OTC) Index

This section presents the thermal comfort indicators to be calculated to analytically assess the impact of the proposed cool pavement solution compared to conventional pavements.

Thermal comfort is a personal reaction that indicates the degree of contentment with the thermal environment conditions. This reaction combines both physiological and psychological responses of individuals and varies in characteristics between indoor and outdoor settings.

One way to examine the physiological response is through a heat equilibrium analysis of the human body, which is represented by Eq. 6.

$$\Delta S=(M-W)\pm(R+C)-E \tag{6}$$

where: ΔS [W/m^2] is the energy variation in the body; M is the heat generated by metabolism [W/m^2]; W [W/m^2], mechanical work, is the heat consumption due to the activity being performed; (R+C) [W/m^2] is the convective and radiant exchange with the environment; and E [W/m^2] is the heat losses due to sweating.

The psychological response is integrated with the physiological response through the derivation of empirical correlations. This process leads to the development of thermal comfort models and various metrics for their assessment.

An example of such an indicator is the PET model. Physiologically Equivalent Temperature (PET) is defined as the air temperature at which the thermal equilibrium of the human body is maintained, with core and skin temperatures equal to those observed under the evaluated conditions (Höppe, 1999). PET has been utilized in different climates globally,

Table 1. Ranges of levels of comfort by PET.

Extremely hot	>41 °C
Very hot	35 °C to 41 °C
Hot	29 °C to 35 °C
Warm	23 °C to 29 °C
Comfortable	18 °C to 23 °C
Cool	13 °C to 18 °C
Lightly cold	8 °C to 13 °C
Cold	4 °C to 8 °C
Very cold	<4 °C

and the results obtained from simulations have shown favorable comparisons with on-site monitoring (Zhao et al., 2021). The PET thermal index comprises 9 thermal sensation scales, offering a detailed insight into an individual's perception (refer to Table 1). As indicated in the work of Binarti et al., the PET model is one of the most widely used and validated indexes in hot climates. For this reason, in this study, the PET model has been chosen to quantify outdoor thermal comfort in the intervention area being examined (Binarti et al., 2020).

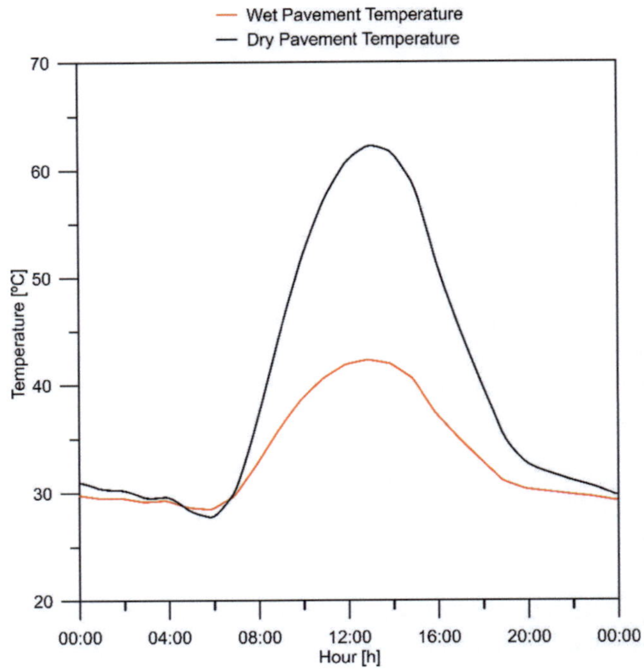

Fig. 7. Thermal performance of pavements in hottest 95th percentile day.

4 Results

This section presents the results obtained from the Digital Twin model analyzing the thermal performance of pavements in Utrera, Seville. Initially, graphical representations of the temperature curves for two types of pavements – a conventional reference pavement and a cool pavement – will be provided. Subsequently, the urban thermal comfort indicators will be presented for both scenarios: the reference pavement and the one enhanced by the proposed solution.

4.1 Thermal Performance of the Cool Pavement

This section presents the surface thermal behavior results of the pavement under sunlight for both a conventional reference pavement scenario and the cold pavement proposed in this study. These temperature models are based on a summer in the Utrera climate of Seville. Figure 7 illustrates the temperature profiles of the pavements' surface on the hottest 95th percentile day during a typical summer in this area.

In Fig. 7, the impact of circulating water beneath the pavement on its surface temperature is demonstrated. The black curve illustrates the temperature of the conventional reference pavement, while the blue curve represents the temperature of the cold pavement with water circulation underneath. A significant temperature difference is observed during hours with higher solar radiation, notably between 10:00 h and 18:00 h. Additionally, Table 2 shows that the maximum temperature of the conventional pavement reaches 62.22 °C, whereas the cool pavement only reaches 42.27 °C, resulting in a maximum temperature difference of 19.96 °C. In terms of average temperatures, the reference pavement averages 40.35 °C, while the cool pavement averages 33.50 °C, indicating a difference of 6.85 °C.

Table 2. Comparison of maximum and average temperature of pavements.

	Reference case (Dry pavement)	Cool pavement (Wet pavement)	Temperature Variation
Ts max [°C]	62.22 °C	42.27 °C	19.96 °C
Ts average [°C]	40.35	33.50	6.85 °C

4.2 Outdoor Thermal Comfort (OTC) Index

This section presents the thermal comfort results based on PET for the analyzed pavements, including both conventional and cold pavements. The study focuses on the thermal behavior from 10:00 h to 19:00 h during a typical summer in Utrera (Seville). For each hour, the pavement falls within one of the pre-defined comfort ranges. The accumulation of percentages for each comfort level is illustrated in Fig. 8.

It was found that for the conventional pavement, only 48% of the hours are in comfort, 25% in warm, 20% in hot, 3% in very hot, and 2% in extremely hot. However, for the cold

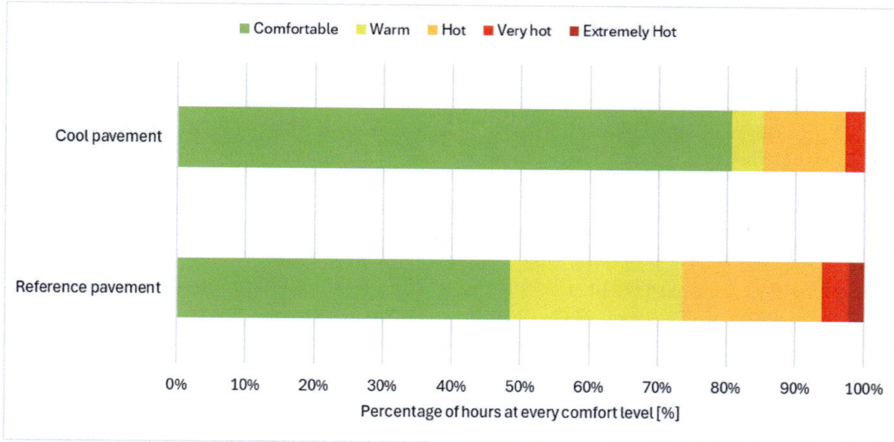

Fig. 8. OTC index of pavements.

pavement, 80% is in comfort, 4% in warm, 11% in hot, 3% in very hot, and 0% in extremely hot. This represents a 40% improvement in the comfort range with the proposed cold pavement solution.

5 Conclusion

In summary, the successful demonstration of a cool pavement technology utilizing sensible and non-evaporative cooling has been achieved. A study on the implementation of the cool pavement solution on a street in Utrera (Seville) has been conducted, detailing the implementation plan for both the cold pavement and rainwater collection and harvesting systems. Furthermore, a mathematical model predicting the thermal behavior of pavements based on environmental conditions has been formulated. An evaluation comparing a conventional pavement scenario with a cold pavement scenario has also been conducted to assess its impact.

The data simulated has yielded very satisfactory results for the innovative concept. In terms of thermal performance, it has been demonstrated that on the hottest 95th percentile day of a typical year in Utrera, a decrease of approximately 20 °C in the maximum temperature of the cool pavement compared to conventional pavement has been computed. Furthermore, an outdoor comfort assessment has been conducted, revealing that the cool pavement offers an increased total number of hours within a comfort range compared to the standard dry pavement, showing an improvement of up to 40%.

Acknowledgments This study has been funded by the projects "CONSTANCY - Resilient urbanisation methodologies and natural conditioning using imaginative nature-based solutions and cultural heritage to recover the street life" (Grant Agreement PID2020-118972RB-I00) and the project "NATURBEAM - Lighting the way to a greener future to restore urban habitability through nature-based solutions" (Grant Agreement TED2021-130416B-I00) by the Spanish Ministry of Science, Innovation and Universities. The second author is supported through the grant agreement C23.I1.P03.S01.01 ANDALUCÍA from "Programa Investigo" funded by the Recovery, Transformation and Resilience Plan of Spanish Government.

References

Agencia Estatal de Meteorología. Informe climático del verano 2023 (2023)

Anupam, B.R., Sahoo, U.C., Chandrappa, A.K., Rath, P.: Emerging technologies in cool pavements: a review. Constr. Build. Mater. **299** (2021). https://doi.org/10.1016/j.conbuildmat.2021.123892

ASHRAE: ASHRAE Handbook-Fundamentals (SI Edition) (2017)

Berland, A., et al.: The role of trees in urban stormwater management. Landscape Urban Plann. **162**, 167–177 (2017). https://doi.org/10.1016/J.LANDURBPLAN.2017.02.017

Binarti, F., Koerniawan, M.D., Triyadi, S., Utami, S.S., Matzarakis, A.: A review of outdoor thermal comfort indices and neutral ranges for hot-humid regions. Urban Clim. **31**, 100531 (2020). https://doi.org/10.1016/j.uclim.2019.100531

Dawson, J.P., Adams, P.J., Pandis, S.N.: Sensitivity of ozone to summertime climate in the eastern USA: a modeling case study. Atmos. Environ. **41**(7), 1494–1511 (2007). https://doi.org/10.1016/j.atmosenv.2006.10.033

Efthymiou, C., Santamouris, M., Kolokotsa, D., Koras, A.: Development and testing of photovoltaic pavement for heat island mitigation. Sol. Energy **130**, 148–160 (2016). https://doi.org/10.1016/j.solener.2016.01.054

Escuela Nacional de Sanidad: Temperaturas umbrales de disparo de la mortalidad atribuible al calor en España en el período 2000–2009 (2015). http://publicaciones.isciii.es

European Comission: Towards an EU research and innovation policy agenda for nature-based solutions re-naturing cities (2015). https://doi.org/10.2777/765301

Feyisa, G.L., Dons, K., Meilby, H.: Efficiency of parks in mitigating urban heat island effect: an example from Addis Ababa. Landscape Urban Plann. **123**, 87–95 (2013). https://doi.org/10.1016/j.landurbplan.2013.12.008

Höppe, P.: The physiological equivalent temperature - a universal index for the biometeorological assessment of the thermal environment. Int. J. Biometeorol. **43**(2), 71–75 (1999). https://doi.org/10.1007/s004840050118

International Energy Agency: Climate resilience electricity security 2021 (2021). www.iea.org/t&c/

Jansson, M., Fors, H., Lindgren, T., Wiström, B.: Perceived personal safety in relation to urban woodland vegetation – a review. Urban For. Urban Greening **12**(2), 127–133. https://doi.org/10.1016/J.UFUG.2013.01.005

Leroy, A.: Urban cooling solutions drawing inspiration from world-wide case studies to address variability in current and future climates (2021). www.carrenoir.com

Lin, Y., Wang, Z., Jim, C.Y., Li, J., Deng, J., Liu, J.: Water as an urban heat sink: Blue infrastructure alleviates urban heat island effect in mega-city agglomeration. J. Clean. Prod. **262**, 121411 (2020). https://doi.org/10.1016/j.jclepro.2020.121411

Oke, T.R., Johnson, G.T., Steyn, D.G., Watson, I.D.: Simulation of surface urban heat islands under ideal conditions at night part 2: diagnosis of causation. Bound. Layer Meteorol. **56**(4), 339–358 (1991). https://doi.org/10.1007/BF00119211

Rodríguez, A.: Una nueva aplicación monitoriza la mortalidad atribuida al calor en España. CSIC **1**(6), 416–419 (2023). https://doi.org/10.1021/envhealth.3c00134

Ruas, A., Clemence, V., Barroca, B., Ulanowski, Y.: Comparison of cooling grey solutions in the sense-city urban demonstrator for adapting cities to heat waves. Urban Clim. **55**, 101874 (2024). https://doi.org/10.1016/j.uclim.2024.101874

Santamouris, M.: Using cool pavements as a mitigation strategy to fight urban heat island - a review of the actual developments. Renew. Sustain. Energy Rev. **26**, 224–240 (2013). https://doi.org/10.1016/j.rser.2013.05.047

Shi, Z., Xu, T., Wang, H.: Sensitivity of Asian climate change to radiative forcing during the last millennium in a multi-model analysis. Glob. Planet. Change **139**, 195–210 (2016). https://doi.org/10.1016/J.GLOPLACHA.2016.02.007

Wang, J., Meng, Q., Tan, K., Santamouris, M.: Evaporative cooling performance estimation of pervious pavement based on evaporation resistance. Build. Environ. **217**, 109083 (2022). https://doi.org/10.1016/j.buildenv.2022.109083

Xie, N., Li, H., Zhang, H., Zhang, X., Jia, M.: Effects of accelerated weathering on the optical characteristics of reflective coatings for cool pavement. Solar Energy Mater. Solar Cells **215** (2020). https://doi.org/10.1016/j.solmat.2020.110698

Zhao, Q., Lian, Z., Lai, D.: Thermal comfort models and their developments: a review. Energy Built Environ **2**(1), 21–33 (2021). https://doi.org/10.1016/j.enbenv.2020.05.007

Traditional Stormwater Management for Urban Resilience: An Analysis of "Ahar and Pyne" System in Sone Floodplains in the Suburbs of Patna, Bihar, India

Anushri Barman, Anjali Sharma, Ashutosh Kumar, and Rishikesh Anand

Abstract

Stormwater management systems use strategies to reduce stormwater flow or runoff, which affects agriculture, society and everyday life. The Ahar and Pyne system is known to be a long-standing, indigenous, and successful flood management and water conservation technique in south Bihar in eastern India, using channels and retention ponds. This traditional system was specifically designed to manage water in South Bihar also, established in response to the region's distinct agro-climatic circumstances. During the British era, this age-old system was ignored, ruining the system and leaving it in poor shape. This research investigates how the "Ahar and Pyne" as a stormwater management system can potentially support sustainable development, fostering urban resilience through case study of Bihta a suburban town of Patna. The current conditions of Ahar and Pyne drainage system is the basis of initial examination, with system enhancement envisioned in future. In order to depict the existing scenario, the drainage system of the Bihta has been mapped using GIS then its existing scenario has been compared to the Ahar and Pyne system, which is the historic drainage system to support agriculture of Bihar. A thorough documentation of the system has been provided through first hand observations completed on-site. The results provide an overview of the present condition of the drainage system and a path forward for the region's sustainable urban development planning with urban resilience. This concept has potential to mitigate floods and water logging, provide cost-effective irrigation in in the hinterland of urban areas, recharge ground water, manage droughts apart from rejuvenation of flora and fauna. The findings highlight the benefits and potential applicability of this traditional method, integrating it into modern blue-green infrastructure systems.

Keywords

Urban resilience · Stormwater management · Sustainable development · Ahar-Pyne system · South Bihar · Blue-green Infrastructure

1 Introduction

Traditional stormwater management techniques play a crucial role in creating urban resilience by leveraging time-honoured methods to address modern challenges. These techniques, rooted in local knowledge and practices, offer sustainable solutions for managing water resources in urban environments. Examining these traditional techniques of stormwater management is necessary as they provide valuable insights into sustainable water management, often using locally available materials and resources with minimal environmental

A. Barman (✉) · A. Sharma
Department of Architecture and Planning, National Institute of Technology, Patna, Bihar 800005, India
e-mail: anushribarman@nitp.ac.in

A. Sharma
e-mail: anjalisharma@nitp.ac.in

A. Kumar
Department of Architecture and Planning, Birla Institute of Technology, Mesra, Jharkhand 835215, India
e-mail: arashutosh@bitmesra.ac.in

R. Anand
(MURP), Department of Architecture and Planning, National Institute of Technology, Patna, Bihar 800005, India
e-mail: rishikesha.pg22.ar@nitp.ac.in

B. He et al. (eds.), *Towards the Framework of Livable and Resilient Cities* Advances in Science, Technology & Innovation, https://doi.org/10.1007/978-3-031-97849-4_8

impact. They enrich our understanding of how ancient knowledge can be adapted to meet present and future urban challenges, promoting more inclusive, safe, resilient, and sustainable cities. Integrating these traditional practices with contemporary urban planning can enhance resilience to climate change, improve water security, and foster sustainable development. Traditional stormwater management methods are increasingly being integrated with modern blue-green infrastructure to create more sustainable and resilient urban environments. This approach combines conventional grey infrastructure with natural green systems to mitigate environmental impacts and enhance urban spaces (Singh et al., 2023). Traditional water knowledge can contribute to flood-resilient urban design and planning to build resilience to urban floods (Asad et al., 2021).

1.1 Importance of Stormwater Management Systems in Sustainable Built Environments

Stormwater management as the management of quality and quantity of water (Borgaonkar & Marhaba, 2021) encompasses systematic planning, implementation, and maintenance of strategies aimed at mitigating the adverse impacts of stormwater runoff on the environment, infrastructure, and human settlements. Stormwater management is crucial for sustainable urban environments, addressing challenges like groundwater depletion and flooding exacerbated by climate change and rapid urbanisation (Mishra, 2019).

It involves a range of measures such as infiltration, detention, retention, and conveyance systems to control the quantity and quality of stormwater runoff. These measures are essential for reducing flooding, erosion, and water pollution in urban and rural areas. Sustainable approaches involve source control of pollutants, decentralised hydraulic retention, and barrier systems for hazardous substances. Integration of stormwater facilities into local landscaping can create attractive blue-green environments (Boller, 2004). Research in this field encompasses urban stormwater quantity, quality, hydrology, treatability, water harvesting, ecosystem services, and related policy and economics (Davis, 2015).

Effective stormwater management requires a comprehensive understanding of hydrological processes, land use patterns, and infrastructure design to ensure sustainable water resource management. By adopting a holistic approach comprising both engineering solutions and ecological principles, stormwater management aims to achieve resilient and environmentally sustainable urban water systems. The system initially envisioned for flood control and nominal improvement in water quality relied upon on-site treatment, mostly stormwater ponds, which are presently "a critical best management practice (BMP) tool for preventing nutrient runoff from urban regions discharging into natural water bodies causing eutrophication" (Ni-Bin Chang, 2014).

"Stormwater management is becoming a challenge in various stages of its implementation, starting from the planning stage," because of expansion of urban areas, "change in existing soil permeability characteristics" due to "construction activities, decrease in vegetation, climate change, change in rainfall and subsequent runoff patterns" (Rentachintala et al., 2022). Various stormwater management strategies employed across different nations include "Sustainable Urban Drainage Systems (SUDS), Low Impact Development (LID), Best Management Practices (BMP), Water Sensitive Urban Design (WSUD), and Sponge City Programmes (SCP)." The "SUDS, WSUD and LID concepts have related aims such as managing urban water through a sustainable approach, preserving the flow conditions close to nature, maintaining water quality and quality of receiving waters, and conserving water resources overall" (Rentachintala et al., 2022). Sustainable stormwater management practices play a pivotal role in conserving water resources by reducing reliance on freshwater sources and tackling water scarcity. Additionally, they aid in alleviating the risk of flooding, a major threat to water resources. Furthermore, stormwater management practices often integrate green infrastructure solutions, such as rain gardens, permeable pavements, and vegetative swales, to mimic natural hydrological processes and enhance ecosystem services.

1.2 Traditional Stormwater Management in India

Traditional stormwater management in India has been an integral part of the country's historical and cultural landscape, demonstrating a profound understanding of local hydrology and sustainable practices and developing various techniques to harvest and manage rainwater. A prominent method was the construction of stepwells (also known as "baolis" or "vavs,"). These intricately designed wells served both as water storage structures as well as community spaces, by capturing and storing rainwater, replenishing groundwater levels (Sharma et al., 2019). Additionally, India's agrarian societies built "tankas" or underground storage tanks that collected runoff from rooftops and courtyards, providing a crucial water source during dry periods. Another significant practice was the creation of "johads," small earthen check dams that captured and percolated rainwater into the soil, enhancing groundwater recharge and supporting agriculture. Another unique system known as "karez" (technology of water distribution through subterranean aqueducts) was introduced by Islamic rulers in India in their territories

(Sharma et al., 2019), examples of which can be found in Bihar as well. These traditional systems were ingeniously adapted to the regional climatic conditions and topography, showcasing a harmonious blend of engineering and ecological wisdom that modern urban planning can draw valuable lessons from.

Water scarcity in India has become a major issue because of unrestricted exploitation, pollution, inequitable distribution of water resources. Revival, adaptation and incorporation of traditional water management systems into contemporary practices is crucial in addressing water shortages; some directly while others with optimisation for increased efficiency in modern applications to ensure sustainable water resource management (Murthy et al., 2022).

1.3 Stormwater Management in Flood Management and Water Conservation

Stormwater management is crucial for effective flood management and water conservation in India, a country frequently grappling with contradictory issues of monsoon-induced flooding as well as water scarcity, often simultaneously. Proper stormwater management systems help mitigate the risk of floods by efficiently channelling excess rainwater away from urban and rural areas, reducing the impact on infrastructure and minimizing property damage. Techniques such as constructing retention basins, percolation pits, and rain gardens slow down runoff, allowing water to seep into the ground and replenish aquifers. This not only alleviates the immediate flooding but also supports long-term water conservation by enhancing groundwater recharge. In regions facing water scarcity, captured stormwater can be treated and reused for various purposes, including irrigation, industrial use, and even potable supply after appropriate treatment. Moreover, effective stormwater management helps maintain the ecological balance of local waterways by preventing the contamination and erosion that unmanaged runoff can cause. In the context of India's rapid urbanisation, implementing robust stormwater management practices is essential for creating sustainable and resilient cities. A more recent approach of stormwater management in watershed planning and management; "a cost effective and environmentally beneficial in terms of controlling runoff and protecting water quality" (Barman & Jha, 2021).

Traditional stormwater management systems in India have been integral to sustainable water practices for centuries, showcasing a deep-rooted wisdom in water conservation (Koul et al., 2012). These systems have not only facilitated irrigation but also supported domestic water supply, livestock management, and environmental services like controlling micro-climatic conditions.

Systems like Ahar-Pyne effectively channel and store rainwater, reducing flood risks and ensuring water availability during dry periods. Additionally, the Ahar-Pyne system in South Bihar continues to provide irrigation to significant areas, emphasizing the importance of indigenous technologies in water management (Koul et al., 2012). As modern challenges like climate change and rapid urbanisation emerge, there is a growing recognition of the need to revive and integrate traditional stormwater management systems with contemporary approaches to ensure sustainable water practices in India (Singh et al., 2023).

The objectives of this paper are to (1) identify the areas around Patna where Ahar and Pyne systems exist in suburbs of Patna, (2) comprehensively analyse historical use of Ahar and Pyne system for irrigation and flood control through literature review, (3) investigate the effectiveness of the traditional Ahar and Pyne system as a stormwater management solution and potential to support sustainable development and urban resilience in the region as well as provide a sustainable source of water through its integration with contemporary urban planning.

2 Methodology

The methodology for this research focuses on identifying the need to utilise traditional stormwater management techniques to address current urban challenges and enhance urban resilience. The primary aim is to analyse how the ancient Ahar-Pyne stormwater management system, originally developed for irrigation and flood control, can be adapted for modern urban settings. This is achieved through a set of clearly defined objectives: (1) investigating the effectiveness of the Ahar-Pyne system as a stormwater management solution through a detailed literature study; (2) identifying areas where these systems are still present and functional through site investigations and analyses; and (3) demonstrating the advantages and contemporary relevance of this traditional technique by integrating it into modern blue-green infrastructure systems. This systematic approach not only explores the historical significance of the Ahar-Pyne system but also evaluates its potential to contribute to sustainable urban development and resilience against the challenges of urbanisation and climate change.

Data for the study has been collected through both primary and secondary sources. The methodology employed in this study begins with an initial examination of the current conditions and drainage system of the Bihta area. This involved a detailed assessment of the area's existing infrastructure, hydrological patterns, and the challenges faced during monsoon seasons. Both GIS-based as well as footprint mapping techniques were utilised to accurately identify and evaluate the existing drainage system, providing a

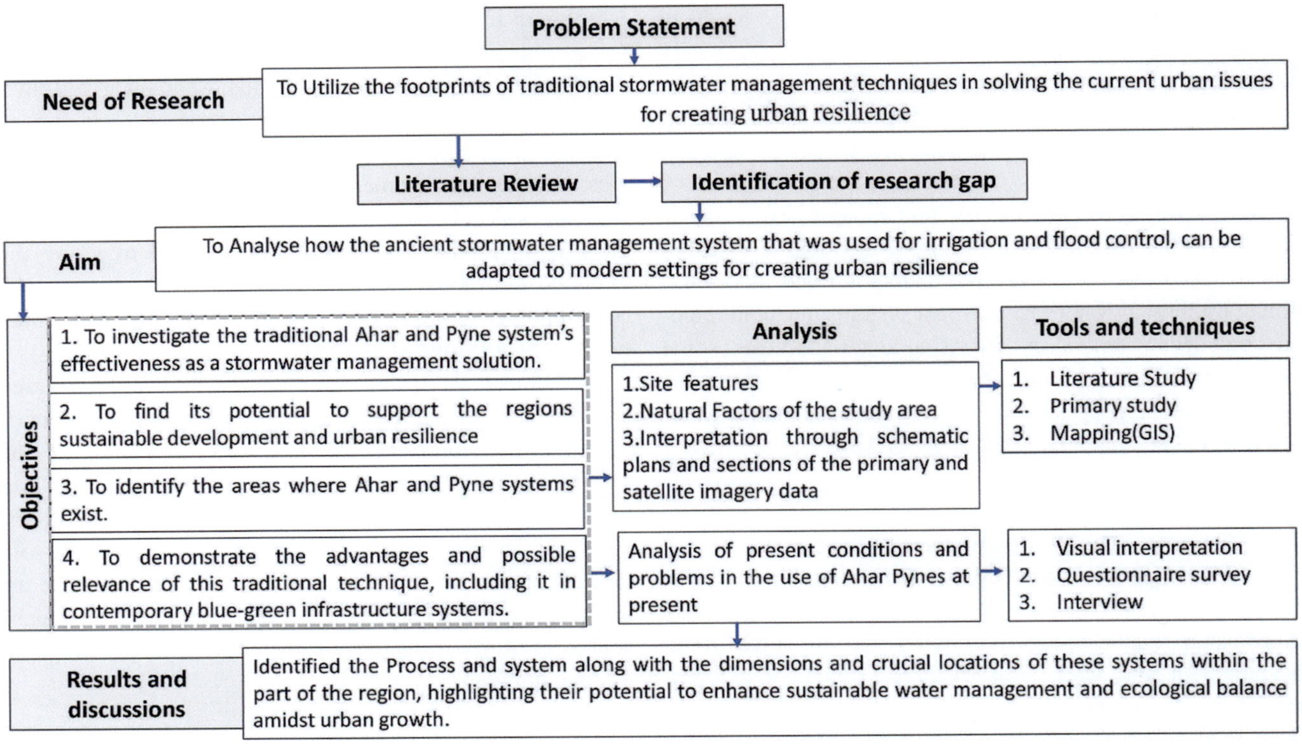

Fig. 1. Methodology for Study (Source: Anjali Sharma)

comprehensive visual representation of the area's water flow and storage capabilities. This data was then used to conduct a comparative analysis between modern drainage infrastructure and the historic Ahar and Pyne systems of Bihar. The research also included thorough documentation through on-site observations, ensuring that real-world conditions and variables were meticulously recorded. This comprehensive approach allowed for a robust analysis of the Ahar and Pyne system's applicability and benefits in the context of contemporary urban resilience and sustainable development in the Bihta region. A flow diagram for the methodology of this study is given at Fig. 1.

2.1 The Study Area

The site considered for study is part of Bihta block where Ahar-Pynes exist. Bihta, a town and block in the Patna district of Bihar, India, located about 30 km west of the city of Patna, is emerging as a satellite town of the city and part of Greater Patna. Bihta is witnessing rapid urbanisation with a real estate boom and significant developments such as IT parks, institutional projects, and proposals for an international airport.

As Bihta area undergoes rapid urbanisation, land use is expected to change significantly and drainage patterns in the area will manifest impacts as a result. Agricultural land along with the traditional Ahar and Pyne systems, are being converted into commercial, institutional, and residential areas without adequate regulatory control. While this necessitates careful planning to manage the shifting drainage needs and to prevent potential flooding or water logging issues in future, this unchecked transformation is leading to the loss of valuable agricultural spaces and the disruption of existing water management infrastructure, posing challenges to sustainable development and ecological balance in the region. The Sone River flows near Bihta and merges with the Ganges River at Hardi Chapra near Maner town. An embankment road separates the Sone floodplain from the main city of Bihta, effectively averting periodic flooding from the Sone River. The location of the study area is shown in Fig. 2.

The study has precisely identified the process and system along with the dimensions and crucial locations of these systems within the Bihta area as shown in Fig. 6, highlighting their potential to enhance sustainable water management and ecological balance amidst urban growth.

The study site at Bihta, situated along the highway to Arrah and adjacent to the Sone River floodplain, encompasses an area flanked by an embankment road. This area features agricultural fields and is witnessing the establishment of several proposed institutes and emerging industrial units. Urbanisation is advancing alongside ongoing agricultural activities in the studied region. Local farmers assisted in identifying

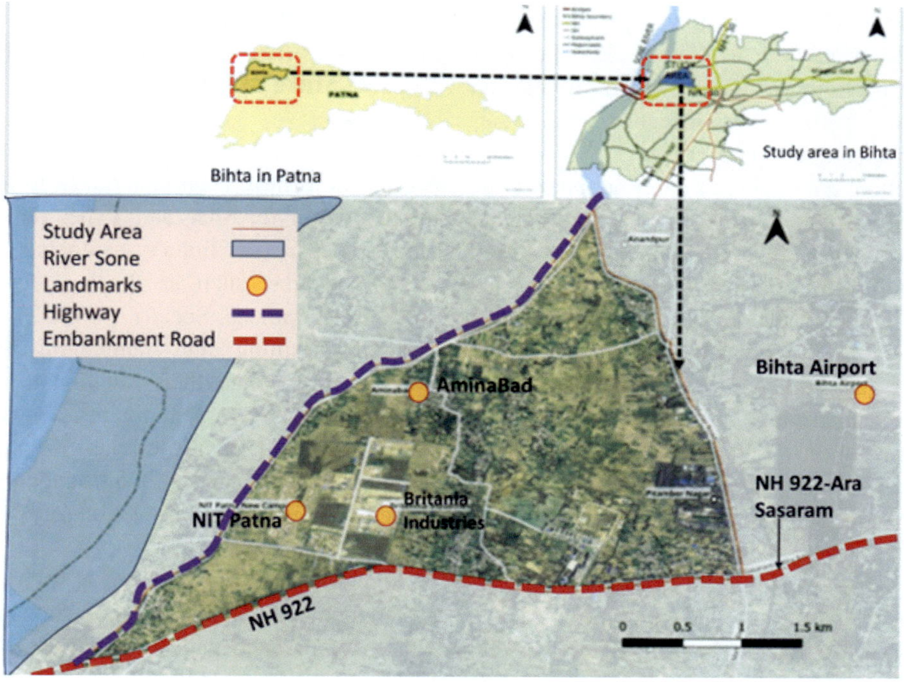

Fig. 2. Location of the study area (Source: Anushri Barman)

the Pynes, with their branching and flow patterns explained through open-ended interviews. On-site observations were recorded, detailing the flow patterns, sectional dimensions of the Pynes, and the associated ecosystem.

3 Results and Analysis

The results include an in-depth understanding of the Ahar-Pyne system through a combination of literature review and ground truthing conducted at the identified site in Bihta.

3.1 Ahar and Pyne Water Management System in Bihar

Ahar and Pyne, a 5000-year-old indigenous irrigation system originated in south Bihar, isa traditional water management network unique to Bihar and comprises a series of interconnected channels (Pynes) and ponds (Ahars). "The word Ahar–Pyne came from Magadhi language of Bihar" (DHAN foundation, 2017). The meaning of Ahars, as informed by the local people in the study area, is to retain water ("*aa*" means to come and "*har*" means to capture). An Ahar resembles "a rectangular catchment basin with embankments on three sides." Ahars were occasionally erected "at the end of small rivulets or artificial channels called Pynes to ensure the supply of water" (Singh et al., 2023). Pynes are small, artificial irrigation channels constructed by the communities that conduct, utilise, and redirect flood water flowing in rivers and streams through gravity and finally discharge the stormwater into nearby rivulets and Ahars. Ahars are large, shallow reservoirs designed to capture and store rainwater and runoff. They resemble retention ponds in terms of blue-green infrastructure in contemporary water management systems (Anjali & Anjali, 2022). A system of Pyne is typically established to draw water from the rivers and divert it to agricultural areas for construction (Fig. 3). "Ahar-Pyne is a traditional floodwater harvesting system" historically employed for irrigation, water conduction, and storage. This method not only facilitates efficient irrigation but also ensures water availability during dry periods or drought. By collecting and storing water during the monsoon season, the Ahar and Pyne system provides a reliable water source for agricultural use throughout the year, significantly aiding in drought mitigation and sustainable water management. "Ahar and Pyne assist in controlling floods by distributing surplus water into its system." It also controls drought as the water in the Ahars is made available for an entire year (DHAN foundation, 2017). This ingenious system exemplifies how traditional practices can effectively address contemporary water challenges, promoting resilience and sustainability in local communities (Singh et al., 2023).

The biggest Pynes are 20–30 km long, feeding several distributaries and irrigating over a hundred villages (Singh et al., 2023). Ahar and Pynes were jointly used and managed by farmers who had to synchronise their operations. Pant (1998) has described Ahar-Pyne as "historically the most important source of irrigation in Bihar" which provides an excellent example of "participatory irrigation" and water management system even today (Pant, 1998).

Apart from irrigation, the important contribution of Pyne system is towards flood water management (Singh et al., 2023). Situated in the downstream of Chotanagpur plateau and flood plains of river Ganga, Bihar is prone to post monsoon floods. However, the abundance of storage structures like Ahar and large-scale collection of torrential floodwaters into Pynes minimises the gush and speed of floodwaters passing through the area. The water management system resembles the Traditional Dong Systems of Assam for flood and stormwater management (Barman & Jha, 2021). The reservoirs (Ahar) and channels (Pyne) plays a critical role in ensuring water and food security. By harvesting river water, these systems provide reliable water security. This availability of water ensures food security and aids in recharging groundwater in the region. Reviving the Ahar Pyne system can be a cost-effective technique to irrigate crops, making water more accessible through innovative ideas and technologies. This system also works as a protective mechanism for flood and drought. The smaller Pynes originate in Ahars and carry water from Ahars to agricultural fields. The larger ones have their origin in rivers from where water is diverted through these Pynes by constructing embankment in the river beds. This leads them some way upstream above the level of the land are intended to be irrigated as shown in the Fig. 3 and Fig. 4. It is usually 3 to 5 kms before the water of the Pynes reaches the level of fields. Some of the longest Pynes are 16 to 32 kms long, while others called *dasian* Pynes (Pynes with 10 branches) irrigate several thousand acres of land in hundreds of villages (India Water Portal, 2024).

The regular upkeep includes cleaning and desilting the Ahar and Pyne systems and maintaining the water conveyance network to ensure their functioning. Consequently, usual maintenance tasks, such as regular silt clearance and the repair of small branches of the Ahars and Pynes, are carried out by local communities themselves under the *goam* system, beginning prior to the onset of the monsoon. Therefore, collective action by the community plays a role in both the operational and maintenance work.

3.2 Institutional Arrangements for Conservation of Ahar Pyne System in State of Bihar

Various initiatives by both government and non-government organisations have been undertaken to rejuvenate the traditional system of Ahar and Pyne, recognizing its importance as part of India's ecological heritage. Government programs, such as those under the Bihar Watershed Development Society (BWDS), have focused on cleaning, desilting, and maintaining these ancient water systems to enhance their functionality and ensure water security. Non-government organisations have complemented these efforts through awareness campaigns, community training sessions, and the introduction of modern techniques to improve the efficiency of the traditional systems.

Some of the projects aimed at revival of this traditional system include Development of Humane Action (DHAN) which collaborated with Indian Tobacco Company Limited (ITC), based on its corporate social responsibility (CSR), emphasising community actions for local development, and the project "Water and Livelihoods Security: Reviving Traditional Ahar and Pyne System Irrigation in Munger Districts of South Bihar" that was initiated with a mission to address poverty through interventions related to water and agriculture specifically in less developed areas of the district of Munger as shown in Fig. 5 (DHAN foundation, 2017).

These Ahar-Pyne systems can be integrated with various stormwater management strategies such as Sustainable Urban Drainage Systems (SUDS), Low Impact Development (LID), Best Management Practices (BMP), Water Sensitive Urban Design (WSUD), and Sponge City Programmes (SCP) employed across different nations. SUDS, WSUD and LID concepts have related aims such as managing the urban water through a sustainable approach, preserving the flow conditions close to nature, maintaining the water quality and the quality of receiving waters, and conserving water resources.

The conservation of the Ahar and Pyne system is essential not only for its historical and ecological value but also as a demonstration of effective community engagement and participation in conservation activities. This collaborative approach has proven successful in promoting sustainable water management practices and enhancing agricultural productivity, thereby securing both water and food resources for local communities. Bihar Watershed Development Society (BWDS) was established to implement the IWMP for the Watershed Development project – 2008 (Revised edition – 2011) by DoLR, Ministry of Rural Development, Government of India. Since 2015–16, IWMP has been renamed PMKSY – WD. The major aims of BWDS

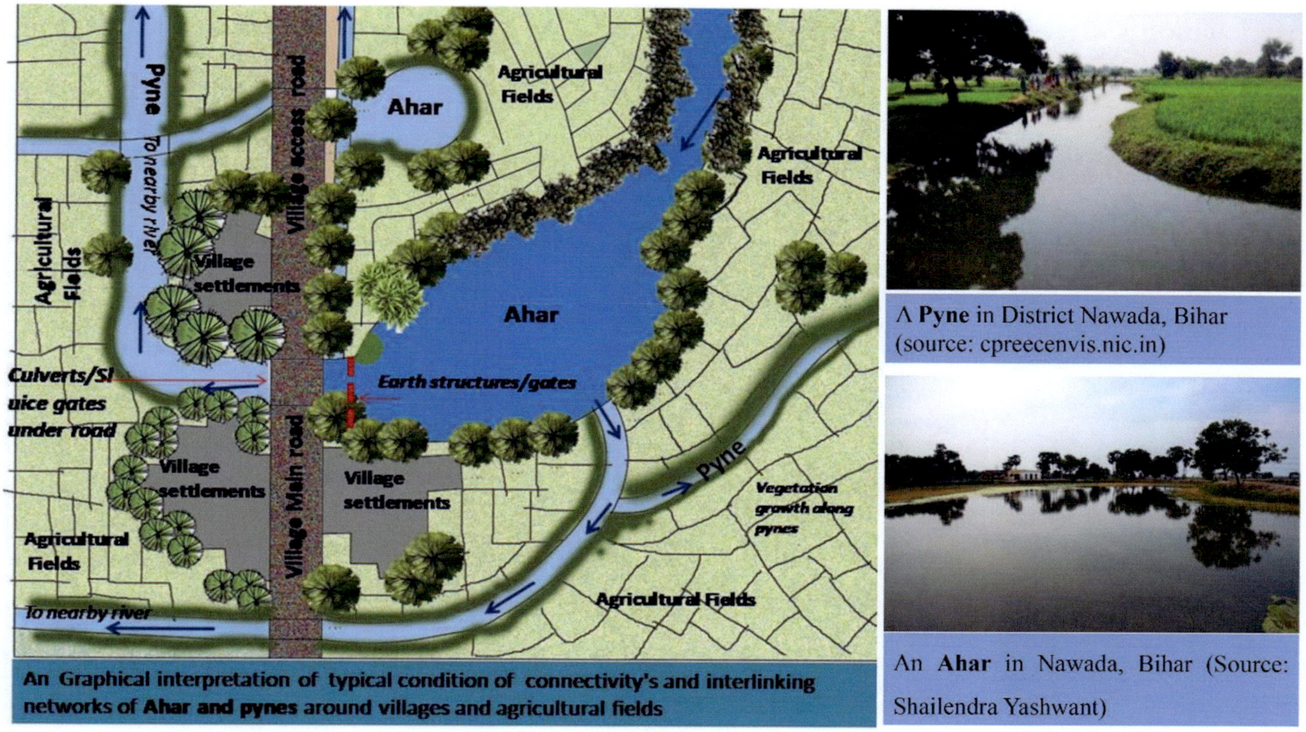

Fig. 3. Typical Arrangements of Ahar and Pynes with Post Monsoon photographs in Bihar, India (Source: By authors)

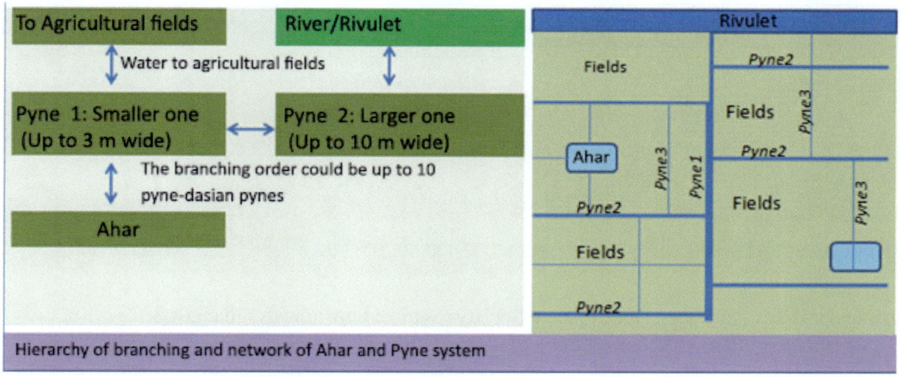

Fig. 4. The interpretation of branching order and hierarchy of sizes of Pynes (Source: Ashutosh Kumar)

are to enhance the prospect of rain-fed degraded land areas through integrated watershed management, strengthening community-based local institutions for supporting livelihoods and watershed sustainability, and enhance the efficiency of watershed-based projects through cross-learning and incentive mechanisms. Under this scheme, several Ahar and Pyne systems have been rejuvenated in different areas of South Bihar. The intent of BWDS is to improve the productivity of crops, livestock, fisheries, and biomass production systems, as well as to support an integrated mechanism for agriculture through the optimal, integrated, sustainable, and effective use of natural resources in project areas. Ahars, Ponds and Pynes are being renovated in various districts of Bihar under the National Renewable Mission (NRM).

3.3 Results of On-site Survey

The results of the study underscore the importance of a detailed study and analysis of the Ahar-Pyne system in the suburb of Bihta as shown in the Fig. 7. As the city undergoes rapid urbanisation, this traditional water management system presents a valuable opportunity to integrate blue-green infrastructure into the developing urban landscape.

Fig. 5. Repair and renovation of Ahar and Pyne in Munger, Bihar (Source: DHAN Foundation Facebook page)

Fig. 6. A renovated Pyne under Bihar Watershed Development Society (Source: https://bwds.bihar.gov.in)

The survey, conducted during the pre-monsoon season in the driest month for easier access, revealed that some Pynes still contained water and supported ecosystem growth. The identified Pynes exhibit varying sectional dimensions, reflecting their adaptation to different parts of the landscape. These Pynes, regardless of their size, converge with larger stormwater channels, which intersect the highway at multiple points. The smaller branch Pynes extend into agricultural lands, effectively carrying water from the upstream Sone River during non-flooding conditions. When the agricultural fields become inundated with water, these smaller branch Pynes transport the excess water back to the larger Pynes, which then channel it out to the main river, the Sone.

During dry seasons, the system operates in reverse: the Pynes serve as irrigation channels, distributing water throughout the area. This dual functionality ensures efficient water management, supporting both flood control and irrigation needs, thereby sustaining agricultural productivity and maintaining ecological balance. The Pynes in the studied areas flow alongside agricultural lands, traversing the fields and carrying away excess water to the Sone River. Whenever the Pynes cross the embankment road, sluice gates have been observed, controlling the water flow and ensuring efficient management of excess water. This system helps prevent flooding in the fields and maintains optimal water levels for agriculture.

The stormwater management mechanism of these Pynes is crucial for maintaining ecological and agricultural balance. The continuous network of Pynes acts as an effective stormwater conveyance system, similar to bioswales in Sustainable Urban Drainage Systems (SUDs). These traditional channels facilitate the natural retention and management of both water quality and quantity. Vegetation growth

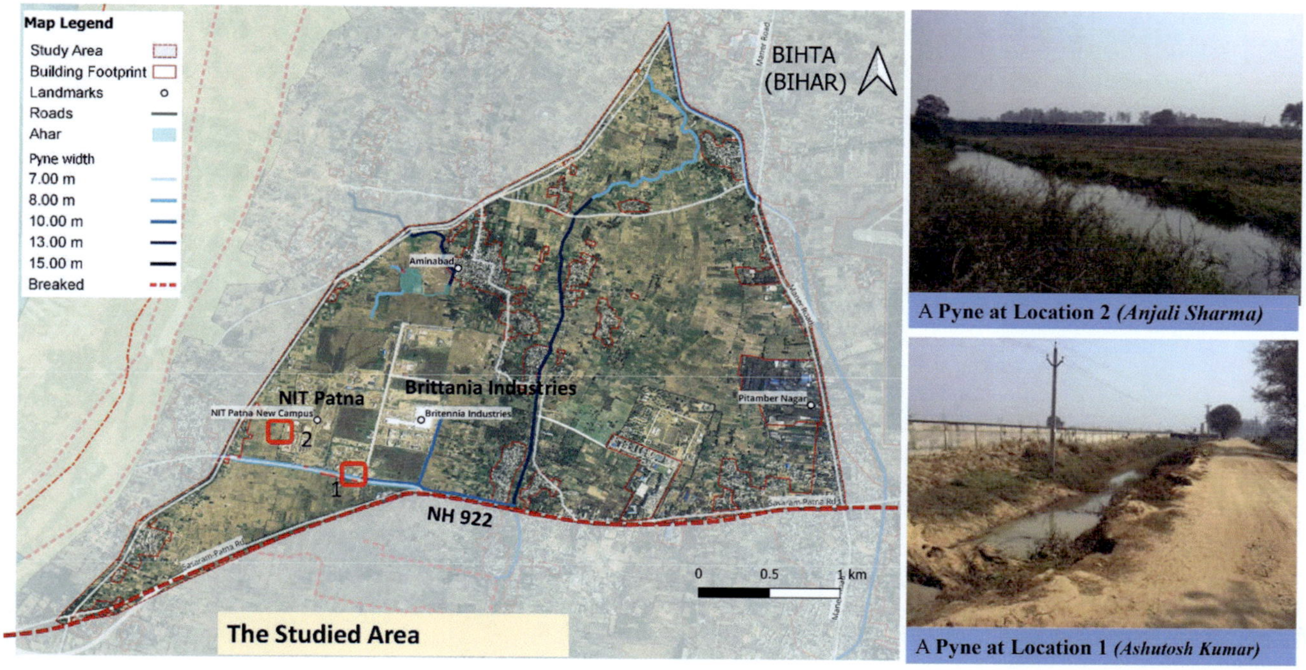

Fig. 7. The study area along NH 922 and embankment Road (Source: Rishikesh Anand)

alongside the Pynes plays a significant role in this process, enhancing water infiltration, reducing runoff, and filtering pollutants. By leveraging these natural systems, the Pynes not only mitigate flooding and ensure water availability for irrigation but also support the local ecosystem, making them an essential component of sustainable water management in the region.

The Pyne 1: In Fig. 6, different dimensions and the hierarchy of stormwater channels are discussed, highlighting their significance in managing water flow. The stormwater channels along the NH to Arrah measure approximately 10 m and 15 m in width. These measurements were recorded at the latitude and longitude coordinates 25.574461, 84.840628, near Raidi village. The channel plays a critical role in carrying water from the adjacent agricultural fields through a series of interconnected Pynes along the village roads and fields. This hierarchical system ensures efficient drainage and water management, preventing flooding and supporting irrigation needs. The channels' varying dimensions reflect their capacity to handle different volumes of water, with larger channels accommodating higher flow rates and smaller Pynes facilitating localised water movement.

Pyne 2: The second Pyne observed and recorded runs along the village road from Arrah NH, connecting to the embankment road at the location 25.575338, 84.835296, near village Modahi. As shown in Fig. 8, vegetated Pynes were found on both sides of the road, measuring 10 m and 8.5 m in width respectively, and depths ranging from 2.5 to 3 m. Although the Pynes were dry at the time of the survey, locals assured that these Pynes efficiently convey stormwater from the fields during the monsoons, eventually meeting Pyne 1. This system highlights the crucial role of the Pynes in managing stormwater and supporting agricultural activities through effective drainage and water distribution during the rainy season.

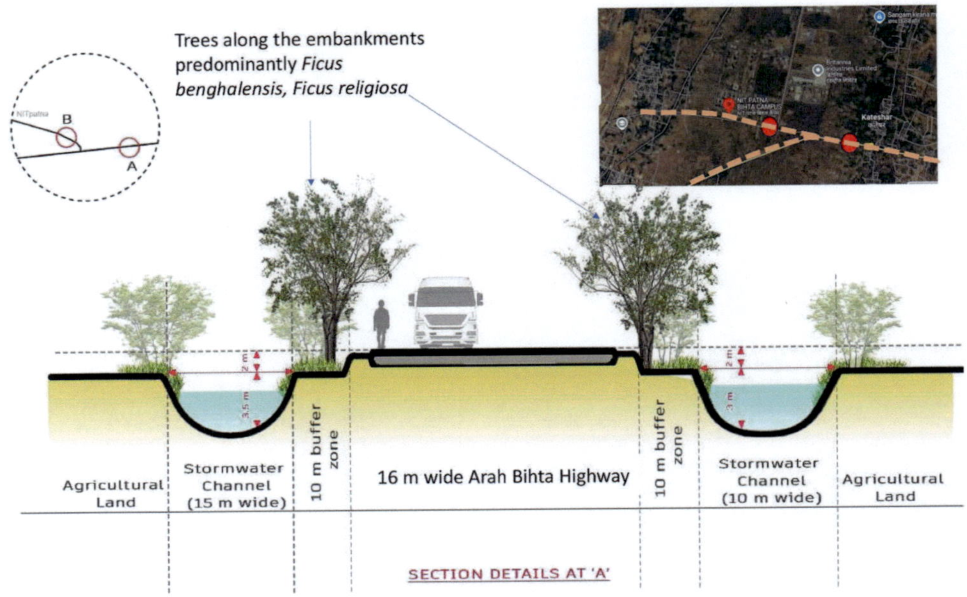

Fig. 8. Typical section through different Pynes along Highway (Source: Rishikesh Anand)

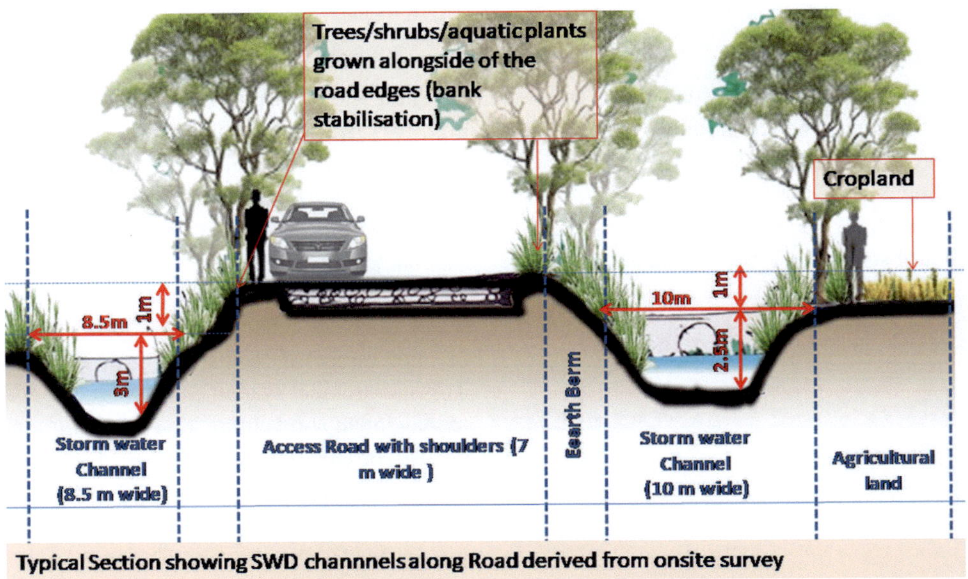

Fig. 9. The Pynes along the access roads to the villages (At location B) (Source: Anushri Barman)

3.4 The Ecosystem of the Pynes

The Pynes host a rich ecosystem, sustained by the presence of water. Acting as bioswales, they not only manage stormwater and recharge groundwater but also support biodiversity, including the growth of aquatic and bank plants. This ecological succession is driven by seasonal variations in water flow and storage. Along the Pynes, plant species such as Calotropis, Carex grasses, Ricinus communis, Saccharum, and Colocasia thrive, while water lilies dominate the water bodies. The junction of two Pynes, where water accumulates, is particularly notable for its abundance of beautiful water lilies, creating a vibrant and ecologically significant habitat. The growth of edge plants and grasses also contribute to the stabilisation of the earthen embankments of the Pynes and Ahars (Fig. 9). This is clearly visible on maps, where the green strips in the landscape can easily be identified as the presence of Ahars and Pynes. The locals mention that they sometimes engage in leisure fishing in the Ahars and Pynes, although this activity is limited to only a few areas. These water bodies also attract birds, particularly egrets, as the area

Fig. 10. Ecosystem of Pynes in Bihta (Source: Anushri Barman)

remains wet throughout the year. The presence of water and the associated vegetation create a conducive environment for both fishing and birdwatching, adding to the ecological and recreational value of the region (Fig. 10).

The third Pyne identified within the agricultural fields in the studied area is located at 25.592352, 84.842019, near Aminabad. This Pyne carries water from the agricultural fields and channels it to the Sone River. A sluice gate is observed in this Pyne where it crosses the embankment road, indicating a controlled water management system. Although the Pyne was partly dry during the survey, with only some water present, it plays a crucial role in stormwater management and irrigation. The Pyne measures approximately 7 to 8 m in width, showcasing its capacity to handle substantial water flow during the monsoon season, thus supporting both agricultural productivity and flood control.

3.5 The Present Status of the Ahar and Pyne in Bihta Region

Due to upcoming development projects, the Ahar and Pyne systems in these areas face significant threats. Connections between these systems have been severed in several places due to construction debris. Culverts are frequently blocked, leading to localised water logging. These disruptions jeopardise the efficiency of the drainage system in the area, potentially exacerbating waterlogging issues following future development. Careful integration of these traditional water management systems is crucial to preserving their age-old effectiveness. Integrating these systems into future stormwater management proposals within upcoming development plans can enhance overall water management efficiency, mitigate flood risks, and sustain the ecological balance of the region. This proactive approach not only safeguards these historical systems but also ensures their continued contribution to sustainable urban development. In Bihta area, the present scenario of the Ahar-Pyne system is fraught with several issues. The Pyne channels are increasingly narrowing due to soil erosion along the banks, soil dumping, and the accumulation of agricultural waste. These channels are often blocked with soil or filled with construction waste, further exacerbating the problem. There are numerous points where the continuity of the Pyne channels has been broken, disrupting the flow and functionality of the system. Additionally, many culverts are in a state of disrepair, hindering the effectiveness of the Ahar-Pyne system in managing water and contributing to urban resilience. These challenges highlight the urgent need for restoration and integration of the traditional system into modern development plans to ensure sustainable water management in the region. The Pynes of Bihta area are under tremendous anthropogenic pressure due to extensive construction activities. This rapid urbanisation has severely impacted the traditional Ahar-Pyne drainage system. Unlike historical times, when the community actively maintained these systems with a strong sense of belonging and responsibility, there is now a significant lack of community involvement in their upkeep. This neglect has led to the deterioration of the Ahar-Pyne infrastructure, compromising its ability to effectively manage water and exacerbating the environmental challenges faced by the region. Reviving community engagement and stewardship is essential to preserving and restoring this valuable heritage for sustainable urban development.

4 Discussions

The Ahar-Pyne system can be compared to various traditional drainage systems both in India and abroad. For instance, the traditional "Dong and Khal" (Barman & Jha, 2021) system of Assam operates with similar functions and processes. While primarily designed as irrigation systems, these traditional stormwater water management frameworks can also be repurposed for stormwater drainage, drought management, ecosystem balance, and habitat corridors. This multifunctionality contributes significantly to healthy and resilient urban development. By drawing parallels with similar systems, such as those found in Assam, we can gain valuable insights into the potential of integrating traditional practices into modern urban planning to enhance sustainability and resilience. These systems exemplify how traditional knowledge can be leveraged to address contemporary environmental challenges, promoting a holistic approach to urban development. A hybrid approach may be adopted while planning for expansion of urban areas or creating new ones, to operate at neighbourhoof levels and join each other at city levels where Ahars and Pynes can be integrated with blue-green infrastructure for an effective and efficient rain water drainage of the area.

4.1 Ahar-Pyne System as Flood Management in Future Development

As per Koul, the Ahar-Pyne system, a traditional water management system in South Bihar, has been identified as a potential solution for flood management in the region (Koul et al., 2012). However, the system's effectiveness is influenced by various factors, including heavy rainfall, a large number of rivers, topography, sand mining, and silt accumulation (Choudhary, 2020). To address these challenges, modern engineering methods have been proposed, such as deepening river beds and constructing permeable screens (Ray, 1954). Despite these challenges, there is hope for the future of flood management in Bihar, with a need for proactive initiatives (Tasnimuddaulah, 2020).

4.2 The Ahar Pyne System in Irrigation

The Ahar-Pyne system operates as a sustainable irrigation solution, offering an alternative to concrete irrigation canals maintained by local communities themselves. Unlike engineered channels that incur high maintenance costs, the construction of Ahars and Pynes is cost-effective or even free, using natural materials and local expertise. This nature-based approach not only reduces initial construction expenses but also promotes ecological balance by integrating with the natural landscape. Along the Pynes, vegetation growth serves a dual purpose, enhancing biodiversity and acting as a natural filter. This vegetation helps regulate the quality and quantity of water flowing into the irrigated lands, ensuring that crops receive cleaner water while minimizing sedimentation and nutrient runoff. In essence, the Ahar-Pyne system exemplifies sustainable water management practices that are economical, community-driven, and environmentally beneficial.

4.3 The Ahar-Pyne System as Ground Water Recharge

The Ahar and Pyne system plays a crucial role as a groundwater recharge mechanism, particularly significant in mitigating the impacts of urbanisation-induced groundwater depletion. As urban areas expand, the demand for groundwater rises, often leading to unsustainable extraction rates and subsequent environmental repercussions. Ahars and Pynes offer a sustainable solution by capturing and storing stormwater, which would otherwise runoff and contribute to flooding. By directing this water into the ground through these systems, they replenish aquifers and elevate groundwater levels. This integrated approach aligns with the principles of rainwater harvesting and sustainable stormwater management, ensuring that urban development supports rather than strains natural water resources. Looking forward, integrating Ahars and Pynes with modern water harvesting systems on buildings and sites can further enhance the efficiency of water use in urban environments, creating a more resilient and sustainable built landscape that addresses both current and future water challenges effectively.

4.4 The Ahar-Pyne System as Drought Management

The Ahar-Pyne system serves a critical role in drought management, particularly during dry seasons. Ahars function as retention ponds, efficiently storing stormwater for future use, thereby ensuring a stable water supply for agricultural and community needs. Simultaneously, the interconnected Pynes play a crucial role in conveying water from Ahars to fields, acting as vital transportation channels. They facilitate the distribution of river water into irrigation channels, while ensuring that crops receive sufficient water even during periods of scarcity (Cheikhaoui et al., 2024). This integrated system not only enhances agricultural productivity but also supports sustainable water management practices, crucial for mitigating the impacts of drought and securing water resources for local communities.

A Retention Pond (source: Sudswale)

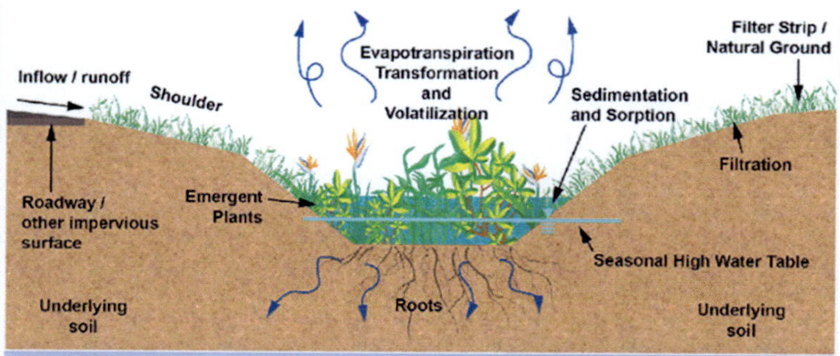

Typical wet swale and pollutant removal mechanisms (source: Ekka and Hunt, 2020)

Fig. 11. Examples as Ahar Pyne could be integrated as sustainable stormwater management (Su et al., 2010)

4.5 The Ahar Pyne as Sustainable Stormwater Management

By integrating landscape planning and innovative interventions, the Ahar region could be transformed into essential components of future blue-green infrastructure (Anjali & Anjali, 2022). These Ahars can be redesigned as retention ponds and detention ponds within this area and can play pivotal roles in managing stormwater effectively, mitigating floods, and enhancing groundwater recharge. These systems could again can be remodified with different innovative swale designs as bio swale, wet swale, swale with infiltration layers for runoff treatment as discussed by Ekka et al. (Fig. 11) (Ekka et al., 2021). Retention ponds play a crucial role in sustainable stormwater management by effectively treating stormwater nutrients and sediment (Yazdi & Sample, 2021). They can also reduce peak flows and negative impacts on downstream locations, particularly when implemented at the upstream portion of a watershed (Fig. 11) (Su et al., 2010). To further enhance their effectiveness, innovative stormwater facilities such as retention ponds should be integrated into local landscaping, creating attractive blue-green environments (Boller, 2004). These interventions not only beautify the landscape but also serve crucial ecological functions by providing habitats for wildlife and improving overall water quality. Similarly, the Pynes can be redesigned as bio-swales or vegetated swales, aligning them with global practices of Nature Based Solutions (NBS) for sustainable stormwater management. Such interventions would not only reduce runoff and erosion but also contribute to enhancing biodiversity and promoting a healthier urban environment. This holistic approach ensures that urban spaces like the Ahar and Pynes are not just functional but also contribute positively to environmental resilience and community well-being.

4.6 Contributor in Sustainable Development Goals

The Sustainable Development Goals (SDGs) have been established as a set of 17 interconnected objectives by the United Nations General Assembly in 2015 for addressing global challenges for an enhanced and more sustainable future by 2030. Various areas, including poverty, inequality, climate change, environmental degradation, peace, and justice, are encompassed by these goals. Listed 11th among the 17 SDGs established by the United Nations General Assembly in 2015, Goal 11 focuses on Sustainable Cities and Communities. Its primary aim is "to make cities inclusive, safe, resilient, and sustainable" (United Nations, n.d.). This goal intends to promote inclusive and sustainable urban development by encouraging investments in public transportation, enhancing urban planning and resource management, and implementing policies for climate adaptation and improved disaster resilience. Goal 11, focusing on Sustainable Cities and Communities, is seen as particularly vital as cities are aimed to be made inclusive, safe, resilient, and sustainable. The growing urbanisation challenges are addressed by this goal through improvement of urban planning and resource management, and the implementation of policies for climate adaptation and disaster resilience.

A study by H. J. Sørup et al. developed a framework to assess the sustainability of nature-based solutions (NBS) for urban stormwater management by linking NBS services to the UN Sustainable Development Goals (SDGs). Applying this framework to three Danish case studies, they found the best matches between local indicators and UN's SDGs for amenities related to flood-protection, minimized material use/carbon footprint, and documentation/knowledge sharing. However, they observed poor matches for services related to linking stormwater to water supply, protecting receiving water bodies, and supporting biodiversity, with no match for mitigating the urban heat island effect (Sørup et al., 2019). The capabilities of Ahar-Pyne system make it a perfect match for such a framework. Ahar Pyne system has also been integrated in the site planning of Nalanda University in Bihar, which is an excellent example of consideration and integration of the system in a site plan by architects and planners.

By integrating the Ahar-Pyne system into city planning, the sustainable water management capability of cities can be substantially enhanced, effectively addressing stormwater challenges and contributing to the achievement of SDGs. This ancient Indian stormwater management technique is exemplified as a contribution to this goal. By effectively channelling and storing rainwater, flooding is mitigated, water availability is enhanced, and sustainable urban living is supported, illustrating the integration of traditional knowledge with modern sustainability practices. By incorporating Ahar-Pyne systems into city planning and urban landscapes, we can harness their ability to efficiently manage stormwater, reduce urban flooding, and replenish groundwater levels. This integration not only preserves cultural heritage but also promotes eco-friendly water management practices, ensuring resilient and sustainable urban environments for future generations.

5 Conclusions

This research paper examines the historical and efficient Ahar-Pyne system of Bihar through a case study in the Bihta area, focusing on its benefits, current condition, and integration potential into modern urbanisation. By bridging traditional knowledge systems with contemporary practices, the study highlights innovative solutions for urban resilience. It offers valuable insights into using indigenous methods to address present-day urban challenges, with outcomes relevant to the broader scientific community.

The research highlights the potential of the Ahar-Pyne system to be incorporated into low-impact development strategies, offering a sustainable alternative for urban stormwater management. It demonstrates how this ancient water management system can significantly contribute to achieving Sustainable Development Goals (SDGs) by addressing water management, urban resilience, and sustainability.

The study presents the Ahar-Pyne system as a model for developing blue-green infrastructure, particularly in peri-urban areas like Bihta. This integration enhances stormwater management while providing co-benefits such as biodiversity conservation, urban cooling, and flood mitigation.

By reintroducing the Ahar-Pyne system in urban planning, the research identifies a cost-effective nature-based solution to mitigate challenges posed by rapid urbanisation, such as increased flooding and loss of agricultural land. It highlights the capacity of this system to save energy and reduce costs, proving its viability in urban drainage infrastructure.

The incorporation of the Ahar-Pyne system in the site planning of Nalanda University exemplifies its practical implementation and potential scalability in modern projects. The Nalanda University example emphasizes the potential of incorporating these systems into larger-scale developmental efforts. Upcoming projects in institutional, commercial, or industrial sectors can benefit from the Ahar-Pyne system's proven utility, showcasing how traditional solutions can meet modern challenges in a cost-effective, environmentally sustainable, and culturally respectful manner. Future research and development initiatives can focus on customizing these systems for diverse urban and rural landscapes, ensuring their relevance and efficacy in the face of climate change and urban expansion.

The study advocates for an integrated approach combining traditional water systems with urban green spaces. This synergy creates resilient urban environments that address drainage, heat island effects, and ecological degradation while fostering healthier living conditions.

Emphasizing the importance of community-level participation, the research underscores the role of local engagement in the successful revival and sustainability of traditional systems.

It encourages stakeholders and policymakers to integrate indigenous knowledge systems into urban planning to enhance urban health, environmental resilience, and well-being.

This research serves as a call to action for scientists, urban planners, and policymakers to rethink traditional knowledge systems as innovative, nature-based solutions for urban sustainability. By combining cultural heritage with cutting-edge urban planning practices, it paves the way for resilient cities capable of meeting the demands of modern development while preserving ecological integrity. Future work should focus on refining these integrations, evaluating their long-term impacts, and creating replicable frameworks for global urban contexts.

References

Anjali, D., Anjali, S.: Mitigation of urban flooding using blue-green infrastructure: a case of Dehradun City, India. Disaster Adv. **15**(11), 50–61 (2022)

Asad, R., Ahmed, I., Vaughan, J., von Meding, J.: Traditional water knowledge: challenges and opportunities to build resilience to urban floods. Int. J. Disaster Resilience Built Environ. **13**(1), 1–13 (2021). https://doi.org/10.1108/IJDRBE-08-2020-0091

Barman, A., Jha, R.: Traditional stormwater management system in watershed of wetland, a study and overview of "Dong" systems in wetland fringe villages of Assam. In: HYDRO 2020 INTERNATIONAL 25th International Conference on Hydraulics, Water Resources and Coastal Engineering, National Institute of Technology Rourkela Odisha, India, March 26–28, 2021 (2021). https://www.researchgate.net/publication/350438903

Boller, M.: Towards sustainable urban stormwater management. Water Sci. Technol. Water Supply **4**(1), 55–65 (2004). https://doi.org/10.2166/ws.2004.0007

Borgaonkar, A.D., Marhaba, T.F.: Evaluation of sustainability strategies—A water quantity and quality perspective. In: Handbook of Water Purity and Quality, 2nd edn., pp. 409–454. Elsevier (2021). https://doi.org/10.1016/B978-0-12-821057-4.00011-2

Cheikhaoui, Y., Mohamed, S., Mohamed, A., Saïd, C., Abdelhak, B., Kenza, K.H.: Remote sensing for sustainable crop water management in a changing climate. Environ. Sci. Sustain. Dev. **9**(4), 79–93 (2024)

Choudhary, S.: Factors affecting flood management in Bihar, India. Int. J. Environ. Sci. **11**(1), 24–28 (2020). https://doi.org/10.53390/ijes.v11i1.5

Davis, P.: Journal of sustainable water in the built environment (2015). https://ascelibrary.org/

DHAN foundation: Reviving traditional Ahar & Pyne system irrigation. In: Development Matters (2017). https://dhan.org/

Ekka, S.A., Rujner, H., Leonhardt, G., Blecken, G.T., Viklander, M., Hunt, W.F.: Next generation swale design for stormwater runoff treatment: a comprehensive approach. J. Environ. Manage. **279** (2021). https://doi.org/10.1016/j.jenvman.2020.111756

India Water Portal: Ahar pynes, traditional flood water harvesting systems can help revive agriculture in south Bihar (2024). https://www.Indiawaterportal.org, https://www.indiawaterportal.org/agriculture/farm/ahar-pynes-traditional-flood-water-harvesting-systems-can-help-revive-agriculture-south

Koul, D.N., Singh, S., Neelam, G., Shukla, G.: Traditional water management systems-An overview of Ahar-pyne system in South Bihar plains of India and need for its revival. Indian J. Traditional Knowl. **11**(2), 266–272 (2012)

Mishra, B.K.: Storm water management in the context of climate change and rapid urbanization: a case of Tokyo metropolitan. JOETP **1**(1), 32–44 (2019)

Murthy, R.N.S., Srikonda, R., Vijayalaxmi Kasinath, I.: Traditional Water management systems of India. ISVS E-J. **9**(2) (2022). https://www.ancient-civilizations.com/

Ni-Bin Chang, J.J.: Ecological Modelling and Engineering of Lakes and Wetlands. In: Jørgensen, S.E., Chang, N.-B., Xu, F.-L. (eds.), vol. 26. Elsevier (2014)

Pant, N.: Indigenous Irrigation in South Bihar, India: A Case of Congruence of Boundaries. In: Economic and Political Weekly, pp. 152–3155 (1998)

Ray, K.B.: Flood prevention in the rivers of Bihar, North Bengal and Assam. The Economic Weekly (1954). http://www.epw.in/system/files/pdf/1954_6/41/flood_prevention_in_the_rivers_of_bihar_north_bengal_and_assam.pdf

Rentachintala, L.R.N.P., Reddy, M.G.M., Mohapatra, P.K.: Urban stormwater management for sustainable and resilient measures and practices: a review. Water Sci. Technol. **85**(4), 1120–1140 (2022). https://doi.org/10.2166/wst.2022.01

Sharma, A., Kumar, M., Singh, M.P., Mazhari, H.K.: Medieval (Islamic) cities in India (1206–1764): an environmental review and its contemporary relevance. Indian Hist. Rev. **46**(1), 55–85 (2019). https://doi.org/10.1177/0376983619856166

Singh, Y.V., Banerjee, S., Sagar, D.: Integrated blue green infrastructure approach for storm water management. EPRA Int. J. Climate Resour. Econ. Rev. **1**(2), 1–5 (2023). https://doi.org/10.36713/epra13053

Sørup, H.J.D., Fryd, O., Liu, L., Arnbjerg-Nielsen, K., Jensen, M.B.: An SDG-based framework for assessing urban stormwater management systems. Blue-Green Syst. **1**(1), 102–118 (2019). https://doi.org/10.2166/bgs.2019.922

Su, D., Fang, X., Fang, Z.: Effectiveness and downstream impacts of stormwater detention ponds required for land development. In: World Environmental and Water Resources Congress 2010: Challenges of Change - Proceedings of the World Environmental and Water Resources Congress 2010, pp. 3071–3081 (2010). https://doi.org/10.1061/41114(371)314

Tasnimuddaulah, M.: Geographical review of flood management in Bihar. Studies in Indian Place Names. UGC Care J. **40**(40), 1569–1574 (2020)

United Nations. (n.d.). THE 17 GOALS. Retrieved December 7, 2024 (2024). https://sdgs.un.org/goals

Yazdi, N., Sample, D.J.: Efficacy of a retention pond in treating stormwater nutrients and sediment (2021)

Designing for Disaster: A Holistic Approach to Architectural Resilience

Hadeel El Naggar and Maha A. Abd El-Wahab

Abstract

Facilities, buildings, and communities worldwide seek better resilience and infrastructure to deal with crisis-like situations. The modern-day world needs to build buildings so that relevant people such as police, firefighters, and medical staff can easily access them in an emergency. Furthermore, with climate change and environmental challenges, buildings must be resilient against environmental damage and climate change events.

The aim of this paper is to explore how infrastructural design can enhance resilience and adaptability to mitigate crises such as natural disasters and climate change. It seeks to provide insights into designing infrastructures that can withstand environmental challenges while ensuring quick access for emergency services and long-term functionality.

The methodology adopted in this study includes a systematic analysis of architectural resilience through real-life case studies such as the Dutch mentality Amsterdam housing development and the Brooklyn Waterfront Greenway. These examples demonstrate the effectiveness of resilient and adaptable infrastructure. The study employs design thinking as a framework to develop strategies for disaster recovery and mitigation, emphasizing the importance of robustness, redundancy, and rapidity in design.

Key concepts such as architectural resilience and building adaptability are discussed, highlighting infrastructure designed for various crises with a high level of resilience. With comprehensive analysis, this paper offers insights that policymakers and architects can use to design infrastructure that is adaptable and resilient, thus enhancing crisis management and recovery strategies.

Keywords

Climate Resilience · Architectural Resilience · Adaptabilities · Mitigation · Disaster

1 Introduction to Architectural Resilience

Buildings, facilities, or communities can be more resilient by using resilience to prevent damage and help them recover. Every building is designed to serve a specific purpose, and every tenant has different needs when it comes to support systems. Resilience strategies and designs must consider these requirements (Berger et al., 2022). Increasing tenant resilience entails a combination of structural elements that protect the building from disruptive events and policies and procedures that reduce effects and expedite a return to regular operations. Resilience elements should be anticipated and incorporated into effective building design.

Building locations and layouts should take a defense-in-depth stance against physical assaults, considering layers of defense imposed by different physical characteristics, technology, and organized operational procedures. With this strategy, in an unfavorable incident, threats should be avoided, detected, delayed, addressed, and ultimately denied (Campos, 2020). Aside from off-site accident hazards like hazardous material releases or explosions that could affect the site, site planning and layout should consider potential natural hazards like flooding, wildfires, and strong winds, as well as any compounding effects from multiple hazards.

The location and design of buildings should also provide quick access to vital infrastructure for first responders,

H. E. Naggar (✉)
Architecture Department, American University in Dubai, Dubai, UAE
e-mail: Helnaggar@aud.edu

M. A. A. El-Wahab
Department of Architecture, Faculty of Engineering, Pharos University, Alexandria, Egypt
e-mail: maha.abdelwahab@pua.edu.eg

B. He et al. (eds.), *Towards the Framework of Livable and Resilient Cities* Advances in Science, Technology & Innovation, https://doi.org/10.1007/978-3-031-97849-4_9

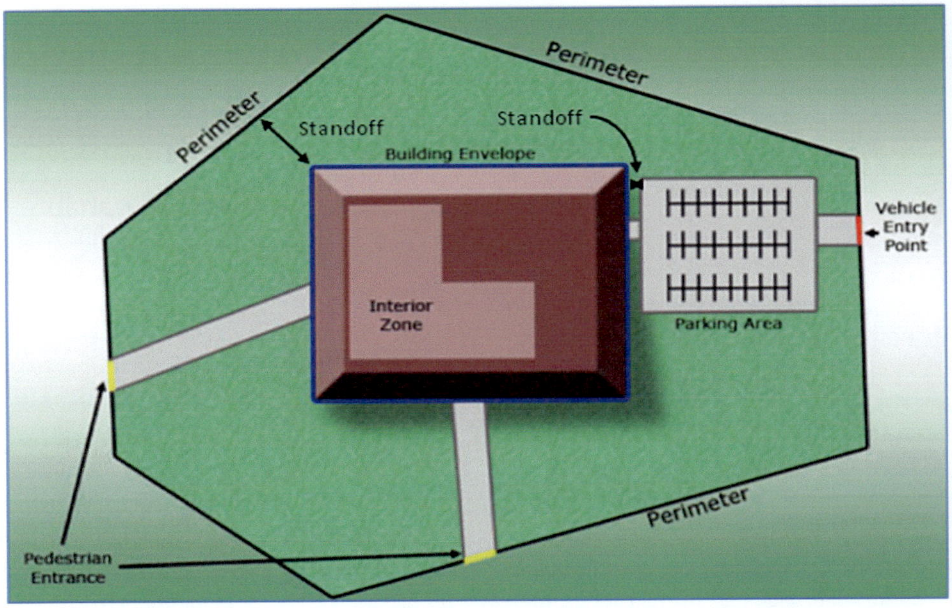

Fig. 1. Example of site layout, source (Campos, 2020)

including police, fire, and medical staff, to support their missions as shown in Fig. 1. To protect against man-made threats, natural disasters, and potential accidents, critical utilities and backup systems should be positioned strategically. The goal is to maintain operability after an incident or quickly restore services as needed. It is advisable to use redundant utility feeds as much as possible, such as gas, water, power, and communications (Brogden, 2020). Site planning should also include marked egress routes and refuge areas for building evacuees, with enough space and physical separation between each location. For ballistic protection, designs that do away with line-of-sight targeting are essential, particularly in urban environments where problems might occur.

When choosing a location for a resilient design, it is essential to consider how resilient the neighborhood is overall, especially regarding utilities, transportation, and building tenant availability after an incident. Protracted outages in the neighborhood can greatly impact a building's recovery resilience. During building design, it is critical to distinguish between operational and life safety/security infrastructure needs because of potential differences in criticality and priority. Building operational infrastructure must be evaluated for sustained operations based on tenant requirements to determine which critical systems must run continuously. Life safety and security systems usually require continuous operation for a predetermined time during unfavorable situations.

According to Malalgoda et al. (2014), a resilient built environment is one in which the people who live or work there, the physical and institutional support systems, and the structures themselves are all designed to withstand, recover from, and lessen the effects of extreme natural and man-made hazards. Furthermore, according to Charlesworth and Fien (2022), the built environment must protect people from natural and man-made disasters. It is anticipated that vital infrastructure and facilities will continue functioning during and after such dangerous occurrences. Urban built assets must, therefore, be planned, developed, run, and maintained in a way that makes them resilient to disasters and guarantees the survival of the city, its residents, and other related physical and social systems. Cutter and Derakhshan (2019) have identified insufficient structural capacity within the built environment as a significant factor contributing to extensive damage during natural disasters. The risk of catastrophe has also been found to be increased by several other factors, including insufficient attention to coastal risks, non-disaster-resistant building designs and constructions in disaster-prone areas, inaccurate hazard assessments, incompatibilities between structural designs and hazard levels, a failure to incorporate risk considerations into town planning, a disregard for building codes and regulations, and illegal occupancy in high-risk areas. Therefore, it is crucial to reduce these risks by implementing hazard-resilient designs, specifications, building techniques, materials, and technologies, as well as by building protective infrastructure and ensuring the security of vital infrastructure. Cost considerations appear to be a factor in some legislators' and professionals' reluctance to acknowledge the increasingly important role that built environment specialists play in disaster resilience (Fallah-Aliabadi et al., 2020). To guarantee the development of a safer built environment, it is crucial to address these concerns at the policy level and allot sufficient funds for disaster risk reduction (DRR) initiatives. The conclusions drawn from empirical data gathered through expert interviews and case studies are presented in the following sections.

2 Methodology

This research adopts a mixed-method approach, combining qualitative analysis and case studies to explore architectural resilience and adaptability in the context of disaster management. The methodology is structured around the following key components:

- Literature Review A comprehensive literature review was conducted to understand the theoretical frameworks and key concepts of resilience in architecture, focusing on robustness, redundancy, adaptability, and resourcefulness. The review also included recent studies on disaster resilience, architectural adaptability, and environmental sustainability. Special attention was given to research related to urban and building resilience, particularly in coastal and flood-prone areas. This review provided the foundational knowledge necessary to understand how buildings can be designed or retrofitted to withstand natural disasters and environmental changes.
- Case Study Analysis Two key case studies were selected to illustrate successful implementations of resilient architecture: -The Brooklyn Waterfront Greenway Project: This project, implemented after Hurricane Sandy, integrates green infrastructure for flood mitigation in New York City. The project focuses on using natural systems like green roofs and swales to manage stormwater and reduce the risk of flooding in urban areas. - Floating Houses in IJburg, Amsterdam: This case highlights how adaptive architecture in the Netherlands addresses rising sea levels and coastal flooding. Floating houses were developed as an innovative housing solution in Amsterdam's IJburg neighborhood, demonstrating a proactive response to the challenges posed by climate change. These case studies were chosen for their relevance to coastal resilience and their application of anticipatory design strategies. Both cases represent innovative approaches that prioritize resilience, environmental sustainability, and adaptability. Each case study was analyzed in terms of design principles, challenges encountered, and the resilience strategies employed, using a combination of primary sources, such as design plans, and secondary data from reports and academic papers.
- Analytical Framework The 4-R Framework—comprising Robustness, Redundancy, Resourcefulness, and Rapidity—was employed to evaluate the resilience of architectural designs. This framework is widely used in disaster management to assess how systems, including buildings, can withstand, recover from, and adapt to various hazards.

This analytical framework was applied to the selected case studies and integrated with the insights from the literature review to provide a comprehensive assessment of architectural resilience. The findings were used to offer recommendations for future architectural designs that can withstand environmental challenges while maintaining adaptability for future needs.

3 Inherent Limitations of Current Architecture Designs

The field of disaster management has several shortcomings despite the growth of disaster studies, especially in the built environment. Specifically, conceptual papers frequently ignore operational concerns; discrete case studies offer little chance of making significant generalizations, and single-sector, single-discipline studies ignore interdisciplinary processes like design essential for solving complex problems. As a result, there is a shortage of systematic research that bridges the knowledge gap between the causes and effects of vulnerability and the creation of resilient settlement infrastructure and landscapes that protect vulnerable communities from disasters. To generate policies and workable solutions from a conceptual roadmap that makes complex realities understandable and facilitates efficient disaster management, Chen et al. (2021) advocate for a fundamental reworking of disaster theory.

Design processes provide a route to this new disaster recovery and mitigation strategy through systems analysis and design thinking (Rak et al., 2021). Design entails applying creativity, spatial innovation, and cross-disciplinary systems thinking to solve problems within the built environment. It consists of iterative observational sequences, design development, fast prototyping, collaboration, and participation to understand people's needs, project implementation and assessment, etc. The ability to create and reform space based on understanding human aspirations, the natural environment, and social, political, and cultural systems is a fundamental skill for architects and other built environment and landscape designers. Design is a methodology and a way of thinking that provides an integrated approach to problems in disaster recovery and mitigation issues.

The terms "design" and "design thinking" are used in the following discussion, emphasizing real-world architectural and spatial applications. These applications cover topics that apply to other fields and disaster recovery and mitigation issues. Housing is one of the most significant losses following disasters, which affects social, economic, and physical infrastructure. However, the people-centered approach of architectural design frequently contrasts with the product-delivery culture that permeates disaster-recovery programs, resulting in a "one size fits all" strategy that ignores local needs, vulnerabilities, and aspirations (Bernabei et al., 2021). Despite the importance of architects' knowledge in people-centered housing and settlement design, there is a shortage of skilled workers to handle challenging reconstruction projects after a disaster. However, through the integration of fieldwork and research, design-led approaches to disaster mitigation and recovery are emerging, providing promising answers to challenging problems.

4 The Need for Holistic Design

While design is widely recognized as a problem-solving tool, its primary goal is usually to address problems. On the other hand, a holistic design paradigm departs from this idea by acknowledging that the system being designed is deeply interwoven in a larger global context. This method gives secondary factors the same weight as the main objective of solving the problem. By integrating sustainability, aesthetics, and user-centered principles, holistic design in architecture surpasses conventional boundaries. Considering how a building affects community dynamics, culture, and climate, architects strongly emphasize the relationship between a building and its surroundings (Flores, 2024). Green practice, which is adopted to mitigate the adverse impacts of the environment on the planet, has become extremely important today (Jiang et al., 2021). When spaces consider the needs and preferences of the users for whom they were created, user-centric design could be a way to cultivate equity and cultural sensitivity. Holistic arrangement promotes creativity, and, in the meantime, steps are taken towards innovative, low-dimensional input regarding multiple dimensions. With this technique, architects and interior designers can measure multiple variables, namely the spatial functionality, interaction with the intended use, and the one in which the environment is aesthetically connected to the context.

This method perceives the design as a means of conveying meaningful information, sometimes implicit and under the broader objectives of project solutions. Similarly, some fundamental aspects of smart urban planning are contemporary projects that consider user comfort, provide for environmental friendliness, and use energy-efficient principles. In essence, holistic design envisions a future when architecture does not serve only functional or decorative purposes but also moves beyond practicality and incorporates the spirit of the environment. Moreover, it enhances the welfare of humanity. A holistic approach is considered in modern architectural projects when choosing material, form, and function along with the environment's and the user's context and climate. Architects use this method to create sustainable, user-focused, and culturally significant spaces, from large-scale urban developments to single buildings. Notable examples are structures that put occupant well-being and environmental sustainability ahead of energy efficiency and sustainability while blending in with their surroundings.

Theoretical and practical approaches to disaster recovery have been developed to enhance immediate response, long-term recovery, and preventive efforts. Figure 2 represents the short-term plan, which lead to transitions phase into a long-term recovery phase. The recent UN World Conference on Disaster Risk Reduction highlighted a lack of effective communication between the scientific community and policymakers. Although scientific literature often emphasizes using scientific knowledge to prevent disasters or inform post-disaster policy responses, a preliminary framework was needed. Thus, a thorough analysis of social science literature utilizing the terms "disaster recovery" and "disaster management" was conducted (Ner et al., 2022). Literature focusing on long-term perspectives, governance, and recovery management was also reviewed and included in this framework.

Social sciences have developed theoretical and practical approaches as shown in Fig. 2 which emphasize resilience to stress and adaptability to support longer-term policies for disaster resilience. One such method is the disaster life-cycle model, which captures the general agreement in contemporary theory that a crisis sets off a crisis response cycle. This cycle—frequently represented as a linear conceptual model—begins with rebuilding, redevelopment, and renewal to enable effective recovery. Alternatively, it proceeds automatically through four stages: mitigation, preparedness, response, and recovery. It comes to an end when people in the impacted area resume their normal activities, and systems are put in place to lessen the effects of catastrophes of a similar nature in the future. It is suggested that there is a progression from one phase of the cycle to the next, even though most of the current disaster recovery literature concentrates on short-term phases and technical response (Meng et al., 2020). It has turned out that the switch to long-term disaster recovery is not as automatic as first thought. Different stages call for different approaches and strategies to be developed to build resilience; however, little is known about the triggers that cause people involved in a disaster context to shift into new approaches that result in long-term recovery and resilience strategies. In the Sendai Framework for Disaster Risk Reduction 2015–2030, the United Nations Office for Disaster Risk Reduction (UNISDR) recently emphasized the need "to facilitate the link between relief, rehabilitation, and development" considering the absence of the anticipated seamless progression; however, a thorough explanation of this link has not yet been supplied.

Across different stages of recovery, the disaster literature identifies various roles and participants. These include vital actors, local actors, and "new actors." While "stakeholders" is a term that is frequently used in disaster literature, Pribadi et al. (2021) use the term "actors" here to emphasize those who are not only included but also have agency and act.

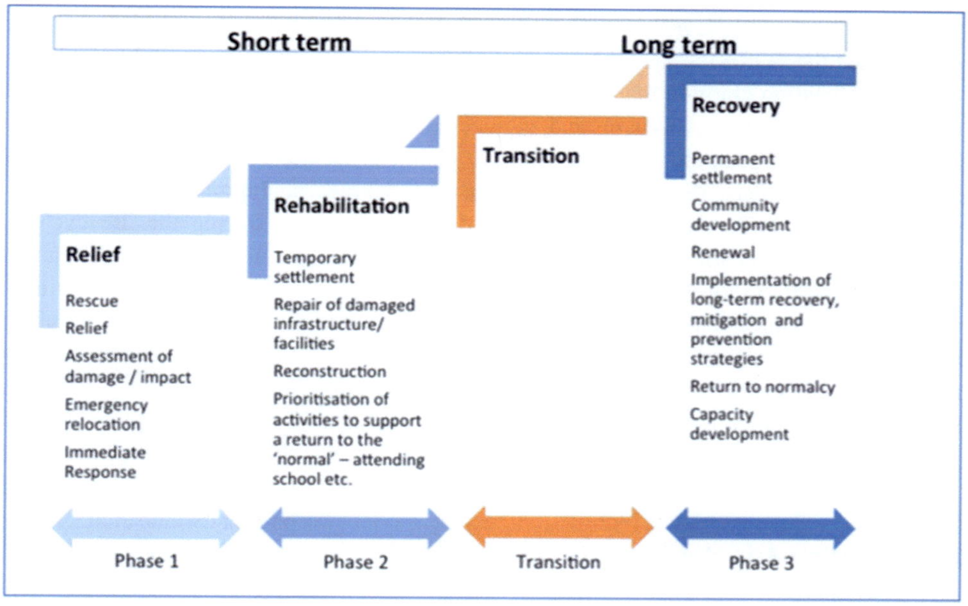

Fig. 2. Disaster transition, source Pribadi et al. (2021)

Notably, grassroots volunteer organizations and community groups are rarely considered significant players. NGOs, bilateral and multilateral development organizations, the scientific community, the insurance industry, the private sector, governments, and local institutions are just a few of the actors in disasters that are described by Christopolos et al. However, they acknowledge that there is not much interaction between these players despite some attempts to build bridges between various viewpoints. Despite this, there is little talk about volunteers outside of NGOs and little community involvement. On the other hand, those like the International Federation of the Red Cross (IFRC) emphasize the importance of local actors and their advantages over external or foreign actors in terms of political rapport, cooperation, and linguistic and cultural ties. This observation implies that current disaster literature fails to fully acknowledge the system, leading to complex interactions.

Aldrich thoroughly analyzed the recovery process following different types of disasters. He concluded that social capital—rather than factors like aid or socioeconomic conditions—plays a significant role in recovery. According to Bemthuis et al. (2020), social capital is the goodwill generated by social relationships that can be used to spur action. It denotes integrating elements like networks, norms, and trust within social organizations. Relationships and personal networks are the sources of social capital; these networks form when people are a part of long-lasting networks, which are vitally necessary but must be created and maintained.

5 Dimensions for Holistic, Resilient Structure

When researchers at MCEER saw a gap in the definition and measurement of resilience within earthquake engineering and disaster loss reduction, they set out to investigate the idea of disaster resilience. "The ability of social units (e.g., organizations, communities) to mitigate hazards, contain the effects of disasters when they occur, and carry out recovery activities in ways that minimize social disruption and mitigate the effects of future earthquakes" was the initial definition of seismic resilience (Imperiale & Vanclay, 2021). The four characteristics of resilience identified by the MCEER resilience framework—now called the "4-R" approach—are robustness, redundancy, resourcefulness, and rapidity. These characteristics can be applied to various units of analysis, including communities, organizations, and engineered systems.

The ability to withstand failure and disturbance and continue to operate effectively in the face of environmental shocks is called robustness. The degree to which backup systems can continue to operate if the central systems malfunction is known as redundancy. Resourcefulness is quickly mobilizing material, financial, informational, or human resources to address disaster challenges (Adekola & Clelland, 2019). Rapidity, which follows the other three resilience-related qualities, promptly resolves disaster-related problems. The technical, organizational, social, and economic domains are the four domains to which resilience applies, as defined by the further development of the 4-R framework (later termed the TOSE framework).

This explanation clarified that the 4-R framework covers organizational and societal domains and engineered systems. Each domain is a separate analytical unit to characterize and measure resilience (Salvato et al., 2020). The degree of resilience can differ significantly between domains. For instance, a lack of technical redundancy and robustness led to the catastrophic failure of the building housing the city's emergency operations center in the 2001 World Trade Center disaster in New York City. The Office of Emergency Management, which oversaw overseeing the disaster response, demonstrated organizational resilience by managing to stay organized and creative even after losing its building.

5.1 Building Adaptability and Architectural Resilience

According to the dictionary, adaptability is the capacity to change or be changed to suit needs or situations. It refers to a system's capacity to internally adjust on its own to adjust to changes in its surroundings in the context of systems engineering. Similarly, adaptable design in product design strives to make products that are simple to alter to suit various needs (Blackman et al., 2017). In the field of the built environment, there is disagreement over the exact meaning of adaptability because it varies based on the situation. In the discourse surrounding architecture, ambiguities such as "adaptability," "flexibility," and "polyvalence" are acknowledged, as they have multiple, frequently overlapping meanings.

Adaptability is further explored by contrasting it with the idea of "maladaptive" buildings—those that cannot meet new demands because of technical or financial inefficiencies. Different academics have defined adaptability differently, considering various building typologies, change types, and motivations. The main reasons for designing adaptable buildings are identified in numerous studies as being building obsolescence and inadequacy resulting from changing operational conditions, user needs, and external factors. Although definitions found in the literature highlight the ability of adaptable buildings to accommodate change as a primary feature, the specifics of change rely on various needs and circumstances.

Changes in components, space layout, size, volume, performance, use or function, and other types of changes are covered in the literature. Changes in building use, however, stand out as the most common category in conversations, probably because users frequently modify space usability to suit their needs. However, more intricate modifications—especially those involving distinct functions—might require expert assistance. It is essential to comprehend these variables that affect change to design flexible buildings throughout their lifespan (Sun et al., 2020). Various change scenarios, from extendibility and scalability to flexibility and recyclability, are summed up under the umbrella generic term of "adaptability," which also includes other general ideas, such as reusability, upgradability, transformability, convertibility, and durability.

During the life cycle of any construction, the use will sometimes vary and may change together with customers' exigencies. As the name suggests, conservative structures are always change-resistant because they cannot adapt to the diverse interrupts and other environmental elements that might affect an architectural body in a particular scenario. Therefore, adaptability should not be just considered as something short-term that is needed for immediate needs but must be considered as a long-term changing requirement of any project (Straub et al., 2020). Incorporating adaptability is not just solving the problem with only one solution but presenting multiple options over a lasting period. This is reinforced by the point that Schmidt et al. made that time can no longer be seen as absolute but multi-dimensional. It also suggests that adaptability forces reassessment in time, values, and attitudes. Recognizing time's role is sufficient to establish buildings as stimulating entities that vary per changing needs while offering long-term sustainability of the built environment. The adding time factor into the process will contribute to the fact that architecture represents the reality of the time since it is a developing thing, having its destiny and constantly changing. It is, therefore, adaptability, which requires an understanding of purpose, applicability, and context and which may not be initially evident in real life. A specific context is created by the decisions made in the design and building stages, which depend on the settings and situations that influence the construction.

On the other hand, a time-based strategy recognizes the inherent instability of contextual conditions (Jia et al., 2020). Practical strategies for adaptability consider how contextual factors change over time. Contextual awareness improves a building's ability to be altered to suit changing circumstances, another feature of flexible buildings that many writers have emphasized. In conventional architecture, the fit is restricted to immediate function and form; in adaptive architecture, the fit is time- and context-based, considering the present and future considerations. An essential goal of adaptability is to increase a building's usable life (Sarker et al., 2020). Because of the imbalance between the conditions of available space and the demands that evolve, many buildings become outdated before their technical lifespan ends, which results in their abandonment. A building's longevity is meaningless if it cannot fulfill the needs of its users. Ultimately, people are the most crucial factor in the equation because buildings must change to meet their changing needs, offer better services, and guarantee comfort and safety. A thorough definition of flexible buildings can be obtained by combining these concepts:

In architecture, adaptability is the ability of a building to change in response to new requirements or changing contextual factors, extending its useful life and maintaining its value to users over time.

5.2 Building Robustness and Resilience

The traditional method of reducing structural risks has been to strengthen the overall robustness of the system and increase the resistance of the elements that make up the structure. However, it is frequently impractical to achieve total protection against all hazards. Once a certain level of acceptable risk is reached, further enhancing structural robustness might not be cost-effective. As a result, incentives and research efforts have changed recently to support resilient design principles. Four essential characteristics of resilience are outlined by Bruneau et al.: robustness, redundancy, resourcefulness, and rapidity. In this framework, robustness is considered a crucial component of resilience and is associated with the performance decline shown in the resilience triangle concept.

According to Mergos (2023), assumptions that member-based design techniques sufficiently affect global resistance—from which structural robustness is derived—are the source of differences between expected and observed structural performance. They suggest defining robustness as a system's intrinsic quality dictated by its geometry and topology. Geometry is the arrangement of load-bearing elements; topology is the structure's configuration of its site and expected exposure to extreme loads. The system's robustness is defined by the fact that both attributes are fixed and cannot be changed without changing the overall structure configuration. As opposed to robustness, which is considered an unchangeable aspect of the system, resilience is thought to be influenced by design choices.

The ability of a structure to withstand, adjust to, and bounce back from extreme events is referred to as resilience. Although resistance is an engineer's attempt to withstand a specified hazard, damage may still happen even with sufficient resistance, requiring corrective measures that temporarily impair the structure's functionality. While recovery refers to the continuous restoration process through corrective actions, adaptation entails having emergency plans to restore functionality after an incident. Because adaptation and recovery are difficult to measure or be affected by structural design, Mergos (2023) offers a more precise definition of resilience that does not include them. Thus, a given hazard magnitude reduced by a structural design with assigned robustness is linked to structural resilience. By quantifying resilience and robustness, structural designers can use this definition as a foundation for structural evaluation that occurs after an event.

5.3 Building Sustainability and Architectural Resilience

The knowledge of life cycles is the starting point for determining a product's suitability and its effect on the environment at the different stages of its life. The relatively obvious goal of tracking any material used on an architectural project is to trace its route from resource extraction (or mining) and processing, production, distribution, and construction through to use and disposal (or life cycle) to analyze and eventually take an informed decision on its environmental impact. The Intergovernmental Organisation of Standardisation (ISO) sets the method used in the Life Cycle Assessment (LCA) that helps evaluate the stage of a material and its environmental impact throughout these stages (Fabbricatti et al., 2020). Eco-objectives can be successfully handled by combining digital analytical tools like BIM and building information modeling during the initial stage of the design phase with lifecycle analysis. However, the problem of establishing where and how the leftover waste should be disposed of is a question that is immediately brought about.

Appraising the whole period of life of a specific allocated material is difficult because few detailed data exist about its second life, which is also crucial for displaying energy utilized in production. The energy embedded in these buildings is being evaluated to find out if it is possible to reuse the pieces already available in the Swedish, German, and Danish building stock. This study used a set of numbers, ranging from embodied energy, greenhouse gas emissions, recycling percentage, and resource utilization (Ahmed, 2023). Using materials in the design process for deconstructing and adaptive reuse will pay off employment for people. Promoting deconstruction as a design principle implemented during the building life cycle stages, namely demolition, is a crucial focus of the design process. Adaptive reuse, a methodological approach that recycles the existing buildings and structures and utilizes them for multiple uses, heightens productivity and brings forth environmental sustainability.

Foresaving the life cycle is the foundation to appraise the over-consumption, utilization, and after-use environmental consequences. Any material employed for an architectural project must be able to go through its complete life cycle to be used again. This fact, along with the effects of such processes on the environment, is invaluable in considering the more suitable materials for our construction and architectural projects. Environmentally friendly construction needs all phases of a material's life cycle to be documented: starting from extraction, processing, etc., through construction, production, use, distribution, and recycling, up to the end of the life cycle (i.e., demolition) for a thorough assessment of the environmental implications. An ISO organization, the International Organization for Standardization (ISO), states that the life cycle assessment has enabled the estimation of the environmental impacts of material containing various components and production stages (Fabbricatti et al., 2020). To meet environmental objectives, architectural designers can use life-cycle analysis (LCA) through various digital analytical tools, including building information modeling at various stages of the building design; however, the most critical question is how to treat the waste after demolition of the building.

Understanding the entire life cycle of the material makes it possible to understand how to treat it in its second life, which is also vital to understanding embodied energy. To recognize the reuse and recycle potential of the material, various research has been done in Denmark, Sweden, and Germany so that the embedded energy of the material could be used positively. Such studies have used various indicators, including recycling percentages, embedded energy, material use, and carbon dioxide production (Ahmed, 2023). Reusing material throughout the design process for deconstruction is becoming extremely critical with time. Periodizing material to be reused throughout the life cycle stages of a building, including demolition, is a critical part of the entire design process of deconstruction. Furthermore, existing structures and buildings for new use come under the term adaptive reuse, which facilitates environmental sustainability and productivity. If products and repurposed materials are integrated into the life cycle of a building, they can be manufactured more quickly with the least cost compared to new ones. However, the lack of an integrated database in making design choices is a huge issue that may result in the abandonment of recovered material. It is crucial to understand which type of reclaimed material can be used in building and what type of drawbacks it may have. The Switzerland, British, and Christ pavilions are some case examples showing the principle of "design for deconstruction " (Elvas et al., 2021)". Using a built-in script under the concept of demolition, such projects can reduce construction waste.

The entire idea of adaptive reuse has been implemented in various famous projects like Herzog & de Meuron's Elbphilharmonie Gradzinski (2019), where material from pre-existing structures was used and converted into steamed public buildings for the regeneration of degraded areas and adaptive reuse has become a global trend that has been shown in number of different projects including Turzyński Quarter in Szczecin, MASS MoCA in the USA, and Docklands in London. Different realizations are an example of how the waste material life cycle is now ending and how various local materials are being used in different projects. The use of discarded and processed waste material is shown in several projects, including the Japanese Pavilion at EXPO 2000 Hanover, Sustainable Emerging City Unit, and Michael Reynolds' Earthship Houses. All these examples show how resources can be effectively reused, and their life cycle can be increased without ending up in a landfill. These examples also show how eco-friendly solutions can be introduced to minimize negative environmental impact throughout the building life cycle (Taherkhani et al., 2021). Using elements such as upcycling, adaptive reusing, recycling, and deconstruction, ingenious architects are increasingly considering environmental principles while developing a design. Experience and knowledge among architects and designers must be shared to advance such practices further. Different types of databases can be used, including advanced tools for evaluating flood risk and properly optimizing solar panel placement, trees in different cities, maps of buildings, and urban observational platforms. Building specifications are now also being made public by companies such as ELEMENTAL, known for actively sharing open-source house designs on platforms like 2030 Palette and Tectónica (Elvas et al., 2021). Such platforms provide solutions to help reduce the environmental impact of construction projects. By developing a groundwork for applying and investigating universally acceptable construction design solutions, such projects promote sustainable methods in architecture and urban planning.

5.4 Building Redundancy and Architectural Resilience

One of the most critical concepts for developing resilient architecture (structure is through proper integration of redundancy. The overall "probability" of getting an event can be minimized by adequately designing and strengthening the overall robustness of the building structure and by integrating fail-safes to avoid the collapse of the entire structure. It is critical to mention that redundancy is also needed for the safety of the entire construction structure because it minimizes the possibility of fatalities and accidents in case of failure. Redundancy is an essential dimension of resilience that architects and construction engineers integrate into their projects for the enhancement of overall safety and also dependability (Nazif et al., 2021). The two types of redundancy are local redundancy and geo-redundancy. When there is an indication of redundancy in the local network, engineers can integrate redundancy through all that architectural design in various campuses, data centers, and cities. Businesses with complicated distributed infrastructure commonly integrate local redundancy and geo-redundancy into their project design for adequate network availability for smoother operations. Redundancy needs to be incorporated into an efficient network architecture to create better resilience and decrease the chances of network outreach. Using local and Geo redundancy makes it easier for businesses to develop physical and online infrastructure resilience (Luke et al., 2021). Redundancy allows engineers and architects to have tools to enhance their capabilities and effectively implement safety. Redundancy is critical for ensuring that catastrophic events power the infrastructure or building. This is specifically applicable to hospitals where the need for continuous power is critical in case of emergencies. Hospitals looking for various renewable energy options because of increased energy costs can use the dimension of redundancy to prevent tighter carbon emission regulations.

In line with Renewable Portfolio Standards, several states require a certain percentage of power from renewable sources like wind and solar. According to Walt Vernon, CEO of Mazzetti, resilience, and sustainability are critical components of future design approaches. This company offers MEP engineering design and technology consulting in the healthcare industry. Design guidelines are changing to prioritize systems that mitigate climate change and build resilience from the ground up. Initiatives like Kaiser Permanente Hospital's plan to design a carbon-neutral building highlight the significance of resilient structures that can run independently without outside energy sources (Lu et al., 2020). Ultimately, the most robust facilities can run independently, guaranteeing continuity under challenging conditions.

6 Resilience and System: Case Example

Age Van Randen coined the phrase "open building" at TU Delft in the middle of the 1980s, drawing inspiration from the concepts of Dutch architect John Habraken. By eschewing the strictly functional design of conventional architectural structures and emphasizing adaptability to social and technological changes throughout the building's lifecycle, it offers a support and infill concept for residential buildings. A multilayer control mechanism known as "open building" allows users to independently manage the infill, which is under their control, and the base building, which is under the control of the owner/investor (Ismail et al., 2017) as a component of the multilayer control structure, furnishing, fixtures, and equipment (FF&E) are also included in this concept.

On the other hand, by incorporating concepts from open buildings, "transformable buildings" provide a dynamic design approach to meet changing needs. This idea, sometimes called the Hendrickx-Vanwalleghem strategy, stresses how a building can adapt its systems or itself to meet changing needs. Disassembly capabilities, interchangeable parts, generative form and dimensioning systems, and building envelopes and infill integration are all features of transformable buildings. In contrast to open buildings, transformable structures could have building infills and kinetic envelopes. "Modularity and standardization" allow building designers to make system adaptation easier. Standardized components known as modules are used in modular construction (Castaño-Rosa et al., 2022). They are produced in factories and can be independently changed or replaced. This method covers materials and whole structural levels in addition to building components. Although standardization and modularity have advantages in terms of better quality, performance, safety, cost savings, and environmental impact, their geometric complexity may restrict design flexibility.

Building codes and regulations are not always flexible enough to account for changes in building structure systems, even with their potential benefits. Various zoning requirements and adaptive regulations are required to facilitate change processes inside building structures. Examples include regulations permitting minor modifications to functional use and adaptive zoning at the neighborhood and building portfolio levels. Resilience indicators are only partially incorporated into the minimum essential standards that are the focus of current building codes. Sustainability and resilience indicators should be systematically considered in regulatory impact assessments, with all stakeholders involved in the design process. Furthermore, instruments such as the Adaptive Reuse Potential (ARP) model can evaluate the possibility of modifying the functional use of already existing buildings, thereby augmenting adaptability within the built environment (Quintana et al., 2021).

By addressing the constraints in the adaptation process and providing a way to reconcile perturbations and changes, a synthesis approach can methodically unify existing concepts related to adaptation to change. The focus of open buildings is how structures can be reused and converted for smaller functional changes within the same building category. Conversely, building transformation categories and partially transformable buildings are included in the category of transformable buildings. Part transformable buildings permit minor modifications with the potential for reversal, such as upgradeability, scalability, or extendibility within the same or closely related class of buildings (Jankovic, 2018). On the other hand, building transformation entails significant modifications that might not be undone, like changing the building's type or infrastructure class, applying the system's change architecture, or going through sustainable demolition procedures. One example of building infrastructure transformation is the Queensway Linear Park in Queens, New York, created from an abandoned railway.

By facilitating the system's form return capabilities, modularity, and standardization act as instruments to accomplish system adaptation objectives. These procedures lead to interchangeable methods and goods, making it simple to swap out old or damaged components without having to replace the entire system. To guarantee a comprehensive approach to managing change, adaptive regulations, and zoning requirements—which function as feedback laws and constraints—must be incorporated into the system's change architecture (Mayar et al., 2023). The system's built-in change architecture does not vary depending on the system domain. Though it is acknowledged that altering organizational or social dynamics can be just as difficult as in technical or physical systems, integrating change architecture for social or organizational systems may seem easier and more convenient at first or later in the system's life cycle.

These ideas are included in the regulations and change architecture. Open buildings often address Convertibility and reusability, which concentrate on minor functional modifications within the same building category. With options for significant and minor adaptations and transformations, such as upgradability, scalability, or extendibility within the same building class, transformable buildings offer greater flexibility for change. Transitioning to a different building class or another type of built infrastructure through sustainable demolition constitutes these significant changes, which may not be reversible (Jankovic, 2018). Standardization and modularity help the system retain its form-return capability by facilitating adaptation to change. The term "adaptive zoning" describes feedback laws that adjust to changes to guarantee that the system handles changes thoroughly and coordinately.

6.1 The Brooklyn Waterfront Greenway

Multiple case studies can be used to show how resilience is being created in various architectural designs and projects. For example, the architectural firm WE designed a highly effective spread heading the Brooklyn waterfront Greenway project in America, New York City showed on the map in Fig. 3. It is critical to know that the project finished in 2015 and aimed to strengthen Brooklyn's overall cost line resilience, specifically after Hurricane Sandy in 2012, which hugely flooded the entire area. The project acted as a proper stormwater management network using various regenerative strategies, including green infrastructure such as green roofs and swales, thus minimizing the probability of flooding (Ribeiro & Pena Jardim Gonçalves, 2019). Considering that the green infrastructure can help keep water bodies clean in case of natural disaster, the management of New York City considered the project extremely valuable compared to conventional grey infrastructure as shown in Fig. 4.

Furthermore, East River pollution and sewer overflow were hugely minimized on the Brooklyn waterfront after Greenway's design, Fig. 4 highlights the effectiveness when harvesting rainwater and treating stormwater. The entire project can be considered a proactive project that minimized the cascading impact of the urban system damaging the area, and climate change further negatively impacted the entire situation. The greatest ability in the entire water system around the area was improved. With these four active strategies, the water treatment system also became easier with the implementation of anticipatory design strategies around coastal cities (Ma et al., 2018). This case example shows how proactive strategies can be used to protect vulnerable coastal areas around the world against rising seaware levels.

6.2 The Dutch Mentality + Amsterdam Housing Developments

Some popular projects hugely used for improving architectural resilience against increasing sea levels and coastal flooding are in the Netherlands. It is critical to know that almost half of the population in the Netherlands lives around sea levels, and it has been taking action to enhance buildings around coastal areas. There was a need to enhance resilience in the Dutch built environment after flooding incidents between 1993 and 1995, which also resulted in he would evacuation and large-scale livestock losses (Askar et al., 2021). The

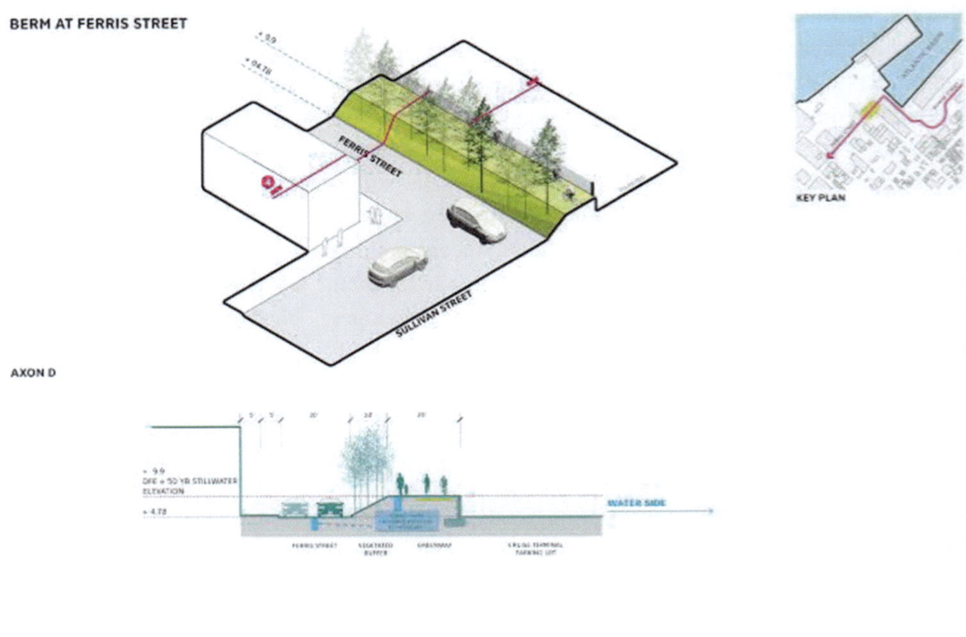

Fig. 3. Street section representing the elevated park toward water side (SCAPE, 2017)

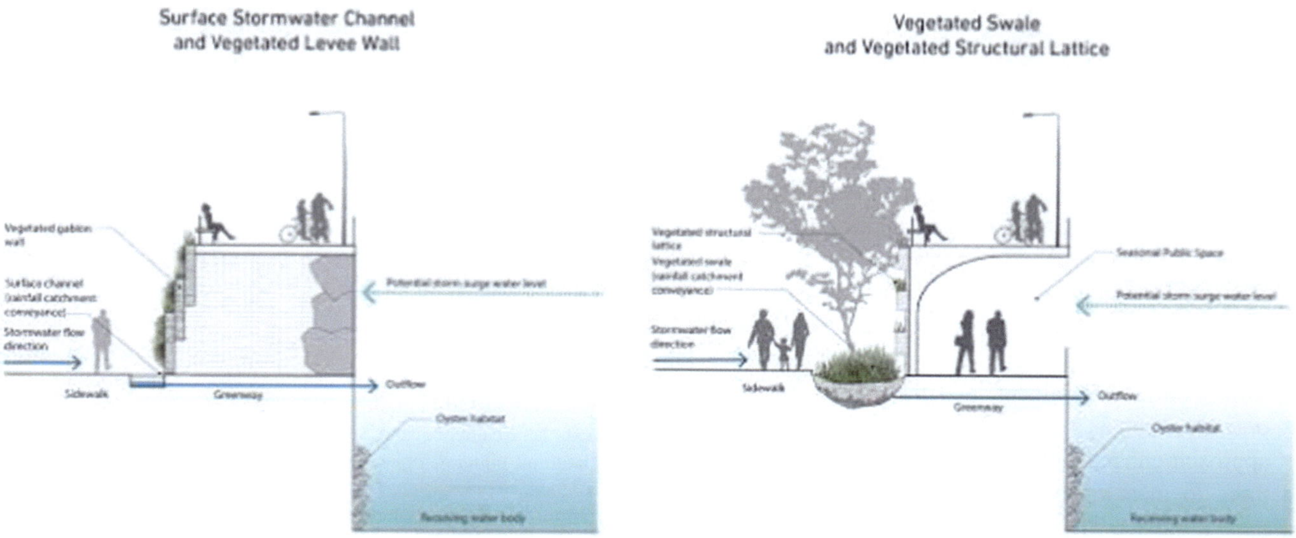

Fig. 4. Multi-purpose barrier typology concepts (SCAPE, 2017)

Fig. 5. Floating Houses at IJburg, source (Shannon, 2018)

Dutch government identified the need for better resilience strategies to prevent further tragedies. Hence, they adopted a new approach to Urban Development with a resilient design dedicated to continuous adoption in line with the changing climate.

One example of anticipatory design in the Netherlands is floating architecture in Fig. 5, as demonstrated by the floating houses in Amsterdam's IJburg neighborhood. However, it is essential to recognize that, despite its creative solutions, floating architecture has drawbacks and restrictions, especially regarding site-specific limitations. The public's perception of climate change and its risks varies more in the United States than in the Netherlands. More effective and coordinated efforts to increase coastal resilience in the United States may be made possible by a shared understanding of the gravity and urgency of climate change, like that in the Netherlands (Stochino et al., 2019). A nationwide adoption and application of more successful anticipatory design techniques may be facilitated by such a high level of awareness and agreement.

7 Conclusion and Findings:

This research has examined the critical need for resilient architectural design in the context of disaster management. Through an extensive review of disaster resilience concepts, adaptability, and sustainability, several key findings emerge:

- **Integration of Resilience in Architectural Design:** Modern architecture must move beyond aesthetic and functional concerns to embrace resilience as a core principle. Resilient designs should anticipate potential threats from natural and man-made disasters, incorporating redundancy, robustness, and adaptability to ensure that buildings not only survive but continue to function in adverse conditions.
- **Holistic Design Approach:** A comprehensive, system-based approach to design is crucial. This involves considering the building's lifecycle, its relationship to the surrounding environment, and the integration of green practices such as sustainability and adaptability. Such an approach enhances the longevity of structures, allowing them to meet evolving needs while reducing environmental impact. The argument made in this report encourages an architectural standard centered around prevention, advocating for integrated design concepts that not only prevent the severe effects caused by disasters but also restore processes for future resilience. From the planning stage onward, architects must anticipate issues like global warming and incorporate infrastructural features that enhance adaptability, sustainability, and overall resilience.
- **Proactive, Not Reactive, Design:** The trend is now shifting towards proactive architectural solutions that prevent severe damage rather than reacting after disasters occur. By focusing on disaster preparedness from the outset, architects can ensure that buildings are designed to withstand future challenges, including those brought on by climate change. Moving from a reactive to a proactive mindset will enable architects to play an integral role in identifying threats and building sustainable environments.
- **Case Studies in Resilience:** Examples such as the Brooklyn Waterfront Greenway and floating architecture in Amsterdam demonstrate how anticipatory design strategies can mitigate disaster impacts. These projects highlight the importance of integrating green infrastructure and sustainable materials to enhance resilience, particularly in vulnerable coastal areas.
- **Challenges in Current Architectural Practices:** The research identifies several limitations in current architectural practices, particularly the lack of interdisciplinary collaboration and the need for flexible regulations to accommodate evolving resilience standards. There is a significant gap between policy frameworks and the practical implementation of disaster-resilient designs.

In conclusion, architectural resilience must become an integral part of modern design. By adopting a proactive, holistic approach that considers the entire building lifecycle, architects can create structures that not only withstand disasters but also contribute to long-term sustainability, safety, and societal well-being. With advanced design elements, architects can craft optimal and resilient shelters that serve both current and future generations. These findings provide a valuable framework for future architectural practices, urging designers to prioritize resilience and adaptability to meet the growing global challenges.

References

Adekola, J., Clelland, D.: Two sides of the same coin: business resilience and community resilience. J. Contingencies Crisis Manage. (2019). https://doi.org/10.1111/1468-5973.12275

Ahmed, I.: Resilience and sustainability in architecture and urban planning: policies, practices, strategies and visions. In: Ahmed, I., Noguchi, M., O'Brien, D., Tucker, C., Vahanvati, M. (eds.) MDPI - Multidisciplinary Digital Publishing Institute (2023). www.mdpi.com.

Askar, R., Bragança, L., Gervásio, H.: Adaptability of buildings: a critical review on the concept evolution. Appl. Sci. **11**(10), 4483 (2021). https://doi.org/10.3390/app11104483

Bemthuis, R., Iacob, M.-E., Havinga, P.: A Design of the resilient enterprise: a reference architecture for emergent behaviors control. Sensors **20**(22), 6672 (2020)

Berger, C., Eichhammer, P., Reiser, H.P., Domaschka, J., Hauck, F.J., Habiger, G.: A survey on resilience in the IoT. ACM Comput. Surv. **54**(7), 1–39 (2022). https://doi.org/10.1145/3462513

Bernabei, L., Mochi, G., Bernardini, G., Quagliarini, E.: Seismic risk of open spaces in historic built environments: a matrix-based approach for emergency management and disaster response. Int. J. Disaster Risk Reduction **65**, 102552 (2021). https://doi.org/10.1016/j.ijdrr.2021.102552

Blackman, D., Nakanishi, H., Benson, A.M.: Disaster resilience as a complex problem: why linearity is not applicable for long-term recovery. Technol. Forecast. Soc. Chang. **121**, 89–98 (2017). https://doi.org/10.1016/j.techfore.2016.09.018

Brogden, L.: From product to process and site to system: Disaster resilience and humanitarian design in architecture education. Eprints.qut.edu.au (2020). https://eprints.qut.edu.au/202988

Campos, P.: Resilience, education, and architecture: The proactive and "educational" dimensions of the spaces of formation. Int. J. Disaster Risk Reduction **43**, 101391 (2020). https://doi.org/10.1016/j.ijdrr.2019.101391

Castaño-Rosa, R., et al.: Resilience in the built environment: key characteristics for solutions to multiple crises. Sustain. Cities Soc. **87**, 104259 (2022). https://doi.org/10.1016/j.scs.2022.104259

Charlesworth, E., Fien, J.: Design and disaster resilience: toward a role for design in disaster mitigation and recovery. Architecture **2**(2), 292–306 (2022). https://doi.org/10.3390/architecture2020017

Chen, D., Wang, G., Chen, G.: Lego architecture: research on a temporary building design method for a post-disaster emergency. Front. Archit. Res. (2021). https://doi.org/10.1016/j.foar.2021.08.001

Cutter, S.L., Derakhshan, S.: Implementing disaster policy: exploring scale and measurement schemes for disaster resilience. J. Homeland Secur. Emerg. Manage. **16**(3) (2019). https://doi.org/10.1515/jhsem-2018-0029

Elvas, L.B., Mataloto, B.M., Martins, A.L., Ferreira, J.C.: Disaster management in smart cities. Smart Cities **4**(2), 819–839 (2021). https://doi.org/10.3390/smartcities4020042

Fabbricatti, K., Boissenin, L., Citoni, M.: Heritage community resilience: towards new approaches for urban resilience and sustainability. City Territ. Archit. **7**(1) (2020). https://doi.org/10.1186/s40410-020-00126-7

Fallah-Aliabadi, S., Ostadtaghizadeh, A., Ardalan, A., Fatemi, F., Khazai, B., Mirjalili, M.R.: Towards developing a model for the evaluation of hospital disaster resilience: a systematic review. BMC Health Serv. Res. **20**(1) (2020). https://doi.org/10.1186/s12913-020-4915-2

Flores, I.: A value-based framework from building stock model to retrofit model. Environ. Sustain. Dev. **9**(1) (2024). https://doi.org/10.21625/essd.v9i1.1044

Gradzinski, P.: Resilience as a sustainable design process in the world climate change. IOP Conf. Ser. Mater. Sci. Eng. **471**, 092054 (2019). https://doi.org/10.1088/1757-899x/471/9/092054

Imperiale, A.J., Vanclay, F.: Conceptualizing community resilience and the social dimensions of risk to overcome barriers to disaster risk reduction and sustainable development. Sustain. Dev. (2021). https://doi.org/10.1002/sd.2182

Ismail, F.Z., Halog, A., Smith, C.: How sustainable is disaster resilience? Int. J. Disaster Resilience Built Environ. **8**(5), 555–572 (2017). https://doi.org/10.1108/ijdrbe-07-2016-0028

Jankovic, L.: Designing resilience of the built environment to extreme weather events. Sustainability **10**(2), 141 (2018). https://doi.org/10.3390/su10010141

Jia, X., Chowdhury, M., Prayag, G., Hossain Chowdhury, M.M.: The role of social capital on proactive and reactive resilience of organizations post-disaster. Int. J. Disaster Risk Reduction **48**, 101614 (2020). https://doi.org/10.1016/j.ijdrr.2020.101614

Jiang, Y., Ritchie, B.W., Verreynne, M.-L.: Developing disaster resilience: a processual and reflective approach. Tour. Manage. **87**, 104374 (2021)

Lu, X., et al.: Quantification of disaster resilience in civil engineering: a review. J. Safety Sci. Resilience **1**(1), 19–30 (2020). https://doi.org/10.1016/j.jnlssr.2020.06.008

Luke, J., Franklin, R.C., Dyson, J., Aitken, P.: Building toward a disaster resilient health system: a study of hospital resilience. Disaster Med. Public Health Prep. **17**, e219 (2021). https://doi.org/10.1017/dmp.2022.204

Ma, Z., Xiao, L., Yin, J.: Toward a dynamic model of organizational resilience. Nankai Bus. Rev. Int. **9**(3), 246–263 (2018). https://doi.org/10.1108/nbri-07-2017-0041

Malalgoda, C., Amaratunga, D., Haigh, R.: Challenges in creating a disaster resilient built environment. Procedia Econ. Fin. **18**, 736–744 (2014)

Mayar, K., Carmichael, D.G., Shen, X.: Resilience and systems—a building structure case Example. Buildings **13**(6), 1520 (2023). https://doi.org/10.3390/buildings13061520

Meng, M., Dabrowski, M., Stead, D.: Enhancing flood resilience and climate adaptation: the state of the art and new directions for spatial planning. Sustainability **12**(19), 7864 (2020). https://doi.org/10.3390/su12197864

Mergos, P.E.: Sustainable and resilient seismic design of reinforced concrete frames with rocking isolation on spread footings. Eng. Struct. **292**, 116605 (2023). https://doi.org/10.1016/j.engstruct.2023.116605

Nazif, S., Mohammadpour Khoie, M.M., Eslamian, S.: Urban disaster management and resilience, pp. 157–185 (2021). https://doi.org/10.1007/978-3-030-61278-8_7

Ner, N.T., Okyere, S.A., Abunyewah, M., Kita, M.: Integrating resilience attributes into local disaster management plans in Metro Manila: strengths, weaknesses, and gaps. Progress Disaster Sci. **16**, 100249 (2022). https://doi.org/10.1016/j.pdisas.2022.100249

Pribadi, K.S., et al.: Learning from past earthquake disasters: the need for a knowledge management system to enhance infrastructure resilience in Indonesia. Int. J. Disaster Risk Reduction **64**, 102424 (2021). https://doi.org/10.1016/j.ijdrr.2021.102424

Quintana, R.M., Fortman, J., DeVaney, J.: Advancing an approach of resilient design for learning by designing for extensibility, flexibility, and redundancy. Uen.pressbooks.pub (2021). https://uen.pressbooks.pub/resilientpedagogy/chapter/advancing-an-approach-of-resilient-design-for-learning-by-designing-for-extensibility-flexibility-and-redundancy/

Rak, J., Girão-Silva, R., Gomes, T., Ellinas, G., Kantarci, B., Tornatore, M.: Disaster resilience of optical networks: state of the art, challenges, and opportunities. Opt. Switch. Netw. **42**, 100619 (2021). https://doi.org/10.1016/j.osn.2021.100619

Ribeiro, P.J.G., Pena Jardim Gonçalves, L.A. Urban resilience: a conceptual framework. Sustain. Cities Soc. **50**, 101625 (2019). https://doi.org/10.1016/j.scs.2019.101625

Salvato, C., Sargiacomo, M., Amore, M.D., Minichilli, A.: Natural disasters as a source of entrepreneurial opportunity: Family business resilience after an earthquake. Strateg. Entrep. J. (2020). https://doi.org/10.1002/sej.1368

Sarker, M.N.I., Peng, Y., Yiran, C., Shouse, R.C.: Disaster resilience through big data: way to environmental sustainability. Int. J. Disaster Risk Reduction **51**, 101769 (2020)

Shannon, E.: Resilient architecture: adaptive community living in coastal locations. Masters Theses (2018). https://doi.org/10.7275/12036675

Stochino, F., Bedon, C., Sagaseta, J., Honfi, D.: Robustness and resilience of structures under extreme loads. Adv. Civ. Eng. (2019). https://www.hindawi.com/journals/ace/2019/4291703/

Straub, A.M., Gray, B.J., Ritchie, L.A., Gill, D.A.: Cultivating disaster resilience in rural Oklahoma: community disenfranchisement and relational aspects of social capital. J. Rural. Stud. **73**, 105–113 (2020). https://doi.org/10.1016/j.jrurstud.2019.12.010

Sun, W., Bocchini, P., Davison, B.D.: Applications of artificial intelligence for disaster management. Nat. Hazards **103**(3), 2631–2689 (2020). https://doi.org/10.1007/s11069-020-04124-3

Taherkhani, R., Hashempour, N., nknnnLotfi, M.: Sustainable-resilient urban revitalization framework: residential buildings renovation in a historic district. J. Clean. Prod. **286**, 124952 (2021). https://doi.org/10.1016/j.jclepro.2020.124952

SCAPE: SCAPE landscape architecture DPC (2017). https://www.scapestudio.com/news/2017/08/integrating-flood-protection-brooklyn-greenway-red-hook/

Sustainability in Campus Building Design from Micro to Macro Scale

Dilruba Ozbay and Sadiye Müjdem Vural

Abstract

This study investigates the criteria for sustainable campus actions and their tangible outcomes in university campuses, which are considered as a small-scale form of cities in terms of their environmental impacts. To contextualize these findings, the term 'sustainability' is employed as a holistic reference basis. The widely used sustainable campus criteria were analyzed together with the sample cases from the campus sustainable performance indicators. The cases were selected from different universities all around the world. As a result of the evaluations made with the help of the data, the meta-analysis table was created and suggestions for future studies were provided especially for promoting sustainable campus applications. On the other hand, negative factors in the process are that applicable indicators are not widely adopted by users in many campuses. These factors have revealed that indicators that require investment and infrastructure costs such as renewable energy production, wastewater recycling and rainwater harvesting are not very actively involved in daily life. The failure of campuses to achieve sustainability goals in terms of physical and systematically relationships lead to the architect's failure to determine the sustainable campus design indicators and to construct the whole. Therefore, it is understood that in the sustainable design of campuses, architects should clearly determine the campus indicators, and the entire design should be created based on the indicators.

Keywords

Value in Sustainability · Sustainable Campus Criteria · Sustainability Criteria and Indicators · Sustainable Campus Design · Sustainability in Architectural Design

1 Introduction

Campuses are considered the small-scale form of cities in terms of environmental impact and are one of the first places where environmental awareness is raised. In recent years, the concept and scope of sustainability have been extensively discussed by academics at universities and campuses. Furthermore, private and public support have been provided for research on this subject, and a public awareness has been tried to be created.

The terms used in the sustainability assessment on campuses have been eliminated to make them clearer and more comprehensible as well as their application areas. In this context, the selected sustainability criteria and the values that contribute to the formation of the factors are considered. Therefore, the study started primarily with defining the criteria in the context of sustainable campuses, evaluating the campus indicators in terms of sustainable campus performance, and analyzing the campuses selected as examples within the scope of this study and making a general 'sustainable campus design' analysis. The study attempts to explain how consistent all economic, social and environmental sustainability decisions and actions are within the concept of 'sustainability in the context of the campus'.

In the context stated above, campuses that adopt sustainable campus evaluation criteria achieve more consistent

D. Ozbay
Astaş and Yapı and Yapı Partnership, Mandarin Oriental Etiler, Mixed-use Project, İstanbul, Turkey

Faculty of Architecture, Eastern Mediterranean University, Famagusta, North Cyprus, Turkey

S. M. Vural (✉)
Department of Architecture, Faculty of Architecture, Eastern Mediterranean University, Famagusta, North Cyprus, Turkey
e-mail: mujdem.vural@emu.edu.tr

B. He et al. (eds.), *Towards the Framework of Livable and Resilient Cities* Advances in Science, Technology & Innovation, https://doi.org/10.1007/978-3-031-97849-4_10

Fig. 1. Princeton University, Scale of Action [2]

results in terms of sustainability. However, it is also understood that the boundaries and definitions of the values have not yet been clearly and adequately defined.

1.1 Sustainable Design Process for Universities

Nowadays, sustainability actions in the campus environment have become important as a starting point for large-scale studies for a greener future. (Fig. 1).

In recent years, many university initiatives have been planned to transform their campuses into 'sustainable' ones. For this transformation, they take many actions in this context. Research on various examples of this subject is developing day by day. In addition, universities have started to set their own sustainability goals, especially since 1990, due to the fact that more importance is given to communication, regulations and partnerships for a sustainable campus in higher education [1]. Recent research on academic studies and motivations also have shown that there are some similarities and differences in almost every university structure for the creation of sustainable campuses.

During recent years, universities have provided sociological awareness on sustainability in terms of their benefits to physical changes. It also focuses on United Nations' sustainable goals including 11 sustainable cities and communities, which provides an understanding of human-ecosystem needs and it supports the production of axioms for this. Goal 11 in the campus education system is also important in terms of its importance in the sustainable campus system and the quality of education. Quality of education results from providing the appropriate tools to monitor the entire sustainability process, from waste management to air quality to social awareness. Institutions that adopt them prepare their communities to manage their resources with the right procedure and to tackle global problems from a micro scale to a macro one. Thus, the participants, the users of the institution, are involved in various discussions and planning actions for the improvement of their own living spaces. Accordingly, Princeton University has aimed to accelerate sustainability action at all scales from personal to global in its 2026 Sustainability Action plan (Fig. 1) [2]. In the light of this process, educational institutions provide the participants with the opportunity to experience in their training, thus providing the development of awareness and sense of duty.

Environmental concerns and sustainability-oriented mission of educational buildings are of great importance for both the renovation of existing buildings and new buildings when considered in physical dimensions. An environmentally, socially and economically sustainable university environment should be provided by minimizing the negative impacts of sustainability activities on campuses. It should provide both university users and society with an instructive and encouraging basis for sustainable living [3].

In short, it would be correct to say that a sustainable university campus should focus on various observations, practical applications, and sustainable technologies [4]. Campus administrators with environmental responsibility should work with their stakeholders on potential environmental problems and their solutions. This will lead university administrators and other staff to an environmentalist approach in education and will ensure the creation of sustainable campuses.

Universities also cooperate because of their common concerns about research and the future to make their campuses sustainable. These collaborations contain various indicators under the name of various sustainability criteria [5].

When sustainability assessment and sustainability indicators are examined, it is seen that sustainability assessment deals with the "process" of the assessment, while sustainability indicators deal with the "technical" aspects of evaluation. All of these are complementary and mutually reinforcing. Their value in decision-making contributes to solving problems related to the difficulty of constructing knowledge as well as problems of interpretation and influencing [6].

Although there are several international efforts to measure sustainability, only a few of them have a holistic approach that considers environmental, economic and social aspects. In most cases, the focus is on one of the three directions. Although it can be said that they can complement each other, the issue of sustainability is more than simply bringing together important issues, it is also about their interconnectedness and dynamics developed within a system.

In order to make a comparative sustainability assessment for higher education institutions, a theoretical model should be created by analyzing the sustainability indicators used in the methodologies. First of all, it means that the criterion must be the object of a representation. It should be clear that the indicator makes it possible to accurately indicate what a particular criterion covers [6]. The sustainability assessment is interpreted under the sustainable indicators of the current situation (including social, cultural, economic, and other sub-criteria…).

Technical methods are applied in order to evaluate the current status of sustainability indicators according to accepted official or unofficial sustainable evaluation criteria and to put them in the current ranking (green ranking systems). Campus criteria, on the other hand, are items created by various organizations as a result of detailed analyzes by taking sustainability evaluation parameters into consideration.

It is seen that many sustainable campus criteria have different aspects as well as common aspects. This causes confusion about what indicators sustainable campuses should have and which criteria should be adopted.

As understood from the studies, generally accepted sustainable campus indicators have not yet been clarified. It is seen that the physical infrastructure on campuses is not adequately provided and the responsibilities of the architect are often ignored at the design stage. It is understood that the implementation of the adopted sustainability criteria is unstable and inadequate in sustainability actions. For these reasons, it is also understood that the process is handled more in terms of certification.

The lack of clear outline of a general 'sustainable campus design' demonstrates a lack of standardization, as well as the lack of 'sustainable campus design criteria', making measurements difficult. To find sustainable solutions, many factors are studied, and recent studies are examined in universities that meet their needs with sustainable campuses in their own cycles.

In this study, the current indicators of sustainability in the campus environment were primarily discussed. The campuses, whose sustainability has been certified with international certificates, were examined. The criteria they adopted and the campus performance indicators were compared. A general 'sustainable campus design' was outlined and 'building design criteria and techniques used in sustainable campus design' were determined.

1.2 Scope of Sustainable Campus Design

Making campuses sustainable makes their potential visible, increases participation, provides campus-based solutions to global problems in the fields of socio-economic-environmental sustainability, and encourages collaboration [7]. Incentives for sustainability should be based on educational strategies [8]. Therefore, sustainable campuses should have innovative steps in every application [3].

When all the 'sustainable campus indicators' of the selected sustainable campuses are examined, it is understood that there is no continuity in the actions.

With the indicators, environments with healthier physical conditions and atmosphere are created and the use of environmentally sensitive and nature-friendly materials is supported. In addition, the use of devices with high energy performance is encouraged and efforts are made to keep the usage costs low [9].

The scope of the sustainable campus for architects includes environment and infrastructure, weather and climate, energy, transportation, waste, water, supply, indoor environment and building indicators. Each sustainable campus indicator includes its own specific information and implementation steps.

In order for universities to have sustainable campuses, it is necessary to set some strict rules in campus sustainable practices and techniques in order to support architectural design with vision, mission and regulations [3].

Architects should use energy-efficient, renewable technologies in their designs and thus contribute to the reduction of greenhouse gas emissions. In summary, instead of fossil fuel-consuming equipment, techniques such as environmentally friendly sustainable equipment should be preferred in designs. In this way, carbon-based emissions from materials will be reduced and a carbon-neutral campus environment will be created.

1.3 Sustainable Campus Indicators

For the purpose of increasing the technical and operational capacity of energy efficiency in environmentally friendly universities, campus indicators need to be evaluated from an architectural design perspective. The evaluations are addressed under two action groups: existing campuses and new campuses.

In the present study, sample campuses were evaluated to determine which sustainable campus indicators covered the international sustainability criteria. The data obtained were used to identify campus sustainability indicators that include more understandable and comprehensive sustainable campus management and design policies that are acceptable to all. It was determined which of the indicators would be included in the architectural scope. Finally, the sustainability potential of the campuses was examined, the indicators that need to be addressed in terms of architecture were evaluated and suggestions were made (Table 1).

The aim of the study is to reveal the main indicators of sustainable campuses. In this study, sustainable campus indicators were categorized into 5 categories as a result of the research conducted.

1.3.1 Campus Physical Operations

To be more sustainable on campuses, sustainability policies applied to all areas and infrastructure on campus must be sustainable in their basic environment and infrastructure. Greenhouse gas and air pollutant emission sources should be identified and measured continuously. Efforts should be made to reduce air pollutants and create a better air and climate environment. Energy procedures should be made mandatory. Conservation measures, namely the transition to cleaner and

Table 1. Sustainable Campus Indicators with Sub-indicators [17]

Sustainable Campus Indicators				
Campus Physical Operations	Institutional Characteristics and Goals	Innovation & Design Process for Green Transformations	Sustainability Engagements for a Sustainable Society	Policy Governance and Administration
- Setting and Infrastructure -Air & Climate -Energy -Transportation -Waste -Water -Buildings -Purchasing -Food & Dining -Indoor Environmental Quality	-Institutional Boundary -Operational Characteristics -Demographic Structure -Institutional Characteristics -Administrative Infrastructure -Education	-Innovative Design Technologies -Effective Equipment Policy	-Continuing Education Center -Orientation Programs -Public Collaborations	-Sustainability Policy -Sustainability Action Plans -Administration

renewable energy sources, including solar, wind, geothermal, and low-impact hydroelectricity, are essential for the efficient protection and utilization of limited natural resources. Sustainable transportation systems and public transportation, including alternative usage options, should aim to reduce waste that pollutes air and water and generates greenhouse gas emissions. Techniques such as reducing water demand, water recovery and greywater management should be implemented. An educational framework should be developed to ensure that studies aimed at integrating sustainability into campus culture are aligned with the curriculum. Campus sustainable purchasing policies should be established to address issues such as purchasing all kinds of sustainable products and equipment, life cycle cost analyses associated with purchases, and the identification of sustainable suppliers and collaborative purchasing practices [10]. An indoor air quality system should be designed to cope with unhealthy and inefficient indoor air conditions resulting from building construction methods, design and materials used. A sustainable food system should be established institutionally to manage food supply, consumption and disposal methods. Additionally, buildings should include campus areas that provide healthy living spaces for users and incorporate designs that protect the environment.

1.3.2 Institutional Characteristics and Aims

While preparing the report for the institutional campuses, the institutional boundaries of the settlements should be determined primarily. All features related to the operational characteristics, demographic structure and academic framework of the campus should be clearly explained. An educational framework should be established to ensure that studies aimed at integrating sustainability into campus culture are aligned with the curriculum. The boundaries of the institution, the operational characteristics it deals with, the demographic structure of the academicians working at the institution and students who constitute the institution, the distinctive features of the institution and the management infrastructure represents the main indicators of the campus characteristics [11].

1.3.3 Innovation and Design Process for Green Transformations

Following the innovation processes with innovative sustainable design enables the quantitative performance improvements to be seen with the energy efficient equipment policies between the current structures and the designed structures [12].

1.3.4 Sustainability Engagements for a Sustainable Society

An environmentally friendly audience should be created in line with sustainability commitments. To maintain their sustainability commitments and further expand their sustainability, they should include studies for information exchange, organizations, online tools for raising individual and corporate awareness through the creation of social partnerships, international commitments, written and unwritten resources and award programs [13].

1.3.5 Policy Governance and Administration

Campuses are defined as microcosms of many environmental issues that concern the whole society [14]. Campuses serve as role models for society through their policies and strategies. When the scope of anthropogenic environmental impacts is understood to be global [15], campuses have become an effective bridge to sustainability.

Qualitative arrangements on issues such as sustainability policies, environmental management plans and systems, environmental audit, recruitment and staff development, ethical investment, local economic development, student access and equality will provide quantitative advantages in the long-term process [16].

2 Materials and Methods

University campuses are expected to be environmentally friendly, have healthier buildings and bring their users to a higher level of welfare. In line with this goal, sustainable campus and its indicators should be evaluated by some certificate programs and guidelines such as Green Metric, Green Unity Toolkit, IARU, LEED.

By this evaluation campuses are thoroughly investigated in the contexts of:

- Identification of Major Systems: Identification and classification of sustainable campuses and structures of different scales, determining basic conceptual approaches and institutional boundaries according to evaluation criteria, measuring the potential of current values at this boundary and defining participation [18].
- Identification of Campus Potentials: Establishing strategic targets and road maps as a result of evaluating the campuses in terms of sustainability [11].
- Creation of a Sustainable Campus Plan: Establishing university-specific goals and principles by designing the evaluation and roadmap [19].

Environmental concerns and sustainability-oriented mission of university campuses are of great importance for both the renovation of existing buildings and new buildings. Environmentally, socially, and economically sustainable universities, by minimizing the negative effects of all their activities, they should provide both university users and society with an instructive and encouraging basis for sustainable living [3].

A comprehensive and long-term process management is a requirement for integrating sustainability into university campuses. The sustainability of universities in UNEP is defined as "the ecological sensitivity, social and cultural justice and economic vitality of the events" [16]. As shown in the statement above, the process for sustainable university campuses requires the determination of performance indicators and the creation of evaluation criteria, and the need for important components such as infrastructural, administrative, and operational thinking.

The inability of many initiatives to be made under the name of sustainability is due to the uncertainties in the reporting and measurement stages of the processes. In addition, the lack of sufficient volunteers, who were the participants in the actions stage, also prevents the actions from becoming compulsory at the level of consciousness. Another problem encountered in the process is that the institutions do not have common goals in their goals and that their working cooperation is not at a sufficient level [20]. Therefore, it is of great importance to establish cooperation between institutions to reach an international scale. From an institutional point of view, it is necessary to aim at transforming the university into a sustainable community that contributes to social, economic and environmental benefits. Committed to institutionalizing best practices in sustainable operations and turning research and teaching into practice to pilot innovative solutions that can be replicated widely using campuses, Harvard's research and teaching mission trains and empowers students to become leaders who are able to use their knowledge to create a sustainable impact in service to the world while on campus (Fig. 2) [21]. A value in sustainability should be created on campuses and these values should be transformed into action [20]. All measures taken must be in line with the system and objectives. Also, a common sustainability ground should be established with concrete actions.

Fig. 2. Harvard University, the corporate road model to a more sustainable future [21].

To avoid being seen as 'doing' in their actions, campuses need to be given measurable value as a result of their actions [22]. The lack of a certain standardization in the reporting of the processes of the campuses is another ambiguity. At this point, integrated reporting processes have been developed. Integrated reporting, according to the definition of the International Integrated Reporting Council (IIRC), operates internationally and undertakes efforts to develop an integrated approach in corporate reporting. It is an integrated system that reports on the value that the business will create in a certain period. It also reports all the elements in the process of creating this value. The integrated report is a concise and understandable report that reveals the value it creates in the short, medium, and long term regarding the management strategy, corporate governance processes, performance and expectations of the stakeholders [23]. The integrated report is an explanatory source that covers both financial and non-financial information of the enterprise and contains information about environmental-social-management performance [24].

The Global Sustainability Standards Board (GSSB) requires integrated reporting processes to be included in their financial reports [25]. For instance, it has been mandatory for institutions to report on value creation in the integrated reporting process to ensure the establishment of standards in EU party countries. On the other hand, financial outputs have become mandatory in the climate-related activity disclosures in the US party countries.

In this context, many guidelines have been developed for sustainable university campus design strategies and practices in the world. Leadership in Energy and Environmental Design (LEED), UI Green Metric, International Alliance of Research Universities (IARU), Green Universities Toolkit V2.0, LEED for Schools V4 are the most important guidelines [26]. The focus of the guidelines is the creation, development and management of university campuses according to the sustainability plan [27]. In order to ensure sustainability on campuses, campus sustainability indicators must first be determined. With the establishment of campus standard criteria, sustainability assessments will be clearer and more understandable.

In this study, sustainability indicators on campuses were first categorized and current application performances on campuses were evaluated. The evaluation was made according to the evaluation reports received from universities and institutions for the criteria in which the indicators were included. Financial authorization, cooperation, reporting and action are essential in globalized action plans. For this reason, all campus applications, from the energy used in buildings to the food consumed in cafeterias, were included in the evaluation of basic performance indicators since considering only physical indicators within the scope of sustainability will prevent the study from providing complete and accurate results. Accordingly, the concept of sustainability in campuses has been addressed in social, environmental and economic contexts in order to understand the concept of 'sustainability' correctly and to examine it from a holistic perspective.

Analytical Hierarchy Process (AHP) methodology is applied to evaluate key performance indicators more concretely. In this way, it is thought that the proposed Key Performance Indicators (KPIs) may be more effective in improving the sustainable campus performance of universities.

Analytical Hierarchy Process (AHP) methodology:
Level 1- Sustainable Campus Performance Indicators with Sub-Indicators
Level 2- Campus Sustainability Criteria
Level 3- Sustainable Campus Action Plan

The objective of our study is to establish a set of criteria for evaluating the sustainability of campus designs. To this end, we first identified the key performance indicators that would be used to assess the sustainability of campus designs considering the information that had been gathered. In the subsequent phase of the study, we determined which of these indicators pertain to the involvement of architects in the design process.

The fact that Problems can never be solved by the way of thinking that first created them [28]. As Einstein quoted, "The problems we face cannot be solved at the level of thinking we have created", classical methods are inadequate to solve problems. This leads us to change our methods. We need to deal with the processes more strategically. The Analytical Hierarchy Process (AHP) focuses on strategic decision-making, prioritization and ranking of criteria with high overall impact to successfully implement sustainable development [4].

The Key Performance Indicators (KPIs) in table (Table 2) in this chapter were evaluated using the AHP methodology. The Analytical Hierarchy Process (AHP) evaluation method, which can evaluate qualitative and quantitative criteria together and allows for more effective and efficient decision-making, was used [29].

Within the context of the limitations of the study, the evaluations were considered from an architectural perspective. The evaluation outputs in the study were given as 0 and 1. The value of 1 was given for outputs that could be supported by concrete data. The value of 0 was given for outputs that could not be supported by concrete data. A summary meta-analysis table was created from the evaluation results of four criteria, 16 universities and 29 indicators. (Table 2).

3 Application Stages of the Sustainable Campus Criteria

Universities contribute to environmental, social and economic problems with their more innovative and sustainable mechanisms [30]. Campuses with different uses stand out as places where sustainable development can be applied.

Evaluating the process from a broader perspective is important, first of all, in terms of determining the method-scope-process stages of the system by working on themes addressed with an inclusive perspective and a holistic approach.

According to the author, the primary purpose of the Sustainable Design Guide is to meet the university's commitment to the protection of natural resources.

Considering this information, 'Sustainable design criteria' should be addressed in three sections:

Section I - Objectives of Sustainable Design Criteria: It is necessary to explain the university's vision, the principles

Table 2. Analysis results [17]

Evaluation of Campus Sustainability Indicators Through Sustainability Evaluation Criteria																													
CRITERIAS	UI GREEN METRIC						IARU									GREENING UNIVERSITY TOOLKIT V2.0								LEED V4					
Indicators / *Universities*	Setting and Infrastructure (SI)	Energy and Climate Change (EC)	Waste (WS)	Water (WR)	Transportation (TR)	Education (ED)	Sustainable campus organization	Campus wide operations	Buildings	Laboratories	Green purchasing	Transport	Communication	Employee and student engagement	Universities as the catalyst for a sustainable society	Defining Sustainability	Initiating Transformations	Indicators	Technologies for Transformation	Policy Governance and Administration	Resources for Change	Greening Your Uni Brochures	Global Exemplars	Sustainable Sites	Water Efficiency	Energy & Atmosphere	Materials & Resources	Indoor Environmental Quality	Innovation & Design Process
Massachusetts Institute of Technology	1	1	0	0	1	1																							
Harvard University	1	1	1	1	1	1																							
University of Winconsin Oskhosh	1	1	1	0	1	0																							
Princeton University	1	1	1	1	1	1																							
Australian National Universitiy							1	1	1	1	1	1	1	1	1														
ETH Zurich							0	1	0	0	1	0	1	1	1														
University of Oxford							1	1	1	1	1	1	1	1	1														
Yale University							1	1	1	1	1	1	1	1	1														
University of Nairobi, Kenya																1	1	1	0	0	1	1	0						
Macquarie University, Australia																1	1	1	1	1	1	1	1						
Tongji University, Shanghai, China																1	1	1	1	1	1	1	1						
Middle East Technical University, Turkey																1	1	1	1	0	1	0	0						
Florida Atlantic University																								1	1	1	1	1	1
University of Miami																								1	1	1	1	1	1
University of Oregon																								1	1	1	1	1	1
Florida International University																								1	1	1	1	1	1
	4	4	3	2	4	3	3	4	3	3	4	3	4	4	4	4	4	4	3	2	4	3	2	4	4	4	4	4	4

required for this vision and the role that future development can play in implementing these principles.

Section II - Technical Criteria: Funding should be provided for future development projects.

Section III - Process Criteria: Suggestions for the implementation of sustainable strategies should be presented within the university's design and construction processes and procedures. In sustainable design guidelines, social responsibility should also be considered in terms of architecture for the protection of limited natural resources such as water, forest, soil and human health, and a sustainable approach should be adopted in designs.

Consequently, in the light of the data obtained, 'Sustainable Design' guidelines should suggest future strategies that can minimize environmental impact in 5 main categories, according to the author. These categories are:

- Planning Sustainable Areas
- Conservation of Water
- Conservation of Materials and Resources
- Increasing Energy Efficiency
- Increasing Indoor Environmental Quality

4 Conclusion

To sum up, it can be said that one of the ways to achieve sustainability is to integrate sustainability into universities with a holistic approach and to have clear and understandable processes. Universities are expected to create campuses that are effective in sustainability in an environment where resources are limited, and conditions are variable. They are also expected to plan for all possible future social-cultural-economic risks. In addition, considering that technology is the locomotive of sustainability, universities should adopt innovative and holistic approaches in their management and actions.

In sustainability studies, universities cannot adopt a holistic approach in which social, economic, cultural and intersectional dimensions are stakeholders in all processes. The lack of a certain standardization in measurement and reporting causes the whole process not to be handled in an integrated and holistic manner integrated with institutional values. It reveals the ignoring or lack of financial empowerment in globalizing action plans. The inability of campuses to reach performances that will show sufficient value and the inability to include sufficient technology in the processes also pose problems. The social responsibilities of architects in sustainable campus design processes are not fully understood. Another reason is the failure to adopt user-oriented designs and theoretically designed processes and to achieve a successful result in efforts to create sustainable campuses. In many cases, due to the lack of a clear understanding of the scope of sustainability, more attention is paid to environmental effects, leading to a more superficial definition of psycho-social effects.

In this context, a 'sustainable approach in terms of social responsibility and architecture' should be included in the technical planning of universities for sustainable campuses with clearer, more systematic, more user and environment focused designs. In the social context, a process that evolves from voluntariness to obligation should be instilled in users. All of these should be adopted by the campus according to sustainability standards. Regarding this adoption, architects should first know their users and be aware of the current values they are responsible for in the environmental context.

Universities should plan, coordinate, monitor and measure their administrative responsibilities and all processes in the context of the institution. All these techniques will ensure that all units of the universities adopt sustainability in sustainable campus management systems and will create a value in sustainability. Since a single scale is needed for the future on a global scale, joint collaborations should also be considered and realized.

Achieving sustainable campuses will only be possible with a campus environment that shapes user habits by ensuring that systems work together as a whole, are measurable, create value and participate. At this point, the active and passive duties of architects and the environmental-social-economic responsibilities of institutions should be to analyze sustainable campus indicators and ensure that users and active participants are guided in line with their sustainable goals. Sustainable campus success can only lead to change if all processes are handled in an integrated manner by architects and institutions, if the actions are value-oriented and enable universities to be more effective in the sustainability process.

One of the conclusions of this study is that the sustainability indicators in universities and the international green building certification criteria that evaluate these indicators are weak in the implementation phase. There is also another conclusion which states that the problems in local campus areas are not addressed and the sustainable areas on the campus are both quantitatively and qualitatively insufficient. According to the conclusions, campuses fail to achieve sustainability goals in terms of physical and system relationships. Architects fail to address sustainable campus design indicators with a sustainable design approach. Actions do not cover all sustainable campus processes. These situations cause the scope of the whole not to be fully and clearly determined. Based on how we can contribute to this issue as architects, it is aimed to find out how we can provide sustainable campus concepts in campus redesigns or improvements by simplifying them and using human behavioral orientations. Therefore, it becomes clear that in the sustainable design of campuses, architects should clearly determine campus sustainability indicators, base the entire design approach on these indicators and create a value or values in terms of the applicability and continuity of understandable actions.

Finally, based on all these analytical evaluations, this study has an architectural perspective for future studies and it tries to determine the framework of a 'sustainable campus' that will be beneficial both for now and for the future by approaching more sustainable designs and adopting sustainable usage habits.

References

1. Tilbury, D.: Environmental education for sustainability: defining the new focus of environmental education in the 1990s. Environ. Educ. Res. **1**(2), 195–212 (1995). https://doi.org/10.1080/135046295001 0206
2. Princeton University. Sustainability Action Plan: Toward 2026 and Beyond (2019). https://cpuc.princeton.edu/sites/g/files/toruqf1221/files/media/sap_ppt_scalable_actions.pdf
3. Velaquez, L., Munguia, N., Platt, A., Taddei, J.: Sustainable university: what can be the matter. J. Clean. Prod. **14**(9–11), 810–819 (2006)
4. Szabo, Z.K., Szádoczki, Z., Bozóki, S., Stanciulescu, G.C., Szabo, D.: An analytic hierarchy process approach for prioritisation of strategic objectives of sustainable development. Sustainability **13**(4), 2254 (2021). https://doi.org/10.3390/su13042254
5. Da Silva, J., Fernandes, V., Limont, M., Rauen, W.B.: Sustainable development assessment from a capitals perspective: analytical structure and indicator selection criteria. J. Environ. Manage. **260**, 110147 (2020). https://doi.org/10.1016/j.jenvman.2020.110147
6. Duque-Uribe, V., Sarache, W., Gutiérrez, E.V.: Sustainable supply chain management practices and sustainable performance in hospitals: a systematic review and integrative framework. Sustainability **11**(21), 5949 (2019). https://doi.org/10.3390/su11215949
7. Leal Filho, W.: Education for sustainable development in higher education: reviewing needs. In: Leal Filho, W. (eds.) Transformative Approaches to Sustainable Development at Universities. World Sustainability Series, pp. 3–12. Springer, Cham (2015).: https://doi.org/10.1007/978-3-319-08837-2_1
8. Luo, H., Yang, J.: Overcoming organisational resistance to sustainability innovations in Australian univerisites. In: Proceedings of the 12th Annual Australasian Campuses Towards Sustainability (ACTS) Conference, pp.1–10. Australasian Campuses Towards Sustainability (ACTS) (2012)
9. Prenger, J.L.: Identifying LEED and Sustainable design processes in the State University System in Florida (2010). http://purl.flvc.org/fsu/fd/FSU_migr_etd-0445
10. Amaral, A.R., Rodrigues, E., Gaspar, A.R., Gomes, Á.: A review of empirical data of sustainability initiatives in university campus operations. J. Clean. Prod. **250**, 119558 (2020). https://doi.org/10.1016/j.jclepro.2019.119558
11. Dawodu, A., et al.: Campus sustainability research: Indicators and dimensions to consider for the design and assessment of a sustainable campus. Heliyon **8**(12) (2022). https://doi.org/10.1016/j.heliyon.2022.e11864
12. Feng, H., Wang, F., Song, G., Liu, L.: Digital transformation on enterprise green innovation: effect and transmission mechanism. Int. J. Environ. Res. Public Health **19**(17), 10614 (2022). https://doi.org/10.3390/ijerph191710614
13. Johnsen, H., Torjesen, S., Ennals, R.: Higher Education In A Sustainable Society: A Case for Mutual Competence Building. Springer, London (2015). https://doi.org/10.1007/978-3-319-159 19-5
14. Dahle, M., Neumayer, E.: Overcoming barriers to campus greening: a survey among higher educational institutions in London, UK. Int. J. Sustain. High. Educ. **2**(2), 139–160 (2001). https://doi.org/10.1108/14676370110388363
15. Vitousek, P.M., Mooney, H.A., Lubchenco, J., Melillo, J.M.: Human domination of earth's ecosystems. Science **277**(5325), 494–499 (1997). https://doi.org/10.1126/science.277.5325.494
16. UNEP. Greening universities toolkit V2.0: transforming universities into green campuses (2014). https://www.unep.org/resources/toolkits-manuals-and-guides/greening-universities-toolkit-v20
17. Özbay, D.: Examining sustainability on university campuses: case study of EMU [Master thesis, EMU]. EMU Institute of Graduate Studies and Research (2022)
18. Eby, R.F.: Campus sustainability plans: a descriptive analysis of sustainability plans from institutions of higher education in the united states [Master thesis, Texas State University] (2020). https://digital.library.txst.edu/server/api/core/bitstreams/652cf2bb-7b7f-44bb-be15-2ae9d7929dfc/content
19. Sugiarto, A., Lee, C.W., Huruta, A.D.: A systematic review of the sustainable campus concept. Behav. Sci. **12**(5), 130 (2022). https://doi.org/10.3390/bs12050130
20. Zhu, B., Zhu, C., Dewancker, B.: A study of development mode in green campus to realize the sustainable development goals. Int. J. Sustain. High. Educ. **21**(4), 799–818 (2020). https://doi.org/10.1108/IJSHE-01-2020-0021
21. Harvard University. Harvard University Sustainability Plan Fiscal Year 2015–2020 (2015). https://zielonecollegium.cm-uj.krakow.pl/cm/uploads/2022/02/Harvard-Sustainability-Plan-Web.pdf
22. Aras, G., Crowther, D.: Corporate governance and corporate social responsibility in context. In: Aras, G., Crowther, D. (eds.) Global Perspectives on Corporate Governance and CSR, pp. 1–42. Gower (2010). ISBN: 978-0-566-08830-8
23. Soyka, P.A.: The international integrated reporting council (IIRC) integrated reporting framework: toward better sustainability reporting and (way) beyond. Environ. Qual. Manage. **23**(2) (2013)
24. Eccles, R.G., Saltzman, D.: Achieving sustainability through integrated reporting. Stanf. Soc. Innov. Rev. **9**(3), 56–61 (2011)
25. Luque-Vílchez, M., Cordazzo, M., Rimmel, G., Tilt, C.A.: Key aspects of sustainability reporting quality and the future of GRI. Sustain. Account. Manage. Policy J. **14**(4), 637–659 (2023). https://doi.org/10.1108/SAMPJ-03-2023-0127
26. Ragazzi, M., Ghidini, F.: Environmental sustainability of universities: critical analysis of a green ranking. Energy Procedia **119**, 111–120 (2017). https://doi.org/10.1016/j.egypro.2017.07.054
27. de A. Romero, M., Kronka Mülfarth, R.C.: University of São Paulo: sustainability masterplan for policies, plans, goals and actions. In: Leal Filho, W., Azeiteiro, U.M., Alves, F., Molthan-Hill, P. (eds.) Handbook of Theory and Practice of Sustainable Development in Higher Education. WSS, pp. 509–530. Springer, Cham (2017). https://doi.org/10.1007/978-3-319-47877-7_34
28. Reynold, M., Holwell, S.: Systems Approaches to Managing Change: A Practical Guide. Springer (2010)
29. Nebati, E., Sağanda, G.N., Erol, H., Subaşı, S.R., Göz, T.E.: Evaluation of employee performance by the analytical hierarchy process (AHP) method. Nigde Omer Halisdemir Univ. J. Eng. Sci. **10**(2), 582–590 (2021). https://doi.org/10.28948/ngumuh.790551 (in Turkish)
30. Diesendorf, M.: Sustainability and sustainable development. In: Dunphy, D., Benveniste, J., Grifiths, A., Sutton, P. (eds.) Sustainability: The corporate challenge of the 21st century, pp.19–37. Allen & Unwin (2000)

Renewable Energy and Sustainable Building Technologies

Building Integrated Photo Voltaic Panels for Achieving Net Zero Energy in Tropical Climatic Conditions

S. B. Karuppannan, B. N. Rao, and K. S. Reddy

Abstract

Globally, the need for power to sustain the buildings of future cities is a critical aspect, and finding solutions for the further way forward is a matter of immediate concern for all of us. If the primary concern is about the structural stability of the building, then the subsequent concern is about its energy usage. Net Zero Energy will be the fundamental approach of future buildings in our cities. To achieve Net Zero Energy, solar power is employed which is available perennially, depending upon the latitude and longitude of the location of the building in any city. Building an Integrated Photovoltaic system aids in achieving the Net Zero Energy condition for buildings. A pilot project to this extent was launched to understand the scope of Building Integrated Photovoltaic. Systematic approach and analysis were also carried out to validate the possibility of Building Integrated Photovoltaic technology in buildings to achieve Net Zero Energy Condition. The PV panels are installed vertically oriented towards the cardinal sides in Chennai city, the capital of the southernmost Tamil Nadu state in India. The thumb rule of tilting the solar panels to 13 degrees facing south and its corresponding solar insolation are also recorded on an hourly basis and daily starting from December 22, 2023. The duration of this study will go for one full solar cycle starting from the winter solstice to the summer solstice and back to the winter solstice. The comparative study will open the doors for Building Integrated Photovoltaic usage in buildings and the best possible orientation corresponding to the location of the building, which will give more insight into how to achieve NZE conditions. Analysis and further inferences are derived from the study.

Keywords

Building Integrated PhotoVoltaic · Net Zero Energy Building · Renewable Energy · Vertical PhotoVoltaic panels · Tropical Climate Insolation · Horizontal PhotoVoltaic Panels

S. B. Karuppannan (✉)
Interdisciplinary, Indian Institute of Technology Madras, Chennai, India
e-mail: ce17d301.suresh@gmail.com; ce17d301@smail.iitm.ac.in

B. N. Rao
Department of Civil Engineering, Indian Institute of Technology Madras, Chennai, India
e-mail: bnrao@iitm.ac.in

K. S. Reddy
Department of Mechanical Engineering, Indian Institute of Technology Madras, Chennai, India
e-mail: ksreddy@iitm.ac.in

1 Introduction

Achieving energy conservation is a huge collective responsibility that lies ahead. This paper explores how building integrated photo voltaic panels can significantly help in achieving the net-zero energy condition for buildings so that the carbon footprint on the energy consumption associated with the energy consumption during the lifespan of the buildings can be reduced to a large extent. Though this effort may appear to be negligible on a massive scale, it is essential that the assumptions based on the past have to be revisited to shape a sustainable future. One such assumption is to keep the solar panels at a tilted angle on the rooftop for receiving solar insolation that will augment power generation. This popular understanding is revisited and various alternatives are explored in this study. Further, the research also considers integrating photovoltaic panels onto the building walls quantifying the solar insolation The rooftop panels can also be integrated into the walls and the solar insolation received will be quantified. In addition, other approaches to achieving the

B. He et al. (eds.), *Towards the Framework of Livable and Resilient Cities* Advances in Science, Technology & Innovation, https://doi.org/10.1007/978-3-031-97849-4_11

net-zero condition for the building are deliberated (Liu et al., 2025).

2 The Need for Net Zero Buildings in Cities

Heat islands in urban areas pose a major issue marring the landscape of the cities. To mitigate this issue, architects and building planners should prioritize innovative building designs. In the concrete jungle of modern cities, urban forests play a crucial role in combating the issue (Bureau of Energy Efficiency, 2017) One factor that contributes to heat islands is solar irradiation which is absorbed and re-reflected by the buildings. Reducing this reflected solar radiation presents a challenging approach.

In the current scenario of concrete jungles, integrating urban forests could help in the reduction of reflected solar irradiation. The increase in the ambient temperature as the solar photons carry varies as per the latitude of the place which may not be controlled but the conversion of the solar irradiation into usable energy power can be a great possibility.

Since the building consumes energy for its daily functions, power generated through the panels integrated into the walls can further reduce the power load on the building. There is enough possibility of achieving net zero energy conditions through vertical PV paneling depending on the contextual location of the building (Jaysawal et al., 2022).

3 Typology of Buildings Vis-à-Vis Energy Consumption

Energy consumption patterns in buildings are largely influenced by different typologies, with HVAC systems and lighting being the main contributors. HVAC systems typically account for 40–60% of a building's energy use, while lighting contributes about 20–30%. In areas such as dark corridors or interior spaces where natural light is scarce, electric lighting becomes necessary. However, using electricity for lighting when the ambient light outside exceeds 10,000 lumens is highly inefficient (Davis Lynn, 2016).

In tropical regions air conditioning is essential. Hence, employing innovative strategies to reduce energy consumption for HVAC and lighting will lead to substantial advantages. Such strategies will not only cut down the reliance on fossil fuels but will also help to mitigate the carbon footprint and urban heat island effects. By optimizing energy usage, buildings can achieve greater sustainability and enhanced efficiency (Parvin et al., 2025).

4 Solar Insolation at Chennai

The area of this research is Chennai, which lies at 13° North Latitude and 80° East Longitude. A typical warm humid climatic condition with a lot of sea and land breezes keeps the place alive during diurnal transition periods. But otherwise, the city gets warmer every year with torrential rains inundating the city. These climatic changes increase energy consumption levels, which pose a global challenge. To address these issues at a local level and at international levels, the study investigates the use of Building Integrated Photovoltaic (BIPV) systems on building facades. But to address the issue at a local level, which can later be replicated at the global level is the usage of BIPV panel systems on the facades of the buildings and to verify if these panels can yield to Net Zero Energy conditions for the buildings in usage.

It is imperative to understand the sun's movement vis-à-vis the location, Chennai.

> Chennai, Capital city of Tamil Nadu State, India @ Latitude: 13.06 and Longitude: 80.23.

The important factor that contributes to the variation in solar insolation in any place is the movement of the Sun between the Summer Solstice and the Winter Solstice. The cycle of one whole year between reaching the Southern Hemisphere Trophic of Capricorn on December 22 at 23.45° South to move towards North to reach 23.45° N, Tropic of Cancer on June 21 every year, creates various sun altitude angles, declination angles, Azimuth angles are shown in the Fig. 1a, Fig. 1b (REUK.co.uk, 2024) and Fig. 3, (Koenigsberger et al., 1975) for better understanding. Sun takes 3 days, 21 h 3 min, and 49 s to move from one latitude to another when it is traversing between solstices (SB Karuppannan et al., 2024).

The potential of solar insolation in Chennai is, released by Indian Meteorological Department (IMD). The solar insolation in TMY at STC (Standard Testing Conditions) is 584.3W/m^2. The average sunshine hour is 8.52 h in the month of January in a TMY (Typical Meteorological Year). The solar insolation throughout the year is shown in Fig. 2 (India Meteorological Department, 2009).

Various factors that contribute to solar irradiation are atmospheric dust, albedo – the reflection from the earth's surfaces, wind movement, and ambient temperature. Chennai has a tropical climate that is warm humid climate, the wind movements create heavy downpours due to oceanic depressions. Hence the value obtained from the calculation is beyond the atmospheric level where the albedo is not present and the wind movement is more associated with the westerlies that are closer to the earth with up to 70–90% humidity as a major influencing factor (SB Karuppannan et al., 2024).

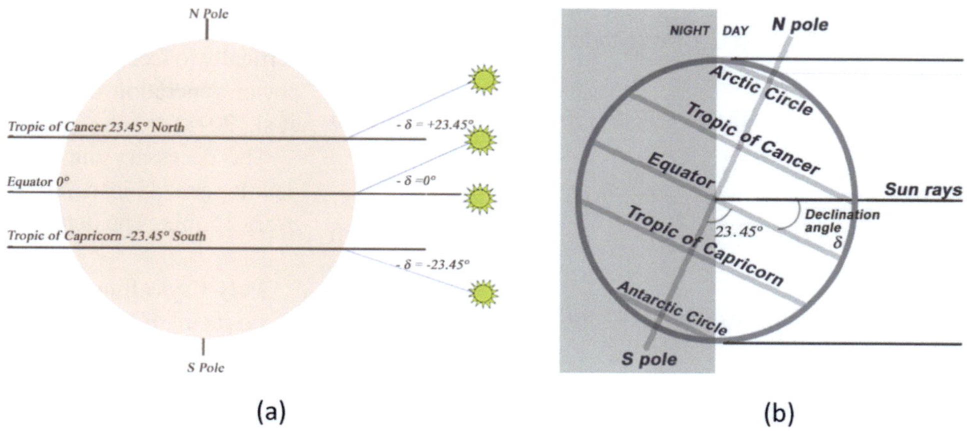

Fig. 1. (a) Sun angles (REUK, n.d.) (b) Sun angles (REUK, n.d.)

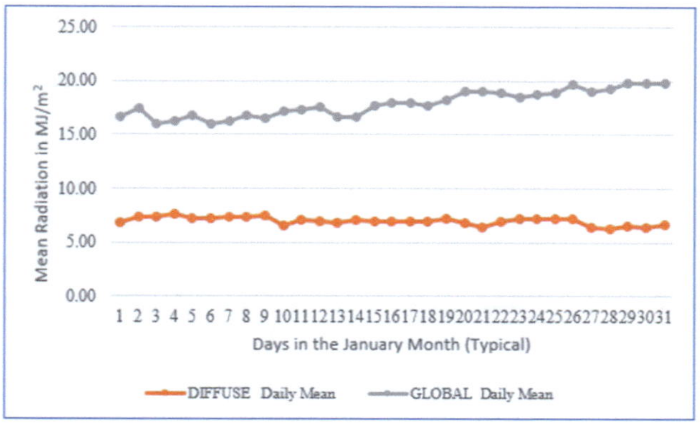

Fig. 2. Indian Meteorological Department Data on Global and Diffuse Radiation in Chennai. *Source: (India Meteorological Department., 2009).*

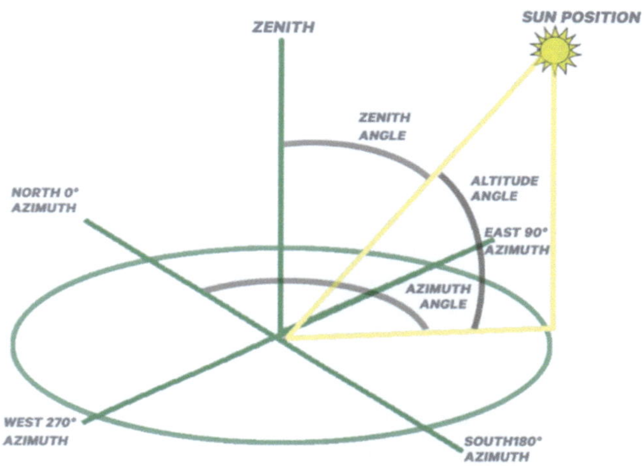

Fig. 3. Solar Angles (Koenigsberger et al., 1975)

Other factors that influence the solar insolation are the solar angles. The incident solar radiation in any particular longitude and latitude plays a crucial role in solar insolation. The influence of various other factors like the ambient temperature, albedo, atmospheric conditions like relative humidity, wind movement, dust particles in the atmospheric air, influencing shades from the environment and various system factors that reduce the power outputs like DC to AC losses, inverter losses, type of PV panel, the efficiency of the PV panel. Considering all these impacts, the factor with which we could calculate the solar insolation are with the help of parameters like the azimuth, altitude, and solar hour angles of the sun (Mamun et al., 2022).

5 The Solar Insolation Calculation

Solar insolation varies with the variation of the altitude of the sun from morning to evening. Further, the day of the year and the position of the sun concerning its Azimuth angle and Altitude angle as shown in Fig. 3 are calculated to arrive at the solar insolation values.

Altitude of the sun on any particular day of the year is calculated by first calculating the Sun's mean longitude angle. The N value used is the particular day of the year starting with Jan 1 as value 1 and December 31 as value 365 (for non-leap year), is presented in Eq. (1). Further Expression of time (EoT) is calculated with the formula (2) and the solar declination angle by formula (3). Local Solar Time (LST) is calculated with the Universal Co-ordinated Time (UTC) by formula (4). With LST, the Hour Angle (H) is calculated (5). With the values of the declination angle and the hour angle, the Sine of the altitude - θ (6), Azimuth angle - α (7), for a particular time as accounted by the LST and for a particular latitude, is calculated. The solar insolation with the values of Altitude and Azimuth, is further calculated (8).

$$B = (360/365) * (N - 81) \tag{1}$$

$$EoT = 9.87 * \sin(2B) - 1.5 * \sin(B) \tag{2}$$

$$\delta = 23.45 * \sin((360/365) * (N - 81)) \tag{3}$$

$$LST = (UTC + (longitude/15) + EoT \quad (4)$$

$$H = (LST - 12) * (360/24) \quad (5)$$

$$\sin(\theta) = ((\sin(\phi)) * (\sin(\delta))) + (((\cos(\phi)) * (\cos(\delta)) * (\cos(H))) \quad (6)$$

$$\alpha = \arcsin((\sin(\theta))/(\cos(latitude))) \quad (7)$$

$$I = S * \cos(\theta) * \cos(\alpha) \quad (8)$$

For 15th February 2024, the solar insolation is calculated based on the above formula. As per the calculation, the derived solar insolation is 1338.76 Wm^2. However, an inference of finding the solar constant at 1361 Wm^2 and the derived value of 1338.76 W/m^2 is to be drawn. A presumption that the above calculation is at above atmospheric conditions that are devoid of the Albedo and atmospheric conditions, as there are no values for albedo is calculated within the calculation. (SB Karuppannan et al., 2024)

Hence A direct value for finding the solar insolation and its power generation through PV panels will resolve for more possibilities. The calculation of this derivative method will call for various other factors and determinants to decide upon the final solar insolation. Hence an experimental setup is established to calculate the solar power generation for further analysis and inferences.

6 The Experimental Setup

4 types of panels were installed facing the cardinal directions. The specifications for the panels are not revealed as the objective of this study is to evaluate the generation of solar power to achieve a Net Zero Energy. The panels were placed vertically to face the cardinal directions to assess if the overall power generation could attain NZE level (Diez-Mediavilla et al., 2018).

The necessary micro inverter, with circuit breakers and data loggers, is installed to fetch the solar power generation every 15 min. The details of the panels are as follows.

Poly-Crystalline DCR Solar Panels 335 W – 1panel kept on facing EAST
Poly-Crystalline Non-DCR Solar Panels 335 W – 1panel - kept on facing NORTH
Mono-Crystalline Non-DCR Solar Panels 550 W – 1panel - kept on facing WEST
Mono-Crystalline Non-DCR Solar Panels 535 W – 1panel - kept on facing SOUTH
Mono-Crystalline-545 W – 1 panel – kept @ 13° inclination facing SOUTH

The installation of the 4 PV panels in cardinal directions was completed and the power generation data has been recorded from 22nd Dec 2023. The plan to fetch the data for one whole solar year is being planned. However, for establishing the logical standpoint of achieving the Net Zero Energy condition in a building, the data has been analyzed within a specific interval period. Each of the panels kept a cardinal East, West, North, and South direction data were collected for 15 days between the 1st of February to the 15th of February 2024. The calculation is to check for a Net Zero condition of a 1BHK house fabricated for the experimental purpose. The details of the PV panel power generation are shown in Fig. 4, Fig. 5, Fig. 6 and Fig. 7 (SB Karuppannan et al., 2024).

The 4 panels facing the cardinal East, West, North & South are supported by a system of DC side isolation switch, Inverter, AC side isolation switch, Power generation meter, and main fuse box to the AC main supply. The unused excess

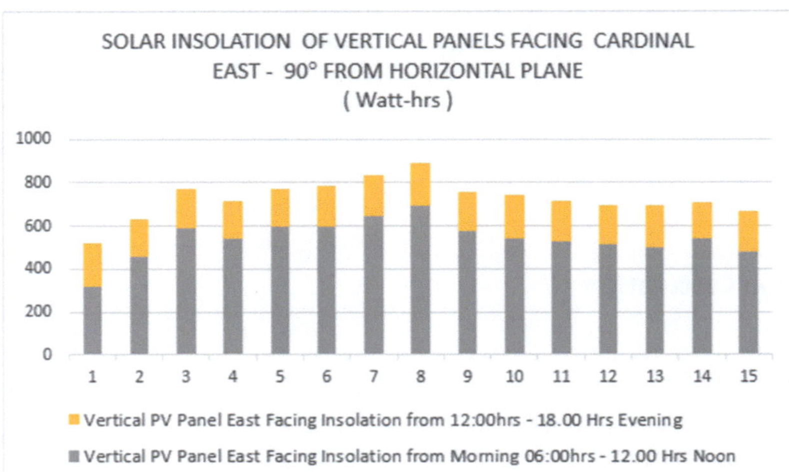

Fig. 4. Poly-Crystalline DCR Solar Panels 335 Watts. Data from the experimental setup *(SB Karuppannan et al.,* 2024*)*

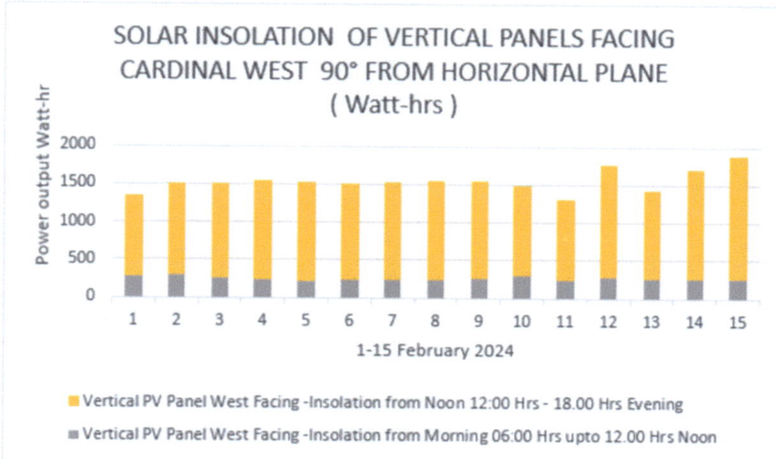

Fig. 5. Mono-Crystalline Non-DCR Solar Panels 550 Watts. Data from the experimental setup *(SB Karuppannan et al.,* 2024*)*

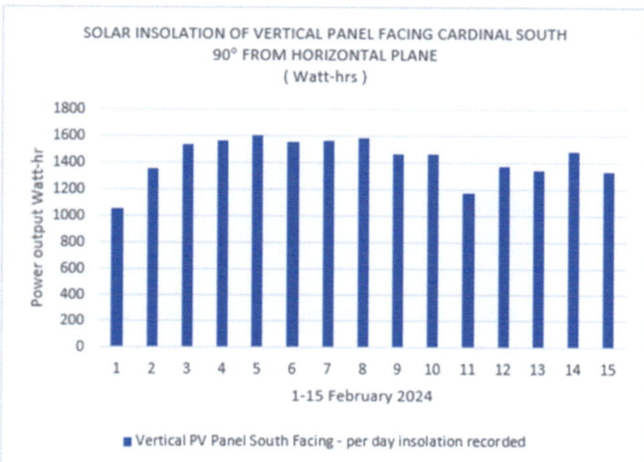

Fig. 6. Mono-Crystalline Non-DCR Solar Panels 535 Watts. Data from the experimental setup *(SB Karuppannan et al.,* 2024*)*

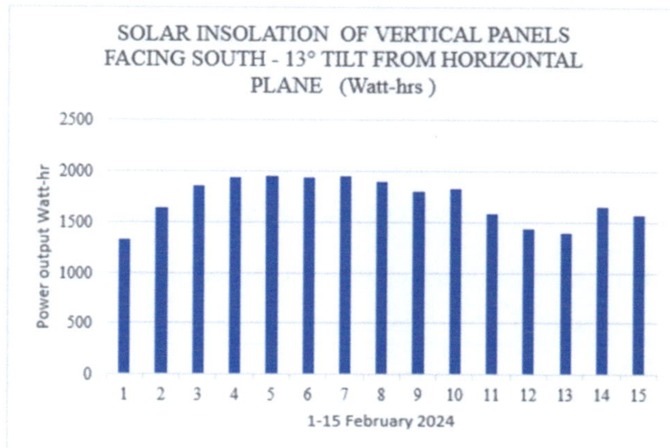

Fig. 8. Mono-Crystalline Non-DCR Solar Panels 550 Watts. Data from the experimental setup (SB Karuppannan et al., 2024)

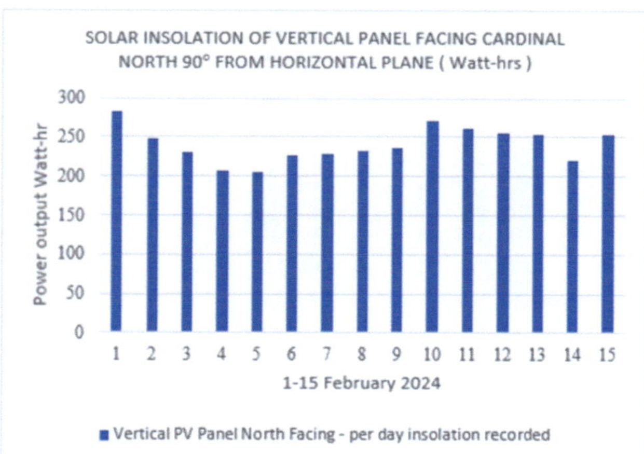

Fig. 7. Poly-Crystalline Non-DCR Solar Panels 335 Watts. Data from the experimental setup *(SB Karuppannan et al.,* 2024*)*

is supplied to the grid to be drawn later. This is a typical grid-connected system used for this experimental setup.

The vertical placement of the PV panel is a new attempt to establish the possibility of achieving Net Zero Energy conditions. However, the existing PV panels in Chennai are kept at 13° tilt facing the South direction.

The 13° tilt (facing South), is a thumb rule of tilt as per the latitude of the place. The power generation of the 13° tilt (facing South) PV panel is shown in Fig. 8.

For consideration, the total power generation of all 4 panels kept vertically facing cardinal directions, and the top 13 tilt (facing South) panel is shown in Fig. 9. The tabular column shows the actual values of the power generation from these panels.

From the total power generated between 1–15 February 2024 the average power generation per square meter of the panel area, per day, is as shown in the Table 1 (SB Karuppannan et al., 2024).

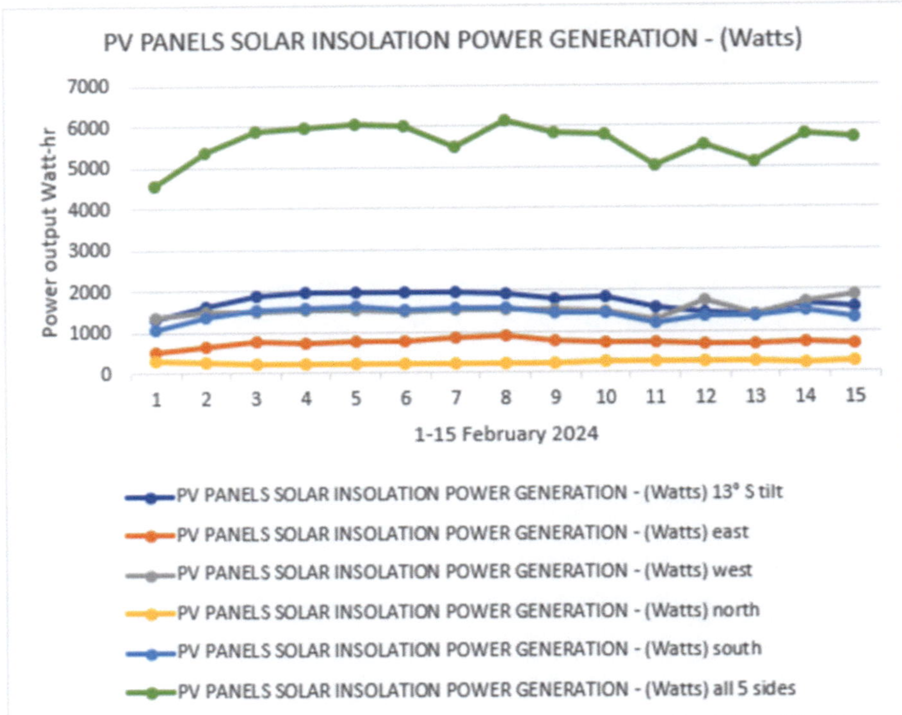

Fig. 9. All PV Panel's power output. Data from the experimental setup *(SB Karuppannan et al.,* 2024*)*

Table 1. Power Generation of PV panels (Karuppannan et al., 2024a)

Date (FEB 2024)	13° S tilt	east	west	north	south	all 5 sides
TOTAL	25872.48	10909	23107.5	3612	21478	84350.23
Watts/m^2/day	958.24*	404.04*	855.83*	133.78*	795.48*	3124.08*

* -Average value of 1–15 days of February power generation

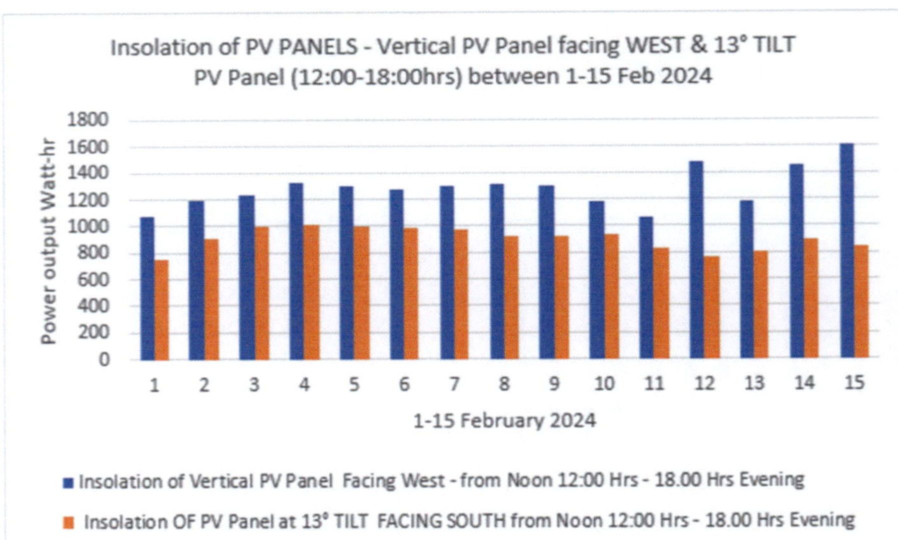

Fig. 10. Mono-Crystalline Non-DCR Solar Panels 550 Watts vs Mono-Crystalline Non-DCR Solar Panels 550 Watts. Data from the experimental setup (SB Karuppannan et al., 2024).

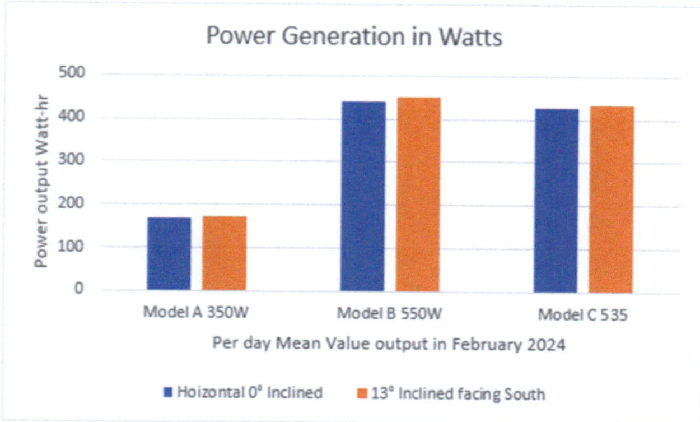

Fig. 11. Comparative Study of PV panels between 0° & 13° tilt. Data sourced from the experimental setup (SB Karuppannan et al., 2024) and PVWatts (National Renewable Energy Laboratory, 2024).

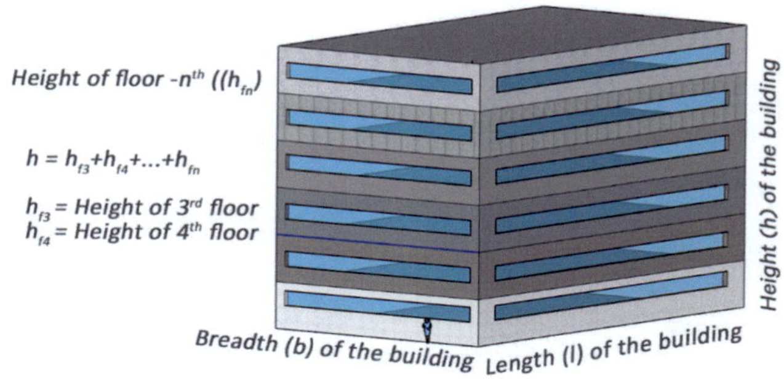

Fig. 12. Typical building block view SB Karuppannan et al., 2024)

Table 2. Calculation of reduction in %age of Power generation between 0° and 13° Inclined PV panels (Karuppannan et al., 2024a)

	Power Generation in Watts		% Age Less in Power
	0° TILT	13° TILT	0° TILT
Model A 350W	168	171	98.25
Model B 550W	442	452	97.79
Model C 535W	426	435	97.93
	Average reduction % of Power generation		97.99

An interesting finding from the study is as follows: For a PV panel facing Cardinal West, the solar insolation between 12:00 h to 18:00 h is more than the solar insolation power conversion of the rooftop 13° tilt facing south panel in the same duration. Figure 10.

A Comparative study of values of simulation for horizontal 0° tilt paneling vis a vis 13° tilt (facing South) with simulation studies using PVWatts application by NREL USA, is shown in Fig. 11 (NREL, 2024).

Table 2 (Karuppannan et al., 2024a) shows the average power generation by 0° tilt PV panel when compared with the 13° tilt (facing South) PV panel is 97.99%.

7 Building Sector

The building sector is the 2nd largest sector of industrial development and growth. It resonates growth in economy and also in other aspects of society.

For a greener future, the land cover should possess more vegetation than the mixture of cement, rubbles, and sand which not only permeate the percolation but also reflect solar radiation. This causes an increase in ambient temperature. In keeping with the rising population, the vertical growth of the city is unavoidable. Various measures in all these aspects of concern are already being implemented to address the issue. However, the high raised structures to comply with the existing and future demands of a vertically growing city.

Hence the niche of considering any building that is a High-rise – as per the CMDA (Chennai Metropolitan Development Authority) to have up to 18.5 mts height or even more to be called a high-rise building, the targeted building for the use of BIPV panels on the walls should be above 3 levels, that is G + 2 levels. This means the ground level plus the higher floors of 2 floors are considered for the panels to be placed. There are rationales for the G + 2 condition of wall paneling for the BIPV. The vegetation and trees in Chennai are predominantly shrubs and trees which are generally below or slightly above 9 m in height. Considering this the G + 2 level of buildings is assumed for BIPV panels (*Tamil Nadu*, 2019).

8 Roof Surface Area Vis a Vis Wall Surface for BIPV Panels

Considering most of the buildings constructed in Tropical climatic areas are mostly cuboids. Considering a cuboid form for a building, the length, breadth, and height parameters are assumed and calculated. The roof area (9), The vertical wall area (10), and (11), give an idea about the area available for fixing PV panels. Now considering the wall area to let go 20% as openings – as per CMDA, Chennai Metropolitan Development Authority, the local authority for building development regulations, the remaining area of the wall can be considered for PV paneling. To strike a balance between aesthetics and practicality, 60% of the terrace floor area is taken into account, added with the 60% of the wall area available on all four sides, as given by the formula (12) and illustrated in Fig. 12 (SB Karuppannan et al., 2024).

$$a_{roof} = lb \tag{9}$$

$$a_{wall1}, a_{wall3} = b(h * n) \tag{10}$$

$$a_{wall2}, a_{wall4} = l(h * n) \quad (11)$$

$$a_{total} = a_{roof} + a_{wall1}, a_{wall3} + a_{wall2}, a_{wall4}$$

$$a_{total} = 0.6\{lb + 2[b(h * n) + l(h * n)]\}; \quad (12)$$

Ideally, the PV panels can be fixed with their footprint area into the available area (13) calculated. There is a common practice in Chennai to orient the PV panel to 13° facing South. It is considered as a thumb rule to follow the latitude angle as the angle of inclination facing towards the South, for maximising the Solar Power Generation. This notion has to be evaluated (SB Karuppannan et al., 2024).

9 Tilted vs. Horizontal PV Panel

The tilted PV panel at 13° facing south occupies more area than the 0-tilt PV panel. Figure 13 shows the perentage of the increase in floor area occupied by the tilted 13° PV panel. The excess area occupied by the PV panel for exposing every panel to solar radiation brings a space in between the row of PV panels. This space is wide enough so as not to let the shadow of the front PV panel cast a shadow onto the rear PV panel. As the angle of inclination is 13°, the sine value gives the opposite height of the panel raised above its horizontal surface. A 45° projection will give the width of the space between the PV panels, as shown in Fig. 13 (SB Karuppannan et al., 2024).

The length of the PV panel is considered as *l*, the breadth as *b*, and the height raised from the ground level is *h*.

$$\sin 13^\circ = h/lp \quad (13)$$

$$h = 0.224951 * l_p \quad (14)$$

$$d_1 = 0.224951 * l_p \quad (15)$$

$$d_2 = 0.224951 * b_p \quad (16)$$

$$l_1 = 0.970296\, l_p \quad (17)$$

$$l_2 = 0.970296\, b_p \quad (18)$$

Given that the actual PV panel is of length *l*, the increase in area that is available for more PV panels that can be fixed is approximately 16.99%. This means that an increase in power production can happen by installing panels horizontally at 0° inclination. However, the reduction in the power yield is 97.99% which was already calculated from the PVWatts App (National Renewable Energy Laboratory, 2024) and as calculated in Eq. (9), when comparing the 0° horizontally inclined PV panel with the 13° inclined facing South PV panel. Hence, it is crucial to optimize the angle of the panel for maximum output.

There is a 97.99% yield of power generation when the panel is placed horizontally but with an increased area of nearly 19.5% floor footprint area for additional PV panels, will see an increase of 116.99% with the increased production of power, thus yielding an increased power of 114.64% yield, compared to the 13° inclined PV panels for the same floor area.

Thus, it can be concluded that, keeping the solar panel at 0° inclination will yield close to a 15% increase in power generation, of course with the cost incurred on the capital investment and maintenance of the PV panel. This increase in cost can be justified by the cost incurred at the source of power generation and supply. With the solar irradiation that

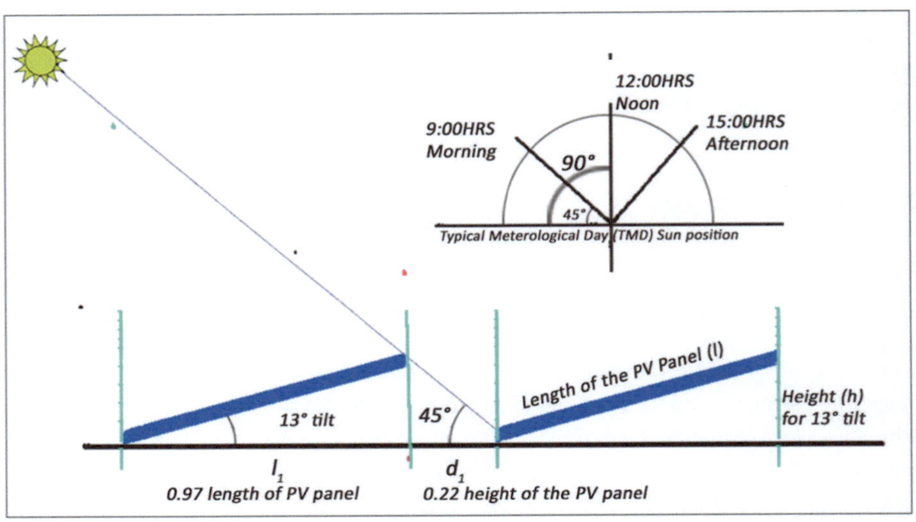

Fig. 13. PV Panel inclination (SB Karuppannan et al., 2024)

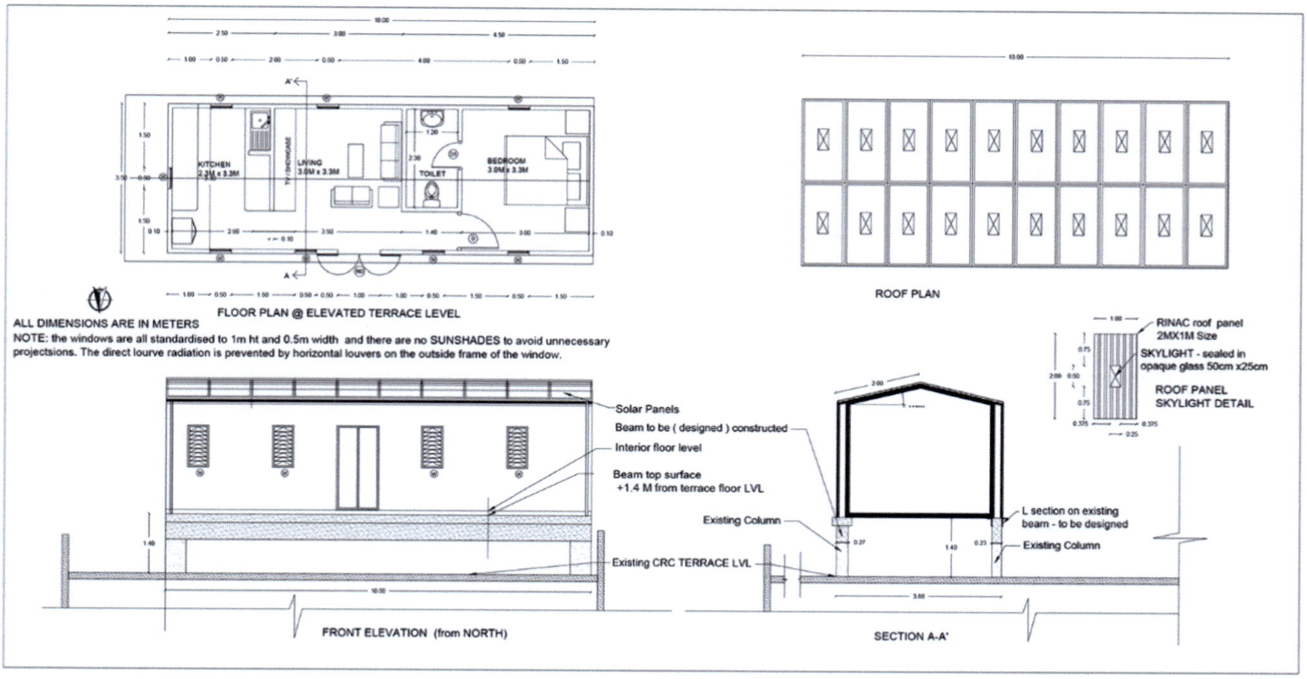

Fig. 14. 1BHK building details as a case-study proposal (SB Karuppannan et al., 2024)

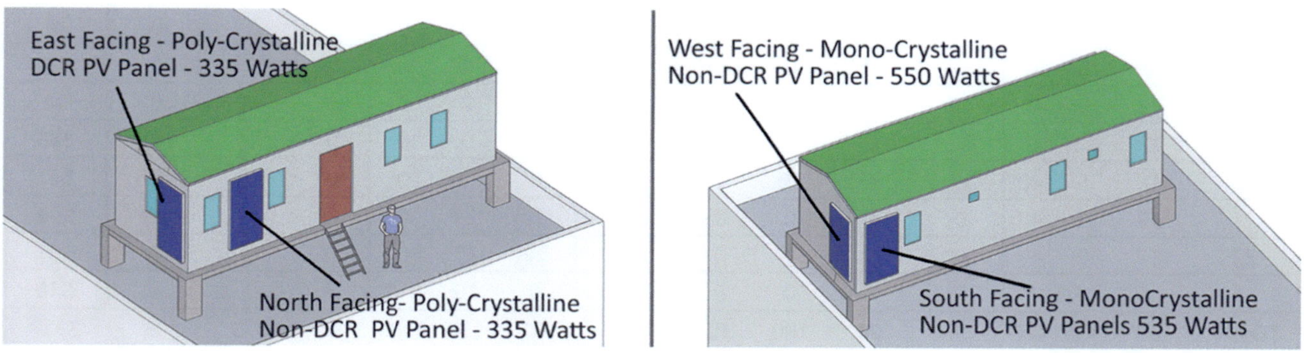

Fig. 15. Placement of the PV panels in four cardinal directions in the proposed 1BHK buidling (SB Karuppannan et al., 2024)

is abundant in a tropical climatic condition, this 13° tilt seems relatively insignificant when compared to power generation (SB Karuppannan et al., 2024).

10 Experimental Setup -1BHK Building

An experimental setup of 1BHK consisting of one living room, one bedroom, and one kitchen as the smallest dwelling unit is built using precast sandwich panels. The installation is done at a + 9 m level from the formed ground level. The following is the plan and the constructed views of the building. Figure 14 shows the plan and elevation of the building, that was constructed. With a floor slab in RCC and the wall panels with the PUF panels sponsored.

The 1BHK house is fabricated with PUF panels sponsored by one of the leading manufacturers of PUF-based sandwich panels. However, the various aspects of the studies are narrowed down to the total power consumption as required in the building vs the total power produced by the building using the 4 vertical panels and one 13° tilt (facing South) panel. 1BHK simulation model is shown in Fig. 15. (SB Karuppannan et al., 2024).

The location of the 1BHK is at the second-floor terrace level, in the CRC (Class Room Complex), inside the campus. The building is at + 9mts level from the formed ground level, in the site location at Chennai.

The building is 10mts long and 3.5 mts wide, with an internal carpet area of 3 mts× 3.3 mts – for the Bedroom; 2.3 mts × 3.3 mts – for the kitchen; 3.0mts x 3.3mts - for the Living room; 1.2 mts× 2.3 mts – for Toilet facility. The windows are openable sideways and of the size 0.5mts width × 1.0 mts height. There is a provision of 2 ventilators for the

1 BHK POWER CONSUMPTION DATA PER DAY (in Wh)															
	Living room			Kitchen					Bed room			Toilet			
hours	Lights	Fan	TV	Lights	Exhaust	Fan	Mixer	Grinder	Light	Fan	A/C	Light	Geaser	Exhaust	TOTAL
0	0	0	0	0	0	0	0	0	0	15	500	0	0	0	515
1	0	0	0	0	0	0	0	0	0	15	500	0	0	0	516
2	0	0	0	0	0	0	0	0	0	15	500	0	0	0	517
3	0	0	0	0	0	0	0	0	0	15	500	0	0	0	518
4	0	0	40	0	0	30	0	0	0	15	500	0	0	0	589
5	0	0	40	0	0	30	10	0	0	15	500	0	0	0	600
6	0	0	0	0	15	0	0	0	0	15	0	12	0	0	48
7	0	0	0	0	15	0	0	0	0	0	0	12	75	15	124
8	0	0	0	0	0	0	0	0	0	0	0	0	0	15	23
9	0	15	0	0	0	0	0	0	0	0	0	0	0	0	24
10	0	15	0	0	0	0	0	0	0	0	0	0	0	0	25
11	0	0	40	0	15	30	0	0	0	0	0	0	0	0	96
12	0	0	40	0	15	30	10	0	0	0	0	0	0	0	107
13	0	0	40	0	0	0	0	0	0	0	0	0	0	0	53
14	0	0	0	0	0	0	0	0	0	0	0	0	0	0	14
15	0	15	0	0	0	0	0	0	0	0	0	0	0	0	30
16	0	0	0	0	0	0	0	0	0	0	0	0	0	0	16
17	0	0	0	0	0	0	0	0	0	0	0	0	0	0	17
18	12	0	40	12	0	0	0	0	12	0	0	0	0	0	94
19	12	15	40	12	15	30	10	20	12	0	0	0	0	0	185
20	12	0	40	12	15	30	0	0	12	0	0	0	0	0	141
21	12	0	40	0	0	0	0	0	12	15	0	12	75	15	202
22	12	0	40	0	0	0	0	0	12	15	500	0	0	0	601
23	0	0	0	0	0	0	0	0	0	15	500	0	0	0	538
TOTAL	60	60	400	36	90	180	30	20	60	150	4000	36	150	45	5317

Fig. 16. Tabular Data of Power Consumption for a 1BHK Building (SB Karuppannan et al., 2024)

toilet and kitchen respectively. The PV panels are placed on to the wall panels, facing the cardinal directions. Figure 15 show the simulated model of the building views from the east and southwest corners. The walls are made of sandwich panels with corrugated metal sheets sandwiched on either side of the PUF. The panels thickness is 0.10 mts, width 1.0 mts and height 3.0 mts. The panels are connected by metal channels that are riveted and bolted for better stability. The total power consumption per day is estimated to be 5317 watts, Fig. 16(SB Karuppannan et al., 2024).

11 Solution for Net Zero Condition for 1BHK

The total average of the Power generation from all 5 sides is 5665.27 Watts per day as shown in Table 3. With the 4 vertical panels and the roof panel being kept at 0 tilt will still yield a power generation of 5630.59 Watts, which is above the needed power requirement of the 1BHK building (SB Karuppannan et al., 2024).

12 Inferences

The daily power consumption requirement of 1 BHK building is 5317 watts/day. The total production of power from the five panels is 5665.37 Watts/day. The limitation of this finding is, that this data is taken for an average of 15 days reading between 1–15 Feb 2024. This model is a combination of estimating the requirement on one end with the PV panel power generation in actuals on the other end. As the system is grid-enabled, the energy produced in the daytime is sent to the grid that is more than the usage during the daytime and drawn

Table 3. Total Power generation of all 5 panels (Karuppannan et al., 2024a)

PV PANELS POWER GENERATION-(Watts) calculated for a period 1–15 Feb 2024						
	13° S	east	west	north	south	all 5 sides
Mean Average/day	1724.83	727.27	1540.50	240.80	1431.87	5665.27

from the grid during the nighttime usage. Energy production will oscillate considering the climatic condition of Chennai. As the study commenced on December 22, 2024, and is still in progress, this sample data is taken to cross-verify if the condition for Net Zero Energy is satisfied with the usage of 4 vertical panels along with one 13° tilted (facing South) panels.

However, the point of study also reveals that even if the 13° inclined towards the south may also be replaced with the PV panel that can be kept horizontal at 0° inclination, as a practice in the future. The findings reveal that the horizontally kept PV panel would still be able to achieve the Net Zero Energy condition, which is a novel finding that would pave the way for more panels kept on the roof at 0° inclination. Another point of reference is the panel's watts power is not uniform, whichever is kept in the cardinal 4 directions. So the total power generation per square meter would be an appropriate measure to calculate for Net Zero Energy condition for any other building typology. Also, the final interpolation and analysis that would happen by the end of the study in a year duration would reveal an accurate total power generation by all 5 panels. The total energy production for achieving Net Zero conditions in buildings by keeping the PV panels vertically and horizontally would yield as much power generation in tropical climate conditions for achieving Net Zero Energy conditions.

13 Way to Reach NET Surplus Energy Condition

It is feasible to attain the net surplus energy condition, in tropical climatic conditions, as was taken for the above experimental setup. Conversely, to achieve the Net Zero building one can also arrive at how many PV panels are required in each of the directions, uncompromising the aesthetics of the building. If aesthetics is incorporated in the building, one can also draw upon the maximum number of panels that can be used in a building so that the Net Surplus Energy can also be achieved and quantified. This approach will transform the Net Surplus condition as a possibility through which each building benefitting out of excess power production.

14 Limitations of the Study

The study did not attempt analyse the details of the building envelope conduction, convection, and re-radiation aspects, which will affect the indoor comfort condition. These factirs will trigger the usage of the various power supply systems that will account for the total power consumption. A comprehensive study on these on these aspecdts has not been carried out. In-depth atmospheric calculation to arrive at the solar insolation has not been done for the following reasons. This calculation will lead to in-depth research to calculate the various factors that are bound to yield solar insolation potentially rejecting the practical side of power generation through PV panels. However, a comparative analysis of the solar insolation achieved at +9mts above the ground level is a major proof of the result, which will pave the door for further research in this area. Life Cycle Analysis and the cost factor comparison have not been completed as of date. With cost justification, the work may bring more innovative implementation process for green buildings, considering the structural stability, aesthetics, and other aspects.

15 Conclusion

Experimental research observation confirms that NZE is achievable with the sample period of study, between the 1 and 15 of February 2024. The solar radiation values in tropical climate locations like Chennai are not as acute as it is normally during the other times of the year. This suggests that an NZE condition may be easily achievable even during other times of the year, that is throughout the year in a location like Chennai. At this juncture, it is noteworthy to mention that the solar insolation data is being collected from 22 December 2023 until 23rd December 2024. An important factor is the structural stability for wind loads, seismic loads have to be ensured without which the BIPV panels of the building become insignificant both for power generation and for aesthetic purposes. The atmospheric conditions like dust accumulation on the surface of the panel, albedo, and Temperature of the panel due to the surrounding atmospheric condition will also impact the power generation of the panels The significance of a location's latitude between the summer and winter solstice, in its ability to produce renewable energy, will not only pave the way for a Net Zero Energy building condition but also will produce surplus energy for future needs. By this strategy, we are evolving towards the development of the greener cities of the future, for the future generation.

Acknowledgements I would like to thank RINAC, Bengaluru for the supply of PUF sandwich panels and the fabrication of 1BHK. DayStar Solar, Chennai for the support in installation, maintenance, and facilitating the data of power generation from the PV panels.

The authors acknowledge the support from the Ministry of Education (MoE), Government of India.

References

Bureau of Energy Efficiency. Energy Conservation Building Code (ECBC) User's Manual. Bureau of Energy Efficiency New Delhi India (2017)

Diez-Mediavilla, M., Rodríguez-Amigo, C., Dieste-Velasco, M.I., García-Calderón, T., Alonso-Tristán, C.: Daily data of Global Vertical Insolation in the four cardinal orientations in Burgos, Spain. Data Brief **21**, 2489–2491 (2018). https://doi.org/10.1016/j.dib.2018.11.099

Lynn, D.: Energy Efficient Lighting Guidance Document for New Construction and Retrofits: The State of North Carolina (2016)

India Meteorological Department. Solar radiant energy over India. India Meteorological Department (2009). https://imdpune.gov.in/library/public/Solar%20Radiant%20Energy%20Over%20India.pdf

Jaysawal, R.K., Chakraborty, S., Elangovan, D., Padmanaban, S.: Concept of net zero energy buildings (NZEB) - a literature review. Clean. Eng. Technol. **11** (2022). https://doi.org/10.1016/j.clet.2022.100582

Koenigsberger, O.H., Ingersoll, T.G., Mayhew, A., Szokolay, S.V.: Manual of tropical housing and building. Part one: climatic design. Build. Sci. **10** (1975)

Liu, R., Duan, W., He, G., Wang, Q.: Development of wall-integrated solar energy technologies. Energies **8**(4) (2025). https://doi.org/10.3390/en18040952

Mamun, M.A.A., Islam, M.M., Hasanuzzaman, M., Selvaraj, J.: Effect of tilt angle on the performance and electrical parameters of a PV module: comparative indoor and outdoor experimental investigation. Energy Built Environ. **3**(3), 278–290 (2022). https://doi.org/10.1016/j.enbenv.2021.02.001

National Renewable Energy Laboratory. PVWatts® Calculator (2024)

Parvin, K., Hossain, M.J., Arsad, A.Z., Ker, P.J., Hannan, M.A.: Building energy technologies towards achieving net-zero pathway: A comprehensive review, challenges and future directions. J. Build. Eng. **100** (2025). https://doi.org/10.1016/j.jobe.2025.111795

REUK.co.uk. Solar Declination (2024). https://www.reuk.co.uk/wordpress/solar/solar-declination/

Karuppannan, S.B., Rao, B.N., Reddy, K.S.: Experimental setup data on poly-crystalline DCR solar panels_herl2024 (2024). https://doi.org/10.17632/pj6p3hfy7v.1

Tamil Nadu, vol. II (2019). https://fsi.nic.in/isfr19/vol2/isfr-2019-vol-ii-tamilnadu.pdf

Effect of Degradation and Size Selection for the Photovoltaic System Installed in a Nearly Zero Energy Building in Tropical Climate

Rosa Francesca De Masi, Antonio Gigante, Michele Parrotta, Silvia Ruggiero, and Alessandro Russo

Abstract

From the perspective of the energy transition, the diffusion of photovoltaic systems and the study of their energy performance considering climate change fulfil a crucial role. The aim of this paper is to evaluate the convenience of installing photovoltaic systems in a tropical climate considering the degradation of system performance and the polluting emissions scenario that would affect climate change in the short term (2030s), in the medium term (2050s) and in the long term (2080s). A calibrated model of an existing nearly zero energy building (nZEB), with residential use, is used as case study. It was simulated for the weather conditions of Santo Domingo, Dominican Republic, with the aim of analysing the building's performance, considering different software for estimating photovoltaic production and considering different opportunities for system design and sizing.

The results indicate that the annual producibility in the reference year is approximately 7900 kWh with a monthly difference between the hottest and coldest periods of the year of around 100–200 kWh. In terms of energy balance, the case study can be considered a Positive Energy Building (PEB) since the electrical generation exceeds the energy demand in each month. Finally, considering different system sizes, degradation of photovoltaic performance and the evolution of climatic conditions, the building can be classified as nZEB and as PEB for a satisfactory number of months.

Keywords

ZEB · PV degradation · future climate projection · size selection · tropical climate

1 Introduction

The transition to renewable energy sources is a global priority in addressing the challenges of climate change and energy security. In this context, solar energy emerges as one of the most promising solutions, offering significant advantages in terms of environmental and economic sustainability. The Dominican Republic, belonging to the tropical climate and characterized by plentiful solar irradiation for most of the year, offers an ideal context for the development and adoption of photovoltaic technologies.

Integrating these systems in the Dominican Republic is a crucial step toward diversifying energy sources, ensuring sustainable access to electricity, and fostering economic and social development. This transition is projected to enable the country to exceed 1.5 GW of installed capacity by 2030 (Pena et al., 2020).

To maximize the potential of solar energy in the region, the integration of photovoltaic systems into buildings emerges as a promising strategy. Not only does this approach make efficient use of available surfaces, but it can also provide additional benefits, such as cooling buildings and reducing carbon emissions associated with energy production.

As noted by Hamzah et al. (2023), the tropical climate provides an ideal setting for the implementation of Building Integrated Photovoltaics (BIPV). The ample solar irradiation

R. F. De Masi · A. Gigante · M. Parrotta · S. Ruggiero · A. Russo (✉)
Department of Engineering, University of Sannio, DING, Benevento, Italy
e-mail: a.russo39@studenti.unisannio.it

R. F. De Masi
e-mail: rfdemasi@unisannio.it

A. Gigante
e-mail: gigante@unisannio.it

S. Ruggiero
e-mail: sruggiero@unisannio.it

B. He et al. (eds.), *Towards the Framework of Livable and Resilient Cities* Advances in Science, Technology & Innovation, https://doi.org/10.1007/978-3-031-97849-4_12

typical of such regions not only supports the generation of clean electricity but also allows for the seamless integration of photovoltaic systems into building designs. This combination enhances aesthetic appeal while fostering sustainability, making BIPV a crucial strategy for incorporating renewable energy into urban environments without compromising architectural integrity.

Montané et al., (2021) found that the energy demands of a Dominican house amount to about 12,703 kWh of electricity per year, where 42% is required for space cooling with the purpose to counteract the high temperatures of the tropical climate and ensure proper indoor comfort conditions. Similarly, for a residential home, the installation of a photovoltaic system could annually avoid producing up to 4.66 tons of carbon dioxide equivalent.

However, the effectiveness of PV systems can be compromised by changing climatic conditions. The degradation of PV performance in response to environmental variations is a critical factor that requires special attention, especially in tropical settings such as Santo Domingo. High temperatures, humidity, and the presence of atmospheric dust, for example, can adversely affect the efficiency and durability of solar panels, thereby reducing their ability to generate power optimally (Bansal et al., 2021).

In order to characterize the degradation of PV performance, the rate of degradation can be taken into account. From the scientific literature, for example, Aboagye et al., (2022) found experimentally that for monocrystalline silicon modules, the degradation rate varies between 0.76%/year and 1.39%/year in different climates in Ghana. Similarly, considering polycrystalline silicon and amorphous silicon, it varied between 1.31%/year and 1.57%/year and 1.62%/year and 1.67%/year, respectively.

Complementing these findings, the study conducted by Islam et al., (2024) illustrates that the reduction in electrical performance of solar PV modules in tropical climates is characterized by intricate degradation phenomena, such as the emergence of hotspots and the effects of environmental stressors. The outcomes indicate average annual degradation rates of 0.3% for open circuit voltage (Voc) in monocrystalline modules at UMPSA and 1.124% at Pasir Mas. This underscores the importance of employing a range of assessment methods, including visual inspections, I-V curve analysis, and thermal imaging, to gain a deeper understanding of the electrical and thermal performance challenges. Such insights are essential for formulating effective mitigation strategies and improving the longevity and performance of solar energy systems in tropical settings.

On the other hand, proper design of the building-system design, taking into account future climate projections, and well-planned maintenance can mitigate the negative impacts of climate change by fully exploiting the potential of renewable technology. In detail, for a tropical climate, Ahmed et al., (2023) found that the existence of a panel cooling system can improve performance by 20.47% in terms of power output.

All told, the literature review highlights a lack of study on the influence of climate changes on the energy balance of building designed to be nearly zero energy in the present condition mainly in countries situated in tropical wet and dry climate zone by also taking into consideration the effect of performance degradation of in-situ renewable plants. Thus, the objective of this paper is to introduce a replicable approach for analyzing the weight of photovoltaic system in the transition of cooling-dominated climate by using as case study a residential nearly zero energy building located in Santo Domingo. Consideration on the optimal sizing selection are discussed by taking into account the performance degradation due to short term (2030s), medium term (2050s) and long term (2080s) future climate projections. The results of the case study are aimed to enrich knowledge regarding solar energy systems in tropical climates, providing useful information for the development of practical policies and implementation strategies.

2 Methodology

The analysis involved building dynamic simulation, by means of EnergyPlus simulation engine with DesignBuilder graphical interface, of a building located classified as nZEB for the tropical climate of the city of Santo Domingo. The case study was modeled through the use of a previously calibrated model based on experimental data obtained for the city of Benevento (Ascione et al., 2019). Briefly, from a systems point of view, the building is supplied by a compact heat pump system that provides the needs for heating, cooling, domestic hot water and mechanical ventilation. In addition, there are integrations with geothermal probes and solar collectors, as well as circulation pumps and thermal storage tanks, to pretreat the air entering the rooms in order to optimize the exploitation of renewable resources on site. Relative to the proposed analysis, the building is connected to a photovoltaic system installed on the roof, connected to the grid and to 6.5 kWh storage batteries through the SOLAX SK-TL(SU)5000 inverter with a maximum efficiency of 97.6%. The PV system consists of 16 SunFortePM096B00 BenQ monocrystalline silicon panels with a peak power of 330 W and an efficiency of 20.3%.

The methodological approach is based on the study conducted by De Masi et al., (2024). However, in this case, the analysis involved a different climate such as the tropical climate represented by the city of Santo Domingo.

Thus, in a first step, the building was simulated with the climate file of the city of Santo Domingo, available in the aforementioned software, analyzing the energy balance

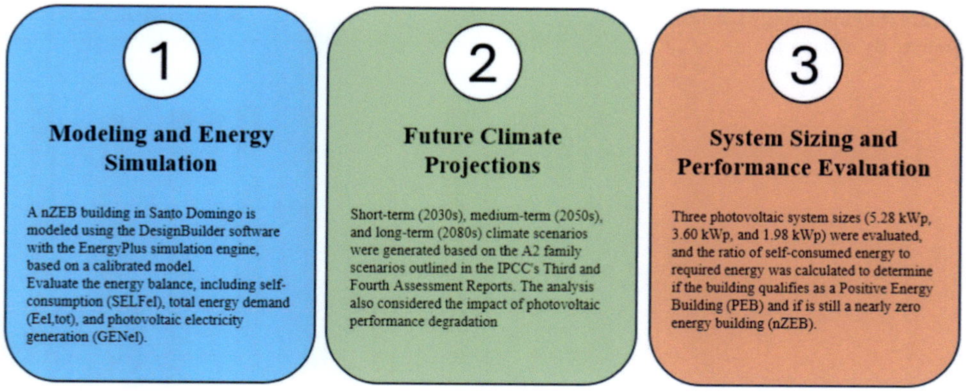

Fig. 1. Methodological approach. (Source: By author)

in detail. In particular, the percentage of self-consumption (SELFel) versus total energy demand (Eel, tot) and photovoltaic electricity generation (GENel) is evaluated. In a second step, short term (2030s), medium term (2050s) and long term (2080s) scenarios were considered. The medium term and long term climate files were obtained as future climate projections using the CCWorldWeatherGen generator from the current climate file used for the short term. These projections are based on the A2 family scenarios as outlined in the IPCC's Third and Fourth Assessment Reports, which consider a world characterized by high population growth, economic development, and a reliance on fossil fuels, leading to significant greenhouse gas emissions and climate change impacts. The degradation rate was set as 0.8%/year in accordance with (Garabintos Lara, 2022). The Performance Ratio (PR) was used to evaluate the performance of the PV system. The standard formulation is reported in Eq. 1:

$$PR = \frac{Y}{\frac{G_i}{G_{STC}}} \quad (1)$$

where Y is the specific efficiency (Energy produced per rated power), kWh/kW; G_i is the global tilted irradiance, kWh/m^2; G_{STC} is the incident radiation at standard test conditions (STC), kW/m^2. For the long term analysis, the system was planned to be replaced every 20 years. In addition, in order to consider the environmental impact related to the degradation of PV performance, the penalty index on the building's operational carbon dioxide equivalent emissions (IP_{OE}) was used according to the following Eq. 2:

$$IP_{OE} = \frac{SELF_{PV,nDEG} - SELF_{PV,DEG}}{SELF_{PV,nDEG}} \quad (2)$$

where $SELF_{PV,nDEG}$ represents the on-site electricity consumption without taking into account the degradation rate of PV production, kWh/year; $SELF_{PV,DEG}$ is the on-site electricity consumption when taking into account the degradation rate of PV production, kWh/year.

In a third step, three different PV system sizes were considered depending on the availability of the south-facing roof. Specifically, the maximum, minimum, and average possible value of 5.28 kWp, 1.98 kWp, and 3.60 kWp were evaluated, respectively.

For all proposed scenarios, the ratio of self-consumed energy to required energy was calculated on an annual basis. According to national and international standards, but also to research papers, the ratio should generally exceed 50%. Finally, in order to define whether the building can be classified as a Positive Energy Building (PEB), the number of months in which monthly production surpasses consumption was calculated.

The methodological approach is summarized in Fig. 1.

3 Results

The results obtained, in terms of electricity consumption (Eel, tot) generation from PV (GENel) and self-consumption (SELFel) are shown in the histogram in Fig. 2. The building exhibits lower values in the winter periods, with minimum generation in November amounting to 559 kWh, and higher values in the summer period, peaking at approximately 784 kWh in August. Regarding electricity consumption, it is possible to see a marked difference in the summer season due to cooling requirements. Specifically, the monthly summer demand is about more than 4 times the energy demand in other months. Similarly, the trends of self-consumption are similar to those of requirements.

Figure 3 shows the changes in operational emissions considering PV degradation. By the 2050s, approximately 30 years from now, the PV system is expected to reach the end of its useful life, while in the 2080s, it is assumed to have been replaced but to maintain the same performance as the system currently under analysis for the short-term (2030s). In the medium term analysis, the anticipated penalties are notably higher, potentially resulting in a surplus of operational emissions exceeding 6% due to the degradation of the PV system.

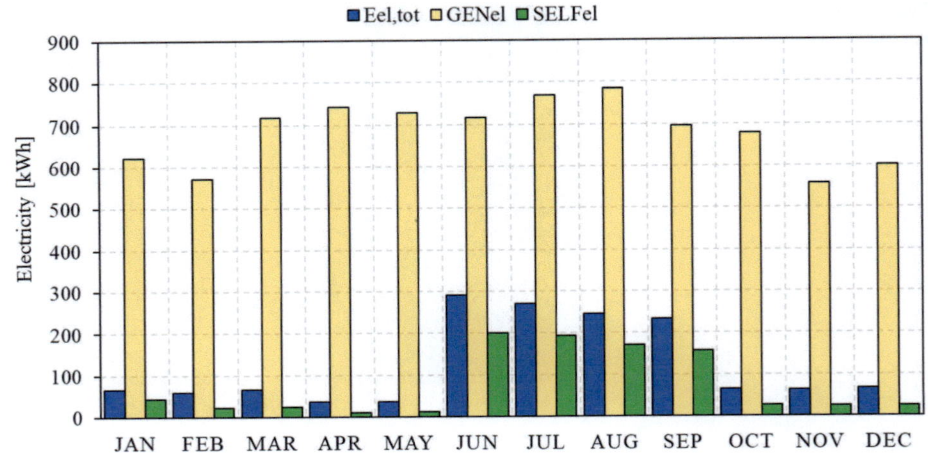

Fig. 2. Building energy balance. (Source: By author)

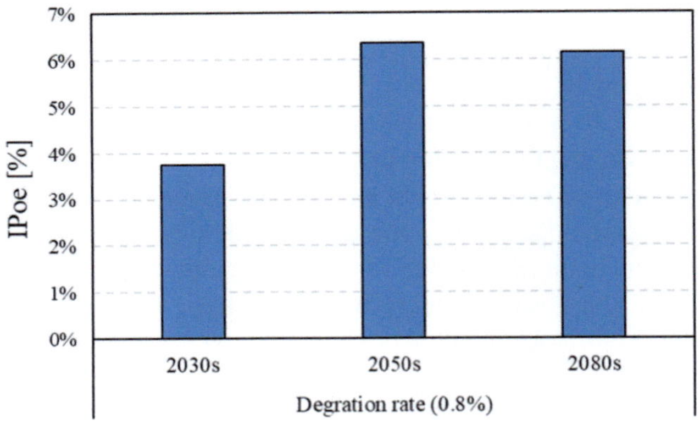

Fig. 3. Penalty index on the building's operational carbon dioxide equivalent emissions. (Source:by author)

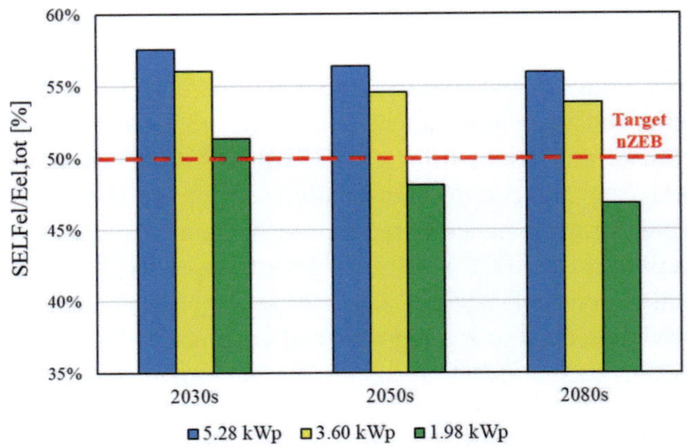

Fig. 4. The nZEB target. (Source: By author)

Conversely, the long term projections suggest slightly diminished penalties compared to the medium term. This could be attributed to several factors, including the possibility of reduced winter consumption as outdoor temperatures rise and increased PV productivity due to higher levels of solar radiation incident on the panels over time. Furthermore, the results considering the size of the photovoltaic system placed in Santo Domingo are reported, considering performance degradation. The electrical self-consumption analysis (Fig. 4) reveals that the building consistently meets the nZEB target when the rated PV power aligns with either the actual (5.28 kWp) or average (3.60 kWp) values. Consequently, despite potential shortcomings in operational emissions, the actual output of the PV system ensures continuous efficiency throughout its operational life. However, in the short term projection, only the minimum installable power achieves the nZEB classification, with a ratio of 51.35%.

Across all future projections, there is a noticeable percentage decrease in on-site renewable consumption. Nonetheless, when considering the same rated power, the results remain comparable. This equilibrium is directly linked to the impacts of global warming. Given the city's context, a shift toward a climate characterized by increased cooling energy demands is anticipated. Therefore, it is reasonable to expect that variations in temperature and solar irradiance will similarly affect the percentage changes in renewable energy production, self-consumption, and building electrical demand in each scenario.

Finally, Fig. 5 shows the results of the analysis to understand the potential Positive Energy Building classification of the building under consideration. For the maximum installed capacity (5.28 kWp), the number of months in which electrical production exceeds consumption is always 12, for each climate scenario considered. This finding is also verified when considering the average installed power (3.60 kWp). Looking at the minimum power (1.98 kWp), however, it can be seen that the number of months decreases slightly with the medium and long term projections. This result is related to the integration of the negative effects related to degradation and the positive effects related to the expected increase in solar radiation with climate change.

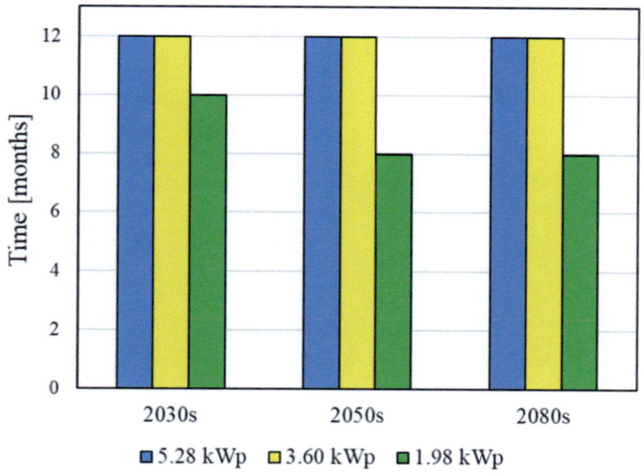

Fig. 5. Number of months in which generation exceeds demand. (Source: By author)

4 Conclusions

This study provided a detailed assessment of the benefits of installing a photovoltaic system in buildings located in Santo Domingo. Through dynamic simulations and short, medium, and long term future climate projections, energy balances were analyzed, considering optimal system sizing and performance degradation due to PV panels aging. The results indicate that despite the degradation, the building maintains an adequate level of electrical self-consumption to be classified as a nearly Zero Energy Building, especially with higher or average rated PV powers. However, there is an increase in operational emissions due to panel degradation, especially in the intermediate periods. This suggests the need to consider strategies for maintenance or replacement of the panels over the lifetime of the system. In addition, the analysis showed a change in the building's energy dynamics due to global warming, with increased energy demand for cooling. Finally, the potential classification of the building as a Positive Energy Building was evaluated, showing that with higher or average installed PV powers the building can produce more energy than it consumes throughout the year, regardless of the climate scenarios considered. However, with lower powers there is a decrease in the number of months when production exceeds consumption, suggesting that the size of the installation is a determining factor in achieving PEB classification. Overall, this study provides important insights for the design and optimization of PV systems in tropical urban settings, highlighting the need to consider not only the current performance of the systems, but also their degradation over time and future climate conditions.

Acknowledgements Thanks to project PRIN 2022 titled HERITAGE – Holistic Energy Renovation and adaptive reuse of historical buildings and sites for an Inclusive, resilienT, and Globally sustainable built Environment. Code 20223M4R8B.

References

Aboagye, B., Gyamfi, S., Antwi, E.O., Djordjevic, S.: Characterisation of degradation of photovoltaic (PV) module technologies in different climatic zones in Ghana. Sustain. Energy Technol. Assess. **52**, 102034 (2022). https://doi.org/10.1016/j.seta.2022.102034

Ahmed, M.S., Karal, R., Das, B.K., Das, A.: Experimental investigation of cooling, wind velocity, and dust deposition effects on solar PV performance in a tropical climate in Bangladesh. Case Stud. Therm. Eng. **50**, 103409 (2023). https://doi.org/10.1016/j.csite.2023.103409

Ascione, F., Borrelli, M., De Masi, R.F., De Rossi, F., Vanoli, G.P.: A framework for NZEB design in Mediterranean climate: design, building and set-up monitoring of a lab-small villa. Sol. Energy **184**, 11–29 (2019). https://doi.org/10.1016/j.solener.2019.03.083

Bansal, N., Jaiswal, S.P., Singh, G.: Comparative investigation of performance evaluation, degradation causes, impact and corrective measures for ground mount and rooftop solar PV plants – a review. Sustain. Energy Technol. Assess. **47**, 101526 (2021). https://doi.org/10.1016/j.seta.2021.101526

De Masi, R.F., Gigante, A., Ruggiero, S., Vanoli, G.P.: Experimental and numerical approach for the evaluation of PV-system performance on energy and environmental behavior of nearly zero energy buildings: case study in Mediterranean climate. Renew. Energy **227**, 120609 (2024). https://doi.org/10.1016/j.renene.2024.120609

Garabitos Lara, E.: Techno-economic model of nonincentivized self consumption with residential PV systems in the context of Dominican Republic: a case study. Energy Sustain. Dev. **68**, 490–500 (2022). https://doi.org/10.1016/j.esd.2022.05.005

Hamzah, A.H., Go, Y.I.: Design and assessment of building integrated PV (BIPV) system towards net zero energy building for tropical climate. e-Prime – Adv. Electr. Eng. Electron. Energy **3**, 100105 (2023). https://doi.org/10.1016/j.prime.2022.100105

Islam, M., Jadin, M.S.B., Al Mansur, A., Alharbi, T.: Electrical performance and degradation analysis of field-aged PV modules in tropical climates: a comparative experimental study. Energy Convers. Manag. X **24**, 100719 (2024). https://doi.org/10.1016/j.ecmx.2024.100719

Montané, M., Ruiz-Valero, L., Labra, C., Faxas-Guzmán, J.G., Girard, A.: Comparative energy consumption and photovoltaic economic analysis for residential buildings in Santiago de Chile and Santo Domingo of the Dominican republic. Renew. Sustain. Energy Rev. **146**, 111175 (2021). https://doi.org/10.1016/j.rser.2021.111175

Pena, J.C, Gordillo, G.: Photovoltaic energy in the Dominican Republic: current status, policies, currently implemented projects, and plans for the future. Int. J. Energy Environ. Econ. **26**(4) (2020). ISSN: 1054-853X

Increasing Renewable Energy Share in Urban Areas by Valorizing Local Wastes: Coupling Anaerobic Digestion and Small Gasification Units

Ebithal Abdelfatah-Aldayyat, Ruben González, Ivan Orlando Cabeza, Jhessica Mosquera Tobar, and Xiomar Gómez

Abstract

Concerns about global warming emissions and their impact on climatic conditions have placed great pressure on the search for energy solutions based on renewable sources. The urban planning of cities often focuses on creating efficient alternatives to ensure the well-being of its citizens. However, waste management and recycling of materials are often overlooked aspects of urban planning and are relegated to areas outside the city limits. Waste can be reintroduced into the economic cycles, but achieving this goal is not an easy task given the nuisance such activities may cause to the local population when carried out close to the generation point. The present study analyzes the feasibility of transforming conventional wastewater treatment plants (WWTP) into biorefineries dedicated to processing urban waste. The possibility of using gasification units to convert sludge, pruning, and garden waste into energy was explored. The energy recovery potential of this approach was studied and compared with the performance of a single centralized unit when adding energy crops grown in the city's surrounding areas. Excessive energy is required to valorize sewage sludge when using thermal technologies due to the high demand for the drying stage. A great extension of land and water resources is necessary to obtain extra energy from this integrated process, which is not always available in many city configurations. However, an increase in energy production can be obtained that is more than double that of the single gasification of sludge.

Keywords

Energy recovery · Waste valorization · Gasification · Anaerobic digestion · Circular economy

E. Abdelfatah-Aldayyat · X. Gómez (✉)
Department of Chemistry and Applied Physics, Chemical Engineering Area, University of León, Campus de Vegazana, 24071 León, Spain
e-mail: xagomb@unileon.es

E. Abdelfatah-Aldayyat
e-mail: eabdea00@estudiantes.unileon.es

R. González
Department of Electrical, Systems and Automatic Engineering, School of Industrial, Computer and Aeronautical Engineering, University of León, Campus de Vegazana, 24071 León, Spain
e-mail: rgong@unileon.es

I. O. Cabeza
Energy, Materials and Environment Laboratory, Faculty of Engineering, Universidad de La Sabana, Campus Universitario Puente del Común, Km. 7 Autopista Norte de Bogotá, Chía, Cundinamarca, Colombia
e-mail: ivan.cabeza@unisabana.edu.co

J. M. Tobar
Instituto de Biotecnología (IBUN), Universidad Nacional de Colombia, Carrera 45 # 26-85, Bogotá, Colombia
e-mail: jhmosquerat@unal.edu.co

1 Introduction

Population growth and urbanization are the main reasons behind the extraordinary amount of waste generated by domestic and industrial sectors. Large amounts of waste are derived from diverse sources. Sewage sludge is an unavoidable product of wastewater treatment and is often disposed of in landfills (Sharma et al., 2022). Anaerobic digestion (AD) is a technology widely used in wastewater treatment plants (WWTPs). Anaerobic microflora transforms organic substrates by a series of sequential reactions, starting with the hydrolysis of the particulate material and following with the solubilization and formation of organic acid compounds, which finally give rise to biogas. Biogas is composed mainly of CH_4 and CO_2, and it is usually valorized as a fuel for energy production in combined heat and power (CHP) units

B. He et al. (eds.), *Towards the Framework of Livable and Resilient Cities* Advances in Science, Technology & Innovation, https://doi.org/10.1007/978-3-031-97849-4_13

or as a natural gas green substitute when up-graded (Minutillo et al. 2019; González et al. 2023a). AD process is very selective and requires long reaction times (Wang et al. 2019), thus explaining the large volume of digesters. In addition to biogas, a liquid slurry stream is also derived from the digestion process commonly known as digestate.

The management of digested sludge, also known as biosolids, is another challenge WWTPs must face. The material contains several nutrients and can be ideally recycled as an organic fertilizer. Unfortunately, this option is not always feasible since the presence of heavy metals, pathogens, and organic pollutants may set restrictions on this practice (Jamali et al. 2007). Dewatering and drying processes are treatment stages commonly found in WWTP operations, allowing for the removal of water content, greatly reducing the mass of sludge requiring final disposal, and facilitating subsequent handling operations and sludge land application. Many WWTPs already incorporate dewatering and drying units since storage costs are significantly reduced (Bennamoun 2012). Therefore, considering thermal valorization of sludge may be a suitable option since the major energy investment regarding sludge drying is already being assumed.

Sludge thermal treatments are attractive for valorizing this material into biofuels (Manara and Zabaniotou, 2012). The integration of different valorization processes provides the opportunity for energy recovery and contaminant removal (Leng et al. 2020). The digested sludge can be subsequently valorized by thermochemical processes such as incineration, gasification, or pyrolysis. Thermochemical processes produce highly oxygenated products that require upgrading using hydrogen treatment before they can be converted into fuels, but hydrogen sources are not naturally available. Gasification is a thermal process that produces syngas as the main fuel stream. The process operates at high temperatures and requires a partial oxidative atmosphere. This atmosphere can be attained by using air or a mixture of air and steam. In this latter case, this gives rise to the production of syngas with higher hydrogen content. The air requirements are lower than those of the combustion process, with equivalence ratios usually found in the range of 0.15–0.6 (Petersen and Werther 2005; Freda et al. 2012). Gasification is an attractive alternative to incineration of complex wastes, as it avoids the production of sulfur and nitrogen oxides, as well as fly ash (Marrero et al. 2004). The gasification process has already been used to produce fuels from organic materials for more than 100 years and is considered a cost-effective process for biomass valorization (Wang et al. 2019).

AD can be used as a pretreatment by integrating different biomass conversion processes to achieve desirable environmental and economic outcomes, optimizing products or byproducts, and developing closed-loop concepts with better energy balance. Several authors have studied the integrated approach of combining AD and thermal processes as a way of valorizing digested slurries when land application of this material is not a feasible option (Antoniou et al. 2019; Gnanendra et al. 2012; Monlau et al. 2016). The main difficulties in attaining a successful integration are associated with the global thermal demand of the process and logistics of collecting operations. However, introducing an additional lignocellulosic material may aid in reaching a positive net energy output (González-Arias et al. 2020). Incorporating short rotation forestry (SRF) into the equation may substantially benefit the energy balance. SRF is characterized by intensive cultivations occurring every few years (Tasca et al. 2019). The residues of short-rotation coppice, like poplar, have been previously studied for producing heat, electricity, or biofuels (Coldwell 2001; Mitchell 1995). The present work considered the feasibility of using a gasification unit to convert digested sludge along with lignocellulosic biomass derived from city green areas and energy crops was considered. The main aim of the study was to explore the feasibility of using a centralized gasification unit for treating lignocellulosic biomass derived from the city's surroundings and green areas. The approach evaluated different scenarios, considering the gasification of digestate as the base case and evaluating the effect of adding garden wastes and poplar biomass in the following scenarios.

Excessive energy is required to valorize sludge using thermal technologies because of the high demand for sludge drying. In addition, due to climatic changes and the expected higher temperatures cities will experience, an increase in urban tree cover is needed to aid in reducing the urban heat island effect. However, this measure also brings along an increment in the mass of garden waste needing proper treatment. The present manuscript attempts to evaluate the feasibility of integrating the combined treatment of sludge and garden waste into a single gasification unit.

2 Material and Methods

2.1 Description of Scenarios

A detailed description of the WWTP can be found in Ellacuriaga et al. (2024). The WWTP has a capacity for treating wastewater for an equivalent population of 150,000 inhabitants. An anaerobic digestion unit treats a mixture of primary and secondary sludge to produce biogas. The methane content of biogas was assumed to be 60% and digestate was subsequently submitted for water dehydration and sludge drying. Figure 1 shows a simplified diagram of the three scenarios evaluated. The WWTP of the city of León already integrates a rotary drier as part of the sludge treatment line. The plant operators are exploring novel treatment alternatives, allowing the plant to attain energy neutrality. Therefore, a gasification unit was initially considered a suitable option for treating

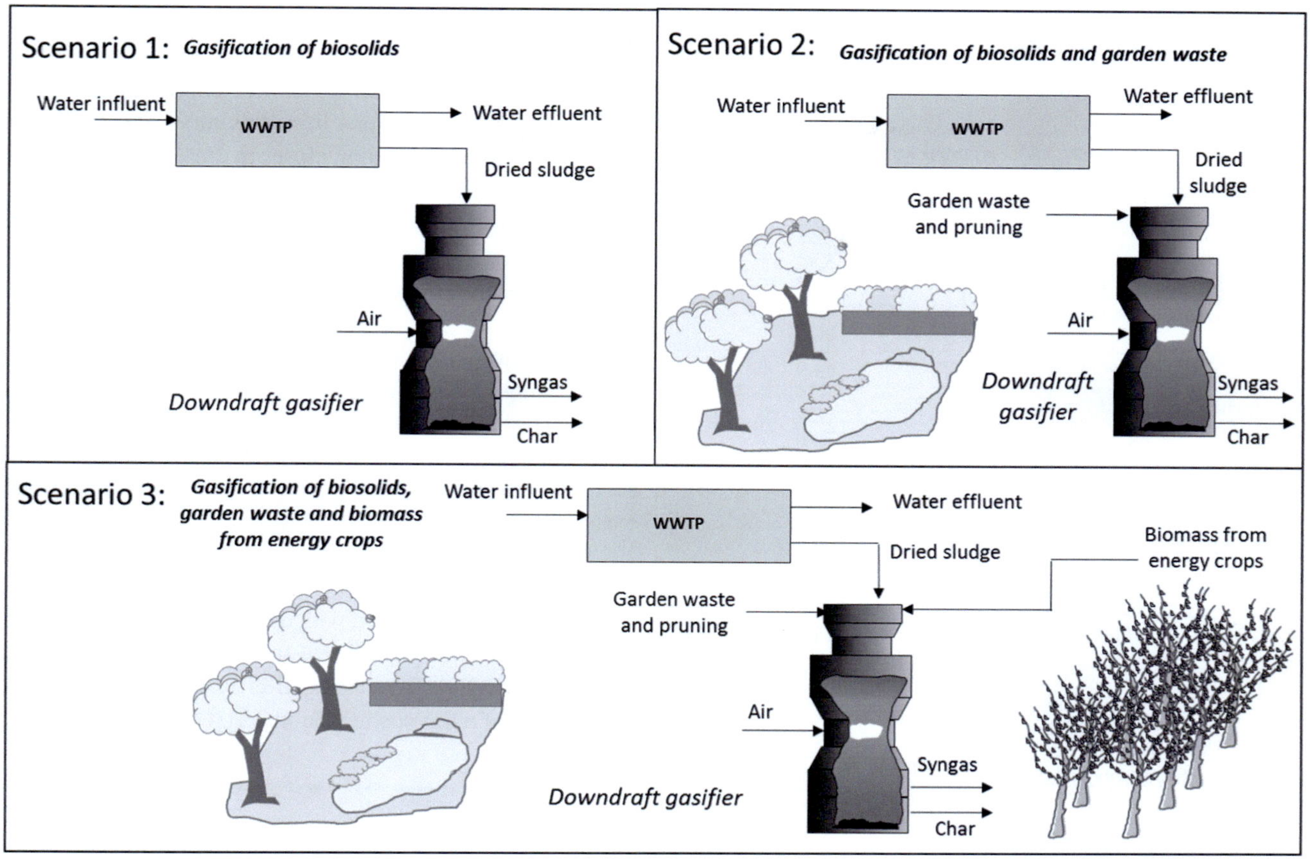

Fig. 1. Schematization of the different scenarios studied where gasification of biosolids is studied in Scenario 1, the co-gasification with garden waste is assumed in Scenario 2. Scenario 3 involves the co-gasification of poplar biomass with garden waste and biosolids. Source: By author.

dried sludge conforming, thus, the base scenario (Scenario 1). Garden and pruning waste derived from residential areas and maintenance activities from community parks and green public areas were assumed to be treated in a gasification unit in conjunction with biosolids. The surface area of the city of León is 39 km^2, with 230.5 ha dedicated to green areas and community parks (www.europapress.es/castilla-y-leon). An estimated 1.5 kg/m^2 of pruning and herbaceous waste was assumed, and 50% of this material was considered suitable for gasifying applications. Central gasification of this material was assumed to be carried out in the same location of the WWTP along with dried biosolids, thus conforming to scenario 2. This material was selected based on finding other valorization options rather than composting for a material with high lignocellulosic content. Scenario 3 considered using energy crops grown close to the WWTP location, in addition to the previous raw materials. Poplar was chosen as the energy crop, given the climatic conditions of the area and the fact that this tree is commonly found close to the river basin.

2.2 Poplar Growth Characteristics and Biomass Characteristics

The city of León is located at 42°36′0.1″ North 5°34′13.2″ West. The selection of suitable land for poplar cultivation was based on the classification of land as herbaceous and crop land cover. A detailed selection of feasible sites was made using Google Earth. León climatic conditions were obtained from https://datosclima.es/Aemethistorico/Tempestad.php. Poplar trees belong to the genus Populus of the Salicaceae family and are of the fastest-growing trees characterized by a very light and soft wood with a high cellulose content (Krutul et al. 2019), making ideal candidates for ecosystems intended for biomass accumulation, carbon storage or as energy crops. Poplar characteristics are listed in Table 1. Where elemental analysis and higher heating value (HHV) are reported from data found in the scientific literature. The results of elemental analysis of biosolids and garden waste are also shown in this table.

Table 1. Characteristics of biomass assumed as input in the gasification process. Reported values are derived from the average values found in references listed in the table for each biomass material. (Source: By author)

Parameter	Sludge	Garden waste	Poplar biomass
Elemental analysis (%)			
C	51.3	45.4	48.4
H	8.2	5.7	5.9
N	6.7	0.9	0.7
S	1.4	0.2	0.1
O	32.4	37.35	44.9
Proximate Analysis (%)			
Volatiles	63.4	70.4	79.6
Fixed carbon	1.9	19.7	16.4
Ash	29.0	9.9	4.0
HHV (MJ/kg)	14.5	18.3	18.2
References	Magdziarz and Werle 2014 Akbari et al. 2000	Ipiales et al. 2022 Roberts 2015 Siddiqui et al. 2018	Paniagua et al. 2019 Chen et al. 2014

2.3 Model Description

SuperPro Designer V10 Software was used to model the different scenarios studied. The model here used for evaluating the performance of the WWTP was based on the previous work of Ellacuriaga et al. (2024), where a detailed description of the plant can be found. The methane specific production was assumed to be 255 mL CH_4/g VS. The digestion unit was assumed to operate at a hydraulic retention time (HRT) of 21 days at the mesophilic regimen (37 °C). A substrate conversion of 98% was assumed. When evaluating lower HRTs, a decrease in conversion was considered, assuming a linear reduction in biogas production, with a 75% decrease at an HRT of 15 days. The methane content in biogas was set at 62% based on Ellacuriaga et al. (2024). The electrical efficiency of the CHP engine was 38% (Shinde et al. 2021). The thermal demand of the digester was estimated as the sensible energy required for increasing sludge temperature from 15 °C to the digestion temperature (37 °C) using the methodology described in González et al. (2020a).

Digested sewage sludge is used as input for the gasification unit once the sludge has been dewatered and dried. SuperPro Designer considers the thermal demand required to reach the gasification temperature and operations associated with the transformation of organic material as separate units. Therefore, a precise estimation of the energy demand in the gasification process can be obtained. This software has been specially conceived to aid in the modeling and simulation of biological processes. The software contains an extensive database of different equipment types, thus allowing accurate estimation of mass and energy balances. The gasification model is based on the Gibbs free energy minimization model. However, deviations from this model were allowed to achieve a gas composition closer to those values obtained experimentally.

The gasification temperature was based on equilibrium equations after setting the temperature of the material at 780 °C before being introduced into the gasifier. Carbon conversion was set at 85%. Sludge gasification was assumed to be carried out in a fluidized bed gasifier with a tar-cracking reactor using a low-cost absorbent, as reported by Jeong et al. (2019). The higher heating value of sludge was 14.5 MJ/kg—based on average values reported by Mun et al. (2013a), Mun et al. (2013b), and Magdziarz et al. (2014)—and the hydrogen content in syngas was in the range of 24 - 30% (Mun et al. 2013a). The simulation model allowed the evaluation of the net energy balance by considering the major items, such as the thermal demand associated with digestion, drying and gasification. Biogas and syngas were the main energy-containing products. The effect of reducing HRT from 21 to 15 days and the addition of a lignocellulosic material was also analyzed.

3 Results

3.1 Results from Scenario 1: Gasification of Biosolids

Figure 2 shows the model configuration based on SuperPro designer software. The scheme shows the section regarding sludge production and subsequent digestion by considering a mesophilic process. The amount of biogas produced was estimated using biogas yield parameters based on Ellacuriaga et al. (2024). The drying demand for reducing the water content of sludge from 80% water content to 20% was 64 590 MJ/d. The amount of methane derived from the digester was 2610 m^3/d, which is equivalent to an energy content of 93 440 MJ/d. About 70% of the energy contained in biogas was needed to dry the material and make it suitable for a posterior thermal valorization option. Many WWTP count with a biogas engine as an energy valorization option for producing heat and electricity. Electric efficiency and thermal energy recovery depend on the size of the biogas engine, with efficiency values usually in the range of 34% to 42% (González et al. 2023a, 2020b); the greater the scale of the engine; the higher the electrical efficiency attained. If biogas is used for electricity production in a combined heat and power (CHP) unit, then not all the energy contained in biogas would be available for producing heat since only a fraction, which is about 50%, will be obtained as heat and some of this heat will be low grade, with a temperature below a 100 °C (González

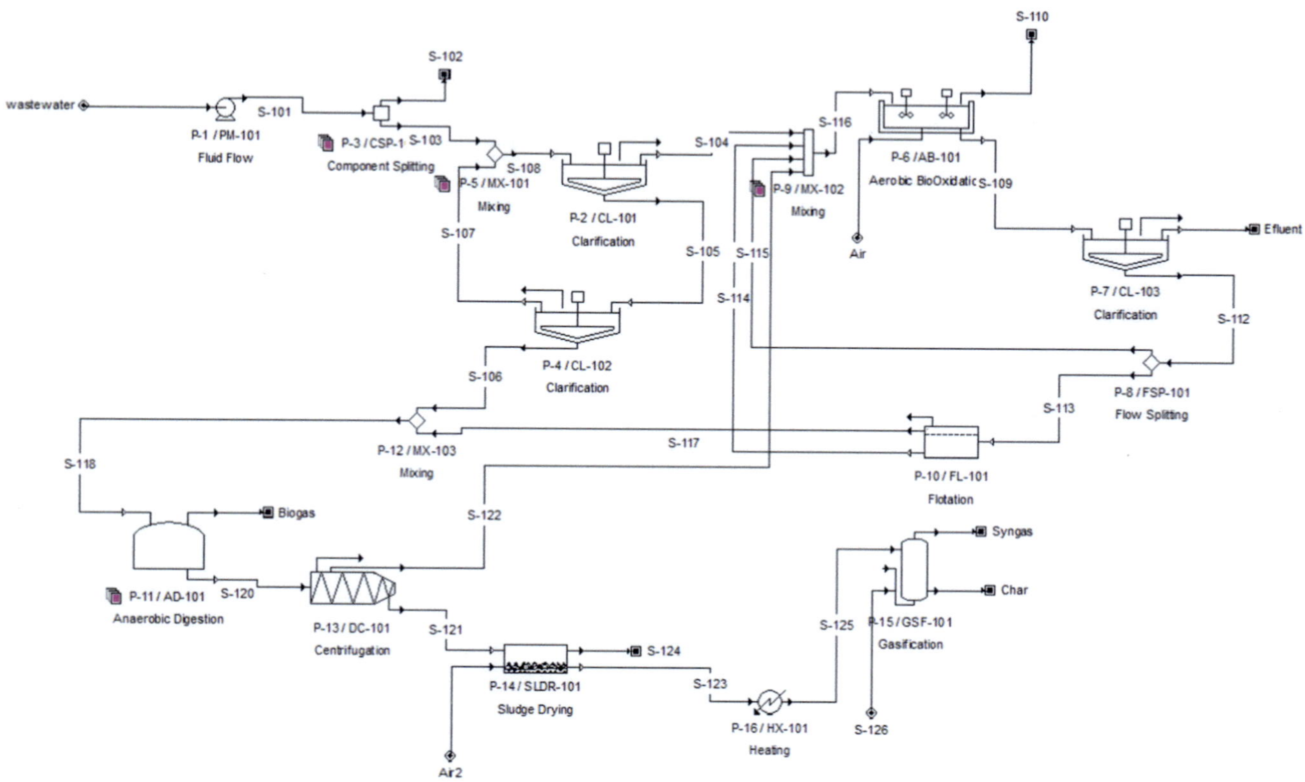

Fig. 2. Scheme derived from SuperPro Designer V10 evaluating the coupling of a WWTP and thermal valorization of digestate through gasification. Source: By author.

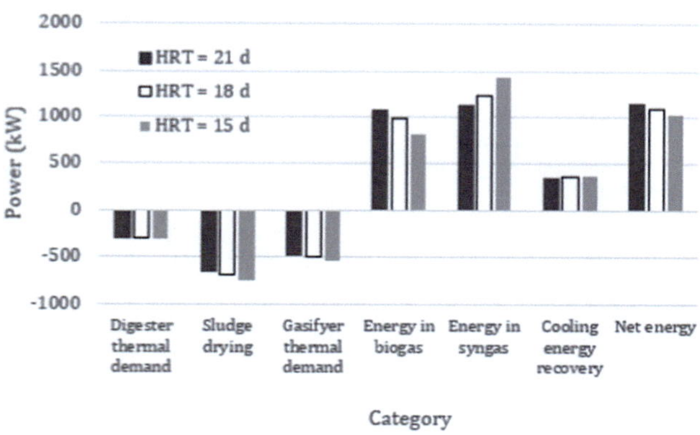

Fig. 3. Energy balance associated with the integration of digestion and gasification of sewage sludge. The balance was calculated for different digestion times, from 21 to 15 days. Source: By author.

et al. 2023b). Therefore, an auxiliary fuel is usually needed to cover the extra thermal demand.

The main thermal demands were associated with keeping the digester temperature constant, sludge drying, and the gasification process's thermal demand. Energy estimations were calculated by considering an HRT of 21 days. Since land application of sludge was no longer considered, the effect of reducing digestion time to 15 days in the net energy balance was assessed. The simulation was run considering a digestion time of 18 and 15 days, assuming a decrease in substrate conversion and, therefore, in biogas evolution. It is observed from Fig. 3 that decreasing the HRT, although it favors the size of the digester since smaller units are now needed, causes a detrimental effect on the net energy balance. The decrease in biogas production was assumed to be linearly dependent on substrate conversion, with a value of 87% when digestion time was set at 18 d and 75% for a digestion time of 15 d. These assumptions translate into a lower biogas evolution and a higher amount of sludge available for gasification, thus increasing the energy associated with syngas evolution. Even though the energy contained in syngas increases by 18.5%, the net energy balance is adversely affected due to the higher thermal demand associated with sludge drying and the lower biogas production.

The gasification of sludge allows the production of a valuable fraction called syngas. Based on the modeling characteristics of the software, the thermal demand of the process can be estimated by determining the heat necessary to reach a temperature close to 600 ºC in the gasification unit. In this case, 480 kW (41 472 MJ/d) are needed. The hydrogen content of syngas was 27%, a value in the range of those reported by Mun et al. (2013a). The chemical energy contained in syngas accounts for 116 345 MJ/d. In addition, this stream is a

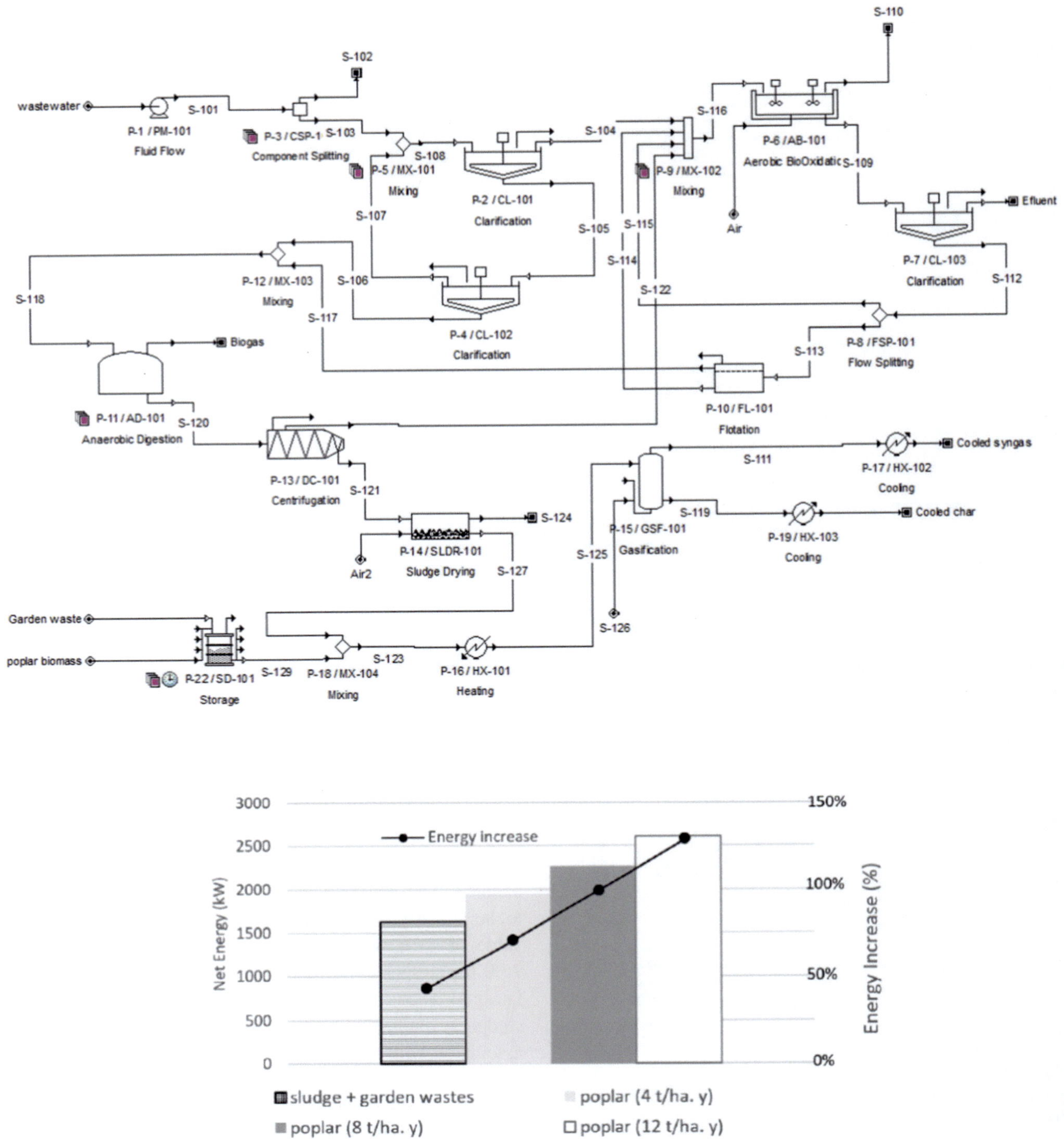

Fig. 4. Sludge handling and gasification section. Garden waste and poplar biomass addition were considered supplements to increase energy production. - Results from the energy balance. Source: By author.

high-temperature fluid that needs cooling for further valorization in a subsequent stage. In the present document, a subsequent valorization stage of syngas was not evaluated. However, cooling is needed either if CHP units are considered or if syngas may be used in any other process taking place either in situ or ex situ. Therefore, some energy recovery is possible from cooling syngas and the char stream.

3.2 Results from Scenario 2 and 3: Co-gasification of Garden Waste and Poplar Biomass with Biosolids

Figure 4 shows the schematic representation of the integrated model (digestion and gasification section). In scenario 2 and 3 the addition of a lignocellulosic material is considered to aid in increasing the amount of energy derived from the gasification reactor. The scheme shows the general case where poplar biomass and garden wastes are added as extra biomass. Net energy values obtained from the balance are also represented in Fig. 4.

Adding high-quality biomass aids in reducing problems associated with tar formation. However, even with the addition of poplar biomass, the ash content of the material is still high. Increasing the proportion of poplar biomass to the mixture up to 80% would reduce the ash content just to a value of 8.2%. Adding such an amount of polar biomass is unfeasible since an area equivalent to 41 km^2 would be necessary for a minimum productivity of 4 t/ha per year. This value may be reduced to 14 km^2 if a productivity of 12 t/ha year is assumed. Considering a circumferential area of 5 km^2 around the WWTP location, the area available for poplar cultivation would be just 37% of the total. Under such premise, expected benefits in the net energy balance were estimated by assuming 4, 8, and 12 t/ha year as biomass productivity values. The net energy benefit almost doubles in all scenarios compared to the base case.

The assumption of adding garden waste along with poplar biomass translates into an input of raw material for the gasifier of 5100 t/year with a syngas production of approximately 780 m^3/h with an LHV of 9.6 MJ/m^3. The higher oxygen content of biomass reduces the amount of air needed to achieve 85% carbon conversion, thus increasing the energy content of syngas. However, this estimation was based on model assumptions when considering air requirements for attaining a desired equivalence ratio. Gasifiers commonly give as an outcome a syngas with lower energy density. The increase in energy derived by adding garden wastes accounts for 29.4%, disregarding the energy needed for drying the material and transport. However, this value reaches only 24% in the net energy balance since gasifier energy demand also increases. Further work will deal with the assessment of logistic operations and water irrigation needs for the growth of energy crops, which would be necessary for reaching a set biomass productivity, given the low precipitation rate of the location. The use of treated wastewater as a water supply and char application, along with sludge addition as a way for recycling nutrients, will also be considered.

4 Conclusions

The energy benefits derived from handling sludge using a gasification unit were analyzed. The energy balance of the integrated approach far outweighs the excessive thermal demand associated with sludge drying. However, when considering the production of electricity, the extra benefit obtained is not enough. Therefore, extra biomass material is required. In the present case, pruning and garden waste were considered, along with poplar biomass derived from energy crops grown in the surroundings. The energy production can be highly increased, reaching a value close to 80% when the lowest poplar productivity is considered. Further studies will be carried out to cover the logistics of this approach and cleaning syngas requirements for valorizing this fuel in CHP engines. The mixture of raw materials is characterized by a high ash content despite the dilution effect associated with adding poplar biomass.

References

Akbari, H., Akbari, H., Fanaei, F., Adibzadeh, A.: Optimization of parameters affecting the hydrothermal carbonization of wastewater treatment plant sewage sludge. Biomass Convers. Biorefinery 1–12 (2022). https://doi.org/10.1007/s13399-022-03427-8

Antoniou, N., Monlau, F., Sambusiti, C., Ficara, E., Barakat, A., Zabaniotou, A.: Contribution to circular economy options of mixed agricultural wastes management: coupling anaerobic digestion with gasification for enhanced energy and material recovery. J. Clean. Prod. **209**, 505–514 (2019). https://doi.org/10.1016/j.jclepro.2018.10.055

Bennamoun, L.: Solar drying of wastewater sludge: a review. Renew. Sustain. Energy Rev. **16**(1), 1061–1073 (2012). https://doi.org/10.1016/j.rser.2011.10.005

Chen, T., et al.: Key thermal events during pyrolysis and CO2-gasification of selected combustible solid wastes in a thermogravimetric analyser. Fuel **137**, 77–84 (2014). https://doi.org/10.1016/j.fuel.2014.07.077

Coldwell, S.: Power from the land. Chem. Eng. **725**, 29–31 (2001)

Ellacuriaga, M., González, R., Gómez, X.: Feasibility of coupling hydrogen and methane production in WWTP: simulation of sludge and food wastes co-digestion. Energy Nexus 100285 (2024)

Freda, C., Cornacchia, G., Romanelli, A., Valerio, V., Grieco, M.: Sewage sludge gasification in a bench scale rotary kiln. Fuel **212**, 88–94 (2018). https://doi.org/10.1016/j.fuel.2017.10.013

Gao, P., Liu, Y., Huang, X., Abulaiti, A., Yang, S.: Effect of wet torrefaction on the physicochemical characteristics and gasification behavior of biochar. Ind. Crops Prod. **197**, 116544 (2023). https://doi.org/10.1016/j.indcrop.2023.116544

Gnanendra, P.M., Ramesha, D.K., Dasappa, S.: Preliminary investigation on the use of biogas sludge for gasification. Int. J. Sustain. Energ. **31**(4), 251–267 (2012). https://doi.org/10.1080/1478646X.2011.559550

González, R., García-Cascallana, J., Gutiérrez-Bravo, J., Gómez, X.: Decentralized biogas production in urban areas: studying the feasibility of using high-efficiency engines. Eng **4**(3), 2204–2225 (2023a). https://doi.org/10.3390/eng4030127

González, R., García-Cascallana, J., Gómez, X.: Energetic valorization of biogas. A comparison between centralized and decentralized approach. Renew. Energy **215**, 119013 (2023b)

González, R., et al.: Performance evaluation of a small-scale digester for achieving decentralised management of waste. Waste Manage. **118**, 99–109 (2020a)

González, R., González, J., Rosas, J.G., Smith, R., Gómez, X.: Biochar and energy production: valorizing swine manure through coupling co-digestion and pyrolysis. Carbon **6**, 43 (2020b). https://doi.org/10.3390/c6020043

González-Arias, J., et al.: Integrating anaerobic digestion of pig slurry and thermal valorisation of biomass. Waste Biomass Valorization **11**, 6125–6137 (2020). https://doi.org/10.1007/s12649-019-00873-w

Ipiales, R.P., Mohedano, A.F., Diaz, E., de la Rubia, M.A.: Energy recovery from garden and park waste by hydrothermal carbonisation and anaerobic digestion. Waste Manage. **140**, 100–109 (2022). https://doi.org/10.1016/j.wasman.2022.01.003

Jamali, M.K., Kazi, T.G., Arain, M.B., Afridi, H.I., Jalbani, N., Memon, A.R.: Heavy metal contents of vegetables grown in soil, irrigated with mixtures of wastewater and sewage sludge in Pakistan, using ultrasonic-assisted pseudo-digestion. J. Agron. Crop Sci. **193**(3), 218–228 (2007). https://doi.org/10.1111/j.1439-037X.2007.00261.x

Jeong, Y.S., Choi, Y.K., Park, K.B., Kim, J.S.: Air co-gasification of coal and dried sewage sludge in a two-stage gasifier: effect of blending ratio on the producer gas composition and tar removal. Energy **185**, 708–716 (2019). https://doi.org/10.1016/j.energy.2019.07.093

Krutul, D., Antczak, A., Kłosińska, T., Drożdżek, M., Radomski, A., Zawadzki, J.: The chemical composition of poplar wood in relation to the species and age of trees. Ann. Warsaw Univ. Life Sci. SGGW For. Wood Technol. **105**, 125–132 (2019). https://doi.org/10.5604/01.3001.0013.7728

Leng, L., et al.: Bioenergy recovery from wastewater produced by hydrothermal processing biomass: progress, challenges, and opportunities. Sci. Total. Environ. **748**, 142383 (2020). https://doi.org/10.1016/j.scitotenv.2020.142383

Magdziarz, A., Werle, S.: Analysis of the combustion and pyrolysis of dried sewage sludge by TGA and MS. Waste Manage. **34**(1), 174–179 (2014). https://doi.org/10.1016/j.wasman.2013.10.033

Manara, P., Zabaniotou, A.: Towards sewage sludge based biofuels via thermochemical conversion – a review. Renew. Sustain. Energy Rev. **16**(5), 2566–2582 (2012). https://doi.org/10.1016/j.rser.2012.01.074

Marrero, T.W., McAuley, B.P., Sutterlin, W.R., Steven Morris, J., Manahan, S.E.: Fate of heavy metals and radioactive metals in gasification of sewage sludge. Waste Manage. **24**(2), 193–198 (2004). https://doi.org/10.1016/S0956-053X(03)00127-2

Minutillo, M., Perna, A., Sorce, A.: Combined hydrogen, heat and electricity generation via biogas reforming: energy and economic assessments. Int. J. Hydrogen Energy **44**(43), 23880–23898 (2019). https://doi.org/10.1016/j.ijhydene.2019.07.105

Mitchell, C.P.: International energy agency bioenergy agreement: progress and achievements 1992–1994. Biomass Bioenergy **9**(1–5), 1–439 (1995)

Monlau, F., et al.: Toward a functional integration of anaerobic digestion and pyrolysis for a sustainable resource management. Comparison between solid-digestate and its derived pyrochar as soil amendment. Appl. Energy **169**, 652–662 (2016). https://doi.org/10.1016/j.apenergy.2016.02.084

Mun, T.Y., Kim, J.W., Kim, J.S.: Air gasification of dried sewage sludge in a two-stage gasifier: Part 1. The effects and reusability of additives on the removal of tar and hydrogen production. Int. J. Hydrogen Energy **38**(13), 5226–5234 (2013a). https://doi.org/10.1016/j.ijhydene.2012.10.120

Mun, T.Y., Kim, J.S.: Air gasification of dried sewage sludge in a two-stage gasifier. Part 2: calcined dolomite as a bed material and effect of moisture content of dried sewage sludge for the hydrogen production and tar removal. Int. J. Hydrogen Energy **38**(13), 5235–5242 (2013b). https://doi.org/10.1016/j.ijhydene.2013.02.073

Paniagua, S., Prado-Guerra, A., García, A.I., Calvo, L.F.: Bioenergy derived from an organically fertilized poplar plot: overall TGA and index estimation study for combustion, gasification, and pyrolysis processes. Biomass Convers. Biorefinery **9**, 749–760 (2019). https://doi.org/10.1007/s13399-019-00392-7

Petersen, I., Werther, J.: Experimental investigation and modeling of gasification of sewage sludge in the circulating fluidized bed. Chem. Eng. Process. **44**(7), 717–736 (2005). https://doi.org/10.1016/j.cep.2004.09.001

Roberts, D.: Characterisation of chemical composition and energy content of green waste and municipal solid waste from Greater Brisbane, Australia. Waste Manage. **41**, 12–19 (2015). https://doi.org/10.1016/j.wasman.2015.03.039

Sharma, I., et al.: Exploring the potential for biomethane production by the hybrid anaerobic digestion and hydrothermal gasification process: a review. J. Clean. Prod. **362**, 132507 (2022). https://doi.org/10.1016/j.jclepro.2022.132507

Shinde, A.M., Dikshit, A.K., Odlare, M., Thorin, E., Schwede, S.: Life cycle assessment of bio-methane and biogas-based electricity production from organic waste for utilization as a vehicle fuel. Clean Technol. Environ. Policy **23**, 1715–1725 (2021). https://doi.org/10.1007/s10098-021-02054-7

Siddiqui, H., Thengane, S.K., Sharma, S., Mahajani, S.M.: Revamping downdraft gasifier to minimize clinker formation for high-ash garden waste as feedstock. Biores. Technol. **266**, 220–231 (2018). https://doi.org/10.1016/j.biortech.2018.06.086

Tasca, A.L., Bacci di Capaci, R., Tognotti, L., Puccini, M.: Biomethane from short rotation forestry and microalgal open ponds: system modeling and life cycle assessment. Bioresour. Technol. **273**, 468–477 (2019). https://doi.org/10.1016/j.biortech.2018.11.038

Wang, A., Austin, D., Song, H.: Investigations of thermochemical upgrading of biomass and its model compounds: opportunities for methane utilization. Fuel **246**, 443–453 (2019). https://doi.org/10.1016/j.fuel.2019.03.015

Green Hydrogen Production Potential from Agricultural Biomass in Togo's Coastal Region

Mawunyo Simon Pierre Kitegi, Kperkouma Wala, Madjouma Kanda, Folega Fousseni, and Tunde David Edunjobi

Abstract

The objective of this study was to assess the possibility of generating green hydrogen from leftover maize and rice. Information regarding the cultivation and land size of these crops was collected from the National Institute of Statistics and Economic and Demographic Studies (INSEED) in Togo, utilizing the latest data from the year 2021. Aspen Plus software was used to simulate the production process. The study showed that the best conditions for producing green hydrogen differ depending on whether rice or maize residues are used. The best temperature for rice residues is 600 °C with an SBR of 2, while for maize residues, it is 650 °C with the same SBR of 2. Yoto and Zio prefectures were found to hold the greatest potential for producing green hydrogen. The coastal region of Togo has a green hydrogen potential of 31.35 Gt equivalent to 1235.37 Gwh, which is much higher than the region's current energy needs. The results of this research indicate that green hydrogen may be a feasible choice for Togo's shift to sustainable energy and the promotion of circularity in the country. Our results suggest that maximizing hydrogen yield from agricultural waste would require identifying the ideal conditions for each crop type during the gasifying process. Hydrogen from agriculture waste may be sufficient to cover the energy needs for the region. Introducing green hydrogen production could help the nation fulfill its obligations to the Paris Agreement and the NetZero initiative.

M. S. P. Kitegi (✉)
CERVIDA-DOUNEDON, University of Lomé, Lomé, Togo
e-mail: simonpierrekitegi@gmail.com

K. Wala · M. Kanda · F. Fousseni
Department of Botanic, Faculty of Sciences, University of Lomé, Lomé, Togo

T. D. Edunjobi
Ladoke Akintola University of Technology, Ogbomoso, Nigeria

Keywords

Green hydrogen · Agriculture waste · Biomass · Bioenergy · Maize residues

1 Introduction

Around 5.13 million deaths annually worldwide are linked to the presence of ambient air pollution caused by the use of fossil fuels, and could potentially be prevented by eliminating the use of these fuels (Lelieveld et al., 2023). This situation could worsen if nothing is done. It is clear in this scenario that in order to protect our lives, we need to shift to renewable energy sources as outlined in SDG 7. Due to the lack of properties of renewable energy resources to decarbonise all the sectors (Martins et al., 2022), researchers have come up with green hydrogento help achieve a more environmentally-friendly world (Haggi et al., 2022). In south Africa, Green hydrogen is helping to reduce carbon emissions in the energy sector (Bessarabov & Human, 2015). One of the many renewable energy sources used to produce green hydrogen is agricultural waste, typically burned with consequences on human health (Awasthi et al., 2010). While agriculture waste contribute to feed animals, and soil health most of farmers use to burn them (Ahmed & Ahmad, 2020). In order to move towards a more environmentally friendly world, it's crucial to reduce pollution that negatively impacts our lives. Producing green hydrogen from agriculture waste (Abdelouahed et al., 2022) is one method. In Togo, agriculture represents nearly 40% of the country's national wealth and 65% of the workforce (Sonhaye, 2022). A large majority of the residues are incinerated, just like in numerous other African nations. Togo's 2021 national greenhouse gas inventory report shows that non-CO_2 greenhouse gas emissions, such as methane (CH_4) and nitrous oxide ($N0_2$), rose by 25.89

B. He et al. (eds.), *Towards the Framework of Livable and Resilient Cities* Advances in Science, Technology & Innovation, https://doi.org/10.1007/978-3-031-97849-4_14

and 25.86%, respectively, between 1995 and 2018. This uptick in emissions is largely due to the expansion of cultivated lands and the extensive burning of agricultural residues (UNFCCC, 2021). The critical issue of air pollution issue in Togo (Yawovi Amouzouvi, 2019) can be associated to agriculture waste burning and energy consumption (Gnoufougou & Bigou-Lare, 2014). 60% of the total carbon dioxide emissions in 2018 originated from the Forestry and Other Land Use sector, 30 percent of the total is allocated to transportation. Most of the methane emissions came from the agricultural sector, while significant contributions were also made by residential biomass combustion and waste sectors (Agbossou et al., 2022). Clean and efficient cooking, efficient charocal production and vehicule efficiency are the main crucial measures to mitigate the production of GHG in Togo (Agbossou et al., 2022). In addition to the pollution problem, Togo is also grappling with a significant issue of energy access. In the maritime region, there is an electricity consumption of 2499 Kwh per person in 2021, with a population of 1144891 residents aged 20 to 75. The total electricity consumption should be 5626755897 Kwh, but only 1012417 Kwh was actually sold (INSEED, 2023). In order to help reduce ongoing GHG emissions in Togo, agricultural waste can be utilized to generate green hydrogen, providing an extra benefit of addressing the issue of energy accessibility (Abad & Dodds, 2020; Kukharets et al., 2021). In a future energy system filled with renewables, green hydrogen could play a vital role as a storage solution for intermittent renewable energy, leading to increased utilization and improved grid reliability. Green hydrogen is regarded as a potentially effective clean energy carrier capable of zero greenhouse gas emissions (Abdelsalam et al., 2024). It can be generated through different renewable energy sources like solar, wind, hydro, and geothermal energy (Abdelsalam et al., 2024; Panchenko et al., 2023).The creation of green hydrogen encounters obstacles such as elevated expenses and the necessity for infrastructure advancement (Jumah, 2024; Zainal et al., 2023). Nonetheless, with the decline in costs and enhancements in efficiency of renewable energy technologies, the price of green hydrogen is anticipated to drop (Łosiewicz, 2024). In developing countries, the possibility of producing and exporting green hydrogen may offer economic prospects, yet it also brings up sustainability concerns regarding land and water usage (Cremonese et al., 2023). Green hydrogen can effectively work alongside renewable energy sources, which are known for their natural unpredictability and instability. Green hydrogen can enhance the use of renewable energy and maintain the power grid's dependability by serving as a prolonged energy reservoir (Chun et al., 2014). Furthermore, green hydrogen is essential for sectors like logistics and heavy industry, where achieving deep decarbonization with electrification is difficult. This helps towards the overall objective of reaching carbon neutrality. Moreover, hydrogen energy offers various choices for energy and fuel sources, enhancing national energy security. It creates jobs and boosts economic development, serving as a versatile answer in the shift towards a sustainable energy future (Han et al., 2023). The production of green hydrogen from agricultural waste offers a hopeful path for sustainable energy production, especially in developing nations such as Togo. The use of biomass waste for hydrogen production corresponds with the concepts of circular economy and provides a remedy for the issues of waste management and renewable energy production (Boddula et al., 2024). Agricultural byproducts, like leftover crops, can act as raw material for generating green hydrogen via different methods. Conventional techniques such as gasification and pyrolysis are utilized, yet they encounter issues like the need for elevated temperatures and pressures, significant capital costs, and possible emissions of pollutants (Boddula et al., 2024). Nonetheless, creative methods such as photo electrocatalytic biomass green hydrogen production are becoming recognized as more sustainable options. This approach converts biomass into hydrogen fuel directly by utilizing sunlight along with an extra electrical impulse, presenting a genuinely environmentally friendly solution (Boddula et al., 2024). Curiously, although producing green hydrogen from agricultural waste holds potential, there is an absence of targeted literature addressing Togo's situation. The sole research available concentrated on generating green hydrogen from solar energy, biomass, and wind (Kitegi et al., 2022). The research concentrated on the total biomass present in the nation. This research gap offers a chance for upcoming studies to investigate the possibilities of producing green hydrogen from agricultural waste in Togo, taking into account the nation's unique agricultural methods, waste disposal trends, and energy requirements. This paper intends to simulate the production of green hydrogen from agriculture waste in the Togo's coastal region. For a successfullness of the study it important to have the proximate and ultimate analysis of the crops to be considered in the analysis (Gondal et al., 2018; Irfan et al., 2022). Up to date the proximate and ultimate analysis has been done in West Africa for only Rice and Maize (Titiloye et al., 2013). This research will focus solely on them to simulate the production of green hydrogen from them in the coastal area of Togo. Rice and maize residues were assessed in this study, and their potential for H_2 production was evaluated through simulations of biomass gasification on Aspen Plus software.

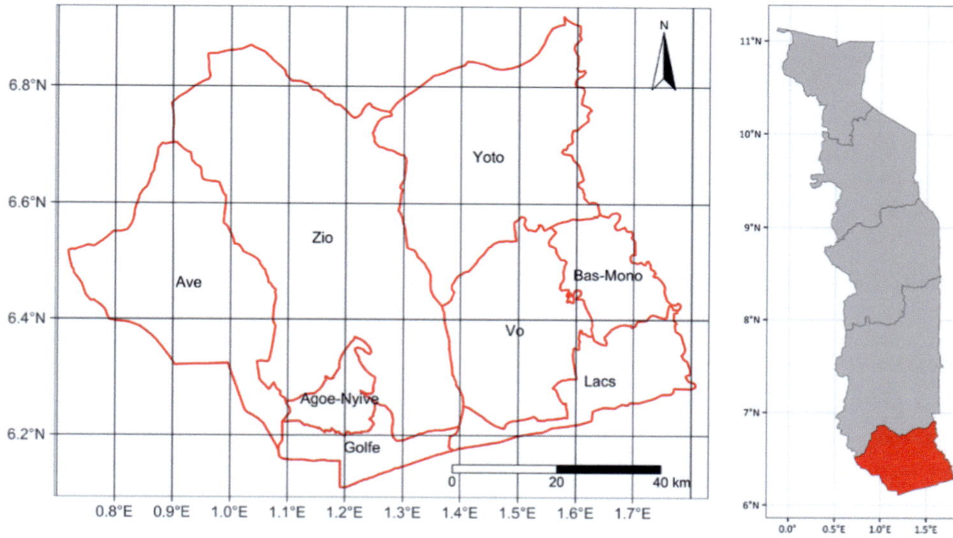

Fig. 1. Map of the Togo coastal area. Source (Authors)

2 Materials and Methods

2.1 Study Area and Data

Situated in the southernmost area of Togo, the Maritime region is between 6°01 and 6°10 latitude North and between 0°70 and 1°85 longitude East. It shares borders with the region des Plateaux to the north, the Atlantic Ocean to the south, the Republic of Benin to the east separated by the Mono River, and Ghana to the west. It spans 6,100 square kilometers, representing about 11% of the nation's total land area. The Maritime region consists of 8 prefectures (Fig. 1) (INSEED, 2023). The data used have been the rice and maize production data of 2021(0), the most recent year available, at the INSEED, the national institute of statistics. The crops components data needed for the simulation have been collected in the literature. They composed of ultimate analysis presented in Table 3, the proximate analysis presented in Table 2 as well as the compositional analysis presented in Table 4. The crop ratios (Residue to Biomass Ratio (RPR), Recovery Factory (RF) and Low Heating Value (LHV)) used were also collected from the literature and presented in 0. Aspen plus software has been used in this study for two reasons. First it is a well-known software in chemical engineering and the next is that the model used in this research have been validated in previous studied using aspen plus (Irfan et al., 2022) (Table 1).

Table 1. Area cultivated and production per crop in Togo coastal region in 2021 (in hectares)

	Area Cultivated (hectares)		Production (tons)	
	Maize	Paddy rice	Maize	Paddy rice
Ave	20546	1535	15056	2916
Bas-Mono	16901	1282	12694	2370
Lacs	11741	0	8703	0
Vo	22063	0	16643	0
Yoto	46699	963	40766	1721
Zio	49467	5067	48305	10150

Source: (INSEED, 2023)

Table 2. Maize and rice proximate analysis (weight percent -wt.%-) (Titiloye et al., 2013)

	Moisture	Volatile Matter	Ash	Fixed Carbon
Maize straws	9.15	75.02	6.13	9.70
Maize cob	8.72	80.72	2.96	7.60
rice straw	7.23	45.68	45.76	1.33
rice husk	8.59	58.22	24.71	8.48

2.2 Biomass Potential Estimation

Crop production yields a considerable amount of crop residues which is the primary energy source for the rural population. Crop residue production depends on the total cultivated area, crop yield, and type of crops. It is tough to determine the total crop residue production due to the unavailability of relevant data regarding residue production (Jingura & Matengaifa, 2008; Karthikeyan et al., 2010). However, crop residues could be estimated based on different ratios, such as residue to crop ratio (RCR). The study considers the biomass potential to be the amount of recoverable crop residue available for bioenergy production, taking into account that not all crop residue biomass may be usable (Tolessa, 2023). The Eq. (1) can help calculate the recoverable crop biomass potential. Where $Ca_{(i)}$

Table 3. Maize and rice ultimate analysis (wt.%) and HHV (MJ/kg) (Titiloye et al., 2013)

	C	H	N	S	CL	O	HHV	HHV	O/C	H/C
Maize straws	42.65	5.57	1.49	< 0.10	1.05	49.16	16.24	16.23	0.86	1.56
Maize cob	43.32	6.02	0.93	< 0.10	0.27	49.37	16.99	17.05	0.85	1.68
rice straw	26.73	4.05	1.02	0.56	1.09	66.55	12.26	6.30	1.87	1.82
rice husk	34.90	5.15	0.31	0.64	<0.01	59.00	14.08	11.69	1.27	1.77

Table 4. Compositional analysis (wt.%) (Titiloye et al., 2013)

	Cellulose	Hemicellulose	Lignin	Extractives
Maize Straw	43.82 ± 0.36	24.39 ± 0.34	9.03 ± 0.04	22.76 ± 0.25
Maize Cob	38.26 ± 0.02	46.59 ± 2.58	7.16 ± 0.12	7.99 ± 0.91
Rice Straw	31.84 ± 1.82	7.29 ± 1.80	28.39 ± 2.25	32.48 ± 1.96
Rice Husk	37.34 ± 0.14	10.07 ± 0.51	41.08 ± 0.18	11.51 ± 0.28

Table 5. RPR, RF and lower heating values (LHV) of crop residues

	Residue	RPR	RF
Maize	Stalk	1.72	0.80
	Cob	0.40	1.00
Rice paddy	Straw	1.46	0.49
	Husk	0.27	0.81

Source: (Tolessa, 2023)

is crop area under the ith crop, hectare (ha); $Cy_{(i)}$ is the yield of the ith crop, ton per hectare (t.ha-1); $RPR_{(i)}$ is the residue-to-product ratio of the ith crop and $RF_{(i)}$ is recoverability factor of the ith crop. The RPR and RF used are presented in Table 5 (Table 2).

$$RCRP = \sum\nolimits_{i=1}^{t} Ca_{(i)} * Cy_{(i)} * RPR_{(i)} * RF_{(i)} \quad (1)$$

2.3 Hydrogen Potential Estimation

Biomass gasification is a renowned technique for transforming biomass into hydrogen and other products without combustion. Biomass goes through multiple phases in a gasifier, such as drying, pyrolysis, burning, and reduction. This procedure generates syngas, made up mostly of hydrogen (H_2) and carbon monoxide (CO), as well as smaller quantities of methane (CH_4), and other gazes (Koido et al., 2017). Biomass gasification consists of four steps: drying, pyrolysis, oxidation, and reduction, depicted in Fig. 2. Using biomass gasification for energy production is considered a sustainable method because it relies on renewable sources and could reduce greenhouse gas emissions more than fossil fuels (Ali et al., 2023).

Various studies have been conducted to assess the potential of green hydrogen from biomass in the global decarbonization process, considering its various benefits (Gondal et al., 2018; Irfan et al., 2022). It has been determined from these studies that factors play a role in the production of green hydrogen from gasification process. The factors include temperature, properties of the gasifying agent feedstock (such as size, type, proximate and ultimate analysis), heating rate, pressure, steam to biomass ratio (SBR), equivalent ratio (ER), bed material, and catalyst (Irfan et al., 2022). In order to simulate the potential for hydrogen production from agricultural waste, an equilibrium-based gasification model was integrated into the Aspen One engineering suite (Aspen Plus, 14). Aspen Plus is advanced simulation software that can be used to conduct industrial process simulations (Pala et al., 2017). Just like previous studies in the literature, the Peng-Robinson equation property method was used for the entire system in this research. Crop residue and char components were identified as unconventional components, and their attributes were defined using the ultimate and proximate analysis presented in Table 3 and Table 4. Crop residues were gasified between 500–1000 °C and under atmospheric pressure of 101.325 Pa.

Figure 3 illustrates the simulation flowsheet of crop residues gasification. The simulation of steam gasification of agricultural residues was conducted in Aspen Plus. At first, the biomass was broken down into its fundamental components through the DECOMP block, with ash being regarded as an inactive substance. The gasifying agent, generated in the HEATER block at a temperature of 250 °C and a pressure of 101.325 Pa, was then transferred to the GASIF reactor. The steam gasification process in the GASIF reactor relied on phase and chemical equilibrium to create syngas at temperatures ranging from 500 °C to 1000 °C. Solid byproducts were removed with the SEP (cyclone) block, and the vapor mixture was cooled to 25 °C using the COOLER block. The cooled

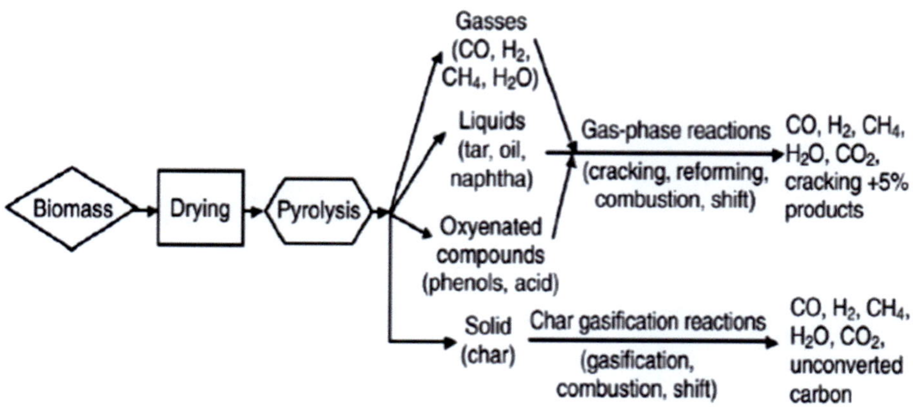

Fig. 2. Schematic steps of biomass gasification (Pradana & Budiman, 2014)

vapor mixture was next divided into gases and water using the SEPRATE block.

Summary of Aspen Plus Simulation Modules:

- DECOMP: Breaks down biomass into elements (C, H_2, N_2, S, O_2) according to ultimate analysis.
- HEATER: Changes water to steam in order to gasify.
- GASIF simulates gasification by calculating phase equilibrium using Gibbs energy minimization.
- SEP: Device used to remove solid particles from gas flow.
- COOLER: Reduces temperature and condenses vapors.
- SEPARATE: Divides gases and liquid (water).

The model that is employed in this study originates from the same done in Pakistan (Irfan et al., 2022). About 141.86 MJ of energy is present in 1 kg of hydrogen, and 1 J is equivalent to 2.77778×10^{-7} kWh (Zheng et al., 2023).

3 Results

3.1 Biomass Potential

Crop residues have the potential to be utilized for energy and chemical extraction as biomass or feedstock sources (Bundhoo, 2019; Jingura & Matengaifa, 2008; Karthikeyan et al., 2010; Lal, 2005). Therefore, accurate assessments of biomass supply from agricultural waste are crucial for determining the potential for H_2 production. The data from crop production 2021 was examined to calculate the recoverable residue biomass potential.

The result presented in 0 and 0 shows that Ave, Bas-Mono, Lacs, Vo, Yoto and Zio have 29.90 Mt respectively; 25.21 Mt; 17.28 Mt; 33.05 Mt; 80.96 Mt and 95.93 Mt of recoverable corn residue biomass. As for rice production, the prefectures of Ave, Bas-Mono, Vo, Yoto have respectively, 2.72; 2,21; 1.61 and 9.48 Mt, and Zio is the leading producer with 9.48 Mt of recoverable for the use of bioenergy production. The combined production of recoverable maize residues from all prefectures is 282.34 Mt, with a total rice residue production of 16.03 Mt. The prefecture of Zio stands out as the first prefecture with the biomass of maize and rice residues (Tables 6 and 7).

Table 6. Maize recoverable residue biomass potential per prefecture and crop (Megatons)

Maize	Straw	Cob	Total
Ave	23.88	6.02	29.90
Bas-Mono	20.13	5.08	25.21
Les Lacs	13.80	3.48	17.28
Vo	26.40	6.66	33.05
Yoto	64.65	16.31	80.96
Zio	76.61	19.32	95.93

Source: Authors

Table 7. Rice recoverable residue biomass potential per prefecture and crop (Megatons)

Rice	Straw	Husk	Total
Ave	2.09	0.64	2.72
Bas-Mono	1.70	0.52	2.21
Les Lacs	0.00	0.00	0.00
Vo	0.00	0.00	0.00
Yoto	1.23	0.38	1.61
Zio	7.26	2.22	9.48

Source: Authors

3.2 Hydrogen Potential

Upon examination of Fig. 4, which were derived from different temperature and SBR variations, it can be observed that the optimum conditions for hydrogen production in the simulations differ based on the specific crop under investigation. The temperature and SBR that result in the highest

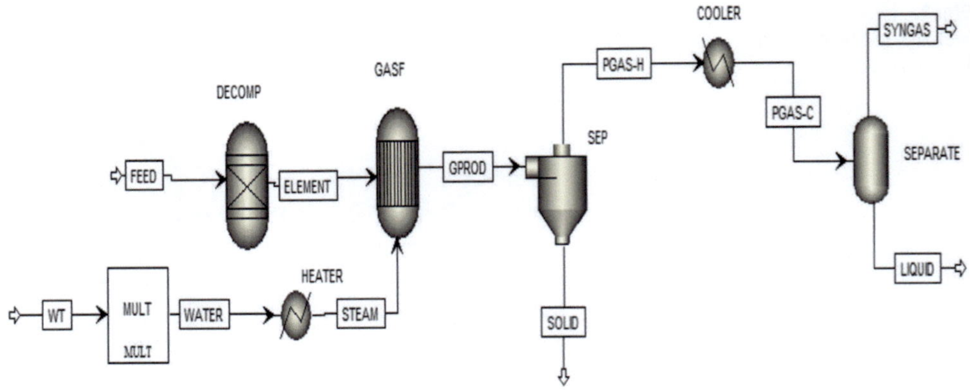

Fig. 3. Aspen Plus flowsheet (Irfan et al., 2022)

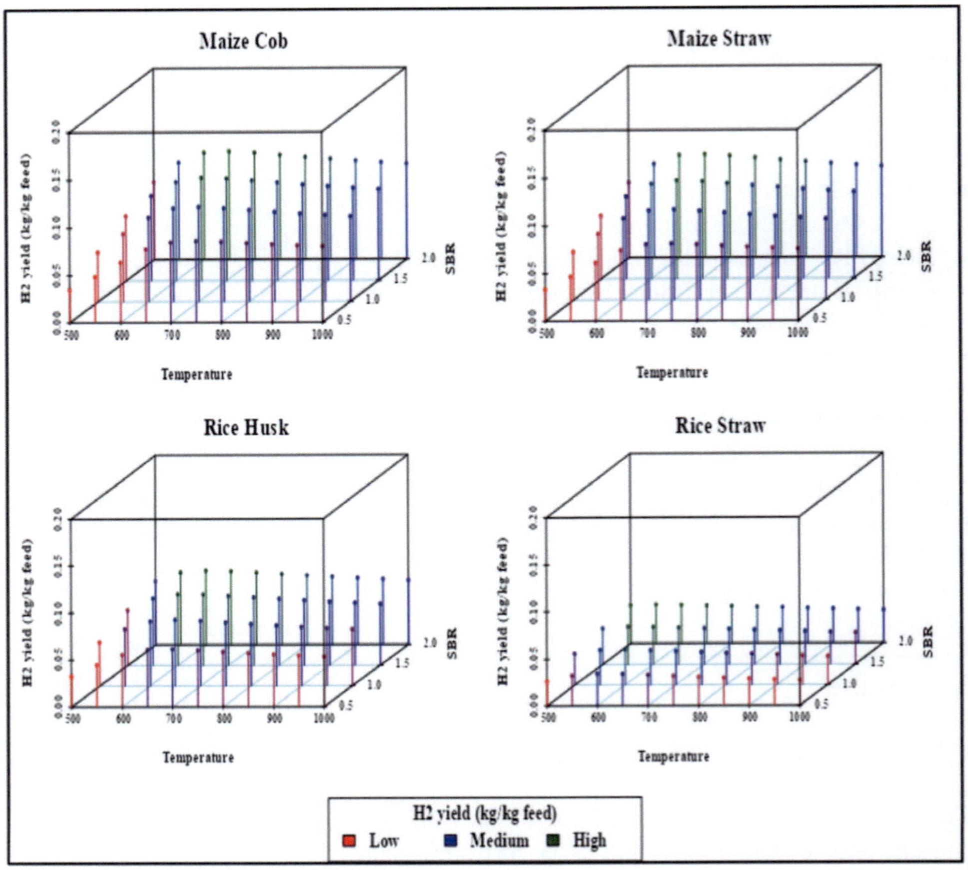

Fig. 4. Effect of Temperature and SBR on H_2 yield (kg/kg feed). Source: Authors

hydrogen yield for maize residues are 650 °C and 2, respectively. For rice husk the highest yield is obtained at 600 °C but at 550 °C for rice straw. The second highest hydrogen yield have been recorded at 550 °C for rice straw, and 650 °C for rice husk. Given that the variation in rice straw production between 550^{0}C and 600 °C is 0.000384864with the common 2 SBR value, we selected 600 °C and 2 SBR as the optimal conditions for rice residues. Therefore, the optimum condition for the rice residue for the study is 600 °C and 2SBR; and 650 °C and 2SBR for maize residues.

The results for the syngas composition, hydrogen yield, and syngas yield are shown on Fig. 4. For the residues at their optimal condition.

The data shown in Fig. 5 indicate that the hydrogen yield for rice straw, rice husk, maize straw and maize cob are 0.039485066, 0.077268133, 0.10821361, and 0.113497772 (kg/kg feed) respectively, with the highest value coming from maize cobs. The rice residues syngas has a higher CO_2 concentration than the maize residues syngas, but with low hydrogen yield. 45,58% of rice straw and 33,48% of rice husk, with 27,87% of maize straw and 27,07% of maize cob.

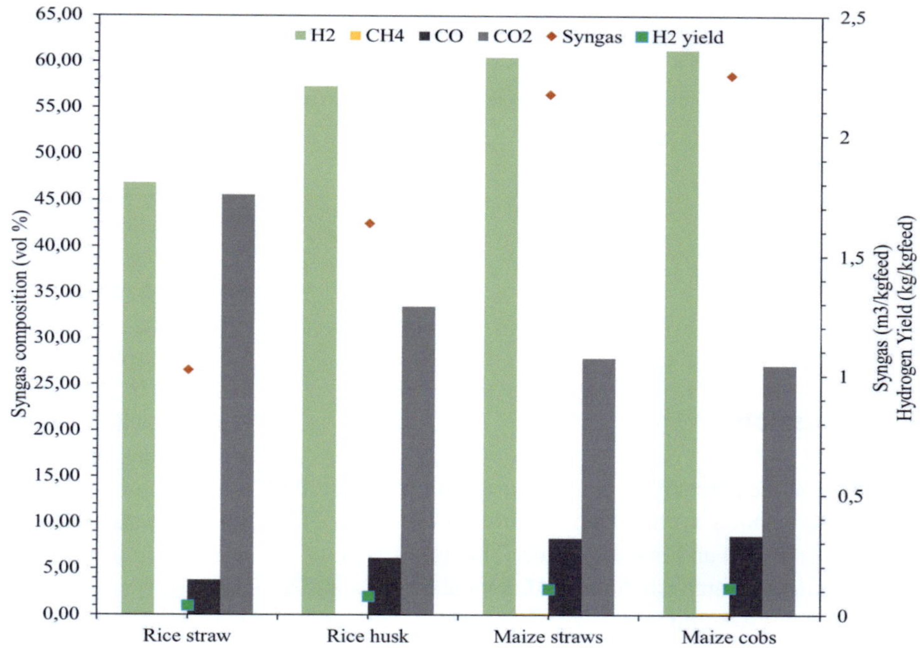

Fig. 5. Syngas composition, H_2 yield, and syngas yield of different residues at 650 °C and 2.0 SBR for maize residues and 550 °C and 2.0 SBR for rice residues. Source: Authors

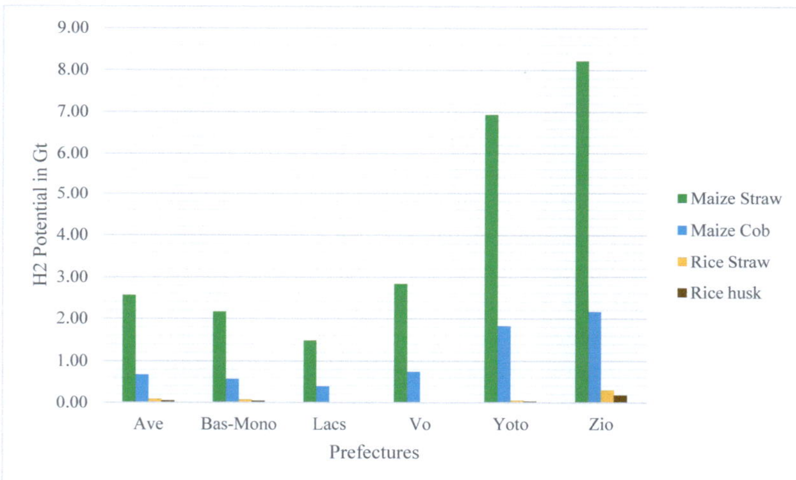

Fig. 6. Hydrogen potential from each prefecture in Togo. Source: Authors

Figure 6 illustrates the hydrogen potential of the various agricultural residues considered in the study (maize straw, maize cob, rice straw, and rice husk) in different prefectures. The contribution of rice residues is very negligible compared to the maize. Zio Prefecture has the highest potential, with 8.22 Gt H_2 from maize straw, 2.16 Gt H_2 from maize cobs, is at the top of the rice residue ranking with 0.29 GtH_2 from rice straw and 0.17 Gt H_2 from rice husks. Yoto prefecture follows with 6.93 GtH_2 from maize straw and 1.83 GtH_2 from maize cobs. The other prefectures, Avé, Bas-mono, Lacs and Vo, have a total potential coming mainly from maize residues in the range of 1 to 3 GtH_2.

Table 8 shows the potential for hydrogen production potential in the Maritime region from different agricultural residues. Maize straw is identified as the most encouraging option, showing a large potential for H_2 production of around 24.18 Gt. Besides, maize cob also demonstrates considerable promise, adding approximately 6.37Gt. On the other hand, rice residues have significantly less potential; rice straw provides approximately 0.5Gt, whereas rice husks add around 0.3Gt. Maize residues, including straw and cob, are highlighted as the main resources for hydrogen production in the area, indicating that initiatives to utilize biomass for H_2 production should prioritize these residues for optimal results.

The result highlight that the hydrogen potential from maize and rice residues for the region can help to meet 1235,37 GWh power need per year.

Table 8. H_2 production potential of region maritime in terms of residue type and total.

Crop residues	H_2 Potential (Gt)	Energy Potential (Tj)	Power (GWh)
Maize Straw	24.18	3430.17	952.83
Maize Cob	6.37	903.65	251.01
Rice Straw	0.5	70.93	19.70
Rice Husk	0.3	42.56	11.82
Total	31.35	4447.31	1235.37

Source: Authors

4 Discussion

The study on the potential of green hydrogen in the maritime area of Togo shows a promising opportunity, with 31.35Gt potential from rice and maize residue. This result is consistent with prior research conducted (Abdelouahed et al., 2022; Ali et al., 2023; Antolini et al., 2022; Irfan et al., 2022; Jardim Borges Da Cunha et al., 2023; Kar & Keles, 2019) emphasizing the significant promise of agricultural waste for generating hydrogen. In addition, our study, pointed out that maximizing hydrogen production from agricultural waste would require identifying the ideal conditions for each crop type. This aspect was overlooked in previous studies (Irfan et al., 2022), where the authors assumed uniform optimal conditions (750 °C and 1SBR) for all crops, leading to reduced hydrogen yield from certain residues especially rice residues. During our research, the best conditions were determined to be 650 °C and 2.0 SBR for maize residues, and 550 °C and 2.0 SBR for rice residues.

Transitioning to green hydrogen will help the region maritime and the nation in tackling their energy issues. In the year 2019, 58.8% of the population in Togo were below the poverty line, with most residing in rural regions. The urban population is projected to increase as individuals transition from rural regions to improving their livelihood, with the current urban population at 42.2%. The estimate is that it will rise to 48.6% by 2030 and hit 60.6% by 2050 (World Bank, n.d.). Changes in how people are distributed in the region or the country due to migration will affect the way energy is used in Togo. Biomass is responsible for supplying up to 70% of the country's energy and serves as the main energy source for homes. Gas and electricity, which are contemporary energy sources for cooking, are rarely utilized. More precisely, biomass, such as firewood, charcoal, and agricultural residues, accounts for 93% of energy consumption in rural households (*SEA*, n.d.). If there are no enhancements in urban living standards and purchasing power, along with specific government interventions, the dependence on natural resources like wood energy is expected to rise in comparison to modern energy sources. Producing green hydrogen from biomass offers a major chance to decrease reliance on conventional natural resources, combat air pollution, and avoid soil degradation from burning residue. Through utilizing collaborations between the government and private sector, the nation can begin the process of shifting towards green hydrogen, offering a way to reduce carbon emissions across all parts of its economy especially in the industry, the transport and the cooking sector. This shift to green hydrogen could also promote economic growth and progress in Togo. Utilizing agricultural residues can help the country generate new green energy sector jobs. Additionally, transitioning to green hydrogen in Togo can improve energy security and stability. Produced within the country from plentiful agricultural waste, green hydrogen provides a more reliable and independent energy source. This could lower the energy import costs of the nation and enhance its trade equilibrium. The creation of green hydrogen production plants would necessitate a proficient staff, resulting in the production of new jobs and the development of skills. This could decrease the high poverty rates in rural areas by offering different income and job opportunities. The implementation of green hydrogen technology has the potential to draw in foreign investments and international assistance. There is a growing worldwide interest in sustainable and renewable energy sources, and Togo's proactive stance on green hydrogen could establish it as a regional leader in the field. This could result in forming alliances with global institutions like the Global Green Growth Institute (GGGI), the Sustainable Energy Fund for Africa (SEFA), the Sustainable use of Energy Resources and Energy Finance (Sunref), the Green Climate Fund (GCF), the Sovereign Green Bonds (SGB) and the African Green Alliance (AGHA), the governmental bodies, and private enterprises, attracting funding, knowledge, and technological advancements. Transitioning to green hydrogen also offers significant environmental advantages. By decreasing the dependence on biomass for energy, Togo can tackle important problems like deforestation and land degradation. Transition to green hydrogen can help Togo to reach the "Energy for All Initiative" and the SDG7. For Togo to effectively shift to green hydrogen, a thorough plan is necessary which involves policy frameworks, financial incentives, and public awareness campaigns. The government ought to introduce measures that promote the growth of green hydrogen infrastructure, like funding for research and development, tax breaks for green energy initiatives, and funding for trial schemes. Financial institutions have the potential to assist by offering cost-effective financing choices for green hydrogen projects. The success of this transition depends heavily on the importance of public awareness and education. Efforts to educate the public on the advantages of green hydrogen, its environmental effects, and its part in the energy shift will boost public backing. Educational programs and training workshops can provide the workforce with the essential skills needed for the green hydrogen industry.

This research provides significant input towards the energy transition and the promotion of circularity in the Maritime region of Togo, despite encountering various obstacles. A significant challenge was the absence of detailed studies on the ultimate and proximate analysis of crops residues that used to be burn in this area. However, the study gains advantages from using information from research carried out in the West African setting, enhancing the strength of its results. Future studies need to concentrate on acquiring coefficients specific to the country in order to improve the accuracy and localized nature of comparisons for instance the proximate and ultimate analysis of each type of crop residues in the context of Togo, thus increasing the relevance and practicality of the findings within Togo's distinct context.

5 Conclusion

The first-ever comprehensive evaluation of green hydrogen production potential from agricultural biomass in Togo's coastal region aims to support the country's energy transition and promote circularity in the country. The study finds that maize and rice residues conversion into green hydrogen can contribute to the energy transition in the country by contribution to circular economy and the country sanitation process. This in-depth research offers important information on the importance of the conversion of agricultural waste into green hydrogen in the energy transition in Togo. By utilizing these leftovers, there is substantial opportunity for generating green hydrogen, as eco-friendly and renewable energy option. The results of this research highlight the significance of utilizing agricultural waste to produce green hydrogen, which could be crucial in fulfilling the energy needs of Togo's coastal area. Using agricultural biomass for green hydrogen production successfully satisfies energy demands and supports global initiatives to shift towards more sustainable and eco-friendly energy systems. Further studies are also needed to advance the knowledge in the sector and contribute effectively to implement green energy transition in Togo.

Acknowledgments The authors are deeply grateful to the World Bank for funding this research through CERViDA-DOUNEDON (Regional Center of Sustainable Cities in Africa).

Conflicts of Interest The authors declare no conflict of interest.

References

Abad, A., Dodds, P.: Green hydrogen characterisation initiatives: definitions, standards, guarantees of origin, and challenges. Energy Policy **138** (2020). https://doi.org/10.1016/j.enpol.2020.111300

Abdelouahed, L., Kourdourli, F., Taouk, B., Estel, L.: Energy evaluation of processes for the production of hydrogen from biomass biodigestion under Aspen Plus. In: Computer Aided Chemical Engineering, vol. 51, pp. 163–168. Elsevier (2022)

Abdelsalam, E., Almomani, F., Alnawafah, H., Habash, D., Jamjoum, M.: Sustainable production of green hydrogen, electricity, and desalinated water via a hybrid solar chimney power plant (HSCPP) water-splitting process. Int. J. Hydrogen Energy **52**, 1356–1369 (2024). https://doi.org/10.1016/j.ijhydene.2023.06.165

Agbossou, A., et al.: Integrated climate change and air pollution mitigation assessment for Togo. Sci. Total. Environ. **844**, 157107 (2022). https://doi.org/10.1016/j.scitotenv.2022.157107

Ahmed, T., Ahmad, W.: Burning crop residue: farmers' choice among various practices. Pak. J. Appl. Econ. **29**, 265–290 (2020)

Ali, R., Elhoweity, M., Zahran, A., Elfadaly, E.: A perspective on hydrogen production from renewable energy and biomass. Int. J. Environ. Stud. Res. **2**(4), 105–119 (2023). https://doi.org/10.21608/ijesr.2023.346724

Antolini, D., Piazzi, S., Menin, L., Baratieri, M., Patuzzi, F.: High hydrogen content syngas for biofuels production from biomass air gasification: experimental evaluation of a char-catalyzed steam reforming unit. Int. J. Hydrogen Energy **47**(64), 27421–27436 (2022)

Awasthi, A., Singh, N., Mittal Cchem Frsc, S., Gupta, P., Agarwal, R.: Effects of agriculture crop residue burning on children and young on PFTs in North West India. Sci. Total. Environ. **408**, 4440–4445 (2010). https://doi.org/10.1016/j.scitotenv.2010.06.040

Bessarabov, D., Human, G.: Solar energy for hydrogen production: experience and application in South Africa (2015). https://www.semanticscholar.org/paper/Solar-energy-for-hydrogen-production-%3A-experience-Bessarabov-Human/e7723809d0f892bab3671ddf4050482ce5dd2299

Boddula, R., Yen-Yi, L., Srinivaas, M.: Sustainable hydrogen production: solar-powered biomass conversion explored through (photo)electrochemical advancements | request PDF. ResearchGate (2024). https://doi.org/10.1016/j.psep.2024.04.068

Bundhoo, Z.M.A.: Potential of bio-hydrogen production from dark fermentation of crop residues: a review. Int. J. Hydrogen Energy **44**(32), 17346–17362 (2019). https://doi.org/10.1016/j.ijhydene.2018.11.098

Chun, D., Woo, C., Seo, H., Chung, Y., Hong, S., Kim, J.: The role of hydrogen energy development in the Korean economy: an input–output analysis. Int. J. Hydrogen Energy **39**(15), 7627–7633 (2014). https://doi.org/10.1016/j.ijhydene.2014.03.058

Cremonese, L., Mbungu, G.K., Quitzow, R.: The sustainability of green hydrogen: an uncertain proposition. Int. J. Hydrogen Energy **48**(51), 19422–19436 (2023). https://doi.org/10.1016/j.ijhydene.2023.01.350

Gnoufougou, D., Bigou-Lare, N.: Technology a long-run relationship investigation of energy consumption and air pollution in Togo (2014). https://www.semanticscholar.org/paper/TECHNOLOGY-A-Long-Run-Relationship-Investigation-of-Gnoufougou-Bigou-Lare/0e44b90c0669e8c6124bfc1b067b708eeff3030c

Gondal, I.A., Masood, S.A., Khan, R.: Green hydrogen production potential for developing a hydrogen economy in Pakistan. Int. J. Hydrogen Energy **43**(12), 6011–6039 (2018). https://doi.org/10.1016/j.ijhydene.2018.01.113

Haggi, H., Fenton, J.M., Brooker, P., Sun, W.: (Invited) renewable hydrogen systems enable deep energy decarbonization of power and transportation sectors. In: ECS Meeting Abstracts, MA2022-01(39), p. 1785 (2022). https://doi.org/10.1149/MA2022-01391785mtgabs

Han, S., Huang, J., Prinzhofer, A., Wang, C.: Advances in global natural hydrogen research and exploration. Adv. Geo-Energy Res. **15**(2), Article 2 (2023). https://doi.org/10.46690/ager.2025.02.01

INSEED. Annuaire statistique 2020–2021 de la region maritime (2023)

Irfan, M., Razzaq, A., Chupradit, S., Javid, M., Rauf, A., Aini Farooqi, T.J.: Hydrogen production potential from agricultural biomass in Punjab province of Pakistan. Int. J. Hydrogen Energy **47**(5), 2846–2861 (2022). https://doi.org/10.1016/j.ijhydene.2021.10.257

Jardim Borges Da Cunha, F., Maia, C., Escobar Palacio, J.C., Ponce Arrieta, F.R.: Theoretical potential of green hydrogen production from biomass in minas gerais. In: Proceedings of the 27th International Congress of Mechanical Engineering. 27th International Congress of Mechanical Engineering (2023). https://doi.org/10.26678/ABCM.COBEM2023.COB2023-0796

Jingura, R.M., Matengaifa, R.: The potential for energy production from crop residues in Zimbabwe. Biomass Bioenerg. **32**(12), 1287–1292 (2008). https://doi.org/10.1016/j.biombioe.2008.03.007

bin Jumah, A.: A comprehensive review of production, applications, and the path to a sustainable energy future with hydrogen. RSC Adv. **14**(36), 26400–26423 (2024). https://doi.org/10.1039/D4RA04559A

Kar, T., Keles, S.: Hydrogen production from renewables: biomass (2019)

Karthikeyan, J., Narra, S., Ay, P.: Energy production from agricultural residues and crop processing wastes: some problems and perspectives (2010). https://doi.org/10.13140/2.1.3330.8965

Kitegi, M.S.P., Laré, Y., Coulibally, O.: (PDF) potential for green hydrogen production from biomass, solar and wind in Togo. ResearchGate (2022). https://doi.org/10.4236/sgre.2022.132002

Koido, K., Iwasaki, T., Koido, K., Iwasaki, T.: Biomass gasification: a review of its technology, gas cleaning applications, and total system life cycle analysis. In: Lignin—Trends and Applications. IntechOpen (2017). https://doi.org/10.5772/intechopen.70727

Kukharets, S., Hutsol, T., Głowacki, S., Sukmaniuk, O., Rozkosz, A., Ткач, O: Concept of biohydrogen production by agricultural enterprises. Agric. Eng. **25**, 63–70 (2021). https://doi.org/10.2478/agriceng-2021-0005

Lal, R.: World crop residues production and implication of its use as a biofuel. Environ. Int. **31**, 575–584 (2005). https://doi.org/10.1016/j.envint.2004.09.005

Lelieveld, J., et al.: Air pollution deaths attributable to fossil fuels: Observational and modelling study. BMJ **383**, e077784 (2023). https://doi.org/10.1136/bmj-2023-077784

Łosiewicz, B.: Technology for green hydrogen production: desk analysis. Energies **17**(17), Article 17 (2024). https://doi.org/10.3390/en17174514

Martins, F., Moura, P., de Almeida, A.T.: The role of electrification in the decarbonization of the energy sector in Portugal. Energies **15**(5), Article 5 (2022). https://doi.org/10.3390/en15051759

Pala, L.P.R., Wang, Q., Kolb, G., Hessel, V.: Steam gasification of biomass with subsequent syngas adjustment using shift reaction for syngas production: an Aspen Plus model. Renewable Energy **101**, 484–492 (2017). https://doi.org/10.1016/j.renene.2016.08.069

Panchenko, V.A., Daus, Y., Kovalev, A.A., Yudaev, I.V., Litti, Y.: Prospects for the production of green hydrogen: review of countries with high potential. Int. J. Hydrogen Energy **48**(12), 4551–4571 (2023). https://doi.org/10.1016/j.ijhydene.2022.10.084

Pradana, Y., Budiman, A.: Bio-syngas derived from Indonesian oil palm empty fruit bunch (EFB) using middle-scale gasification, vol. 10, pp. 1–8 (2014)

SEA (n.d.). https://www.se4all-africa.org/fileadmin/uploads/se4all/Documents/Country_RAGAs/TOGO_RAGA_FR_Released.pdf. Accessed 26 July 2024

Sonhaye, K.N.: L'Intelligence Artificielle, une opportunité pour l'agriculture au Togo. Commun. Technol. Dév. **11** (2022). https://doi.org/10.4000/ctd.7219

Titiloye, J.O., Abu Bakar, M.S., Odetoye, T.E.: Thermochemical characterisation of agricultural wastes from West Africa. Ind. Crops Prod. **47**, 199–203 (2013). https://doi.org/10.1016/j.indcrop.2013.03.011

Tolessa, A.: Bioenergy potential from crop residue biomass resources in Ethiopia. Heliyon **9**, 1–13 (2023). https://doi.org/10.1016/j.heliyon.2023.e13572

UNFCCC. Rapport national d'inventaire des gaz a effet de serre Togo (2021)

World Bank. World Bank [Text/HTML]. World Bank (n.d.). https://www.worldbank.org/en/country/togo/overview. Accessed 26 July 2024

Yawovi Amouzouvi, M.: Contribution of the satellite observation on the analysis and the forecast of the state of air pollution in TOGO. Am. J. Environ. Protect. **8**(1), 31 (2019). https://doi.org/10.11648/j.ajep.20190801.15

Zainal, B.S., et al.: Recent advancement and assessment of green hydrogen production technologies. Renew. Sustain. Energy Rev. **189**, 113941 (2023). https://doi.org/10.1016/j.rser.2023.113941

Zheng, L., Ambrosetti, M., Marangoni, D., Beretta, A., Groppi, G., Tronconi, E.: Electrified methane steam reforming on a washcoated SiSiC foam for low-carbon hydrogen production. Aiche J. **69**(1), e17620 (2023). https://doi.org/10.1002/aic.17620

Intelligent Energy Management Systems: A Systematic Review of Technologies, Challenges, and Opportunities

Nazgol Hafizi and Sadiye Mujdem Vural

Abstract

The growing use of smart technologies in buildings highlights the crucial need to prioritize efficient energy use and management, especially in architectural design. This research explores the convergence of contemporary architectural approaches and advanced technologies, including Artificial Intelligence (AI), the Internet of Things (IoT), and machine learning. Its aim is to review energy-efficient solutions specifically designed for smart residential complexes. The inquiry examines several aspects of energy optimization, including both technology improvements and architectural challenges. The incorporation of smart grid technologies and the integration of renewable energy sources are examined as essential elements of architectural design with the goal of improving energy efficiency. The study aims to provide a thorough evaluation of the present state of incorporating intelligent technologies in energy management for smart residential complexes, focusing on architectural aspects. The objective is to provide information to architects, academics, and policymakers who are interested in efficiently optimizing energy usage in the changing environment of smart residential complexes.

Keywords

Smart Residential Complex · Energy Management · Intelligent Systems · Artificial intelligence · Architectural Design · Smart Technologies

1 Introduction

The fast growth of smart technology has radically altered the setting of architectural design, highlighting the crucial relevance of efficient energy usage and management [1]. In an age of environmental difficulties and a rising desire for sustainable living, residential and office complexes have appeared as significant actors in combining innovative technology and architectural design [2]. The merging of contemporary architectural methods and cutting-edge technology, especially Artificial Intelligence (AI), the Internet of Things (IoT), and machine learning (ML), provides new possibilities to explore innovative energy-optimization techniques in smart architectural complexes [3].

The growth of smart technology in buildings creates an urgent question: how can we prioritize and improve energy efficiency in architectural design to suit the changing requirements of the contemporary home landscape? This study attempts to answer this issue by investigating many aspects of energy optimization, including both technical developments and the inherent limitations of architectural settings [4].

A smart architectural complex is a building development that uses innovative technology to improve inhabitants' overall quality of life, productivity, and sustainability [5]. These complexes use a mix of IoT devices, artificial intelligence, and other technological advances to automate and optimize numerous operations, notably energy usage, security, comfort, and total management of buildings [6].

The relevance of this study stems from its timely nature and possible effect. As urbanization accelerates and energy

N. Hafizi (✉)
Department of Urban Design and Landscape Architecture, Faculty of Design, Arkin University of Creative Arts and Design, Kyrenia, North Cyprus
e-mail: nazgol.hafizi@arucad.edu.tr

S. M. Vural
Department of Architecture, Faculty of Architecture, Eastern Mediterranean University, Famagusta, North Cyprus
e-mail: mujdem.vural@emu.edu.tr

B. He et al. (eds.), *Towards the Framework of Livable and Resilient Cities* Advances in Science, Technology & Innovation, https://doi.org/10.1007/978-3-031-97849-4_15

needs grow, sophisticated, energy-efficient building solutions become more important [7]. This study aims to fill the gap between theoretical technical breakthroughs and actual architectural application, offering useful insights to architects, urban planners, politicians, and technology developers.

This study conducts a comprehensive literature review to examine the present condition of smart technologies in architecture, analyses the integration of AI, IoT, and ML in energy optimization, and assesses architectural obstacles and advances in technology for establishing energy-efficient strategies [8]. It also investigates the use of smart grid technology and renewable energy sources as integral components of architectural design to increase energy efficiency [9]. The study also seeks to identify important issues and potential advancements in the area of energy administration for smart residential buildings [10].

1.1 Objectives of the Research

The major goal of this study is to give a thorough analysis of energy management techniques in smart residential complexes, with a particular emphasis on the integration of intelligent technologies. This study aims to bridge the gap between modern architectural techniques and new technologies such as Artificial Intelligence (AI), the Internet of Things (IoT), and machine learning (ML).

This study aims to analyze how smart technology in residential complexes affects energy management.

1. Analyze the use of AI, IoT, and ML for energy optimization in smart residential structures.
2. Assess architectural obstacles and technical breakthroughs for adopting energy-efficient solutions in smart residential complexes.
3. Identify difficulties and potential advances for energy management in smart residential complexes.

By targeting these goals, this study hopes to add to the expanding body of knowledge on sustainable and intelligent design by providing a comprehensive picture of the junction of sophisticated technology and architectural practices in the framework of energy management.

1.2 Scope and Limitation

This study explores the integration of AI, IoT, and ML in energy management, with a particular focus on architectural design (Fig. 1). While the scope covers many elements of energy optimization, it recognizes certain inherent limits, such as the changing character of technology improvements and the varying architectural settings of smart residential complexes.

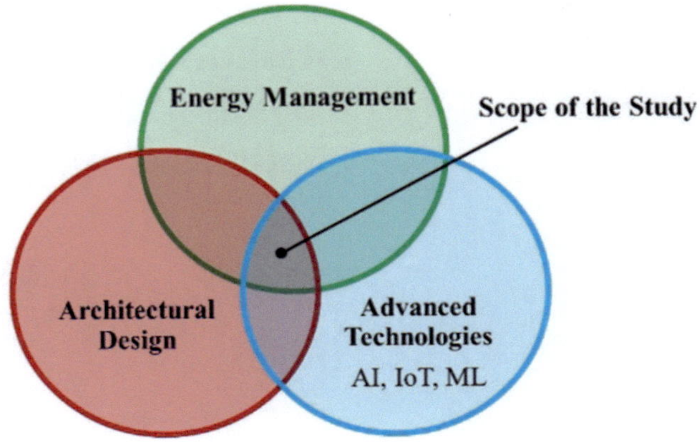

Fig. 1. Scope of the Study (Source: By authors)

2 Methodology

This study uses a thorough literature review technique to examine the incorporation of intelligent technologies for energy management in smart buildings. To guarantee a systematic and complete investigation of the area, we used the Web of Science (WoS) as our major source for gathering relevant academic articles, conference papers, books, and reports all in English and published after 2010.

2.1 Data Sources

The preliminary search on the Web of Science used the following search string:

TOPIC: ("smart building*" OR "intelligent building*") AND TOPIC: ("energy management") AND TOPIC: ("artificial intelligence" OR "machine learning" OR "IoT"). This search produced 845 articles used for bibliometric analysis and visualization using VOSviewer. A further, more extensive search was performed with the query: TOPIC: ("smart residential complex*" OR "smart building*" OR "smart home*") AND TOPIC: ("energy management") AND TOPIC: ("AI" OR "IoT" OR "machine learning"), resulting in 1,247 publications.

Scholarly publications, conference papers, books, and reports on the following important topics were used as primary data sources for this research.

- Smart Building Technology: Investigating the progress of smart technologies and their uses in residential buildings.
- Residential Complex Energy Management: Assessing current ideas and methodology for energy optimization in smart residential complexes.
- AI, IoT, and Machine Learning Integration: Examining the significance and effect of AI, IoT, and machine learning in improving energy management systems in residential settings.

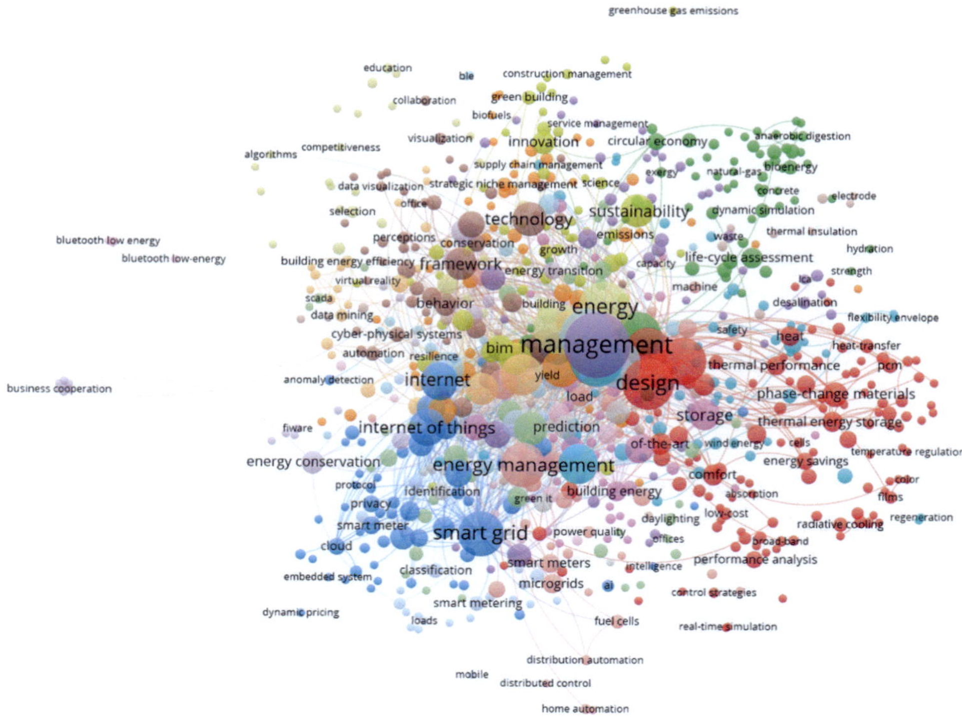

Fig. 2. Network Visualisation of Main Keywords of Building Energy Management and Advanced Technologies (Source: By authors)

During the first Web of Science (WoS) search, the relationship between keywords and their availability in the literature was investigated in order to discover research gaps. Overall, 845 WoS papers have been visualized and analyzed to determine the links between major topics in the literature. The VOSviewer is a software application for creating and visualizing bibliometric networks. This enabled the writers to find clusters of similar phrases and comprehend the links between various fields of research.

Analysis of Fig. 2 reveals several key insights:

- Centrality of "Energy Management": The word "energy management" is essential to the literature on smart residential complexes and sophisticated technologies.
- Clustering of Technologies: It has been noted that separate clusters are emerging around phrases like "Internet of Things," "artificial intelligence," and "machine learning." This clustering indicates that these technologies are often mentioned in connection with energy management in smart buildings.
- Interconnectedness: The extensive network of links between phrases depicts the multidisciplinary character of this topic, where ideas from computer science, engineering, and energy management meet.
- Emerging Themes: Smaller nodes on the network's perimeter might signify rising topics or other areas of inquiry that are gaining interest in the field.

Figure 3 provides a more detailed look at the relationships between advanced technologies and energy management:

- AI, IoT, and ML Prominence: Each of these technologies are shown as big nodes, demonstrating their importance in the literature on smart energy administration.

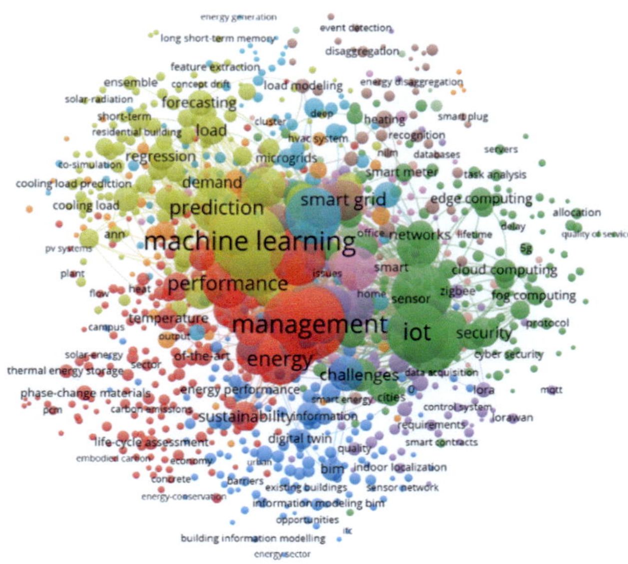

Fig. 3. Network Visualization Main Advanced technologies (AI, IoT, ML) in Detail with Energy Management (Source: By authors)

- Subcategories: Clusters of similar words may be found around each key technology. For example, "deep learning" and "neural networks" are associated with "artificial intelligence," implying related applications or methodology.
- Cross-Technology Links: The linkages between AI, IoT, and ML nodes indicate that these innovations are often utilized together for energy management strategies.
- Application Areas: Concepts like "smart grid," "demand response," and "energy efficiency" are strongly associated with the primary technologies, indicating significant potential uses in the smart residential setting.

2.2 Article Selection Process

The paper selection adhered to a methodical methodology with well-stated criteria. The preliminary screening criteria defined essential characteristics for the search: publications were restricted to those published from 2015 to 2023, composed in English, and included journal articles, conference papers, and reviews. This period was selected to include current research while reflecting the notable technology advancements in smart building energy management over the last ten years.

Articles were required to fulfill many specified criteria for inclusion in the research. Initially, they needed to concentrate mostly on residential structures or complexes, since this corresponded with the study's fundamental aims. Secondly, energy management must be a central issue of discourse rather than a secondary one. Third, the study must clearly illustrate the application of artificial intelligence, the Internet of Things, or machine learning technology. Ultimately, manuscripts were mandated to exhibit transparent approaches and outcomes to guarantee the integrity and dependability of the findings.

Conversely, many exclusion criteria were used to preserve the primary objective and standard of the review. Papers concentrating only on non-residential buildings were omitted, since their energy management needs and solutions sometimes diverge markedly from those of residential applications. Furthermore, purely theoretical publications without practical applications were excluded to guarantee the results had real-world significance. Papers devoid of explicit energy management elements were eliminated, regardless of their discussion on smart technology in residential contexts.

The rigorous selection method narrowed the original pool of 1,247 articles to a final group of 127 publications that constituted the heart of this study (Fig. 4). The procedure included many phases of evaluation, commencing with the elimination of duplicates, progressing to title and abstract review, and culminating in a comprehensive full-text analysis based on the specified inclusion and exclusion criteria.

2.3 Research Gap Identification

This bibliometric study identifies numerous key research needs in the subject of energy management for smart buildings. While the literature is rich in reviews of individual technologies, there is a striking paucity of thorough research addressing the issues of combining AI, IoT, and machine learning, particularly for energy management in residential environments. This gap emphasizes the necessity for research that comprehensively explores the interaction and synergy between these sophisticated technologies in practical applications.

Additionally, the network visualizations indicate a possible under-representation of architectural and design viewpoints in the technology discourse. This highlights the need for an investigation that overcomes the gap between sophisticated technology and architectural design principles in smart buildings. Such research might give useful insights into how technology solutions can be smoothly incorporated into home designs, improving both form and function.

A further significant gap occurs in the area of long-term performance assessment. The present study tends to prioritize technology adoption above long-term performance and sustainability in residential settings. This provides an intriguing chance for ongoing investigations that test the effectiveness, longevity, and developing influence of these technologies in real-world residential settings.

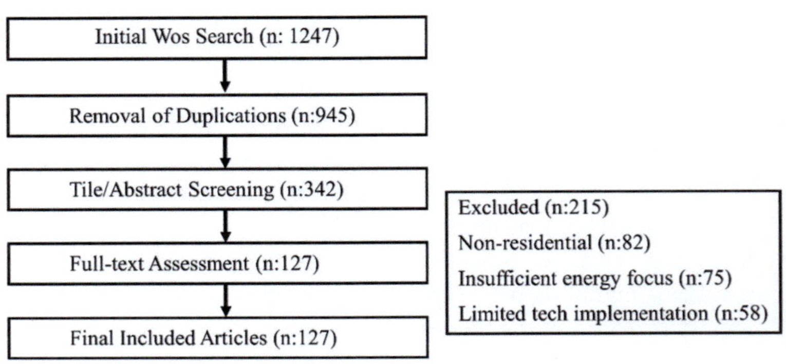

Fig. 4. PRISMA Flow Diagram of the Systematic Review Process (Source: By authors)

Although energy management is crucial to the conversation, there is a notable lack of research focusing on user experience and the acceptability of these innovations in residential environments. This gap emphasizes the necessity of user-centric methods in future research, studying how inhabitants interact with, adapt to, and profit from intelligent energy management systems in their everyday lives.

Finally, the network's varied range of words indicates a lack of standardization in methods for smart energy management. This variation, although illustrative of the field's vibrancy, also points to a significant need for research on scalable and standardized methods that may be efficiently deployed across varied residential environments.

By addressing these highlighted deficiencies, this research hopes to bring useful insights to the area of energy management in smart residential complexes. The study's purpose is to give a more comprehensive knowledge of the integration of sophisticated technologies in this setting, taking into account not only technical factors but also architectural, long-term performance, user experience, and standardization difficulties.

2.4 Limitations

Though the literature review process gives an overview of current information, it is critical to recognize its basic limitations, such as possible publication biases and the fluid nature of technical improvements. This evaluation of the literature provides a basis for subsequent empirical investigation projects, establishing the platform for following publications that will provide unique data and ideas.

3 Literature Review: Systematic Review Methodology

To guarantee a complete and impartial evaluation of the literature, a systematic review process based on the PRISMA criteria was used [12]. The search approach consisted of the following databases: Web of Science, Scopus, and IEEE Xplore. The search phrases "smart residential complex" OR "smart buildings" AND "energy management" AND ("AI" OR "IoT" OR "machine learning") for articles from 2015 to 2023 have been used.

The first search results generated 1,247 papers. After deleting duplicates and applying inclusion criteria (peer-reviewed English publications on energy management in residential settings), 342 papers were chosen for full-text evaluation. Finally, 127 articles were included in the review. In order to present the partial part of the review Table 1 presents the paper selected with significant findings in energy management for smart residential complexes.

3.1 Statistical Overview of the Literature

The systematic study identified important trends and patterns in the research landscape for energy management in smart residential complexes. To offer a thorough perspective, we analyzed the spread of articles over time, the main areas within the subject, and the major technologies covered in the literature.

The time series study of publications from 2015 to 2023 shows a clear rising trend (Fig. 5), demonstrating increased academic interest and research effort in this topic. Figure 4 depicts this pattern, showing a consistent rise in the number of publications every year throughout the previous decade.

The graph shows that the number of publications grew more than fivefold between 2015 and 2023, with a particularly rapid increase beginning in 2018. This trend reflects the rising relevance of energy management in smart residential complexes, as well as significant technical improvements in this sector.

Further examination of the literature indicates significant trends in the areas of study in this discipline. Table 2 shows the distribution of articles according to their principal technical emphasis.

The categorization of publications based on technological emphasis (Table 2) was established by a rigorous content analysis methodology. Each article underwent a three-phase evaluation process:

- Initial screening of keywords and title
- Detailed analysis of the abstract
- Review of the methodology section in the full text

The statistics in Table 1 show that Internet of Things (IoT) technologies dominate the present research landscape, closely followed by Artificial Intelligence (AI) and Machine Learning (ML). A smaller but considerable percentage of the literature focuses on hybrid techniques that include various technologies.

3.2 Key Themes in the Literature

A comprehensive analysis of the literature on energy management in smart residential complexes identified numerous common themes and areas of attention. These topics represent the current level of research, emphasize the most promising technologies and techniques, and indicate future developments in the subject. The following subsections investigate four major topics that dominate the contemporary literature:

- IoT-based Energy Management Systems: This subject looks at how Internet of Things technologies are being

Table 1. Papers with Significant Findings in Energy Management for Smart Homes (Source: By authors)

Ref#	Year	Authors	Title	Key Findings/ Innovations	Impact
[13]	2023	Alanne, K., et al	Study on reducing the grid dependency of urban housing in Nordic climate by hybrid renewable energy systems	Digital twins increased energy efficiency by 10–15% by better predicting and controlling building systems	Shown the practical use of digital twin technology in energy management
[14]	2022	Li, W., et al	Intelligent multi-agent system for smart home energy management	Utilizing the reinforcement learning-based multi-agent framework, we were able to save 22% on energy while increasing user comfort	AI has shown considerable promise for optimizing household energy consumption
[15]	2021	Talei, H., et al	Smart building energy inefficiencies detection through time series analysis and unsupervised machine learning	The proposed comprehensive framework has the potential to reduce energy consumption in smart buildings by 40%	Introduced an integrated method with significant energy savings possibilities
[3]	2020	Antonopoulos, I., et al	Artificial intelligence and machine learning approach to energy demand-side response: A systematic review	In demand-driven applications, AI/ML approaches saved 10% to 25% of the energy used	Measured the effect of AI/ML on consumption management
[16]	2020	Moustaka, V., et al	A systematic review for smart city data analytics	Smart city settings and energy management solutions based on data analytics have lowered energy use by up to 30%	Emphasized the need for big data in urban energy management
[17]	2019	Wang, Z., et al	Data fusion in predicting internal heat gains for office buildings through a deep learning approach	The deep learning method resulted in a 15% decrease in HVAC energy usage while maintaining comfort levels	AI has been shown to be useful in HVAC optimization
[2]	2019	Jia, M., et al	Adopting the Internet of Things for the development of smart buildings	IoT-enabled smart building systems revealed possibilities for 20–25% energy savings	Evaluated the effect of IoT integration on building energy management
[5]	2018	Mocrii, D., et al	IoT-based smart homes: A review of system architecture, software, communications, privacy and security	IoT-enabled houses have the potential to save up to 30% more energy than conventional homes	A thorough evaluation showcasing IoT's potential for domestic energy savings
[18]	2018	Park, E., et al	Smart home services as the next mainstream of the ICT industry: determinants of the adoption of smart home services	User-friendly interfaces and clear explanations of advantages resulted in 40% greater adoption rates of smart home energy solutions	Emphasized the significance of user experience in the adoption of energy management systems
[1]	2017	Beaudin, M., & Zareipour, H	Home energy management systems: A review of modeling and complexity	Innovative residential energy management technologies have shown the potential for 15–20% reductions in energy expenses	Information on the financial advantages of smart home energy systems was given
[19]	2021	Khan, S., et al	Towards intelligent building energy management: AI-based framework for power consumption and generation forecasting	AI-powered demand response tactics cut peak energy use by 30% in smart buildings	AI has had a considerable influence on controlling peak energy consumption

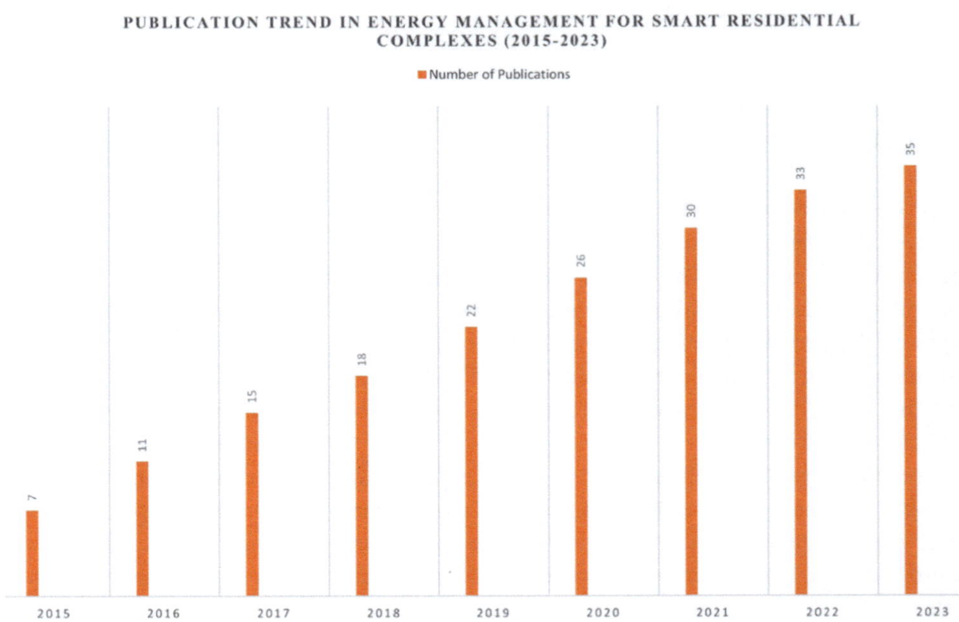

Fig. 5. Publication Trend in Energy Management for Smart Residential Complexes (2015–2023) (Source: By authors)

Table 2. Distribution of Papers by Technology Focus (Source: By authors)

Technology Focus	Number of Papers	Percentage
IoT	52	41%
AI	38	30%
Machine Learning	28	22%
Hybrid Approaches	9	7%

used to provide networked, efficient energy management systems in home settings.

- AI and Machine Learning Approaches: This field focuses on the use of artificial intelligence and machine learning approaches to optimize energy consumption and forecast usage trends.
- Integration Challenges: This issue discusses the challenges and complications of deploying and integrating sophisticated energy management technology in residential complexes.
- User-Centric Approaches: This growing issue emphasizes the significance of user involvement, acceptability, and behavior in the success of smart energy management systems.

Analyzing these major issues allows us to acquire a full overview of the present situation of energy management in smart residential complexes, identify gaps in existing research, and suggest the potential for future studies.

Before Focusing on each key theme identified through literature analysis, the study defines the definition of "Smart Residential Complex" for this research and what are the main components.

A "smart residential complex" is a housing project that incorporates sophisticated technology to improve the occupants' quality of life, efficiency, and sustainability. These buildings utilize a mix of Internet of Things (IoT) devices and artificial intelligence (AI), along with additional digital technologies, to automate and optimize activities such as energy consumption, security, comfort, and overall building management (Fig. 6).

Advanced automation of homes is at the core of a smart residential complex. Occupants may use this function to regulate and control their home surroundings, such as lighting, heating, ventilation, air conditioning (HVAC), and appliances. Because the automation systems are extremely customizable, homeowners may match the technology to their specific tastes and everyday activities [6].

The Internet of Things (IoT) plays a critical role in these complexes, with a network of connected devices and gadgets that can be operated by central platforms or mobile applications. This connection allows for smooth control and monitoring, which improves the usefulness and ease of several home systems.

These complexes run on data, which is collected from a variety of sensors and devices. This strategy allows for informed decision-making on maintenance, energy usage, and other operational issues.

A smart residential complex, in basic words, is a progressive living space that combines cutting-edge technology to create an efficient, safe, and enjoyable setting, prioritizing its occupants' needs and welfare while embracing sustainable practices [2].

After highlighting the definition of smart residential complexes and their main component, the study delves into the each key themes defined earlier.

3.2.1 IoT-Based Energy Management Systems

IoT technology has emerged as a key component in smart domestic energy management. Mocrii et al. [5] did a thorough

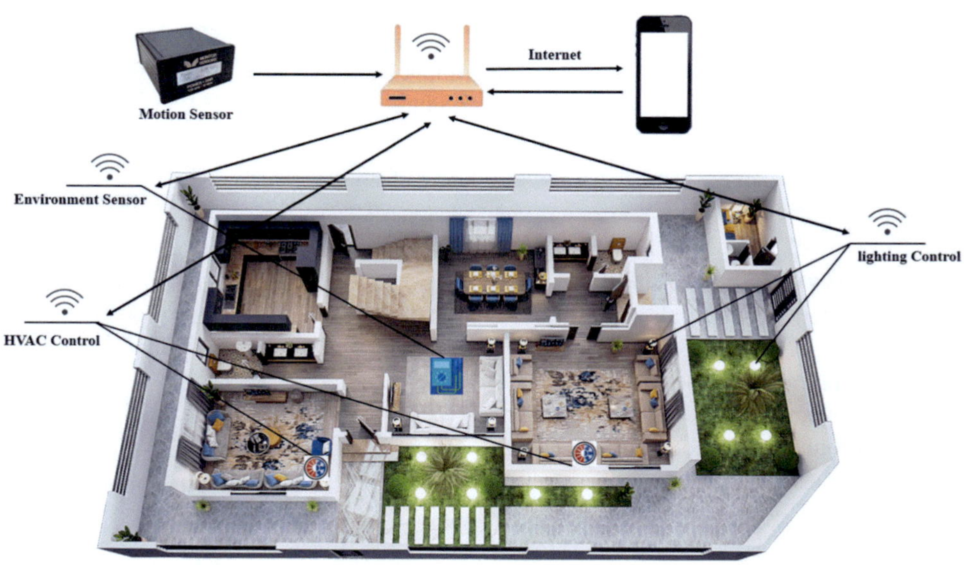

Fig. 6. IoT Application Diagram in Smart Housing (Source: By authors)

evaluation of IoT-based smart homes, emphasizing the power of networked devices in optimizing energy use. According to their analysis, IoT-enabled houses may save up to 30% more energy than conventional homes.

Jia et al. [2] investigated the use of IoT in smart buildings, focusing on enabling technologies and applications. They found major difficulties in IoT deployment, such as data security and interoperability concerns. Their results imply that although IoT has considerable promise for energy management, standardization initiatives are critical for wider implementation.

3.2.2 AI and Machine Learning Approaches

Artificial intelligence and machine learning have demonstrated favorable results in anticipatory energy management. Antonopoulos et al. [3] performed a thorough assessment of AI and machine learning techniques for energy demand response. Their examination of 160 articles found that reinforced training and deep neural networks were the most widely employed approaches, with stated energy savings ranging between 10% and 25%.

Wang et al. [17] proved the efficacy of advanced reinforcement learning in optimizing HVAC systems in smart homes. Their technique resulted in a 15% decrease in energy use while preserving occupant comfort.

3.2.3 Integration Challenges

Despite their promise, these technologies continue to pose major integration challenges. Carlander and Thollander [11] highlighted the challenges to deploying energy-efficient technology in construction projects. Their case study in Sweden identified high initial expenses, a lack of knowledge, and opposition to change as important impediments.

3.2.4 User-Centric Approaches

Recent research has emphasized the significance of user participation in energy management systems. Park et al. [20] investigated consumer acceptability of smart home services and found that perceived ease of use and usefulness were significant determinants of adoption rates. Their findings indicate that user-friendly interfaces and straightforward communication of advantages are critical to the achievement of smart energy management systems.

3.2.5 Research Gaps and Future Directions

The systematic review reveals several key research gaps:

1. Limited research on the long-term performance of integrated systems in real-world environments.
2. Inconsistent methods for assessing energy savings among smart home systems.
3. Limited study on the scalability of AI and ML solutions in big residential complexes.
4. Additional research is needed to reconcile energy economy with user comfort and privacy.

Further study should concentrate on overcoming these gaps, including via longitudinal studies and the creation of standardized assessment measures for smart household energy management systems.

4 Smart Technologies in Buildings

The incorporation of smart technology in buildings, especially in the framework of energy management systems, is transforming how residential complexes utilize and handle energy. These solutions improve not just the use of energy but the whole sustainability, comfort, and efficiency of living areas. This integration entails a mix of hardware and software solutions that function collectively to track, control, and optimize the energy usage of the structure. The key components of this integration can be categorized as follows;

- Smart Sensors and Metres, These sensors gather information on a variety of characteristics such as temperature, humidity, occupancy, and energy use. This real-time data is critical for the EMS to make educated energy distribution and preservation choices. Studies have shown that smart meter implementation can lead to energy savings of 3–5% in residential settings [20].
- IoT-enabled devices, such as smart thermostats and lighting systems, enable automatic and remote control of building systems. According to Jia et al. (2019), IoT-enabled smart building systems have the potential to cut energy consumption by 20–25%. [2]
- Artificial Intelligence and Machine Learning (AI and ML) are being utilized to forecast and optimize building systems. For example, Wang et al. (2019) built a based on artificial intelligence HVAC system that reduced energy usage by 15% while preserving comfort levels. [17]

To illustrate the effectiveness of these technologies, consider the following Table (Table 3):

Table 3. Effectiveness of advanced technologies on energy saving (Source: By authors)

Technology	Average Energy Savings	Source
Smart Meters	3–5%	[20] Sovacool et al. (2017)
IoT-enabled Systems	20–25%	[2] Jia et al. (2019)
AI-driven HVAC	15%	[17] Wang et al. (2019)

4.1 Application of the Smart Technologies in Residential Complexes

Smart technologies integration in residential architecture is transforming how houses are planned, constructed, and managed. These innovations not only increase the comfort and convenience in residential areas, but they also enhance energy consumption and environmental sustainability dramatically. This section investigates numerous smart technology uses in residential architecture, demonstrating their influence on contemporary life.

The usage of smart residence automation technologies is a critical factor. Lighting, heating, ventilation, air conditioning (HVAC), security, etc., are all combined into a single, intuitive user interface with these systems. This connection enables residents to manage key features remotely using smartphones or tablets, increasing ease and effectiveness in operation. Automated lighting and temperature management, for example, may vary in accordance with usage and the moment of day, reducing wasteful energy usage [26].

Smart technologies are a game changer in the field of energy management. Smart meters and energy management systems provide homeowners with real-time insights into their energy use, allowing them to identify possibilities for energy reduction. The use of renewable energy sources such as solar panels and wind turbines, in conjunction with energy storage devices, optimizes household energy usage and minimizes reliance on conventional power networks [2].

Smart technologies have an impact on building materials and construction procedures. Materials having adaptive characteristics, such as phase-changing materials and smart glass, are increasingly being used in contemporary houses to help with natural temperature adjustment. Furthermore, smart technology-enabled novel construction technologies such as 3D printing and modular construction are simplifying the process of constructing, making it faster, more efficient, and less wasteful. Another important component of contemporary home design is the Internet of Things (IoT). IoT-enabled equipment, such as refrigerators, washing machines, and ovens, provide increased interconnection and automation. These gadgets not only increase daily comfort but also help with energy saving by automating operations and giving homeowners useful consumption data[21, 22].

Advanced smart technology solutions have improved house security and safety. Smart locks motion sensors, for example, increase monitoring and notify residents of possible security breaches. The incorporation of these systems with mobile devices allows for real-time updates and remote control, providing an additional layer of protection to contemporary houses [23]. Smart technologies are enabling a new paradigm in dwelling architecture: health and wellness monitoring. Sensors and equipment that monitor indoor air quality, as well as various other external variables, are increasingly being installed in homes to provide a healthy living environment. In addition, by incorporating wearable technology and health monitoring devices into house design, inhabitants' health indicators may be tracked, providing significant health insights and alarms [9].

At last, smart technologies play an important role in encouraging environmentally friendly and sustainable techniques in home design. Home automation systems optimize the utilization of resources, decreasing waste and lowering environmental footprints. Smart water management systems, for example, not only effectively regulate water consumption but also enable rainwater collecting and recycling programs, therefore encouraging water conservation.

4.2 Implementation Challenges and Solutions

Despite its promise, using smart technology in residential complexes presents some substantial hurdles. One of the biggest impediments is the high upfront costs connected with these technologies. The initial expenditure needed for smart systems may be significant, which frequently deters property developers and homeowners. However, these expenditures may be mitigated in the long term. According to a recent cost-benefit study undertaken by Carlander and Thollander (2023), the majority of smart systems get a return on investment within 3–5 years, making them economically feasible over time [11].

Another big worry with the introduction of smart technology is privacy and security. The gathering and transmission of data, which is essential to the operation of smart devices, naturally raises concerns about data security and privacy. To overcome these issues, it is critical to design strong encryption mechanisms and create clear data regulations. These methods may assist increase consumers' confidence and guarantee the safe functioning of smart technologies. [5].

Interoperability is another major difficulty in the smart technology world. The market is presently saturated with a wide range of smart gadgets, many of which employ incompatible protocols. This lack of standardization might result in fragmented systems and diminished efficiency. Nevertheless, the industry is taking steps to solve this problem. The introduction of open standards like as Matter in 2022 marks a concentrated attempt to tackle interoperability concerns, perhaps opening the way for easier integration between varied smart devices [24].

Finally, user acceptance is important to the success of smart technology. Even the most powerful technologies might fall short of their full potential if users do not actively connect with them. A lack of user interaction may dramatically reduce the efficacy of smart systems in practice. To overcome this, developers are working to

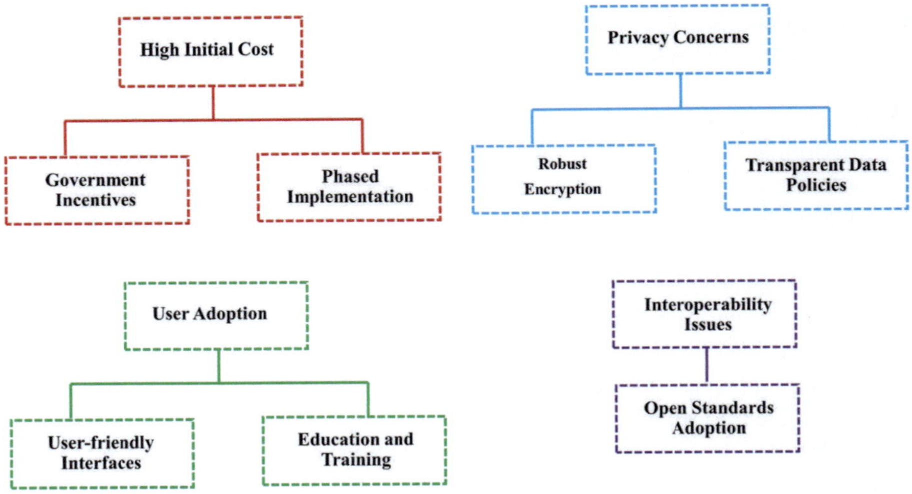

Fig. 7. Challenges and Solutions in Smart Technology Implementation (Source: By authors)

create more user-friendly interfaces and properly communicate the advantages of these technologies to end users. Such efforts have had positive results, with research showing that these tactics may boost adoption rates by up to 40% [18]. Figure 7 depicts the implementation issues of smart technology (Fig. 6).

4.3 Emerging Trends and Future Directions

The context of smart technology in residential complexes is quickly changing, with various new ideas influencing the future. One of the most interesting advances is the usage of digital twins, which are virtual duplicates of actual structures used to improve performance. Alanne et al. (2023) found that digital twins may enhance building energy efficiency by 10–15% [13]. This development enables more exact monitoring and control of building systems, opening the path for more effective energy management.

Along with digital twins, edge computing is gaining traction in the smart building industry. Edge computing minimizes latency by processing data closer to its origin and improves real-time decision-making. This functionality is especially important for time-critical applications like demand responses in energy management systems [2]. The ability to process data rapidly and locally not only enhances the system's adaptability but also overcomes some of the bandwidth and privacy problems related to cloud-based systems.

Another creative development is the use of blockchain technology in energy trading. Peer-to-peer energy trading systems based on blockchain are gaining pace, opening up new avenues for energy management in residential communities. A revolutionary research by Chen et al. (2022) showed that blockchain-based systems might lower energy expenses by 18% while improving renewable energy utilization by 25%. This method supports not just more efficient energy consumption, but also the adoption of renewable energy sources inside residential communities.

Federated learning is emerging as a possible answer to privacy problems in smart building systems. This method enables machine learning models to be trained on numerous decentralized devices without sharing raw data. Kumar et al. (2023) proved the method's efficacy, achieving a 25% increase in energy forecast accuracy while respecting user privacy [24]. Federated learning therefore provides a mechanism to exploit the value of community data without sacrificing individual privacy, which is an important factor in residential settings.

Looking forward, some crucial areas demand more study and development. Long-term performance studies of integrated smart systems in residential complexes are required to better understand their long-term impact and efficacy. There is also an urgent need for the creation of standardized measurements that can objectively assess the efficacy of various smart technologies. This standardization would enable more informed decision-making in the adoption of smart systems.

Furthermore, as these technologies become more widely used in residential settings, it is critical to investigate their social and behavioral consequences. Understanding inhabitants' interactions with and adaptation to smart technologies may influence improved design and implementation tactics, perhaps leading to increased acceptance rates and more effective usage of these systems.

Finally, the potential of quantum computing in optimizing complex energy management systems is a promising area for future study. As quantum computers become more available, their enormous computational capacity might revolutionize how we approach energy optimization in smart buildings, possibly discovering answers to complicated issues that are presently unsolvable with traditional computing approaches.

These rising trends and future research orientations all lead to a future in which residential complexes are not just smart, but also highly efficient, user-friendly, and fully integrated with sophisticated technology. As research advances in these areas, we should expect to see major breakthroughs in energy management, user comfort, and overall sustainability in smart residential complexes.

5 Energy Management in Residential Complexes

In the present day, when energy efficiency and sustainability are key, energy management in residential complexes is becoming more vital. Effective energy management strategies in residential settings are critical in light of rising energy consumption and the need to minimize carbon footprints [1]. This section delves into the numerous facets of household energy management, including technology, techniques, and obstacles.

5.1 Technologies for Energy Management

Advanced technologies are becoming more important in residential complex energy management, where they considerably improve the capacity to monitor and regulate energy consumption. Smart meters and smart home gadgets, for example, enable people to monitor their energy usage in real time, identifying particular areas for possible energy savings. Integration of Internet of Things (IoT) devices extends this by automating the operation of household systems such as heating, ventilation, air conditioning (HVAC), lighting, and appliances. This automation optimizes energy consumption while preserving comfort levels, demonstrating a healthy balance between technology and everyday life.

Nevertheless, deploying these sophisticated energy management systems is not without difficulties. The initial cost of implementing energy-efficient technology and renewable energy systems is a significant barrier. These fees, although advantageous in the long run, might be daunting at first. Furthermore, the variable structure of residential complexes implies that energy demands vary dramatically from one building to the next, making customization of energy management systems both necessary and difficult [11].

The next phase of energy management in residential complexes is set to be exciting, thanks to more incorporation of intelligent technology. Machine learning and artificial intelligence (AI) are projected to play a greater part in anticipating energy needs and optimizing use [26]. Blockchain technology might emerge as a crucial participant in community-based energy trading, allowing locals to sell extra renewable energy. Furthermore, the drive towards net-zero buildings is gaining momentum. These energy-efficient buildings attempt to create as much energy as they use, integrating passive design methods with active technological innovations to minimize energy consumption while maximising output. This future envisions a more ecological and energy-efficient living environment that smoothly integrates technology into daily life.

5.2 AI and Machine Learning Energy Management

AI and machine learning are at the core of current energy management systems in homes. These technologies can process vast amounts of data produced by different home devices and systems. AI can identify trends in energy use and provide optimization based on this information. AI algorithms, for example, may learn a household's usual energy usage habits and alter heating and cooling systems ahead of time to optimize energy use and save expenses.

AI is also useful in detecting anomalies. It may detect unexpected patterns in energy usage that may suggest a faulty appliance or system, encouraging early repair or replacement and reducing energy waste. This predictive maintenance is an important step toward improving energy efficiency and lowering unexpected failures and expenses [26]. Machine learning algorithms excel in predictive analytics, forecasting future energy demands based on previous data. This skill enables improved energy resource planning and optimization. For instance, on occasions when the forecast predicts a lot of sunshine, a smart energy system may store more solar energy, expecting less dependence on the grid.

5.3 IoT in Residential Energy Management

In smart homes, IoT acts as a foundation for gathering real-time data. Sensors and smart meters placed throughout a home collect data on energy use, temperature, and other environmental factors. This data is then sent to a centralized database where it may be analyzed and acted accordingly [8].

When IoT is integrated with AI and ML, its whole potential is realized. Smart thermostats, for example, utilize IoT to monitor home conditions and AI algorithms to learn homeowner preferences and modify settings for both comfort and energy savings. Smart lighting systems, on the other hand, may alter brightness depending on natural light availability, occupancy, and time of day, resulting in considerable energy savings. Another major benefit of combining AI, IoT, and ML is improved user contact with energy systems. AI-powered smart home assistants enable people to handle numerous parts of their home energy system via simple voice requests or mobile apps. Users may operate heating, lighting, and other appliances remotely, which adds convenience while assuring effective energy consumption [10].

5.4 Challenges and Future Developments

While incorporating AI, IoT, and ML into household energy management systems has various advantages, it also has drawbacks, such as data privacy issues, the necessity for effective cybersecurity regulations, and the potential difficulty of maintaining networked systems. Furthermore, there is a constant need for the development of increasingly advanced algorithms that may improve energy efficiency even further.

6 Findings and Conclusion

The complete analysis of energy management in smart residential complexes provides various significant insights into the field's present condition, developing trends, and future research opportunities. These results highlight the transformational potential of intelligent solutions for domestic energy management. The most effective techniques combine various technologies, such as IoT devices for data collecting, artificial intelligence for decision-making and analysis, and intuitive user interfaces for resident participation [27]. The combination of renewable energy sources, smart grid technology, and energy storage systems has also shown promise for improving overall energy efficiency [28]. However, significant obstacles in adopting these systems remain, such as high initial costs, privacy and security concerns, and interoperability issues across various smart devices and systems. [5, 11].

Emerging developments influencing the prospects of energy management in smart residential complexes include the usage of digital twins and edge computing for improved real-time decision-making [25]. The evaluation discovered multiple fields for future research, including long-term performance examinations of incorporated smart systems in real-world settings, the creation of standardized metrics for comparing different technologies, and investigations into the social and behavioral consequences of smart technology adoption [20]. The potential of new technologies, such as quantum computing, in optimizing complicated energy management systems is also largely untapped.

The study emphasizes that user participation is crucial to the success of smart energy management systems. User-friendly interfaces and clear explanations of advantages have been proven to enhance adoption rates by up to 40% [18], emphasizing the need to take human elements into account when designing technology solutions. To summarize, although smart technologies hold enormous potential for improving energy management in residential complexes, their effective deployment necessitates overcoming a variety of technical, economic, and social problems. The sector is continuously changing, bringing new opportunities but also requiring ongoing study to fully realize the promise of these sophisticated technologies in generating more energy-efficient and sustainable home settings [19].

References

1. Beaudin, M., Zareipour, H.: Home energy management systems: a review of modeling and complexity. Renew. Sustain. Energy Rev. **45**, 318–335 (2015). https://doi.org/10.1016/j.rser.2015.01.046
2. Jia, M., Komeily, A., Wang, Y., Srinivasan, R.S.: Adopting Internet of Things for the development of smart buildings: a review of enabling technologies and applications. Autom. Constr. **101**, 111–126 (2019). https://doi.org/10.1016/j.autcon.2019.01.023
3. Antonopoulos, I., et al.: Artificial intelligence and machine learning approach to energy demand-side response: a systematic review. Renew. Sustain. Energy Rev. **130**, 109899 (2020). https://doi.org/10.1016/j.rser.2020.109899
4. King, J., Perry, C.: Smart buildings: using smart technology to save energy in existing buildings, pp. 1–46. Washington, DC, USA: American Council for an Energy-Efficient Economy (2017)
5. Mocrii, D., Chen, Y., Musilek, P.: IoT-based smart homes: a review of system architecture, software, communications, privacy and security. Internet of Things **1**, 81–98 (2018). https://doi.org/10.1016/j.iot.2018.08.009
6. Mirdula, S., Roopa, M.: MUD enabled deep learning framework for anomaly detection in IoT integrated smart building. e-Prime-Adv. Electr. Eng. Electron. Energy **5**, 100186 (2023). https://doi.org/10.1016/j.prime.2023.100186
7. United Nations: 2018 Revision of world urbanization prospects (2018). Retrieved from https://www.un.org/development/desa/publications/2018-revision-of-world-urbanization-prospects.html
8. Marinakis, V., Doukas, H.: An advanced IoT-based system for intelligent energy management in buildings. Sensors **18**(2), 610 (2018). https://doi.org/10.3390/s18020610
9. El Jaouhari, S., Jose Palacios-Garcia, E., Anvari-Moghaddam, A., Bouabdallah, A.: Integrated management of energy, wellbeing and health in the next generation of smart homes. Sensors **19**(3), 481 (2019). https://doi.org/10.3390/s19030481
10. Mazhar, T., et al.: The role of ML, AI and 5G technology in smart energy and smart building management. Electronics **11**(23), 3960 (2022). https://doi.org/10.3390/electronics11233960
11. Carlander, J., Thollander, P.: Barriers to implementation of energy-efficient technologies in building construction projects—Results from a Swedish case study. Resour. Environ. Sustain. **11**, 100097 (2023). https://doi.org/10.1016/j.resenv.2022.100097
12. Moher, D., Liberati, A., Tetzlaff, J., Altman, D.G., Prisma Group.: Preferred reporting items for systematic reviews and meta-analyses: the PRISMA statement. PLoS Med. **6**(7), e1000097 (2009). https://doi.org/10.1371/journal.pmed.1000097
13. Alanne, K.: Study on reducing the grid dependency of urban housing in Nordic climate by hybrid renewable energy systems. Renew. Energy Focus **46**, 1–15 (2023). https://doi.org/10.1016/j.ref.2023.05.006
14. Li, W., Logenthiran, T., Woo, W.L.: Intelligent multi-agent system for smart home energy management. In: 2015 IEEE Innovative Smart Grid Technologies-Asia (ISGT ASIA), pp. 1–6. IEEE (2015). https://doi.org/10.1109/ISGT-Asia.2015.7386985
15. Talei, H., Benhaddou, D., Gamarra, C., Benbrahim, H., Essaaidi, M.: Smart building energy inefficiencies detection through time series analysis and unsupervised machine learning. Energies **14**(19), 6042 (2021). https://doi.org/10.3390/en14196042
16. Moustaka, V., Vakali, A., Anthopoulos, L.G.: A systematic review for smart city data analytics. ACM Comput. Surv. (cSuR) **51**(5), 1–41 (2018). https://doi.org/10.1145/3239566

17. Wang, Z., Hong, T., Piette, M.A.: Data fusion in predicting internal heat gains for office buildings through a deep learning approach. Appl. Energy **240**, 386–398 (2019). https://doi.org/10.1016/j.apenergy.2019.02.066
18. Park, E., Kim, S., Kim, Y.S., Kwon, S.J.: Smart home services as the next mainstream of the ICT industry: determinants of the adoption of smart home services. Univ. Access Inf. Soc. **17**(1), 175–190 (2018). https://doi.org/10.1007/s10209-017-0533-0
19. Khan, S.U., Khan, N., Ullah, F.U.M., Kim, M.J., Lee, M.Y., Baik, S.W.: Towards intelligent building energy management: AI-based framework for power consumption and generation forecasting. Energy Build. **279**, 112705 (2023). https://doi.org/10.1016/j.enbuild.2022.112705
20. Sovacool, B.K., Kester, J., Noel, L., de Rubens, G.Z.: The demographics of decarbonizing transport: the influence of gender, education, occupation, age, and household size on electric mobility preferences in the Nordic region. Glob. Environ. Chang. **52**, 86–100 (2018). https://doi.org/10.1016/j.gloenvcha.2018.06.008
21. The Internet of Things: Making Sense of the Next Mega-Trend. New York: The Goldman Sachs Group, Inc. (2014). www.academia.edu/31067883/The_Internet_of_Things_Making_sense_of_the_next_mega_trend
22. Tracy, P.: "How IoT Is Lowering the Cost of Building Management Systems." Enterprise IoT Insights (2016). August 8. Retrieved from www.rcrwireless.com/20160808/network-infrastructure/inbuildingtech/building-management-system
23. Peterson, A.: 'Internet of Things' Compounded Friday's Hack of Major Websites. The Washington Post (2016). https://www.washingtonpost.com/news/the-switch/wp/2016/10/21/someone-attacked-a-major-part-of-the-internets-infrastructure/
24. Connectivity standards alliance: Matter: The Foundation for Connected Things (2022). https://csa-iot.org/all-solutions/matter/
25. Deng, S., Zhao, H., Fang, W., Yin, J., Dustdar, S., Zomaya, A.Y.: Edge intelligence: the confluence of edge computing and artificial intelligence. IEEE Internet Things J. **7**(8), 7457–7469 (2020). https://doi.org/10.1109/jiot.2020.2984887
26. Russell, S.J., Norvig, P.: Artificial Intelligence: A Modern Approach. Pearson (2016)
27. Hannan, M.A., et al.: A review of internet of energy based building energy management systems: Issues and recommendation. IEEE Access **6**, 38997–39014, 8403212 (2018). https://doi.org/10.1109/ACCESS.2018.2852811
28. Mazzola, S., Astolfi, M., Macchi, E.: The potential role of solid biomass for rural electrification: a techno economic analysis for a hybrid microgrid in India. Appl. Energy **169**, 370–383 (2016). https://doi.org/10.1016/j.apenergy.2016.02.051

Constructions of Green Façade and Effects on Thermal Environment: Simulations with CFD

Hankun Lin, Huijie Wang, Manchun Liu, Xinye Su, Yuwen Wang, Xiaomei Lin, and Siyuan Zhan

Abstract

In response to the challenges posed by high-density urbanism, such as the urban heat island (UHI) effect, increased traffic noise, and exacerbated air pollution that collectively diminish human comfort, vertical greenery systems (VGSs) have emerged as a strategy. Among VGSs, green façades (GFs) are particularly noteworthy for their ability to guide climbing plants to grow vertically, providing shade and thus influencing the thermal environment of buildings. However, the natural growth of these climbing plants makes the shading ratio and its correlation to the thermal environment difficult to control and predict. This paper endeavors to construct four types of GF configurations that are both predictable and controllable in terms of shading methods and thermal impact. Computational fluid dynamics (CFD) simulations were utilized to assess the thermal performance of different GF foliage settings. A coupled ventilation and heat transfer model was applied for the simulations, which included four types of GF constructions and a control model without shading. To evaluate the thermal impact of GFs on both indoor and buffer spaces, we compared air temperature, wind velocity, and wall surface temperature. The key outcomes are summarized as follows: (1) The average temperature in the buffer space was reduced by 0.67 °C to 3.24 °C across different shading methods. (2) The range of thermal optimization for the indoor environment was 1.27 °C to 3.45 °C, and for the buffer space, it was 0.67 °C to 3.24 °C. (3) The wall temperature in models H1 and H2 was reduced by up to 7.68 °C on the south orientation and 13.24 °C on the west orientation. (4) The average indoor wind velocity showed minimal difference between shaded and unshaded conditions. (5) Based on comparisons of air and wall surface temperatures, models H1 and H2 demonstrated the most favorable effects in this study. As part of a broader investigation into the effects and configurations of GFs, these results will inform future research on the thermal impact of GFs on human thermal comfort through field measurements and numerical simulations.

Keywords

Green façade · Thermal environment · CFD · Coupling simulation · Shading construction

1 Introduction

Rapid urbanization has amplified environmental stressors in densely built regions, particularly in hot-humid climates where elevated temperatures and humidity compromise outdoor livability. The dominance of impervious surfaces intensifies urban heat island (UHI) effects, while diminishing green spaces exacerbate air pollution and thermal discomfort (Santamouris & Osmond, 2020). These challenges demand innovative solutions that reconcile ecological benefits with spatial efficiency (Huang et al., 2024). Vertical greenery systems (VGS), including green façades (GFs) and living walls, have gained traction for their dual role in mitigating microclimatic extremes and enhancing urban aesthetics (Zaid et al., 2018).

VGS are typically classified into two categories: living wall systems (LWS), which rely on modular substrates and automated irrigation, and GFs, which utilize climbing plants supported by lightweight frameworks (Manso & Castro-Gomes, 2015). While LWS demonstrate superior thermal insulation, GFs offer cost-effectiveness and simpler maintenance, achieving surface temperature reductions of up to

H. Lin (✉) · H. Wang · M. Liu · X. Su · Y. Wang · X. Lin · S. Zhan
School of Architecture and Urban Planning, Guangdong University of Technology, Guangzhou, China
e-mail: hklin@gdut.edu.cn

B. He et al. (eds.), *Towards the Framework of Livable and Resilient Cities* Advances in Science, Technology & Innovation, https://doi.org/10.1007/978-3-031-97849-4_16

10–16 °C through shading and evapotranspiration (Perini & Rosasco, 2013; Kontoleon & Eumorfopoulou, 2010; Coma et al., 2017). GFs also provide shading for building windows, which can influence daylighting and enhance the visual experience for users (Sharbafian et al., 2024). Existing research on GF and LWS has demonstrated significant potential in enhancing building energy efficiency, improving thermal comfort, and reducing local air pollution across various climate zones (Pérez et al., 2014; De Jesus et al., 2017; Zhao et al., 2023). However, their application in building façade systems remains underexplored, especially in regions where balancing solar protection and natural ventilation is critical.

To better understand the potential of GFs in building façade systems, several simulation studies have been conducted. For example, Alsaad et al. (2022) used computational fluid dynamics (CFD) to assess the effects of living walls on microclimatic conditions. Šuklje et al. (2016) carried out a numerical simulation of the thermal performance of GFs, developing a mathematical model to estimate their cooling potential. Their research demonstrated that GFs can significantly reduce façade surface temperatures by up to 34 K (degrees Kelvin). These studies collectively highlight the potential of VGSs to enhance thermal comfort and energy efficiency in urban environments. Existing research underscores the effectiveness of VGSs in improving urban microclimates and thermal comfort, with GFs showing notable potential for reducing façade surface temperatures.

Despite the growing use of GFs, the combined effects of GFs and shading methods on building façades remain underexplored, especially in regions where balancing solar protection and natural ventilation is critical. This paper aims to address this gap by developing optimized GF construction strategies through integrated CFD simulations. Four GF typologies are evaluated across south, east, and west orientations to quantify their impacts on wind velocity, air temperature, and wall surface temperature. By bridging empirical data with numerical modeling, this paper aims to develop predictable and controllable GF construction strategies, focusing on shading methods and thermal environment optimization.

2 Methodology

As a façade technology, it is crucial to clearly describe the organization, properties, and environmental effects of GFs. To compare the effects of GF constructions more efficiently, this paper utilizes a simulation method rather than field research. We adopted the CFD method to study the effects of different GF forms in front of the building windows. The process involved two main steps: First, we conducted a field measurement of a GF project and used these data to validate the CFD simulation method (Gromke et al., 2015; Lin et al., 2019; Lin et al., 2023). Second, we simulated four different types of GF constructions in different orientations and compared their effects.

2.1 Thermal Environmental Indices

This study focuses on the varying effects of GFs' constructions on the thermal environment. The following sections will introduce a coupling simulation method that incorporates wind and heat transfer in CFD. Several thermal environmental indices were utilized to evaluate the effects of GFs, including wind flow velocity (U), dry-bulb temperature (Ta), and material surface temperature (Ts).

2.2 Setting of the CFD Models

In the CFD simulations, Ansys Fluent 16.0 was utilized to solve the models. The study building model was designed as a two-story structure with double-sided windows. The scale of the model was 4 m × 6 m × 4 m (Fig. 1). The total computational domain scale was 24 m × 108 m × 40 m. The model grid was generated using hexahedral cells, with sizes ranging from 0.05 m, 0.1 m, 0.2 m, and 0.4 m to 1.0 m at the model boundary (Fig. 2). A grid sensitivity test was conducted, and the relative error was found to be less than 3% (Fig. 3). Consequently, the smallest grid size of 0.1 m was selected for the subsequent simulations.

The solution models were based on the 3D steady-state Reynolds-averaged Navier-Stokes (RANS) equations. Mei & Yuan (2022) recommended using RANS models in scenarios with weak buoyancy at the ground level. For solar irradiation and radiative transfer, the solar ray tracing model was employed in the simulation. The coupling pressure-velocity was solved with SIMPLEC scheme. Second-order schemes were ulitilized to solve the U, turbulence, and energy equations.

The Ta and U at the wind inlet boundary of the test model were set at 32 °C and 1.5 m/s, respectively, at a height of 10 m. The location, date, and solar irradiation of the model were set to 5th August in Guangzhou, China, based on the previous validation model (Lin et al., 2019; Lin et al., 2023).

At the setting of the wind inlet boundary, vertical profiles are defined following the best practice guidelines (Franke, 2011). The U, turbulence kinetic energy (k_i), and turbulence dissipation rate (ε) of the neutrally stratified atmospheric boundary layer were employed according to

$$U(z) = \frac{u_*}{k} \ln \frac{z + z_0}{z_0} \quad (1)$$

$$K_i(z) = \frac{u_*^2}{\sqrt{C_u}} \quad (2)$$

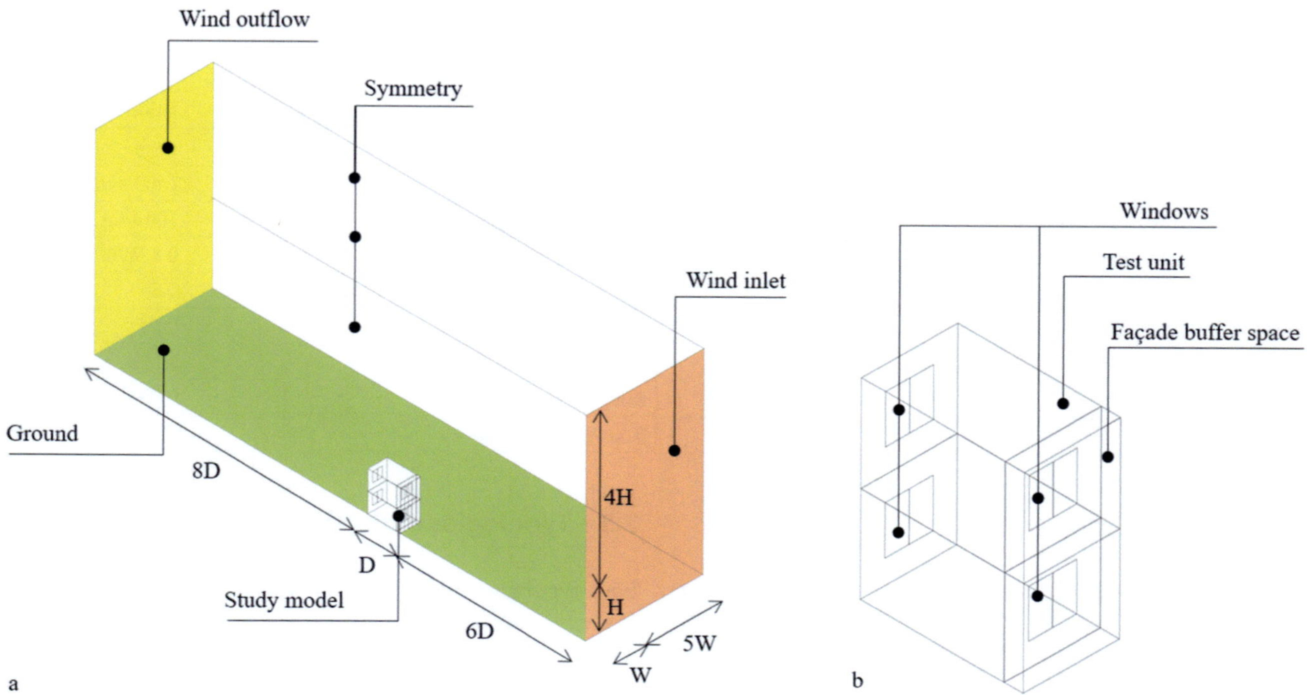

Fig. 1. CFD model of the simple 2 floors building (a. Scale and facet definition; b. Detail information of the study building model)

Fig. 2. Grid of the CFD model (a. Grid of the test building model; b. Grid of the computational domain)

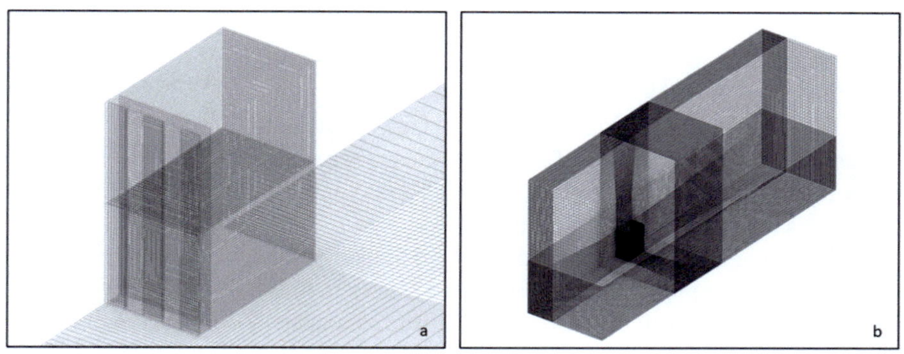

$$\varepsilon(z) = \frac{u_*^3}{k(z+z_0)} \quad (3)$$

where z is the vertical position, k is the van Karman constant (k = 0.42), z_0 is the aerodynamic roughness length (z_0 = 0.5 m), C_u = 0.09, and u_* is the friction velocity. U (1.1 m) was set as 1.5 m/s in the model.

Other boundaries, including symmetry boundaries, outflow boundary, and ground boundary, were assigned as Fig. 1. The definitions and parameters of the walls and the ground were assigned in Table 1.

To incorporate the effects of wind drag coefficient (C_d) and evapotranspiration cooling from plant foliage, the models were enhanced by defining porous zone cells with leaf area density (LAD). The thickness of these cells was set at 0.3 m, aligning with the average foliage layer thickness observed in field measurements of climbing plants. Drawing on the work of Buccolieri et al. (2019), additional terms were introduced

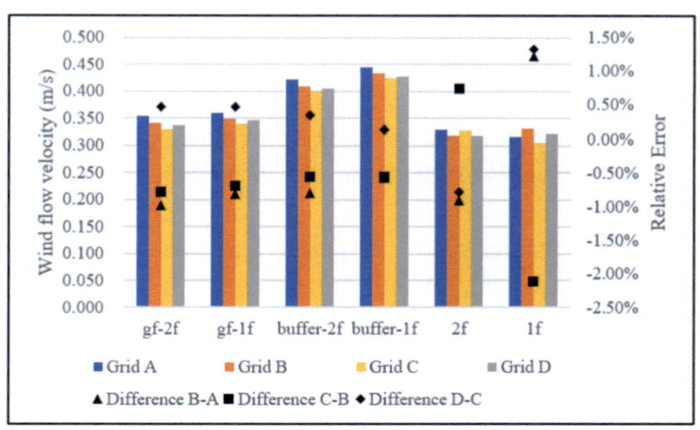

Fig. 3. Grid sensitivity test result the CFD model (Grid A: 0.05m, Grid B: 0.1m, Grid C: 0.2m, Grid D: 0.4m)

into the transport equations for momentum S_{u_i}, turbulence kinetic energy S_k, and turbulence dissipation rate S_ϵ within

Table 1. Boundary definitions and parameters of the study CFD model

Building walls conditions		Ground conditions	
Parameter	Value	Parameter	Value
Thickness	0.3 m	Thickness	10 m
Density	2000 kg/m^3	Density	1600 kg/m^3
Specific heat capacity	900 J/kg K	Specific heat capacity	300 J/kg K
Thermal conductivity	0.8 W/m K	Thermal conductivity	0.8 W/m K
Short-wave radiative absorption coefficient	0.8	Short-wave radiative absorption coefficient	0.85
Emissivity	0.9	Emissivity	0.85
Temperature	20 °C	Temperature	10 °C
Roughness height k_s	0	Roughness height k_s	0.05
Roughness constant C_s	0	Roughness constant C_s	0.5

the foliage computational cells:

$$S_{u_i} = -\rho C_d \text{LAD} U_i \text{U} \tag{4}$$

$$S_k = \rho C_d \text{LAD}\left(\beta_p \text{U}^3 - \beta_d \text{U} k_i\right) \tag{5}$$

$$S_\epsilon = \rho C_d LAD \frac{\varepsilon}{k_i}\left(C_{\varepsilon 4}\beta_p U^3 - C_{\varepsilon 5}\beta_d U k_i\right) \tag{6}$$

where ρ is the density of air (1.225 kg/m^3), $C_d = 0.2$, LAD = 2.5 m^2/m^3, U_i is the velocity component in direction i, β_p and β_d are the fraction of mean kinetic energy and the coefficient accounting for short-circuiting of the eddy cascade separatly ($\beta_p = 1.0$, $\beta_d = 5.1$), and $C_{\varepsilon 4} = C_{\varepsilon 5} = 0.9$ are empirical coefficients (Lin et al., 2019; Lin et al., 2023).

Sharbafian et al. (2024) evaluated the thermal load of the evergreen climbing plant *English Ivy (Hedera Helix)*. In their research, they input specific thermal characteristic parameters for the plant, which included thermal conductivity of 0.22 W/m K, specific heat of 3.08 J/g K, thermal absorptance of 0.9, and solar radiation absorptance of 0.7. This study is significant as it provides empirical data on how GFs can impact the thermal environment of buildings. According to the previous validation model (Lin et al., 2023), the cooling power (P_c) of the foliage layer is estimated to 625 W/m^3 when the LAD is 2.5 m^2/m^3.

Another parameter of the porous zone cells is the heat transfer capacity (C_T) (Heidkamp et al., 2017), according to

$$C_T = c_p \rho z_{veg} \tag{7}$$

where c_p is the specific heat capacity of air ($c_p = 1005$ J/kg K), z_{veg} is the height of the plant. Considering the foliage of GF is distributed vertically, z_{veg} is defined as 3.8 m in the model. The convergency value of different parameters was set as 10^{-5}. The settings of the CFD simulation have been validated in the previous studies (Lin et al., 2023).

2.3 Typologies of the GF Constructions

Four typologies of GF constructions were employed in the study: two vertical and two horizontal. Additionally, a model without GF was included for comparative analysis (Fig. 4).

The vertical GF models were designated as V1 and V2. In the V1 model, the GF was designed as a vertical shading structure that directly covered the front side of a window, providing more direct shade. In contrast, the GF in the V2 model was perpendicular to the window, offering less direct coverage but fostering a greater visual connection between the indoor and outdoor spaces. The horizontal GF models were named H1 and H2. The distinction between H1 and H2 models was similar to that between V1 and V2 models. Details of the different GF models' scales are provided in Table 2.

To simplify the simulation model, elements such as plant pots, support and guiding structures (which are typically integrated with the foliage), and irrigation and drainage pipes were omitted. In the model scenario, various orientations of south, west, and north were investigated in the same basic model.

3 Results and Discussion

3.1 Results of the Thermal Environmental Indices

Three sets of results are displayed in Fig. 5, which record data at a height of 1.1 m on both floors. Both indoor and buffer spaces were considered in the analysis. Figure 6 shows the temperature and wind flow velocity distribution across the models.

The air temperature results indicate that the GFs significantly reduced temperatures in shaded areas (Fig. 5a). The range of thermal optimization for the indoor environment was 1.27 °C to 3.45 °C, and for the buffer space, it was 0.67 °C to 3.24 °C. These variations depended on the typology of the

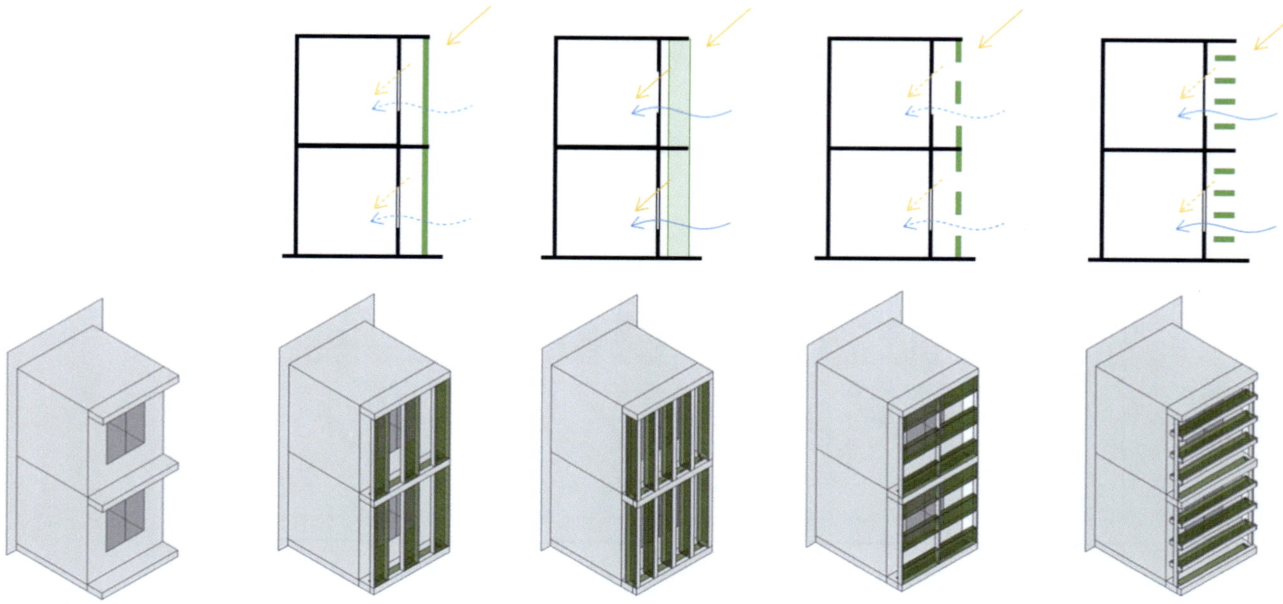

Fig. 4. Typologies of the GF constructions models with an unshaded model

Table 2. Scale of the GF units in different models

Model	Width (m)	Depth (m)	Height (m)	GF Distance (m)
V1	0.8	–	3.8	0.8
V2	–	0.6	3.8	0.75
H1	3.8	–	0.75	0.75
H2	3.8	0.6	–	0.75

GF construction and its orientation. On the 1st floor, model H1 showed the highest reduction in indoor temperature across three scenarios, while model V1 did the same for the buffer space. On the 2nd floor, models V1 and H2 provided the best optimization for both indoor and buffer temperatures.

Wind velocity results showed minimal differences (less than 0.1 m/s) in indoor wind speed across the three scenarios (Fig. 5b). Additionally, there was little variation in velocity between the scenarios. The highest difference between the unshaded and shaded areas reached up to 0.13 m/s in the buffer space of the 1st floor in model V1.

Regarding the material surface temperature of the external walls shaded by GFs, models H1 and H2 demonstrated the most significant reduction in wall temperature in the south and west orientations. The temperature difference was up to 7.68 °C in the south orientation and 13.24 °C in the west orientation (Fig. 5c).

3.2 Analysis of Wind Velocity and Air Temperature at Two Central Lines

In the south scenario, a further analysis of wind velocity and air temperature was conducted at two central lines across the opening windows at a height of 1.5 m on both floors of the test building (Fig. 7). Figure 7 also illustrates the GF area, with a width of 0.8 m, to demonstrate different variations of wind velocity and air temperature around the GFs.

The GF configurations significantly influenced the wind velocity distribution along the selected lines on both floors. Due to the drag effects, wind velocity was primarily reduced in front of the GF area but increased in the indoor space at the position of the opening windows. The difference in wind velocity between the GF models and the bare model was approximately 0.1 m/s to 0.3 m/s in the buffer space and up to 0.2 m/s to 0.9 m/s in the indoor space. On the selected lines, model V2 had the strongest effect on wind velocity, while model H2 had the weakest effect.

The GF configurations also had clear effects on air temperature reduction in both the buffer and indoor spaces across different GF models. At the selected lines, the difference in air temperature between the GF models and the bare model was approximately 0.3 °C to 4.0 °C on the 2nd floor and 0.1 °C to 3.6 °C on the 1st floor. In the indoor space, this difference was approximately up to 2.0 °C to 6.0 °C on the 2nd floor and 2.0 °C to 4.5 °C on the 1st floor. Model H2 showed the greatest temperature reduction on the 2nd floor, while Model H1 showed the greatest reduction on the 1st floor.

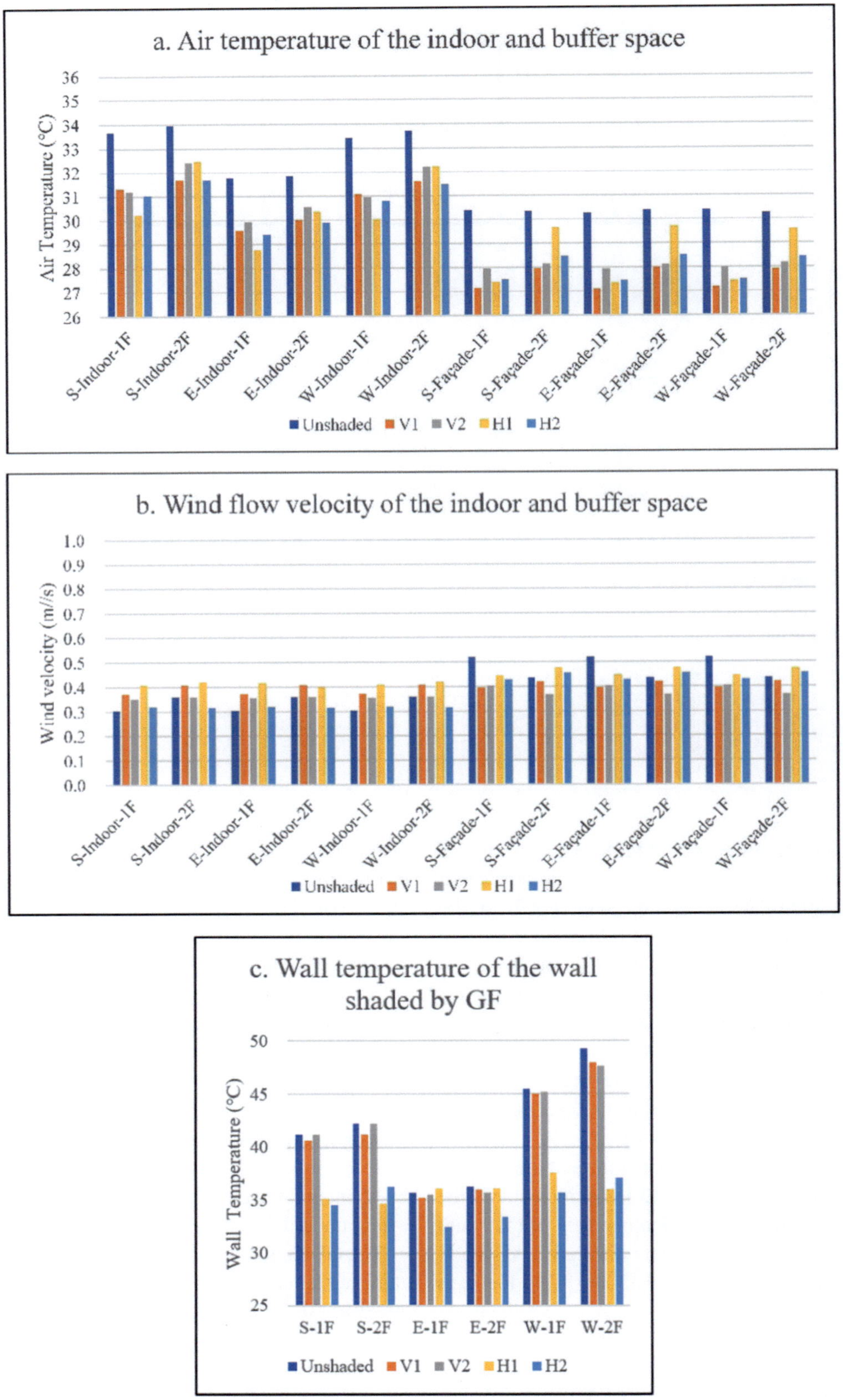

Fig. 5. (a,b,c) Results of the simulation of GF constructions

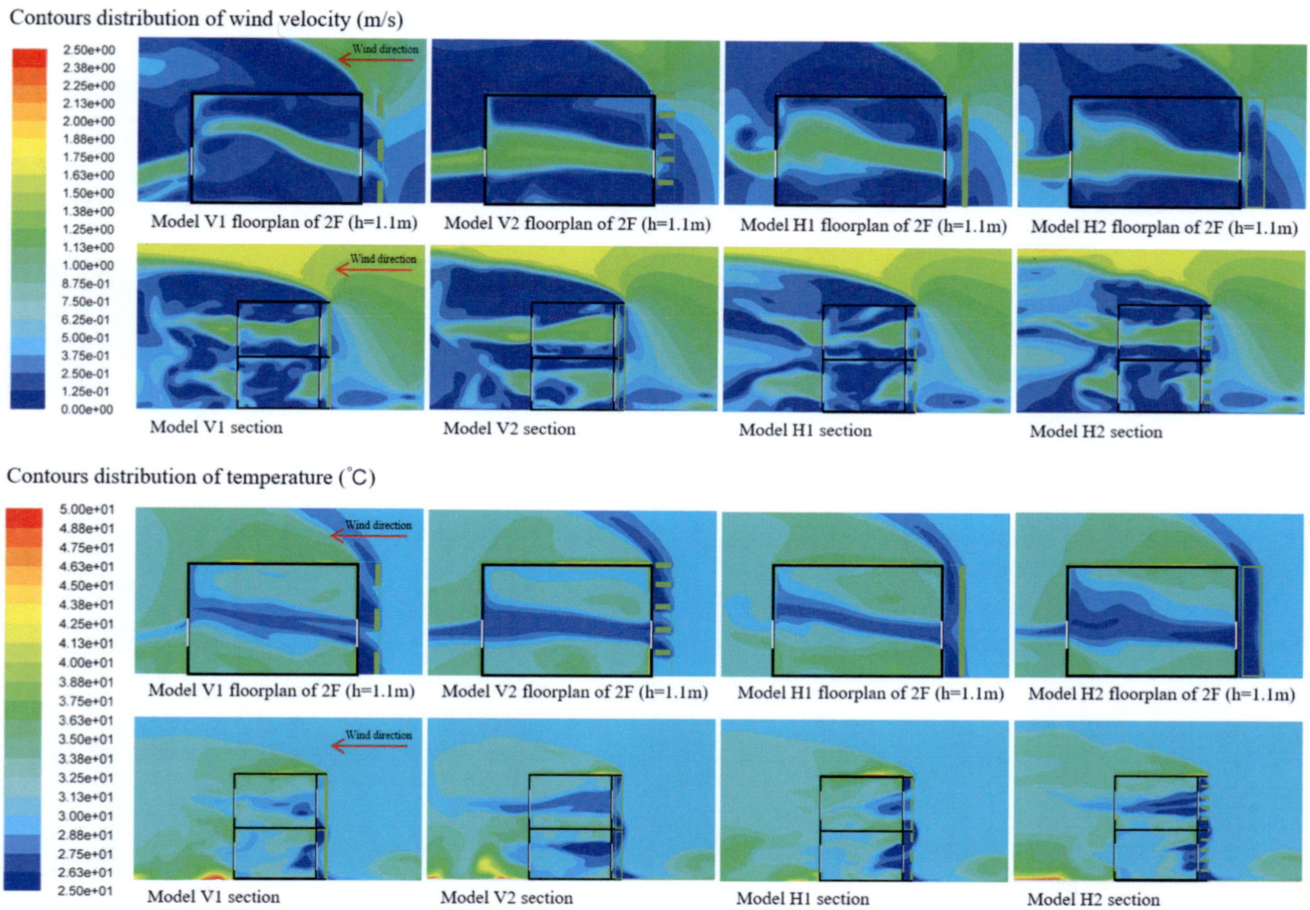

Fig. 6. Contours distribution of air temperature and wind velocity on room floorplans and sections (south orientation)

Since the selected lines were located at the center of the opening windows, the results on these lines enhanced the contrast in wind velocity and air temperature due to the transportation of cooling power at the GF constructions. The results also suggest that the design of the forms and configurations of the GFs significantly affects the performance of thermal regulation.

3.3 Discussion

In the results across five cases and three scenarios, the models with GFs showed a more significant effect on temperature than on wind flow velocity in the indoor environment. The average temperature reduction in the buffer space ranged from 0.67 °C to 3.24 °C across different shading methods. However, the average wind velocity in the indoor space in the simulations showed little difference between shaded and unshaded cases. Additionally, the wind pressure difference on the front and back sides of the rooms was minimal.

The results of wall temperature reduction are comparable with previous research. Koyama et al. (2013) reported a reduction in external wall surface temperature of 3.7 °C to 11.3 °C with different plants facing south during summer. Thus, the results for air temperature, wind velocity, and wall temperature with varying types of GF constructions fall within a reliable range. Moreover, different types of GF constructions have distinct impacts on the thermal environment. Since the wind velocity showed little difference in the case studies, the reduction in temperature becomes the primary judgment factor for comparison. In this regard, models H1 and H2 demonstrated better effects on both environmental and wall temperatures.

In the simulation cases, although GFs reduced the indoor temperature by a maximum of 3.45 °C, the indoor environment temperature still exceeded 29 °C. Additional technologies should be considered for integration during the design phase to optimize the indoor thermal comfort.

4 Conclusions

This paper utilized CFD simulations to compare four distinct construction methods of GFs. The results demonstrated the potential of GFs to optimize the thermal environment in both

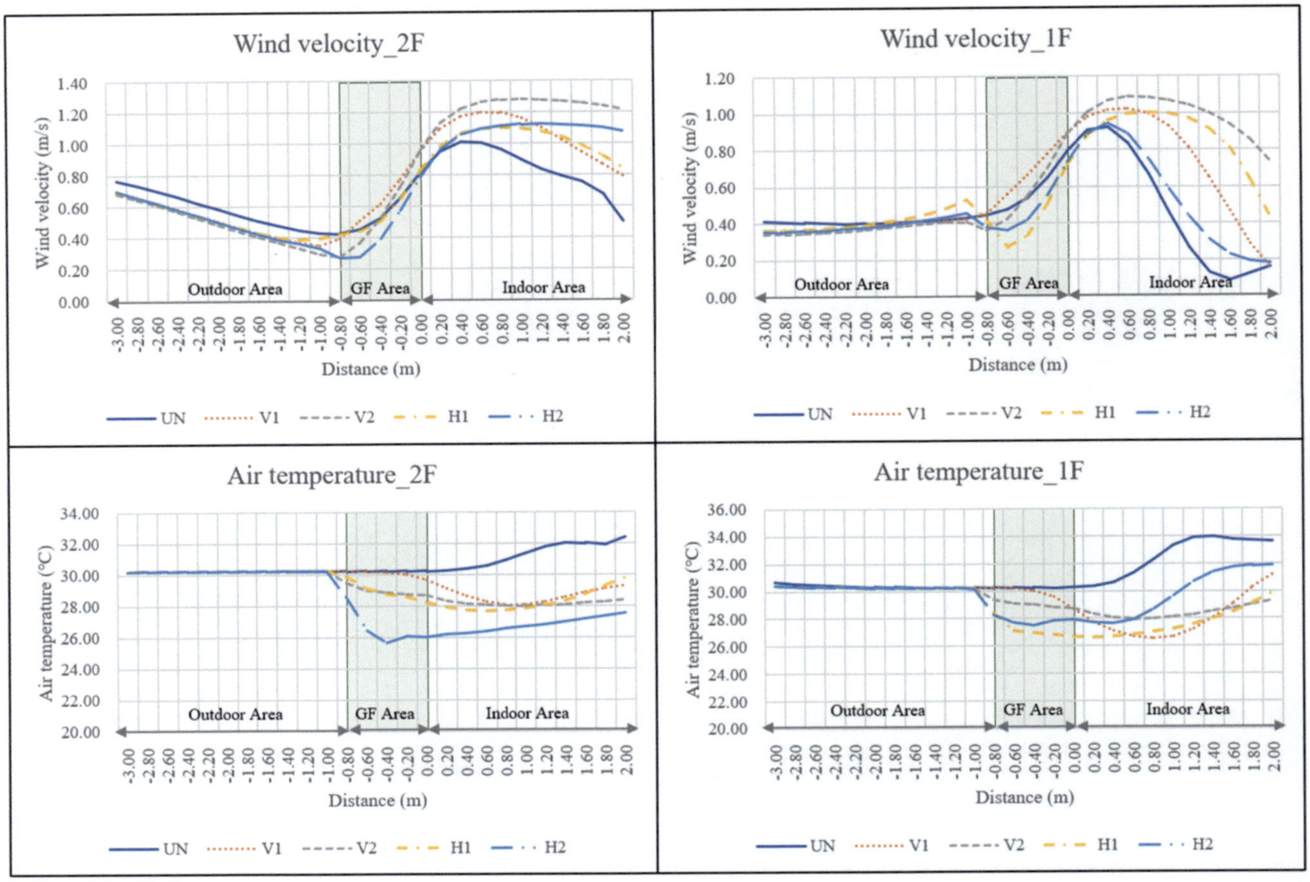

Fig. 7. Results of wind velocity and temperature at two central lines at height of 1.5 m on both floors

indoor and buffer spaces. The key outcomes are summarized as follows:

(1) The average temperature in the buffer space was reduced by 0.67 °C to 3.24 °C across different shading methods.
(2) The range of thermal optimization for the indoor environment was 1.27 °C to 3.45 °C, and for the buffer space, it was 0.67 °C to 3.24 °C.
(3) The wall temperature in models H1 and H2 was reduced by up to 7.68 °C on the south orientation and 13.24 °C on the west orientation.
(4) The average indoor wind velocity showed minimal difference between shaded and unshaded conditions.
(5) Based on comparisons of air and wall surface temperatures, models H1 and H2 demonstrated the most favorable effects in this study.

This study has several limitations that could be addressed in future research. The simulations were conducted at a specific time during the summer in Guangzhou, China. To provide a more comprehensive understanding, future studies should be conducted at multiple times and across different seasons. Additionally, the current CFD model only coupled ventilation with heat transfer; future studies should incorporate the factor of moisture to account for the effects of plant evaporation. Furthermore, this study focused solely on thermal environment simulations. Future research should couple thermal indices with energy performance assessments to offer a more comprehensive evaluation. In conclusion, the results of this study will support future reserches of GFs' effects on building energy and carbon footprint.

Acknowledgements This paper was funded by the Guangdong Basic and Applied Basic Research Foundation (No. 2023A1515010074), the Research Project of Humanities and Social Sciences of the Ministry of Education (No. 23YJC760069), and the GDUT Innovative Training Program for University Students (No. xj2024118451007 and No. xj2024118451015).

References

Santamouris, M., Osmond, P.: Increasing green infrastructure in cities: impact on ambient temperature, air quality and heat-related mortality and morbidity. Buildings **10**(12), 233 (2020). https://doi.org/10.3390/buildings10120233

Huang, H., et al.: Urban heatwave, green spaces, and mental health: a review based on environmental health risk assessment framework.

Sci. Total. Environ. **948**, 174816 (2024). https://doi.org/10.1016/j.scitotenv.2024.174816

Zaid, S.M., Perisamy, E., Hussein, H., Myeda, N.E., Zainon, N.: Vertical greenery system in urban tropical climate and its carbon sequestration potential: a review. Ecol. Ind. **91**, 57–70 (2018). https://doi.org/10.1016/j.ecolind.2018.03.086

Manso, M., Castro-Gomes, J.: Green wall systems: a review of their characteristics. Renew. Sustain. Energy Rev. **41**, 863–871 (2015). https://doi.org/10.1016/j.rser.2014.07.203

Perini, K., Rosasco, P.: Cost–benefit analysis for green façades and living wall systems. Build. Environ. **70**, 110–121 (2013). https://doi.org/10.1016/j.buildenv.2013.08.012

Kontoleon, K.J., Eumorfopoulou, E.A.: The effect of the orientation and proportion of a plant-covered wall layer on the thermal performance of a building zone. Build. Environ. **45**, 1287–1303 (2010). https://doi.org/10.1016/j.buildenv.2009.11.013

Coma, J., Pérez, G., de Gracia, A., Burés, S., Urrestarazu, M., Cabeza, L.F.: Vertical greenery systems for energy savings in buildings: a comparative study between green walls and green facades. Build. Environ. **111**, 228–237 (2017). https://doi.org/10.1016/j.buildenv.2016.11.014

Sharbafian, M., Yeganeh, M., Baradaran Motie, M.: Evaluation of shading of green facades on visual comfort and thermal load of the buildings. Energy Build. **317**, 114303 (2024). https://doi.org/10.1016/j.enbuild.2024.114303

Pérez, G., Coma, J., Martorell, I., Cabeza, L.F.: Vertical greenery systems (VGS) for energy saving in buildings: a review. Renew. Sustain. Energy Rev. **39**, 139–165 (2014). https://doi.org/10.1016/j.rser.2014.07.055

De Jesus, M.P., Lourenço, J.M., Arce, R.M., Macias, M.: Green façades and in situ measurements of outdoor building thermal behaviour. Build. Environ. **119**, 11–19 (2017). https://doi.org/10.1016/j.buildenv.2017.03.041

Zhao, Y., et al.: Beating urban heat: multimeasure-centric solution sets and a complementary framework for decision-making. Renew. Sustain. Energy Rev. **186**, 113668 (2023). https://doi.org/10.1016/j.rser.2023.113668

Alsaad, H., Hartmann, M., Hilbel, R., Voelker, C.: The potential of facade greening in mitigating the effects of heatwaves in Central European cities. Build. Environ. **216**, 109021 (2022). https://doi.org/10.1016/j.buildenv.2022.109021

Šuklje, T., Medved, S., Arkar, C.: On detailed thermal response modeling of vertical greenery systems as cooling measure for buildings and cities in summer conditions. Energy **115**, 1055–1068 (2016). https://doi.org/10.1016/j.energy.2016.08.095

Gromke, C., Blocken, B., Janssen, W., Merema, B., Van Hooff, T., Timmermans, H.: CFD analysis of transpirational cooling by vegetation: Case study for specific meteorological conditions during a heat wave in Arnhem, Netherlands. Build. Environ. **83**, 11–26 (2015). https://doi.org/10.1016/j.buildenv.2014.04.022

Lin, H., Xiao, Y., Musso, F., Lu, Y.: Green façade effects on thermal environment in transitional space: field measurement studies and computational fluid dynamics simulations. Sustainability **11**(20), 5691 (2019). https://doi.org/10.3390/su11205691

Lin, H., Ni, H., Xiao, Y., Zhu, X.: Couple simulations with CFD and ladybug + honeybee tools for green façade optimizing the thermal comfort in a transitional space in hot-humid climate. J. Asian Architect. Build. Eng. **22**(3), 1317–1342 (2023). https://doi.org/10.1080/13467581.2022.2081574

Mei, S.-J., Yuan, C.: Urban buoyancy-driven air flow and modelling method: a critical review. Build. Environ. **210**, 108708 (2022). https://doi.org/10.1016/j.buildenv.2021.108708

Franke, J., Hellsten, A., Schlunzen, K.H., Carissimo, B.: The COST 732 best practice guideline for CFD simulation of flows in the urban environment: a summary. Int. J. Environ. Pollut. **44**(1/2/3/4), 419 (2011). https://doi.org/10.1504/IJEP.2011.038443

Buccolieri, R., Santiago, J.-L., Rivas, E., Sáanchez, B.: Reprint of: review on urban tree modelling in CFD simulations: aerodynamic, deposition and thermal effects. Urban Forestry Urban Greening **37**, 56–64 (2019). https://doi.org/10.1016/j.ufug.2018.07.004

Heidkamp, M., Chlond, A., Ament, F.: Closing the energy balance using a canopy heat capacity and storage concept – a physically based approach for the land component JSBACHv3.11. Geosci. Model Dev. **11**(8), 3465–3479 (2018). https://doi.org/10.5194/gmd-11-3465-2018

Koyama, T., Yoshinaga, M., Hayashi, H., Maeda, K., Yamauchi, A.: Identification of key plant traits contributing to the cooling effects of green façades using freestanding walls. Build. Environ. **66**, 96–103 (2013). https://doi.org/10.1016/j.buildenv.2013.04.020

Cost Analysis for Passive Design Buildings in Cold Climate Regions in China: A Preliminary Review

Penglong Gao and Fara Diva Mustapa

Abstract

The climate change dilemma has become the primary challenge of the 21st century, posing significant threats to both the environment and the economy of human society derived from natural resource consumption and urbanisation. Prior research has highlighted the importance of the designer's involvement in creating sustainable design during the initial design stage. Nevertheless, passive design structures are associated with a significant initial construction cost. This study is to provide a preliminary review of comparative cost analysis for passive and active design buildings in cold regions of China. It is achieved through assessments of life cycle costings (LCC) analysis for both active and passive design buildings. Case studies research strategies were adopted to gather the initial construction and cost use of housing projects, about $500,000 compared to $420,000 for active building, coupled with the standard form of elemental cost analysis, ECA assessment for both active and passive design with similar building typologies characteristics in Zhangjiakou, China. It is also found that the operational cost is reduced from $13,000 to $7,000 annually. The aspects of passive design in buildings, including the building envelope, ventilation design, and utilisation of renewable energy, were examined to quantify the elemental cost analysis of constructing the building. Data were analysed using LCC assessment of the initial cost and cost in the use of passive over active design buildings. The preliminary research indicates that passive design has reduced yearly operating costs with higher initial and end-of-life costs. In contrast, the Active Design Building has higher annual operating costs but lower initial costs. Additionally, the Passive Design's residual values are larger than the Active Design's. This preliminary findings model will serve as a basis for cost estimation and assist in the preparation of feasibility studies for effective passive design housing development.

Keywords

Cost analysis · Energy Efficiency · Feasibility studies · Life cycle costing · Passive design housing

1 Introduction

Ecosystems and human well-being on Earth are being threatened by detrimental climate change resulting from human activities, causing unsustainable practices and increasing energy consumption (Jiang et al., 2017). Specifically, about 36% of the energy the whole world uses has been accounted for by the construction sector, which also can be a significant first step to reducing energy demand during the global effort to mitigate climate change (Santamouris & Vasilakopoulou, 2021). Inflation and exorbitant living expenses have created financial pressures. It makes it necessary to produce cost-effective building concepts that maximise construction and continuing operational costs (Myers, 2022). Fortunately, passive design building, which employs natural sources of energy for heating, cooling, and lighting, is considered one of the effective ways to solve the uneven environmental consequences and economic sustainability of urbanisation (Bhamare et al., 2019). However, its benefits are often linked with high construction costs and uncertainty in cost-in-use savings (X. Sun et al., 2018). Hence, this research aims to demonstrate through cost analysis the possibility of adopting passive design buildings to strategically reduce the demand

P. Gao · F. D. Mustapa (✉)
Department of Quantity Surveying, Faculty of Built Environment and Surveying, Universiti Teknologi Malaysia, Johor Bahru, Malaysia
e-mail: faradiva@utm.my

P. Gao
e-mail: gaopenglong@graduate.utm.my

B. He et al. (eds.), *Towards the Framework of Livable and Resilient Cities* Advances in Science, Technology & Innovation, https://doi.org/10.1007/978-3-031-97849-4_17

for energy and greenhouse gases in cold climate areas of China, particularly in chilly regions such as Northern China. The research aim is achieved through case studies analysis scrutinising the initial cost (construction) and the cost in use (maintenance) of housing development in the Northern China region. Zhangjiakou was chosen as the research scope as it has a very cold climate and hosts many important events, like the Winter Olympics. Hence, it is very crucial to conduct the need for passive building design in inclement areas.

The research outcome is expected to give a preliminary cost model in demonstrating the cost-benefit analysis of passive buildings in Zhangjiakou, China. To give recommendations for sustainable building design strategies for architects, developers and policymakers. Positive evidence is essential to prove passive buildings' market potential, economic feasibility and environmental benefits to have passive building design established as a leading solution in cold regions. This research hopes to provide some insights into urban development and environmental sustainability for frozen cities around the world. Consequently, this research will look at the social and economic perspectives to present some suggestions on the global sustainability goal that will agree with economic development and ecologically sustainable development.

2 Literature Review

2.1 Impact of Regional Climate on Passive Buildings

In China, the diversification of climate zones significantly influences passive building designs, having adaptations that cater particularly to local environmental conditions. Bai et al. (2021) offer a critical review of the appropriateness of current thermal climate zoning and argue the need for revised criteria to better represent local variations in solar radiation and humidity (Bai & Song, 2021). These are important for effective passive design in places such as Zhangjiakou. Extending this review, Bai and Song (2021) comment on dramatic changes in heating and cooling demands all across China brought by climate change, showing that design adaptation is capable of adjusting to declining heating degree days in the north and rising cooling degree days in the southeast is necessary (Bai et al., 2021).

Wei et al. (2022) go a step further to provide a pragmatic observation for architectural adaptation, whereby the shape and orientation of the buildings can be a great determinant of energy use across various climate zones (Wei et al., 2022). In that regard, the results they find make the prescription of specific changes in built form quite amenable for immediate application to cities such as Zhangjiakou. In turn, Liu et al. (2023) discuss some of the climate-responsive urban design strategies that would help reduce the impact of extreme heat (S. Liu et al., 2023). Thereby supporting the argument that detailed architectural planning at both building and urban scales is essential.

In this regard, Li et al. (2022) mention the potential of climate-responsive design in traditional vernacular dwellings of Chaoshan (Li et al., 2022). It could offer useful lessons for optimising the building parameters, such as the orientation and construction of walls, for enhanced thermal performance. These historical insights are vital in the development of modern passive buildings that effectively respond to local climatic conditions. Offering a bridge between traditional wisdom and contemporary sustainable design practice, as indicated by a series of related papers.

2.2 Development of Passive Buildings in China

In China, passive building design has become a recent trend for eco-friendliness with China's fast growth and adaptation despite its time-honoured tradition of lagging behind Western building standards (Z. Liu et al., 2019). In these passive design structures, clever design and placement of the structure serve to maximise benefit cheaply from natural lighting and heat sources, thus dramatically taking down heating demand. This, at the same, enhances indoor environments concerning both a "sense of place," thermally stabilising and creating lower volatile substances pollution and leading to a better welfare state (Gou et al., 2018).

Notably, collaborative endeavours have been established, such as those in the Sino-German Ecopark in Qingdao, to implement large-scale construction projects with the foundation of Passive House standards. The focus was on technical, policy and managerial aspects (Chung & Yang, 2022). Fortunately, it has shown great potential for energy reduction in residential buildings during the winter. The space heating demand can be reduced by up to 90% compared with conventional structures (Han et al., 2023). The operation success and occupant satisfaction of such buildings have further shown China's progressive improvement in global energy efficiency standards. China has dramatically changed its construction paradigms over the last couple of years. China's adoption of passive house designs that minimise heating and cooling energy use represents an important trend for sustainability amidst increasing environmental concerns (Su et al., 2020). Passive houses outperform conventional buildings in terms of energy efficiency and emissions mitigation, as several studies indicate (Y. Sun, 2023). Passive houses almost always achieve significantly lower space heating demands and much lower overall energy use, thus offering an approach towards China's energy and environment goals (Xu et al., 2016). Although there are barriers like immature technology and a lack of professionals, it is foreseeable that passive building will be universally adopted and become an industry standard, ultimately harmonising with global sustainability and significantly reducing CO2 emissions.

Investigation into the passive building technology in China underlines its rapid progress and significant benefits. The research by Han et al. (2022) indicates the possibility and energy effectiveness of the biggest certified passive house office building in China (Han et al., 2022). It makes a remarkable decrease in electricity consumption compared to conventional buildings and brings greater comfort to occupants. This case study is important for reference of future passive house projects and policy formulation. Furthermore, the study by Lin et al. (2021) examines developments in passive building technology in China, discussing the current state of passive building technology, benchmark projects, and emerging initiatives (Lin et al., 2021). These scholars also looked at the policies and challenges opposing passive building development, claiming that it is mandatory to create local incentive mechanisms and supportive regulations. It is also evident from Liang et al. (2022) that supportability in the application of passive energy-conservation theory can be enhanced in passive design principles (Liang et al., 2022). It is done through the incorporation of the supportability of a residential building construction project in Africa with support from China, thereby also making such internationally supported projects more sustainable and resource-efficient. Overall, in general, these studies suggest that both the recognition and application of passive building technology in China are increasing and that energy efficiency and sustainability objectives are also being met.

While the advantages of using passive design in buildings are well-established, significant obstacles hinder its implementation, including higher upfront building costs, technological obstructions, and a lack of knowledgeable practitioners (Pan & Pan, 2021). These economic barriers in many related investigations are frequently inadequately explored, with no extensive examination of these issues and a scarcity of research on their long-term effects on practicality.

2.3 Economic Viability of Passive vs. Active Buildings

Passive design on the economic sustainability of building construction has captured the focus of some recent literature. For instance, Dong et al. (2023) showcase how LCSA has been used in investigating the broader impact of building projects in China (Dong et al., 2023). Their results point out the very significant environmental and economic benefits of buildings with passive designs over conventional ones in mitigating ozone depletion and freshwater eutrophication. This agrees with the study conducted by Lu et al. (2023) using a systematic review of LCCA, whereby it is revealed that, when combined with LCCA, the integration of BIM can optimise building costs and improve economic, environmental, and social performances (Lu et al., 2023). This approach is similar to the cost-benefit analysis exercise to determine the viability of a particular project proposal. Cost-benefit analysis aims to offer a complete assessment of the economic profitability and sustainability of the environmentally passive design buildings, focusing on the benefits in terms of reduced lifetime operating costs and the effects on the ecosystem. (Rocchi et al., 2018).

Elaouzy and El Fadar (2022) have further quantified the same benefits about the fact that passive design strategies can result in saving energy by 6.7–66.2%, but the payback varies on account of different upfront costs (Elaouzy & El Fadar, 2022). Hence, the 3E (Energy, Economic, and Environment) benefits of passive designs must undergo a holistic approach in their assessment. Anand et al. (2023) point out that even though the initial costs would be high, the life-cycle cost of a passive building can very well be much less than that of a conventional building since it will use much less energy (Anand et al., 2023). To be more precise, Han et al. (2023) go even deeper by providing tangible evidence through a long-term monitoring survey in northern China. It demonstrated that passive houses could reduce space heating demand by up to 86.3% over the local conventional building stock (Han et al., 2023). This operational performance, backed by theoretical models, shows the tangible benefits of passive design in realising the sustainable building goals in the cold climate zones of China. Altogether, these studies prove the economic viability and environmental benefits of passive buildings, reinforcing their important place in sustainable urban development.

Although previous studies have successfully identified the advantages of passive design buildings, they often do not have detailed cost breakdowns. They mainly focus on certain building types, such as office or commercial buildings, instead of residential housing. This paper aims to fill this gap by using Elemental Cost Analysis (ECA), which offers a consistent universal standard for assessing costs. ECA allows a clearer identification of which building elements incur higher initial costs, which will justify their selection toward sustainable passive design housing construction. Such an organised approach would contribute towards the interpretation and practicability of the cost-benefit analysis by offering insightful, precise and practicable applications for policymakers, architects and developers.

2.4 Evolution of Passive Design Techniques in Response to Climate Change

Passive design concepts have a significant past originating from the early building methods that were full of sunny days and heat, allowing for improved indoor comfort without

Table 1. Passive design techniques in response to climate change (Saifudeen & Mani, 2024)

Technique	Traditional Application	Modern Adaptation for Climate Change	Specific Application for Cold Climates	Timeline
Building Orientation	Maximized natural sunlight exposure, especially in winter	Optimization using simulation software to maximize solar gain and reduce shading impact	Orientation towards south to capture maximum sunlight	Since early architecture
Insulation	Basic use of local materials like straw, mud, and wood for insulation	High-performance materials (e.g., fiberglass, mineral wool) to reduce heat loss	Thick layers of insulation in walls and roofs to minimize heat loss	20th century to present
Thermal Mass	Use of heavy materials like stone or adobe to retain heat in cold nights	Advanced materials (e.g., phase-change materials) to regulate indoor temperature	Heavy thermal materials in walls and floors for gradual heat release	Modern era adaptations
Glazing	Single-pane glass windows	Double or triple-pane windows with low-emissivity coatings for better insulation	Low-e coated, triple-glazed windows to retain warmth	1970s to present
Natural Ventilation	Strategically placed openings for air flow in warmer seasons	Mechanically-assisted ventilation systems with heat recovery to conserve indoor warmth	Minimally used due to high heat loss; air-tight structures preferred	Evolved in colder regions
Photovoltaic (PV) Systems	None (PV technology was not available)	Integration of solar panels to generate renewable energy	Rooftop PV panels to supplement heating and reduce grid dependence	2000s onward
Adaptive Shading Devices	Static shading (overhangs, deep window recesses)	Adjustable shading devices (e.g., louvers, blinds) for seasonal solar control	Fixed overhangs to block high-angle summer sun and allow winter sun	1980s to present
Dynamic Insulation Systems	Use of thermal curtains, shutters for night-time insulation	Smart insulation materials that adapt to temperature changes	Advanced insulation materials under development	Experimental phase

supplemental equipment, which has been getting by since then. Table 1 shows some techniques of passive design that are developed as a response to climate change.

Generally, constructions of the past that were mostly on the side of caution were using tactics such as positioning, insulation, and natural ventilation to guarantee energy efficiency, specifically in cold climates. The mitigation of climate change has led to the introduction of these techniques, together with high-tech materials and methods, which are more adaptive to the environments of today. High-performance insulation is, for instance, the new benchmarking of passive design and thus is now being used in the regions of minimum temperature records to reduce the amount of heat loss. Likewise, cutting-edge windows that include features such as thermal breaks and dynamically operated blinds also keep heat inside the house, which in warmer climates causes electricity bills to plummet. Due to the zero-energy building goals, renewable energies (solar systems) have become an inseparable part of any building, especially in areas where heating and cooling are most needed. In climate areas with low hinting demand like Zhangjiakou, passive building strategy modifications are stuck to improving thermal resistance, airtight, and mass utilisation to maintain or release heat smartly. The future favour can be seen in the application of passive design, which easily adapts to variable climates, along with novel products such as dynamic insulation and adaptive shading devices that create new uses of the existing technology and make it more perishable-less unity while dealing with changing nature crisis. So far, these new changes have brought forth the part of the design that contributes to meeting sustainability and climate-resilient targets.

2.5 Challenges in Cold Climate Regions for Passive Building Design in China

Passive design buildings in cold climate areas, for example, Zhangjiakou in northern China, are not without setbacks, which influence the workability and efficiency of sustainable architecture. One of the major problems in these cold regions is adequate insulation and energy efficiency without the use of active heating systems (Felius et al., 2020). The reason for this is that passive buildings are susceptible to excessive heat loss through the envelopes during the long freezing periods, thereby causing the loss of substance design features. The research framework, designed to solve the issue, uses LCCA and ECA to objectively evaluate the chosen insulation materials and energy-efficient windows based on the lifetime cost and effectiveness. The method of cost prioritisation enables a detailed examination of materials, with the primary advantage being increased insulation and energy efficiency, hence directly solving the problem in cold climates.

Another significant obstacle that stands in the way of passive-building design adoption is the high cost of installation, which is the main reason for the refusal of investment, usually because only long-term ROI is visible. Addressing this problem, the study doubles as a comparative approach evaluating the initial and lifecycle costs of passive and active building designs. By rolling net present value calculations and significance analyses, constructed framework proves the durability of passive building techniques by comparison with energy price fluctuations (Wang et al., 2020). It represents a clear path to cost recovery. This substantiates the economic viability of passive design by demonstrating how these types of buildings can become energy-saving meerkats, thus making the initial investment, which would have been a burden, more appealing to the property developers and stakeholders (Elaouzy & El Fadar, 2022). Additionally, technological problems and the lack of skilled labour in China bring more difficulties in the implementation of passive. One of the useful methods of this research is the usage of standard tools, such as the RETscreen Expert Software, for predictive modelling. This is a catchy way for those who are new in the field of passive buildings to monitor and evaluate the performance of the buildings. This structured approach of analysing financial returns along with the technical requirements framework proceeds to the investment in the acquisition of skills and the technology needed, which, on a broader scale, namely the execution of passive design, is one of the main factual obstructions.

In remote areas where the resources are scarce, the mischief is to be able to secure the resources of high quality, such as triple glazing windows or advanced insulation, that is vital for the success of energy-saving building design. These material issues are very dear and not easy to supply logistically (Hussein et al., 2021). Shifting to the Elemental Cost Analysis (ECA), this kind of research permits the intricate analysis of each building component's cost and profit relationship. This gives the option of selecting local material options that are eco-friendly without a performance loss, in addition to the regulatory incentives that could incite the requisite local production of these materials that are indispensable for the construction of passive buildings.

Lastly, it is mainly the climate-responsive design and adaptability to the regional conditions, which are, on the one hand, the most necessary but, on the other hand, the most difficult aspects to cope with for passive design in places like Zhangjiakou, where very cold winters require special architectural solutions. Climate-responsive design benefits a passive design system by comparing alternative designs and taking parameters like orientation, envelope, ventilation, followers, and others into account. This unique flexibility allows passive design to still be implemented effectively in different regions with extreme temperatures, thus becoming a more environmentally friendly and site-specific way to build in cold areas.

2.6 Summary

It would be beneficial to verify the basis of the current understanding of passive design buildings with such a view that it is the most significant for cold climate cities, as well as why it is so, and at the same time, point out research gaps and challenges. A Passive design building is based on the principle of minimising the energy consumption of a construction. It is realised through the use of natural resources such as building envelope, insulation, and ventilation, which are the resulting parts of commonly centralised energy systems. These strategies have been applied internationally, including in China, a country where projects like the Sino-German Ecopark have proven very successful, both in terms of energy savings and comfort. At the same time, the passive design of such buildings brings to the fore positive features, but there are still splices and challenges that render a full implementation and adoption impossible.

The literature discloses several essential areas of research that have been overlooked. In particular, economic viability studies in passive buildings are extremely limited, particularly in the research areas in compliance with the needs of cold regions in China. The majority of the existing studies concentrate on energy efficiency but often ignore the holistic perspective of the life cycle cost analysis customised for cold climates. Moreover, there is an overemphasis on the commercial and office building sectors in passive design research that combines with a lack of knowledge about the cost implications of residential buildings like single-family homes. Moreover, operational cost savings, though they are gaining the necessary attention and assumed to be the advantage of passive buildings, the longer period thereof, the lack of depth of analysis in terms of actual financial benefits such as Net Present Value (NPV) modelling is not overlooked at all.

Challenges in adopting passive design are related to the massive first money needed for construction, which might drive away developers. Moreover, the implementation of passive architecture skills that are still in their formative stage adds to the problem of the projects that are expected to be replicable. Finally, the decline in fossil fuel prices and the availability of renewable energy projects present economic reasons for the reduction in uncertainty, financial forecasting, and long-term savings estimation. This paper sets out to tackle these gaps and challenges through a directed lifelong cost, net present value, and sensitivity to economic variables analysis. The issue is that the research failed to be interpreted as a text tailored to academic audiences. Through the analysis of the passive design financial viability in cold climates, this research aims at the development of a cost-benefit model that may enable architects, developers, and policymakers to make well-informed decisions. Thus, it will eventually become a sustainable and energy-efficient urban development case in regions with a harsh climate. Such a study would make it possible for reviewers to relate the study to the broader discussions over sustainable architecture, which is their contribution.

3 Research Methodology

Applying Saunders' Research Method theory, the research employs a mixed-method approach in the academic reference base, incorporating qualitative and quantitative case study analysis as shown in Fig. 1. The research shows Life Cycle

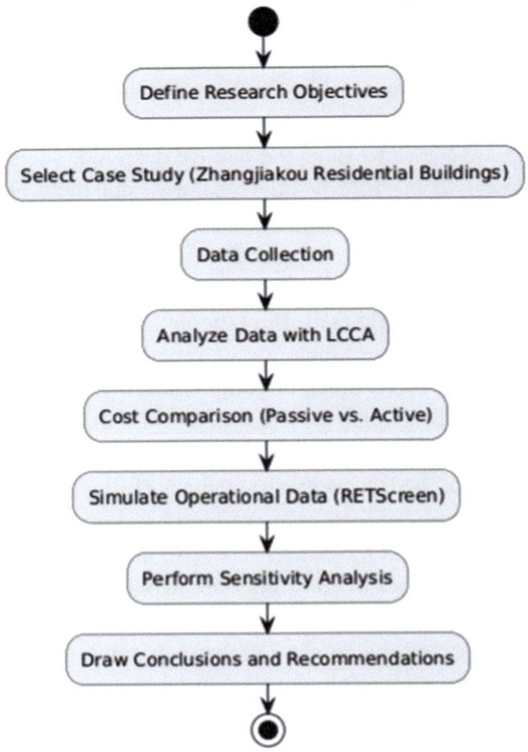

Fig. 1. Flowchart of the research (Generated by authors)

Cost Analysis (LCCA) as the methodology to investigate the economic suitability and cost-benefit analysis of passive and active design buildings. It comprehensively compares two building designs by considering the initial, operational and end-of-life costs. This method involves a methodical approach to collecting data and then analysing it. Covering preliminary costs of buildings and giving particular focus to various building types (typologies) such as residential, commercial, and office structures. Annual operating and maintenance costs and, finally, the costs at the end of the life cycle of buildings designed by passive and active energy principles (Kaming, 2017). Firstly, the research correctly identifies and chooses case study buildings in Zhangjiakou (state building types, and the building year, size, building height, etc. It controls the building obsolescence and design economics aspects), and in similar size and functional buildings, this is importance to have a similar base for analysis which are to make sure that the study has equivalence. Data about construction costs at the initial time are directly taken from project documentation that includes all direct costs for the materials and indirect costs, which include design and labour.

The analysis looks at data from ten projects, focusing on single-family houses. Those houses had a 150 m^2 average floor area per house, 2 floors, 6 m height and 300 m^2 living area in total. Initial construction costs, operation costs and other metrics were calculated using the mean/average values from those 10 projects. The estimated operation cost is based on power consumption, maintenance costs and possible savings due to governmental incentives or subsidies. The analysis is based on historical data and predictive modelling. Predictive data enables the author to predict the costs throughout the project lifespan, considering inflation and the economy. The estimation of end-of-life cost is based on the lifetime of the construction's components.

$$LCC\ formula = C + PV\ Reccurring - PV\ Residual\ Value$$

(Biolek & Hanák, 2019)

Often, the residual value of the construction components is estimated at disposal or demolition (Petroutsatou et al., 2023). Sensitivity analysis is used to evaluate variables like changing energy prices or maintenance costs and their effect on the cost-effectiveness of the passive building. The result is a side-by-side comparison between passive and active design estimating systems. This is a structured process for evaluating building projects from a financial perspective and works well regardless of the building projects' additional sustainable attributes.

This paper delves into the money matters between passive and active design buildings to establish their cost-effectiveness and whether they make economic sense in the long run. It seeks to establish which of the two building designs yields a financially more sustainable model at present and in the future, considering capital investments, operation costs, and exit costs. This research makes use of Life Cycle Cost (LCC) analysis and performance and risk assessment simulation. Net Present Value (NPV) is used to calculate the total savings and costs in their present value over the lifetime of the buildings, incorporating a discount rate that adjusts future value to present value.

Through Net Present Value (NPV) calculations, the Life Cycle Cost (LCC) analysis assesses the financial consequences of both passive and active design structures, accounting for all expenditures incurred over the buildings' life cycles.

3.1 Case Study Area and Description

This study, through incredible heat demand in extreme winters, has chosen Zhangjiakou, China, which is a city located in a cold climate region known for passive design. Zhangjiakou's environmental conditions are mainly the constant low temperatures and work day heating degree days, making energy-efficient building design a necessity for operating. The study conducted covers ten detached houses, each of about 150 m^2 floor area, 2 floors, 300 m^2 cumulative living area, and a 6-m height to keep the analysis homogeneous. Buildings were selected based on the homogeneity in typology and function to compare between passive and active

Table 2. Simulation (Generated by authors)

Building typology	
Floor area	150 m^2
Number of floors	2
Total living area	300 m^2
Window area	60 m^2
Building height	6 m

designs. Passive design includes high levels of insulation, PV systems, as well as controlling the position of the building relative to the sun to let in as much natural light and heat as possible, whereas active thus heating is based on mechanical systems and energy is supplied from the grid. The data on the costs of the projects were obtained from the project documentation that included the initial construction cost, the operation cost, and the end-of-life cost, followed by predictive modelling for future cost estimates. This detailed case study is designed to cover a wide range of issues, including the cost-effectiveness and the environmental sustainability of passive design in cold climate regions, thus providing substantial insights.

3.2 Cost in Use Simulation

To conduct the risk assessment and performance data, a simulation of the RETscreen expert software is performed with the building size being 150 m^2 floor area, two floors, total living area of 300 m^2, and height of 6 m, specifically focusing on single-family home typology, as indicated in Table 2 under building typology. Figure 2 shows the location and climate data in the software for the location of Zhangjiakou are proven from being.

Photovoltaic (PV) systems have competitive energy production costs when compared to alternative power generation technologies. The benchmark price per kWh for producing energy on the central grid is $0.10. Figure 2 shows the benchmark details for the location in the simulation software.

Cash flow analysis of each year and the cumulative data analysis is performed. Over the 20-year lifetime of the project, this cash flow indicates that the majority of the project is in the investment phase. It also reveals an annual negative cash flow, as shown in Fig. 3.

Figure 4 illustrates a 20-year trend of increasing cumulatively negative cash flow, which is brought on by continuing operating and investment expenses that aren't countered by yearly income or cost savings.

Finally, the financial viability is calculated by the software, as shown in Fig. 5. Costs of Energy Production: PV systems are competitive with other methods of producing energy because they substantially lower the $0.10/kWh benchmark. The financial metrics (Fig. 6) here are the incentives to reduce the impressive $75,000 initial investment of over $600,000. Careful financial planning is essential, as seen by the substantial annual costs associated with debt and operations (Santillán-Yuqui et al., 2020). Carbon credits and energy savings do not yield substantial financial returns, underscoring the necessity for more finance or enhanced operational efficiency. Issues with the flow of cash: The importance of securing adequate finance and seizing every chance to improve operational efficiency or secure additional incentives is highlighted by a persistently negative cash flow. The environmental impact of this is that despite the project's financial challenges, there is a good chance that it will lower greenhouse gas emissions and contribute to the attainment of environmental sustainability goals.

4 Data Analysis

4.1 Overview

Figure 1 provides location information for Zhangjiakou. LCC aggregates all costs associated with the building from inception to demolition. These measures are applied to establish a clear overview of the economic effects that will be associated with either building design and help stakeholders make prudent investment decisions in sustainable architecture.

4.2 Life Cycle Cost Analysis

The analysis includes data from 10 multiple projects and focuses on 2-story, 150 m^2 single-family homes, which results in 300 m^2 of living area. The buildings are about a decade old and have an average height of 6 m. This building typology can be used as a helpful reference point throughout this section regarding design economics.

The expenses of Active and Passive Design Buildings are contrasted in Table 3. The Passive Design has reduced yearly operating costs of $7,000 but higher initial and end-of-life costs of $50,000 and $500,000, respectively. On the other hand, the Active Design Building has higher operating costs of $13,000 per year but is less expensive initially ($420,000) and at the end of its useful life ($45,000). Additionally, the Passive Design's residual values are larger ($10,000) than the Active Design's ($5,000).

Tables 3 and 4 display the net present values (NPVs) of active and passive structures at different discount rates. When the discount rate rises, the net present value (NPV) of passively designed buildings decreases, going from $674,156 to $627,477. Additionally, the NPVs for operations and

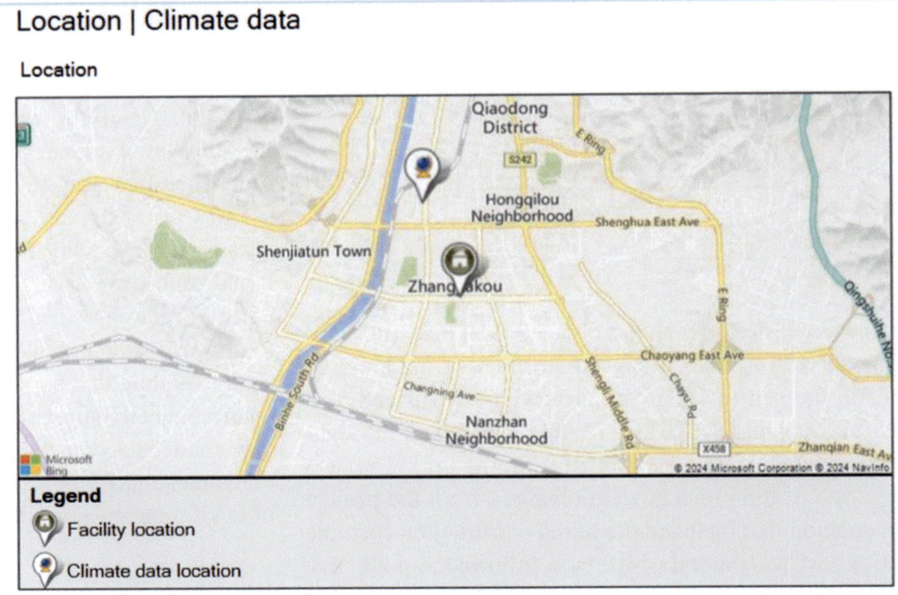

	Unit	Climate data location	Facility location
Name		China - Hebei - Zhangjiakou	China - Hebei - Qiaodong District
Latitude	°N	40.8	40.8
Longitude	°E	114.9	114.9
Climate zone		5B - Cool - Dry	5B - Cool - Dry
Elevation	m	726	720

Fig. 2. Location information (*Climate & Weather Averages in Zhangjiakou, Hebei, China*, n.d.)

end-of-life decline. However, the overall net present values (NPVs) in Active Design Buildings follow a similar pattern as the end-of-life and operational NPVs decline, falling from \$724,849 to \$647,129. The data shows how the value of prospective cost reductions decreases for both building types as discount rates rise, with the Active Design consistently showing bigger net present values (NPVs) due to its higher operational expenses.

4.3 Sensitivity Analysis

A 10% increase in energy costs for passive buildings results in an operating net present value (NPV) of \$108,977 and a total NPV of \$625,454. On the other hand, a 10% reduction brings these numbers down to \$89,163 and \$605,638, respectively, as shown in Table 5. A 10% increase in energy costs in active buildings increases the total NPV to \$700,156 and the operational NPV to \$263,679, while a 10% decrease in energy costs lowers these figures to \$251,486 and \$687,963. For active buildings, the same 10% increase leads to an operational NPV of \$263,679 and a 10% decrease results in \$251,486.

For active buildings, the same 10% increase leads to an operational NPV of \$263,679 and a 10% decrease results in \$251,486. The final Summary of the passive building and active building calculation can be seen in Table 6.

The LCC analysis demonstrates that passive buildings, despite higher initial costs, have significantly lower operational costs that contribute to a lower total NPV over the lifecycle compared to active buildings. The sensitivity analysis further illustrates the robustness of passive buildings against fluctuations in operational expenses, underscoring their economic and environmental benefits. This analysis aids stakeholders in making informed decisions about sustainable architecture investments by quantifying the long-term financial impacts of building designs.

5 Discussion

The complete preliminary analysis provided in this paper serves to underline the fact that passive design buildings are indeed feasible in terms of both economy and environmental sustainability, particularly for cold climates like Zhangjiakou, China. The intertwining of life-cycle cost analysis with net present value assessments in the research has created a potential framework for looking at long-term financial and ecological benefits between passive and active building designs.

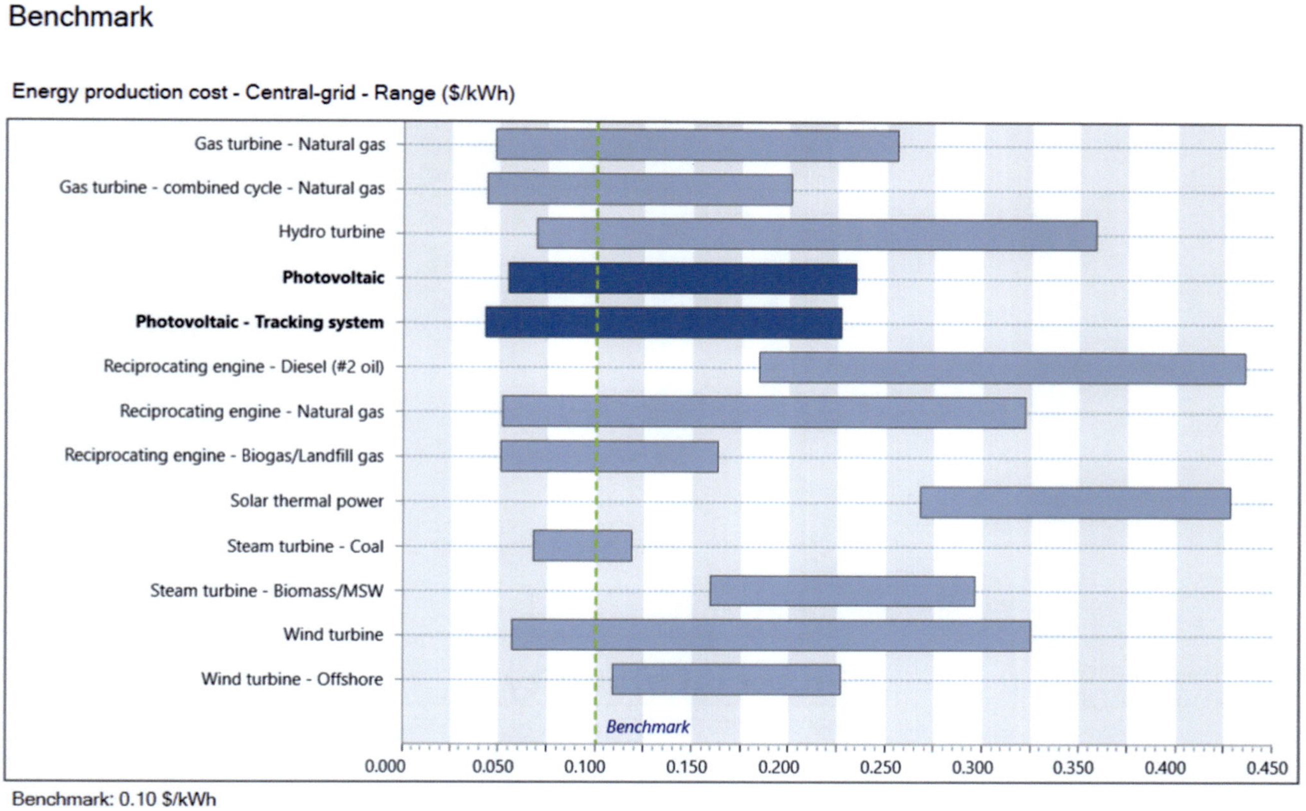

Fig. 3. Benchmark Data for Passive Design Building (Agency, n.d.)

Fig. 4. Annual cash flow analysis (Generated by authors)

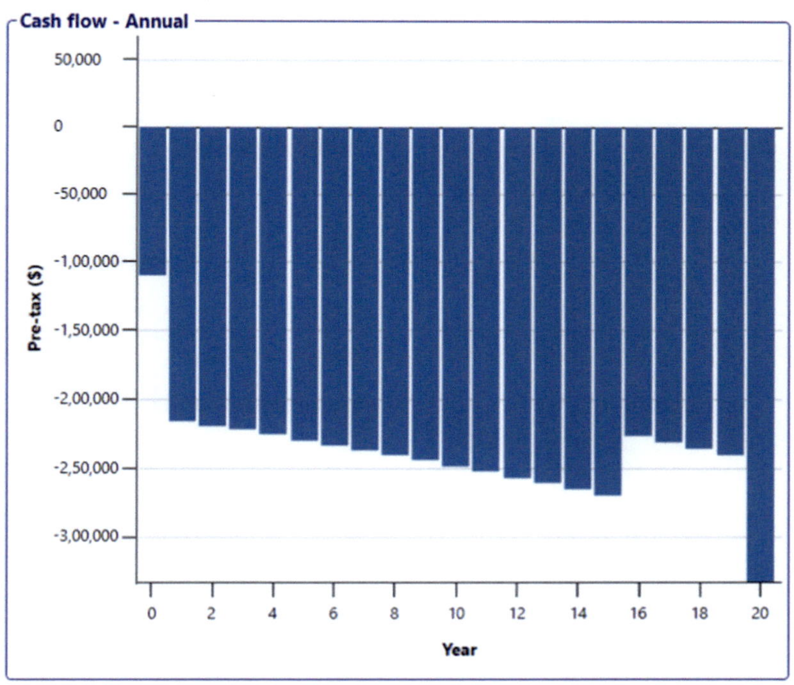

The LCC shows that there is a high initial investment in passive buildings compared to the active ones, $500,000 against $420,000. For passive structures, the cost per square meter is about $3,333, whereas for active buildings, it is about $2,800. The realisation of operational savings is each year: $7,000 against $13,000. Around $46.67 and $86.67 per square meter for passive and active structures, respectively, is the cost per square meter for operating savings. One of the biggest

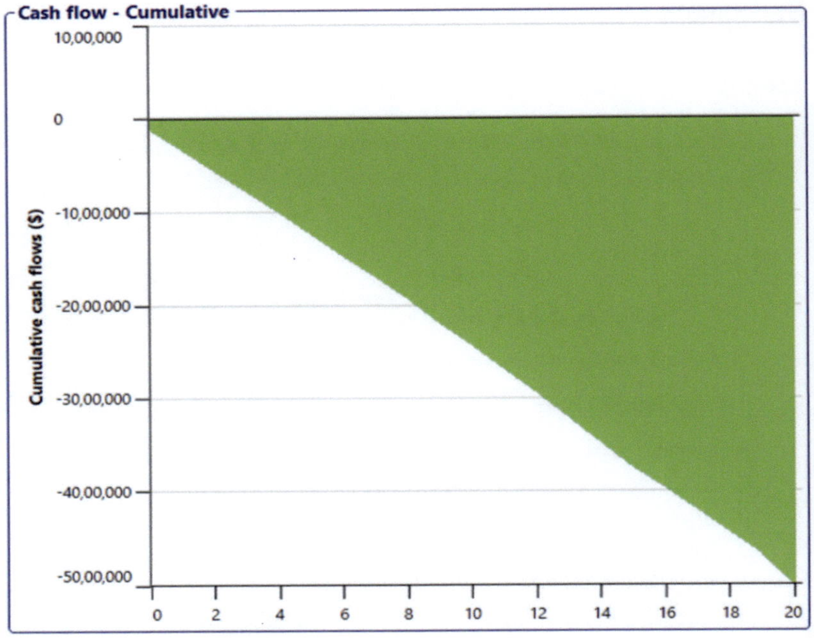

Fig. 5. Cumulative cash flow analysis (Generated by authors)

Fig. 6. Financial viability (Santillán-Yuqui et al., 2020)

Financial viability

Financial parameters

Inflation rate	%	2%
Project life	yr	20
Debt ratio	%	70%
Debt interest rate	%	7%
Debt term	yr	15

Costs | Savings | Revenue

Initial costs

Feasibility study	24.3%	$	1,50,000
Development	1.3%	$	8,000
Engineering	0.89%	$	5,500
Power system	23.5%	$	1,45,000
Balance of system & miscellaneous	50%	$	3,08,664
Total initial costs	100%	**$**	**6,17,164**
Incentives and grants		$	75,000

Yearly cash flows - Year 1

Annual costs and debt payments

O&M	$	1,75,000
Debt payments - 15 yrs	$	47,433
Total annual costs	**$**	**2,22,433**

Annual savings and revenue

User-defined	$	60,000
GHG reduction savings - 5 yrs	$	0

Table 3. Cost Table (Generated by authors)

Building type	Initial cost (average)	Operational costs (average)	End-of-Life cost aka salvage cost (average)	Residual cost	Cost per metre square
Passive Design	5,00,000	7000	50000	10000	$3333
Active Design	4,20,000	13000	45000	5000	$2800

Table 4. NPV for Passive and active design buildings (Generated by authors)

Discount Rate	NPV (Operational)		NPV (End-of-Life)		Total NPV		Cost per metre square	
	Passive	Active	Passive	Active	Passive	Active	Passive	Active
2%	$152,049	$282,742	$22,107	$22,107	$674,156	$724,849	$1014	$1885
3%	$138,698	$257,582	$16,477	$16,477	$655,175	$694,059	$924	$1718
4%	$127,569	$236,331	$12,277	$12,277	$639,846	$668,608	$851	$1577
5%	$118,235	$217,887	$9,242	$9,242	$627,477	$647,129	$789	$1453

Table 5. Sensitivity analysis for passive and active design building (Generated by authors)

Energy Price Change	NPV average (Operational)		Total average NPV		Average cost per meter square	
	Passive	Active	Passive	Active	Passive	Active
10%	$108,977	$263,679	$625,454	$700,156	$725	$1758
Base	$138,698	$257,582	$655,175	$694,059	$924	$1718
−10%	$89,163	$251,486	$605,638	$687,963	$594	$1679/

Table 6. Passive Design Building similar comments (Generated by authors)

Parameter	NPV average (Operational)	Total average NPV (Passive)	Cost per metre square (average building area)	NPV (Operational)	Total NPV (Active)	Cost per metre square (average building area)
Base Case	$138,698	$655,175	$924	$257,582	$694,059	$1718
Discount Rate 2%	$152,049	$674,156	$1014	$282,742	$724,849	$1885
Discount Rate 4%	$127,569	$639,846	$851/	$236,331	$668,608	$1577
Discount Rate 5%	$118,235	$627,477	$789	$217,887	$647,129	$1453
Energy Price + 10%	$108,977	$625,454	$726	$217,954	$654,431	$1758
Energy Price -10%	$89,163	$605,638	$594	$178,327	$614,804	$1679/
Maintenance Cost + 10%	$142,661	$659,138	$951	$263,679	$700,156	$1758
Maintenance Cost -10%	$134,735	$651,212	$898	$251,486	$687,963	$1679

factors in the LCC evaluation of the building over the whole lifespan. These findings are in agreement with the larger literature, showing that passive buildings are consistently more energy-efficient and less costly in the long run.

Besides, the sensitivity analysis deepens the financial appraisal in that it shows that passive designs can survive operational cost turbulence. For instance, at the rise of energy prices by 10%, passive buildings still maintained a lower operational NPV, indicating that passive buildings can cope with economic variabilities better than active ones. This is important because one can never have any idea about the state of energy prices or maintenance costs, for that matter, over several decades. The different discount rates eventually determine the NPV even more strongly, showing that passive buildings still have a large investment cost. They still have a far smaller total NPV over the life cycle as compared to active buildings. This is a very interesting result because it confirms the previously formulated hypothesis that the higher passive building investment at the beginning can pay off very well due to huge operational savings over the whole lifetime of the building (Shim et al., 2018). The end-of-life costs are relatively similar to each building type, a factor conventionally missing in the massive majority of building cost analyses. That is, when taken into account, any difference in initial cost will not be proportionately higher when taking into account end-of-life costs. This feature ensures that the total cost of ownership of passive buildings remains competitive and hence forms another strong financial justification for their adoption.

Summing up, the economic feasibility of passive buildings is rather well supported by the outcome of the preliminary analyses in this paper, and under such circumstances, it promotes them as an alternative to the challenges raised by climate change. Such outcomes could provide a seminal point of reference for policymakers, architects, and developers when considering the integration of sustainable practices into new building projects, especially in areas characterised by extremely cold climates. A strong methodology and extensive analysis of data are the basis of proof for this paper that passive building designs should be used much more due to their environmental benefits and, in addition, due to their clear economic gains. The next stage of this research will detail the elemental cost analysis of passive building design to systematically detail the impact of passive design building elementally. This is important to assess and select the quality and specifications in selecting g building items and elements in contrast to building life spans and cost in use.

This research offers emphasis in the fields of architecture, urban planning, and engineering. The adoption of sustainable construction practices becomes a clear option, especially in cold climates such as Zhangjiakou. The use of findings energy-efficient materials and building orientation are prioritised, implying a decrease in heating energy requirements. The improvements of energy-saving and environmental protection of the city architects. For instance, a careful design of a high-performance façade with the use of even materials, such as transparent insulation, will allow for a controlled thermal exchange and passive cooling design of a building. The engineers might add these valuable outcomes derived from LCCA of the mechanical, electrical and energy systems of a structure in the conceptual phase and speak of very useful investments such as high-quality insulation, energy-saving HVAC systems, and modern glazing methods conducive to long-term money savings. Furthermore, urban planners must motivate and jump on incentives for policies that counterbalance the expensive costs associated with the passive design that such projects go beyond the objectives of sustainability. Adapting passive design strategies to local climates is the best way to not only cut down on costs but also meet the environmental goals that provide a green light to the path to urban development sustainability.

6 Conclusion

This study advocates preliminary assessments for passive buildings, particularly in cold climates such as Zhangjiakou, China, because of their environmental and economic advantages over traditionally designed active buildings. Detailed life cycle cost analysis and computation of simulation of the net present value reveal that although the capital cost of passive buildings is quite high, the increment gets justified because of high operational savings and almost similar expenses in the end-of-life phase. This makes passive buildings a financially feasible and better choice for sustainable urban development. This sensitivity analysis provides evidence for great robustness in passive buildings against oscillations in operational costs and underpins their suitability to possibly uncertain economic conditions and energy price volatility. They are directly well related to the global sustainability goals to reduce carbon footprints and attain energy efficiency in the building sector. This paper convincingly argues the case for embedding passive design principles within the built form, more specifically in regions with extreme climatic conditions. The financial and environmental benefits argued present motivating factors to the policymakers, architects, and developers to consider passive design in their future practices for the creation of sustainable and resilient urban environments. The study henceforth sets the groundwork for a more profound investigation toward the optimisation of building design that begets economic and ecological sustainability.

Acknowledgments Disclaimer. The views and opinions expressed in this article are those of the author(s) and are the product of professional research. It is not a requirement to conform to the official policy or stance of any of the organisations, sponsors, or agencies, nor is it the position of the publisher. The authors of this article bear full accountability for its conclusions, examinations, and content.

Funding Information The author(s) received no financial support for the research, authorship, and publication of this article.

Data Availability Statement The data that support the findings of this study are available on request from the corresponding author.

Competing Interests The author(s) declare that they have no financial or personal relationship that may have inappropriately influenced them in writing this article.

Ethics Approval This research does not rely on human or sensitive data, instead utilising publicly available data sources. Hence, this research did not require any ethics approval.

References

Agency, I.E.: An energy sector roadmap to carbon neutrality in China_Chinese. www.iea.org

Anand, V., Kadiri, V.L., Putcha, C.: Passive buildings: a state-of-the-art review. J. Infrastruct. Preserv. Resilience **4**(1), 1–24 (2023). https://doi.org/10.1186/S43065-022-00068-Z

Bai, L., Song, B.: The effect of climate change on building heating and cooling energy demand in China. E3S Web Conf. **228**, 2005 (2021). https://doi.org/10.1051/E3SCONF/202122802005

Bai, L., Song, B., Yang, L.: The applicability of the current thermal climate zones for building energy-saving design of china. IOP Conf. Ser. Earth Environ. Sci. **668**(1), 12057 (2021). https://doi.org/10.1088/1755-1315/668/1/012057

Bhamare, D.K., Rathod, M.K., Banerjee, J.: Passive cooling techniques for building and their applicability in different climatic zones—the state of art. Energy Build. **198**, 467–490 (2019). https://doi.org/10.1016/J.ENBUILD.2019.06.023

Biolek, V., Hanák, T.: LCC estimation model: a construction material perspective. Buildings **9**(8), 182 (2019). https://doi.org/10.3390/BUILDINGS9080182

Chung, H.J., Yang, J.K.: Passive house performance evaluation and key management factor analysis in the Sino-German Ecopark: a study through fieldwork and interdisciplinary joint research. J. Glob. Area Stud. (JGA) **6**(3), 153–169 (2022). https://doi.org/10.31720/jga.6.3.7

Climate & Weather Averages in Zhangjiakou, Hebei, China. https://www.timeanddate.com/weather/china/zhangjiakou/climate

Dong, Y., Liu, P., Hossain, M.U.: Life cycle sustainability assessment of building construction: a case study in China. Sustainability **15**(9), 7655 (2023). https://doi.org/10.3390/SU15097655/S1

Elaouzy, Y., El Fadar, A.: Energy, economic and environmental benefits of integrating passive design strategies into buildings: a review. Renew. Sustain. Energy Rev. **167**, 112828 (2022). https://doi.org/10.1016/j.rser.2022.112828

Felius, L.C., Dessen, F., Hrynyszyn, B.D.: Retrofitting towards energy-efficient homes in European cold climates: a review. Energ. Effi. **13**(1), 101–125 (2020). https://doi.org/10.1007/S12053-019-09834-7/METRICS

Gou, S., Nik, V.M., Scartezzini, J.L., Zhao, Q., Li, Z.: Passive design optimization of newly-built residential buildings in Shanghai for improving indoor thermal comfort while reducing building energy demand. Energy Build. **169**, 484–506 (2018). https://doi.org/10.1016/J.ENBUILD.2017.09.095

Han, F., Liu, B., Feist, W., Pfluger, R., Dermentzis, G.: Operational performance evaluation of a Passive house residential building in northern China based on long-term monitoring in winter. Indoor Built Environ. **32**(7), 1464–1486 (2023). https://doi.org/10.1177/1420326X231169874/ASSET/IMAGES/LARGE/10.1177_1420326X231169874-FIG16.JPEG

Han, F., et al.: Verifying of the feasibility and energy efficiency of the largest certified passive house office building in China: a three-year performance monitoring study. J. Build. Eng. **46**, 103703 (2022)

Hussein, M., Eltoukhy, A.E.E., Karam, A., Shaban, I.A., Zayed, T.: Modelling in off-site construction supply chain management: a review and future directions for sustainable modular integrated construction. J. Clean. Prod. **310**, 127503 (2021). https://doi.org/10.1016/J.JCLEPRO.2021.127503

Jiang, L., Jiapaer, G., Bao, A., Guo, H., Ndayisaba, F.: Vegetation dynamics and responses to climate change and human activities in Central Asia. Sci. Total. Environ. **599–600**, 967–980 (2017). https://doi.org/10.1016/J.SCITOTENV.2017.05.012

Kaming, P.F.: Implementation of life cycle costing for a university building. Indonesian J. Life Cycle Assess. Sustain. **1**(1), 29–38 (2017). https://doi.org/10.52394/IJOLCAS.V1I1.5

Li, J., Wu, Z., Zhang, Y.: Research and application of climate-responsive design of traditional vernacular houses in Chaoshan, China. Environ. Sci. Sustain. Dev. **7**(1), 47–65 (2022). https://doi.org/10.21625/essd.v7i1.867

Liang, Y., Ren, S., Zhang, L., Ren, S., Zhang, L.: Research on sustainable design of China aid residential construction project in Africa based on passive energy-saving theory. Adv. Civil Funct. Struct. Ind. Architect. 614–620 (2022). https://doi.org/10.1201/9781003305019-85

Lin, Y., Zhao, L., Yang, W., Hao, X., Li, C.-Q.: A review on research and development of passive building in China. J. Build. Eng. **42**, 102509 (2021). https://doi.org/10.1016/j.jobe.2021.102509

Liu, S., Yin, S., Hua, J., Ren, C.: Climate-responsive architectural and urban design strategies for adapting to extreme hot events. In: Adapting the Built Environment for Climate Change: Design Principles for Climate Emergencies, pp. 253–272 (2023). https://doi.org/10.1016/B978-0-323-95336-8.00002-0

Liu, Z., Zhou, Q., Tian, Z., He, B., Jin, G.: A comprehensive analysis on definitions, development, and policies of nearly zero energy buildings in China. Renew. Sustain. Energy Rev. **114**, 109314 (2019). https://doi.org/10.1016/J.RSER.2019.109314

Lu, K., Deng, X., Jiang, X., Cheng, B., Tam, V.W.Y.: A review on life cycle cost analysis of buildings based on building information modeling. J. Civil Eng. Manag. **29**(3), 268–288 (2023). https://doi.org/10.3846/JCEM.2023.18473

Myers, D.: Construction Economics: A New Approach. Construction Economics A New Approach (2022). https://doi.org/10.1201/9781003287513

Pan, W., Pan, M.: Drivers, barriers and strategies for zero carbon buildings in high-rise high-density cities. Energy Build. **242**, 110970 (2021). https://doi.org/10.1016/J.ENBUILD.2021.110970

Petroutsatou, K., Ladopoulos, I., Polyzos, M.: Construction Equipment's Residual Market Value Estimation Using Machine Learning. Springer Proceedings in Business and Economics, pp. 195–203 (2023). https://doi.org/10.1007/978-3-031-24294-6_21

Rocchi, L., et al.: Sustainability evaluation of retrofitting solutions for rural buildings through life cycle approach and multi-criteria analysis. Energy Build. **173**, 281–290 (2018). https://doi.org/10.1016/J.ENBUILD.2018.05.032

Saifudeen, A., Mani, M.: Adaptation of buildings to climate change: an overview. Front. Built Environ. **10**, 1327747 (2024). https://doi.org/10.3389/FBUIL.2024.1327747/BIBTEX

Santamouris, M., Vasilakopoulou, K.: Present and future energy consumption of buildings: challenges and opportunities towards decarbonisation. E-Prime Adv. Electr. Eng. Electron. Energy **1**, 100002 (2021). https://doi.org/10.1016/J.PRIME.2021.100002

Santillán-Yuqui, C.R., Narváez-Zurita, I., Ormaza-Andrade, J.E., Narváez-Zurita, C.I.: Presupuesto de operación y su incidencia en la toma de decisiones gerenciales en empresas manufactureras. Cienciamatria **6**(2), 584–614 (2020). https://doi.org/10.35381/CM.V6I3.398

Shim, J., Song, D., Kim, J.: The economic feasibility of passive houses in Korea. Sustainability **10**(10), 3558 (2018). https://doi.org/10.3390/SU10103558

Su, X., Tian, S., Shao, X., Zhao, X.: Embodied and operational energy and carbon emissions of passive building in HSCW zone in China: a case study. Energy Build. **222**, 110090 (2020). https://doi.org/10.1016/J.ENBUILD.2020.110090

Sun, X., Gou, Z., Lau, S.S.-Y.: Cost-effectiveness of active and passive design strategies for existing building retrofits in tropical climate: case study of a zero energy building. J. Clean. Prod. **183**, 35–45 (2018). https://doi.org/10.1016/j.jclepro.2018.02.137

Sun, Y.: Passive house: a practice of low carbon and sustainable development in china. Appl. Comput. Eng. **3**(1), 122–127 (2023). https://doi.org/10.54254/2755-2721/3/20230371

Wang, R., Lu, S., Feng, W.: A three-stage optimization methodology for envelope design of passive house considering energy demand, thermal comfort and cost. Energy **192**, 116723 (2020). https://doi.org/10.1016/J.ENERGY.2019.116723

Wei, Y., Wang, S., Dang, H., Liu, P.: RETRACTED: climate adaptability analysis on the shape of outpatient buildings for different climate zones in china based on low-energy target. Atmosphere **13**(12), 2121 (2022). https://doi.org/10.3390/ATMOS13122121

Xu, W., Sun, D., Liu, Z.: Performance criteria system for passive nearly zero energy buildings in China. Indoor Built Environ. **25**(8), 1181–1184 (2016). https://doi.org/10.1177/1420326X16674376

TCAD Modeling of Bifacial Silicon Solar Cells

Rayimjon Aliev, Murodjon Komilov, Muzaffar Djuraev,
Nozima Abbosova, and Jahongir Ziyoitdinov

Abstract

This work is devoted to studying the effect of the doping degree of the rear isotype layer in the silicon bifacial solar cell (SC) using the "Sentaurus TCAD" modeling software package. The "Sentaurus TCAD" system allows to create a model of SCs with different structures and illumination conditions. The tools of the software system allow to change the physical parameters of the base and isotype layer of the SC. The traditional design of a monofacial sensitive SC with the structure n^+-p-p^+ and p^+-n-n^+ was investigated for comparative analysis. For the traditional design of a monofacial sensitive SC, the best results are achieved with a doping concentration of the rear isotype p^+ (n^+) layer of 10^{19} cm^{-3} for both structure types n^+-p-p^+ and p^+-n-n^+. Results showed that at the same time, the efficiency of SC with the n^+-p-p^+ structure is 5% higher than the efficiency of SC with the p^+-n-n^+ structure. When the bifacial sensitive SC was illuminated from the back side, the highest value of the short-circuit current was achieved at a doping concentration 10^{19} cm^{-3} for the p^+ layer in the n^+-p-p^+ structure and 10^{19} cm^{-3} for the n^+ layer in the p^+-n-n^+ structure. At the same time, the efficiency of SC with the n^+-p-p^+ structure is 17% higher than the efficiency of SC with the p^+-n-n^+ structure. When bifacial sensitive SC was illuminated from the front side, the highest efficiency value was achieved at a doping concentration of 10^{18} cm^{-3} and 10^{19} cm^{-3} for heavily doped p^+- and n^+- layers of both n^+-p-p^+ and p^+-n-n^+ structures. The assessment showed that in the case of simultaneous illumination of the front and back surfaces of bifacial sensitive SC, the short-circuit photocurrent density is increased by 1.57 times. Thus, the output power increases by 1.56 times.

Keywords

Silicon · Bifacial · Solar cell · "Sentauruse TCAD" · Efficiency

1 Introduction

The photovoltaic (PV) systems used in the renewable energy sources have various solar power conversion efficiency [1–4]. Mainly it depends on the physical characteristics of the semiconductor material from which the solar cell (SC) is made [4]. The works on the front surface texture and different metal nanoparticles are one of the perspective directions to increase the efficiency of SCs and solar panels (SP) [5–9]. Another promising alternative to increase the output power of SCs is the bifacial sensitive SCs [5].

Based on publicly available data, the installed capacity of bifacial solar PV modules increased from 97 MW in 2016 to over 2,600 MW in 2018. Gradual growth has allowed reaching up to 20,000 MW in 2024, which will account for 17% of the global photovoltaic module industry. There are forecasts for a significant increase of up to 60% of such photovoltaic systems by 2029 [1]. The proper rear surface reflector design is a major obstacle to the efficient use of bifacial SCs. In [10, 11], the impact of exterior rear surface reflectors, such as reflection from walls, windows, and parking places, on bifacial solar cells for residential deployments has been studied. However, bifacial solar panels can function well even without a reflector in many situations.

Analysis of photoelectric conversion of solar radiation by bifacial SC under illumination from different sides shows that these SCs differ significantly. The photoelectric efficiency of the rear side is less than the front side for bifacial SC.

R. Aliev (✉) · M. Komilov · N. Abbosova · J. Ziyoitdinov
Andijan State University, Andijan, Uzbekistan
e-mail: alievuz@yahoo.com

M. Djuraev
Andijan Medical Institute, Andijan, Uzbekistan

B. He et al. (eds.), *Towards the Framework of Livable and Resilient Cities* Advances in Science, Technology & Innovation, https://doi.org/10.1007/978-3-031-97849-4_18

The level of doping of the base and isotype layer with impurity atoms provides optimal conditions for the collection of photogenerated electron-hole charge carriers for the case of illumination from the front side. However, this condition is not optimal for rear-illuminated cases, in which a different depth of effective photogeneration of nonequilibrium charge carriers occurs. According to the physical processes of charge transfer, the isotype layer (n^+ layer in the p^+-n-n^+ structure and n^+layer in the n^+-p-p^+ structure) on the back side of a conventional solar cell creates an internal pulling electric field for photogenerated minority charge carriers, ensuring a reduction in volume recombination losses. However, there are many issues in the field of developing efficient bifacial SCs as such charge transfer processes for bifacial SC have not been studied in detail. The issues regarding the optimization of internal electric fields in bifacial SCs with various p^+-n-n^+ and n^+-p-p^+ structures have not been considered. The problems of optimizing the doping level of isotype n^+ and p^+ layers in p^+-n-n^+ and n^+-p-p^+ structures for bifacial SCs have not been solved. The presented work is devoted to modeling the influence of the doping level of isotype n^+ and p^+ layers in p^+-n-n^+ and n^+-p-p^+ structures on the main photoelectric parameters of silicon bifacial SCs using the "Sentaurus TCAD" software package.

2 Research Methodology

Currently, modern digital modeling methods are often used to perform studies of semiconductor structures. In this work, to achieve the research goal, the "Sentaurus TCAD" software system was used. This system allows to creating a geometric model of solar cells with different structures and illuminates its surfaces under different conditions. The tools of the software system allow you to vary the physical parameters of the base and isotype layer of the solar cell. So, the "Sentaurus TCAD" software platform was used as a research method, the theoretical and methodological aspects of which are described in detail in [12–14]. The basic physical and mathematical description of the processes of photoelectric energy conversion and the necessary equations for calculating the main parameters of the studied structures are described in detail in the sources [15–17]. The created models of two types of solar cells were silicon structures p^+-n-p^+ and n^+-p-p^+. The main parameters of the construction components of the SCs model are given in Table 1. The objective of the study is traditional designs of one-sided and two-sided lighting. Single-sided sensitive SC has a grid electrode on the front surface and a solid metal electrode (Al) on the rear surface. Double-side (bifacial) sensitive SC have a grid electrode on the front and rear surfaces made of the same metal (Al). Illumination of the SC surface was carried out by a solar radiation simulator which provides condition AM1.5.

3 Results and Discussion

Based on the results of the calculation, the I-V characteristics of the studied structures were formed, which are shown in Figs. 1, 2 and 3.

Each figure shows I-V curves corresponding to different doping concentrations of the rear isotype p^+ (n^+) - layer in the range from 10^{16} cm^{-3} to 10^{19} cm^{-3}. For a more detailed analysis of the physical nature of the photovoltaic process, individual cases of calculating the characteristics of one-side and two-side sensitive SCs under conditions of frontal and rear illumination are considered. In particular, in Fig. 1a and Fig. 1b show the I-V curves for two types of one-side sensitive SCs with structures n^+-p-p^+ (Fig. 1a) and p^+-n-n^+ (Fig. 1b).

The software of the "Sentaurus TCAD" platform allows to calculate the main photovoltaic parameters of solar cells under lighting conditions from the front side. Therefore, using well-known expressions of the physics of semiconductor photovoltaics and the data shown in Fig. 1a and Fig. 1b the main photovoltaic parameters of two types of solar cells with n^+-p-p^+ and p^+-n-n^+ structures are determined. For easy comparison of data from two types of solar cells, the results of the calculations are summarized in Table 2.

A comparison of the values of the main photovoltaic parameters of two types of one-side sensitive solar cells indicates the following. With an increase in the doping concentration of the p^+– layer in the n^+-p-p^+ structure from 10^{16} cm^{-3} to 10^{19} cm^{-3}, the open-circuit voltage gradually increases from 0.468 V to 0.570 V. In this case, the value of the short-circuit photocurrent also increases from 16.34 mA to 11 mA. The efficiency of photoelectric energy conversion of such a solar cell is 13.11%.

It can be noted in Fig. 1a and Fig. 1b, as well as based on the results of the calculation of the main photovoltaic parameters (Table 2), the efficiency of photoelectric energy conversion of

Table 1. Main physics parameters of Silicon p^+-n-n^+ and n^+-p-p^+ structures. (Cited by author)

№	Structure and conductivity of the layer		Doping concentration	Thickness of layer
1	p^+-p-n^+	n^+-n-p^+	(cm^{-3})	(mkm)
2	n^+	p^+	10^{17}	1
3	p (base)	n (base)	10^{15}	18
4	p+	n+	$10^{16} \div 10^{19}$	1

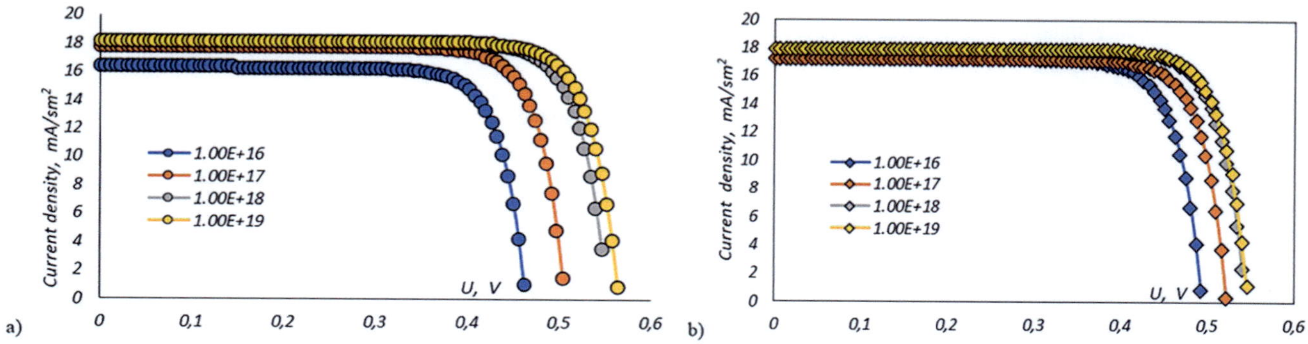

Fig. 1. (a,b) I-V characteristic of a silicon single-sided solar cell with the structure n^+-p-p^+ (a) and p^+-n-n^+ (b) for conditions of lighting from the front side. (Cited by author)

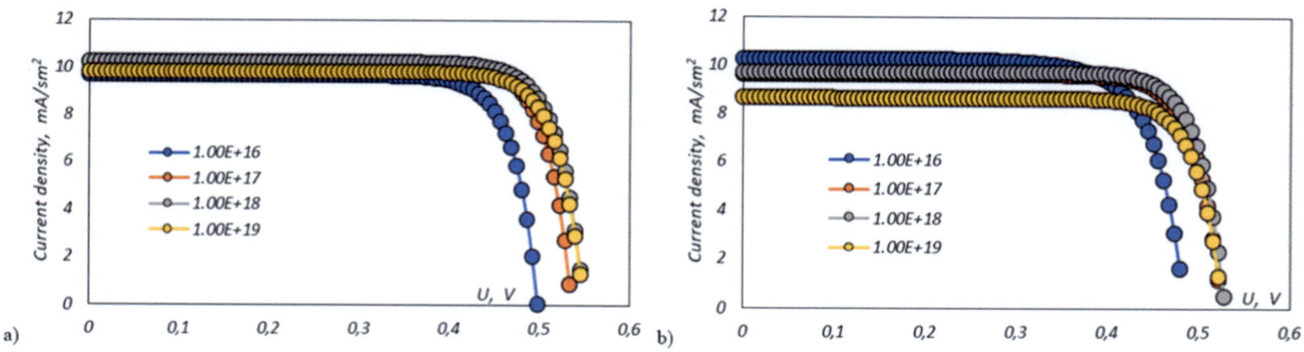

Fig. 2. (a,b) I-V characteristic of a silicon bifacial solar cell with the structure n^+-p-p^+ (a) and p^+-n-n^+ (b) under rear lighting conditions. (Cited by author)

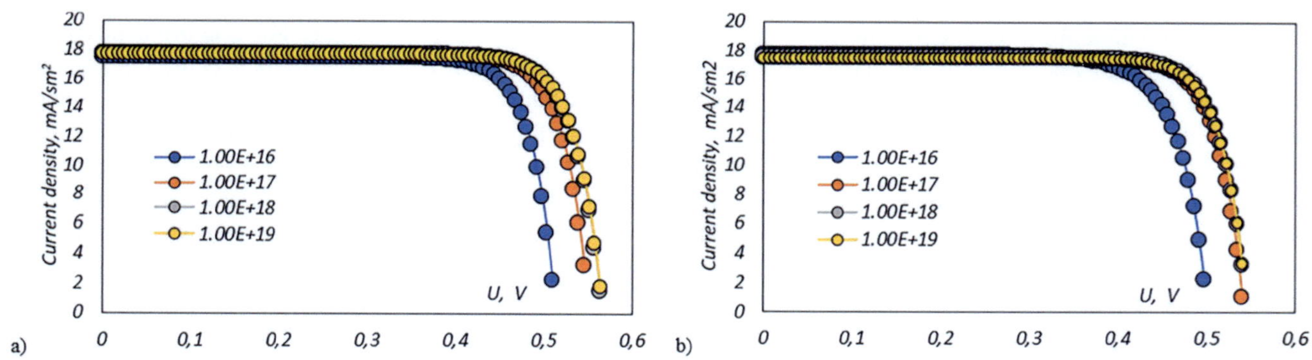

Fig. 3. (a,b) I-V characteristic of a silicon bifacial solar cell with the structure n^+-p-p^+ (a) and p^+-n-n^+ (b) under frontal lighting conditions. (Cited by author)

solar cells with the n^+-p-p^+ structure (13.11%) is 1.05 times higher than the efficiency of the p^+-n-n^+ structure (12.54%).

As mentioned in the introduction the back side of a bifacial sensitive solar cell is illuminated by secondary reflected radiation. As a result, additional electrical energy is generated in the solar cell volume. To estimate the possible additional energy due to illumination of the rear side, the I-V characteristic was calculated under conditions of illumination of the rear side of the solar cell with the original radiation.

The results of calculating the I-V characteristics for n^+-p-p^+ and p^+-n-n^+ structures are shown in Fig. 2a and Fig. 2b.

The method described above and the data shown in Fig. 2a and Fig. 2b define the main photovoltaic parameters of two types of solar cells with structures n^+-p-p^+ and p^+-n-n^+ for the case of illumination of their backside. The calculation results are summarized in Table 3.

A comparison of the values of the main photovoltaic parameters of two types of bifacial-sensitive solar cells indicates the following. With increasing doping concentration of the p^+– layer in the n^+-p-p^+ structure from 10^{16} cm^{-3} to 10^{18} cm^{-3}, the open-circuit voltage gradually increases from 0.504 V to 0.552 V. Moreover, with the increase of

Table 2. Doping concentration of p^+ (n^+) - layer and main photovoltaic parameters of Silicon solar cells with n^+-p-p^+ (p^+-n-n^+) - structure at frontal illumination. (Cited by author)

№	$N_p(N_n)$ (cm^{-3})	U_{oc} (V)	J_{sc}(mA/sm^2)	P_{mpp} (wt)	FF	FIK (%)
n^+-p-p^+						
1	1016	0.468	16.34	5.97	0.780	9.43
2	1017	0.510	17.78	7.23	0.797	11.42
3	1018	0.552	18.09	8.10	0.811	12.80
4	1019	0.570	18.11	8.29	0.804	13.11
p^+-n-n^+						
5	1016	0.498	17.22	6.73	0.785	10.63
6	1017	0.528	17.21	7.24	0.797	11.44
7	1018	0.546	17.85	7.86	0.806	12.42
8	1019	0.552	17.91	7.94	0.803	12.54

Table 3. Doping concentration of p^+ (n^+) - layer and main photovoltaic parameters of Silicon solar cells with n^+-p-p^+ (p^+-n-n^+) structure at back illumination. (Cited by author)

№	Np (sm^{-3})	U_{oc} (V)	J_{sc}(mA/sm^2)	P_{mpp} (wt)	FF	FIK (%)
n^+-p-p^+						
1	1016	0.504	9.68	3.86	0.791	6.09
2	1017	0.540	10.18	4.41	0.803	6.97
3	1018	0.552	10.22	4.54	0.806	7.18
4	1019	0.552	9.82	4.36	0.805	6.90
p^+-n-n^+						
5	1016	0.486	10.27	3.73	0.747	5.89
6	1017	0.528	9.62	4.07	0.801	6.43
7	1018	0.534	9.72	4.14	0.798	6.55
8	1019	0.528	8.64	3.66	0.803	5.79

doping concentration from 10^{18} cm^{-3} to 10^{19} cm^{-3} the open circuit voltage remains constant. The magnitude of short-circuit photocurrent in the range of doping concentration of the p^+– layer in the n^+-p-p^+ structure from 10^{16} cm^{-3} to 10^{18} cm^{-3} gradually increases from 9.68 mA to 10.22 mA. With increasing doping concentration from 10^{18} cm^{-3} to 10^{19} cm^{-3}, the short-circuit photocurrent decreases to 9.82 mA. The maximum efficiency of photoelectric energy conversion of such a solar cell is 7.18%.

From these graphs in Fig. 2a, Fig. 2b and the results of calculating the main photovoltaic parameters (Table 3) it follows that the efficiency of photoelectric energy conversion of solar cells with the n^+-p-p^+ structure (7.18%) is 1.1 times higher than the efficiency of the p^+-n-n^+ structure (6.55%). The described results apply to the case of lighting the solar cell from the rear side. The characteristics of solar cells are more important when they are illuminated from the front.

Therefore, the I-V characteristics of two types of solar cells were calculated under frontal. The calculation results of I-V characteristics for n^+-p-p^+ and p^+-n-n^+ structures are shown in Fig. 3a and Fig. 3b. Using the method described above and the data shown in Fig. 3a and Fig. 3b the main photovoltaic parameters of two types of solar cells with n^+-p-p^+ and p^+-n-n^+ structures are determined for the case of illumination of their front side. For a convenient comparison of data from two types of solar cells, the results of the calculations of main photovoltaic parameters are summarized in Table 4. A comparison of the values of the main photovoltaic parameters of two types of bifacial solar cells shows the following results. With an increase in the doping concentration of the p^+– layer in the n^+-p-p^+ structure from 10^{16} cm^{-3} to 10^{19} cm^{-3}, the open-circuit voltage gradually increases from 0.516 V to 0.570 V. In this case, the value of the short-circuit photocurrent also increases from 17.48 mA to 17.70 mA. The efficiency of photoelectric energy conversion of such a solar cell is 12.83%.

A different dependence is observed for other types of solar cells. The magnitude of short-circuit photocurrent in the range of doping concentration of the n^+– layer in the p^+-n-n^+ structure from 10^{16} cm^{-3} to 10^{18} cm^{-3} decreases from 17.73 mA to 17.44 mA. It was noted that for both types of solar cells, the efficiency of photoelectric conversion does not change with increasing doping concentration p^+ (or n^+) of the layer from 10^{18} cm^{-3} to 10^{19} cm^{-3}. All parameters of monofacial and bifacial sensitive SCs, determined under different conditions depending on the choice of their structure (n^+-p-p^+ and p^+-n-n^+), as well as their mutual differences, can be interpreted within the framework of the generation-recombination model

Table 4. Doping concentration of p^+ (n^+) - layer and main photovoltaic parameters of Silicon solar cells with n^+-p-p^+ (p^+-n-n^+) structure at frontal illumination. (Cited by author)

№	Np (sm^{-3})	U_{oc} (V)	J_{sc}(mA/sm^2)	P_{mpp} (wt)	FF	FIK (%)
n^+-p-p^+						
1	1016	0.516	17.48	7.23	0.801	11.42
2	1017	0.552	17.67	7.91	0.810	12.5
3	1018	0.570	17.70	8.12	0.805	12.83
4	1019	0.570	17.68	8.12	0.806	12.83
p^+-n-n^+						
5	1016	0.504	17.73	6.78	0.758	10.71
6	1017	0.528	17.07	7.64	0.802	12.07
7	1018	0.546	17.44	7.72	0.808	12.21
8	1019	0.546	17.45	7.71	0.809	12.18

of charge transfer. In this case, volume and surface recombination losses of photo-generated minority charge carriers play an important role. The assessment shows that in the case of simultaneous illumination of the front and back surfaces of bifacial sensitive solar cells, the short-circuit photocurrent density is increases 1.57 times. In this case, the output power increases 1.56 times.

4 Conclusion

In this work, the photovoltaic characteristics of bifacial sensitive solar cells with the n^+-p-p^+ and p^+-n-n^+ structure were studied using the "Sentaurus TCAD" modeling method. The influence of the doping level of the rear isotype p^+ (n^+) - layer on the I-V characteristics and the main photovoltaic parameters under the conditions of front and rear illumination was determined.

It is shown that for the traditional design of a one-sided sensitive SC, the best results are achieved with a doping concentration of the rear isotype p^+ (n^+) layer of 10^{19} cm^{-3} for both types of structure n^+-p-p^+ and p^+-n-n^+. At the same time, the efficiency of solar cells with the n^+-p-p^+ structure is 5% higher than the efficiency of solar cells with the p^+-n-n^+ structure.

When illuminated from the back side of a bifacial sensitive solar cell, the highest value of the short-circuit current is achieved at a doping concentration of 10^{19} cm^{-3} for the p^+ layer in the n^+-p-p^+ structure and at a doping concentration of 10^{19} cm^{-3} for the n^+ layer in the p^+-n-n^+ structure. At the same time, the efficiency of solar cells with the n^+-p-p^+ structure is 17% higher than the efficiency of solar cells with the p^+-n-n^+ structure.

When illuminated from the front side of a bifacial sensitive SC, the highest efficiency value is achieved at a doping concentration of 10^{18} cm^{-3} and 10^{19} cm^{-3} for heavily doped p^+- and n^+- layers of both n^+-p-p^+ and p^+-n-n^+ structures.

The assessment shows that in the case of simultaneous illumination of the front and back surfaces of bifacial sensitive solar cells, the short-circuit photocurrent density is increases 1.57 times. In this case, the output power increases by 1.56 times.

Acknowledgments This work has been carried out in the Renewable Energy Sources Laboratory at Andijan State University under basic research Project number FZ 2020092973 of the Ministry of High Education, Science and Innovation of Uzbekistan.

References

1. Aliyev, R., Bozarov, O., Kodirov, D., Kaxxorov, J., Xonbutayeva, D.: Study on photovoltaic characteristics of bifacial solar panelsю. In: E3S Web of Conferences, ICECAE 2024, p. 01016, vol. 497 (2024). https://doi.org/10.1051/e3sconf/202449701016
2. Seroka, N.S., Taziwa, R., Khotseng, L.: Solar energy materials-evolution and niche applications: a literature review. Materials **15**, 5338 (2022)
3. Deline, C.A., Ayala Pelaez, S., Marion, W.F., Sekulic, W.R., Woodhouse, M.A., Stein, J.: Bifacial PV System Performance: Separating Fact from Fiction, (Sandia National Labs) PVSC-46, Chicago, IL (2019)
4. Kodirov, D., Mirzabaev, A., Isakov, A.J., Mirzabekov, S., Makhkamov, T.: Problems of integration of the photovoltaic power stations with the grid systems. IOP Conf. Ser. Earth Environ. Sci. **614**(1), 012016 (2020)
5. Mo'ydinova, M.A., Qosimjonova, X.B.: Raqamli modellashtirish orqalitextura va optic qatlamli fotoelektrik qurilmalarni o'rganish. Sci. Innov. Int. Sci. J. **2**(1), 39–41 (2022). https://doi.org/10.2581/zenado.6596286
6. Gulomov, J., Accouche, O., Ziyoitdinov, J., Mirzaalimov, A.: Optimizing CZTSe Solar Cell Architecture: Comparative Study of ZnO, TiO2, and MoO3 as Electron Transport Layers. Digital Object Identifier (2024). https://doi.org/10.1109/ACCESS.2024.3351675
7. G'ulomov, J.J., Accouche, O.: Gold Nanoparticles Introduced ZnO/Perovskite/Silicon Heterojunction Solar Cell. IEEE Access **19**, 119558–119565 (2022). https://doi.org/10.1109/ACCESS.2022.3221875
8. Kodirov, D., Tursunov, O., Ahmedov, A., Khakimov, R., Rakhmataliev, M.: Economic efficiency in the use of solar energy: a case study of agriculture in Uzbekistan. IOP Conf. Ser. Earth Environ. Sci. **614**, 012031 (2020)
9. Gulomov, J., Aliev, R.: Numerical analysis of the effect of illumination intensity on photoelectric parameters of the silicon solar cell

with various metal nanoparticles. Nanosystems Phys. Chem. Math. **12**, 569–574 (2021)
10. Durusoy, B., Ozden, T., Akinoglu, B.G.: Solar irradiation on the rear surface of bifacial solar modules: a modeling approach. Sci. Rep. **10**(1), 13300 (2020)
11. Yun, M.J., Sim, Y.H., Lee, D.Y., Cha, S.I.: Kirigami-inspired automatically self-inclining bifacial solar cell arrays to enhance energy yield under both sunny and cloudy conditions. iScience **25**(7), 104649 (2022)
12. Sentaurus™ Device User Guide O-2018.06. 1530 (2018)
13. Idrishenyuk, V.I., Pectsov, E.F.: Osnovy priborno-texnologicheskogo modelirovaniya v Sentaurus TCAD. M.: Moakovskiy texnologicheskiy univarsitet, 139 p. (2018)
14. Web page of company Synopsys. https://www.synopsys.com/content/dam/synopsys/silicon/datasheets/sentaurus_ds.pdf
15. Mirzaalimov, A., Mirzaalimov, N., Ziyoitdinov, J., Komilov, M., Temirov, S., Tojiboyeva, Sh.: Development of 3D format photoelectric energy devices based on silicon solar cells. Glob. J. Eng. Technol. Adv. (2024). https://doi.org/10.30574/gjeta.2024.20.2.0153
16. G'ulomov, J.J., Ziyoitdinov, J.N., G'ulomova, I.M.: The effect of temperature and base thickness on photoelectric parameters of amorphous silicon solar cells. J. Nano Electron. Phys. **14**(5), 05010, 1–6 (2022). https://doi.org/10.21272/jnep.14(5).05010
17. Gulomov, J., Accouche, O., Neji, B., Ziyoitdinov, J.: Suntulip: new tool to simulate solar cells. IEEE Access **11** (2023). https://doi.org/10.1109/ACCESS.2023.3315593

Climate Adaptation Strategies and Environmental Quality

Climate Change Adaption Projects: Three Case Studies

Lorenzo Massimiano, Paolo Fusero, and Maura Mantelli

Abstract

To mitigate the damage caused by climate change effects, cities worldwide are exploring innovative solutions to transform their built environments, adapting them to the increasing intensity of extreme weather events. This study analyzes some of the strategies and actions implemented to achieve this goal. Specifically, it focuses on three international case studies: the "Rainproof" project in Amsterdam, which employs nature-based solutions (NBS) to manage intense rainfall and prevent flooding situations; the "Climate Resilient Neighborhood" in Copenhagen, which develops resilient neighborhoods capable of dynamically adapting to climate change; and the "Urban Forest Strategy" in Melbourne, which integrates NBS with information and communication technologies (ICT) to mitigate the impacts of global warming on urban areas. Through a comparative analysis of these case studies, the strengths and weaknesses of each strategy are identified. The paper aims to foster a discussion—based on the examination of these three best practices—that contributes to the scientific discourse on urban resilience, in alignment with the carbon neutrality objectives of the European Green Deal.

Keywords

Climate change · urban design · Nature based solutions · Information and communication technologies · Best practices

L. Massimiano (✉) · P. Fusero · M. Mantelli
Department of Architecture, University "G.d'Annunzio" of Chieti and Pescara, Pescara, Italy
e-mail: lorenzo.massimiano@unich.it

P. Fusero
e-mail: paolo.fusero@unich.it

M. Mantelli
e-mail: maura.mantelli@unich.it

1 Introduction

In recent decades, adaptation to climate change has emerged as a pressing global priority. The increasing frequency and severity of extreme climate events, such as floods, heatwaves, and storms, along with rising global temperatures, have underscored the urgent need for effective strategies to mitigate the impacts of climate change on both the environment and society (IPCC, 2007; Field et al., 2012). As urban areas are particularly vulnerable to climate-related risks due to their high population density and infrastructure concentration, there is a growing recognition of the necessity to develop adaptation and mitigation strategies specifically tailored to urban contexts. These strategies aim to enhance the resilience of communities, safeguard critical infrastructure, and reduce potential economic losses (Seto et al., 2014; Revi et al., 2014). In response to these challenges, international organizations and governments have intensified their efforts to fund and support climate resilience initiatives. For example, the European Union's Green New Deal represents a substantial financial commitment to fostering sustainable urban development and reducing greenhouse gas emissions across member states (Pettifor, 2020). This policy framework emphasizes the integration of innovative solutions and collaborative approaches to address the unique challenges that urban areas face due to climate change. Despite significant advancements in climate science and the increased understanding of urban resilience, numerous gaps remain in our ability to effectively implement adaptation strategies that fully address the complexities of urban ecosystems. Research has increasingly highlighted the potential role of Nature-Based Solutions (NBS), Information and Communication Technologies (ICT), and Community Engagement (CE) in fostering adaptive capacities and building sustainable, resilient cities (Davis et al., 2017; Kabisch et al., 2017). These components have shown promise in enhancing ecosystem services, improving disaster

B. He et al. (eds.), *Towards the Framework of Livable and Resilient Cities* Advances in Science, Technology & Innovation, https://doi.org/10.1007/978-3-031-97849-4_19

response and early warning systems, and mobilizing community support for climate action. Starting from these assumptions, the study aims to contribute to the scientific discourse on climate change adaptation strategies by examining the significance of these three key components: NBS, ICT, and CE. These elements are not always consciously utilized in adaptation plans. Moreover, they are often addressed separately, without creating the necessary synergies needed to maximize their effectiveness. Through an analysis of three international case studies, the study explores how different urban contexts have implemented these elements to mitigate the impacts of climate change, highlighting the characteristics and benefits of this approach. It becomes clear that NBS, ICT and CE, while each having a significant impact when implemented individually, offer even greater benefits when used in synergy. In fact, integrating these elements makes it possible to leverage the synergies between technological, environmental, and social solutions, creating a more holistic approach to climate adaptation (Frantzeskaki et al., 2017). This multidimensional approach can provide valuable and concrete insights for the scientific community and policymakers to develop and adapt effective strategies for urban resilience in response to the challenges of climate change. The study will follow this structure: in the first part, the paper will present an overview of the urban contexts and adaptation strategies employed in the three case studies. The second part will provide a detailed analysis of NBS, ICT, and CE solutions, identifying their strengths and challenges. In the third part the study will conclude with recommendations and suggestions for future research directions in the field of urban climate change adaptation.

2 Methodology and Case Studies

The research adopts a methodology based on the analysis of three international case studies related to climate change adaptation in urban contexts: Amsterdam "Rainproof", Melbourne "Urban Forest Strategy" and Copenhagen "Climate Adaptation Plan". The selection of the three case studies was conducted through a systematic procedure developed over several stages, briefly described in the diagram below (Fig. 1). Initially, studies related to climate adaptation strategies implemented at the urban level were identified via a Google Scholar search using keywords such as "climate change," "adaptation plan," "urban planning," "resilience," and "city/ies." A list of 151 cities was extracted from the research paper *Opportunities for Biodiversity Conservation as Cities Adapt to Climate Change* (Butt et al., 2018). This list was then cross-referenced with the 94 cities belonging to the C40 Cities Climate Leadership Group, a global network dedicated to enhancing urban resilience in line with the Paris Agreement (C40, 2024). The comparison yielded 54 cities common to both lists. Cities from Asia, Africa, and South America were subsequently excluded, as they were considered too different in terms of climatic and socio-economic contexts relative to the objectives of this study. This step reduced the list to 30 cities. From this list, cities recognized as being at the forefront of the field—based on their winning of the C40 Climate Leadership Award, sponsored by Siemens—were further selected. Ten cities were awarded in 2013 (Siemens, 2013), and nine in 2014 (Siemens 2014).

As a result, a shortlist of seven cities was created: Amsterdam, Barcelona, Copenhagen, London, New York, San Francisco, and Melbourne. For each of these seven cities, relevant documents available on official municipal websites, concerning climate change adaptation strategies, were analyzed. Following this detailed analysis, three final case studies—Amsterdam, Copenhagen, and Melbourne—were selected, as they exemplified innovative and comprehensive approaches to urban climate adaptation. This was recognized in several scientific papers, which were identified by adding the names of the cities to the previously mentioned keywords during the search process (Geneletti and Zardo, 2016; Sørensen and Torfing, 2022; Axelsson et al., 2021; Hartigan et al., 2021). At this point, an analysis of the case

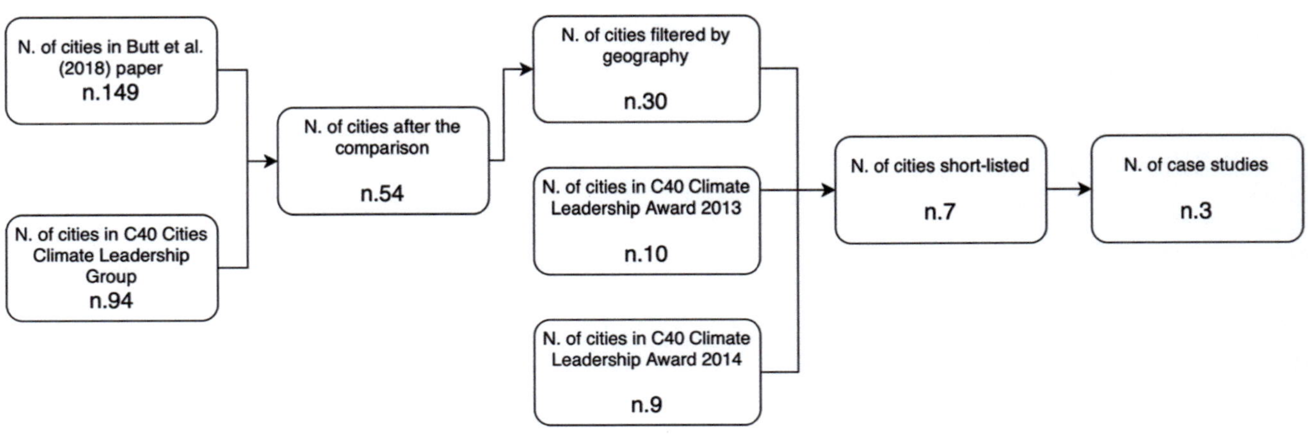

Fig. 1. Flowchart of the case studies selection phases (Source: By author).

Table 1. List of cities involved in climate adaptation plan, aligned with the selection process described in the flowchart (Source: By author).

Source	Number	Cities
Opportunities for Biodiversity Conservation as Cities Adapt to Climate Change (Butt et al., 2018)	151	Addis Ababa, Adelaide, Ahmedabad, Albertirsa, Albuquerque, Alexandria, Alameda, Amsterdam, Ancona, Antwerp, Aspen, Athens, Auckland, Austin, Baltimore, Bandar Lampung, Bangalore, Barcelona, Basel, Bellingham, Berkeley, Berlin, Birmingham, Blacktown, Bogota, Bologna, Boston, Boulder, Brighton-Hove, Brisbane, Bullas, Buenos Aires, Cairo, Calgary, Can Tho, Cape Town, Caracas, Changwon, Charleston, Chattanooga, Chiang Rai, Chicago, Curitiba, Danang, Darebin, Darwin, Delhi, Denver, Dhaka, Dresden, Durban, Dublin, Edinburgh, Evanston, Exeter, Freemantle, Ghent, Glasgow, Gold Coast, Gorakhpur, Hanoi, Hamburg, Hamilton, Hat Yai, Heidelberg, Helsinki, Ho Chi Minh City, Hobsons Bay, Homer, Hong Kong, Houston, Indore, Istanbul, Jakarta, Johannesburg, Kalamaria, Karachi, Keene, Lagos, Leicester, Lima, Lisbon, Liverpool, London, Los Angeles, Louisville, Madrid, Malmo, Mandurah, Manchester, Melbourne, Mexico City, Miami, Milan, Montreal, Moscow, Munich, Nairobi, Nagoya, Nelson, New Orleans, New York, Oslo, Ottawa, Oxford, Paris, Patras, Philadelphia, Pittsburgh, Pomaz, Portland, Port Phillip, Quito, Qui Nhon, Redland, Riga, Rio de Janeiro, Rome, Rotterdam, Salt Lake City, San Diego, San Francisco, Santiago, Sao Paulo, Seattle, Semarang, Seoul, Shanghai, Shenzhen, Singapore, Sfantu Gheorghe, Stockholm, Stuttgart, Surat, Sydney, Tallinn, Tatabanya, Tehran, Tokyo, Toronto, Tshwane, Venice, Vienna, Warsaw, Washington, Wellington, Wuhan, Yokohama
C40 Cities Climate Leadership Group	54	Abidjan, Accra, Addis Ababa, Ahmedabad, Amman, Athens, Auckland, Austin, Baghdad, Bangkok, Barcelona, Beijing, Bengaluru, Berlin, Bogotá, Boston, Buenos Aires, Cape Town, Chennai, Chengdu, Chicago, Curitiba, Dalian, Dakar, Dar es Salaam, Dhaka (North and South), Dubai, Durban (eThekwini), Ekurhuleni, Fortaleza, Freetown, Fuzhou, Guadalajara, Guangzhou, Hangzhou, Hanoi, Heidelberg, Ho Chi Minh City, Hong Kong, Houston, Istanbul, Jakarta, Johannesburg, Kuala Lumpur, Lagos, Lima, Lisbon, London, Los Angeles, Madrid, Melbourne, Medellin, Mexico City, Miami, Milan, Montreal, NCT of Delhi, Nairobi, Nanjing, New Orleans, New York City, Oslo, Paris, Philadelphia, Phoenix, Portland, Quezon City, Quito, Rio de Janeiro, Rome, Rotterdam, Salvador, San Francisco, Santiago, São Paulo, Seattle, Seoul, Shanghai, Shenzhen, Singapore, Stockholm, Sydney, Tel Aviv - Yafo, Tokyo, Toronto, Tshwane, Vancouver, Warsaw, Washington, DC, Wuhan, Yokohama, Zhenjiang
C40 Climate Leadership Award (2013)	10	Bogota, Copenhagen, Melbourne, Mexico City, Munich, New York City, Rio de Janeiro, San Francisco, Singapore, Tokyo
C40 Climate Leadership Award (2014)	9	Amsterdam, Barcelona, Buenos Aires, London, Melbourne, New York City, Portland, Seoul, Shenzhen
Shortlisted	7	Amsterdam, Barcelona, Copenhagen, London, New York, San Francisco, and Melbourne

studies was conducted, focusing on three key areas: the implementation of Nature-Based Solutions, the use of Information and Communication Technologies, and citizen participation through Community Engagement. The table below provides the comprehensive list of cities considered at each stage of the case studies selection process, highlighting those that advanced through various filtering criteria and were ultimately shortlisted for in-depth analysis (Table 1).

2.1 Amsterdam "Rainproof"

The "Rainproof" project was launched in 2014 in the city of Amsterdam, Netherlands. Most of the information about the project is available in the publication distributed by the Amsterdam Municipality (Amsterdam Municipality, n.d.). Following the exceptional downpour that hit Copenhagen in 2011, causing damage amounting to one billion euros due to 150 mm of rain falling in just 90 min, the water company Waternet initiated a meticulous vulnerability study of the city of Amsterdam. In collaboration with the engineering firm Tauw, "stress tests" were conducted, highlighting a significant flood risk for the city. These results demonstrated the urgent need to prepare Amsterdam to face increasingly intense and frequent downpours. For this reason, the "Amsterdam Rainproof" program was launched by the municipality, aiming to implement innovative solutions to manage especially heavy rainfall and protect the city from the increasingly severe impacts of climate change. The project focuses on two specific risk factors, both related to water: flood risk and inundation risk. These aspects led to the formulation of two main objectives: increasing the permeability of urban soil and managing water flow by slowing down, collecting, and reusing rainwater. The intervention is distributed throughout the city area, although with different levels of pervasiveness: from a minimum of 60% of adoptable solutions in the historic center neighborhoods to a maximum of 95% in neighborhoods developed after the Second World War, which are more capable of accommodating the prevention measures proposed by the project.

Fig. 2. The polder roof on the Vivaldi office building (Source: Dakdokters Studio)

The Amsterdam Rainproof project employs various types of NBS to manage heavy rainfall and mitigate flood risks. These solutions include: transforming standard roofs into "green-blue" roofs (Fig. 2), increasing parks, permeabilizing soils through vegetation strips along roads, and implementing natural rainwater collection systems. The green strips run parallel to the roads and are made using plant species that require low maintenance and can withstand floods. In the Bellamybuurt neighborhood, green permeable surfaces were increased by 1.241 m^2. The building's drainage system was then modified to optimize water flow: rainwater pipes were redirected to allow water to flow towards the green area and nourish it. The specific goal is to make Amsterdam's soil capable of withstanding 60 mm/h of rainwater, aiming to become completely rainproof by 2050 (Dai, Wörner and Van Rijswick, 2018). This approach aims to sustainably manage heavy precipitation, prevent flood risk, and reduce the burden on the sewer system. Particularly interesting is the approach used for roofs. Green roofs with rooftop gardens have been implemented, as seen in the Polder Roof project in the Zuidas neighborhood. Additionally, "green-blue" roofs have been adopted, exemplified by the Buiksloterham district, where designer Frank Anselma has pioneered a roof type featuring a hydraulic system distributed across multiple levels: during rain, the rainwater falling on the roof is absorbed by the vegetation and collected in rain barrels on the structure; once the roof reaches its capacity, excess water is directed to the rainwater pond on the lower roof. In case of heavy rain saturating the pond, the water is directed to a lower tank acting as an emergency overflow with a capacity of 6.000 L. At this point, only if the tank is also full, the water is discharged into the sewer system. All these strategies serve to slow down rainwater runoff, reduce the burden on the sewer system, and promote water infiltration into the ground, thereby preventing floods and improving water quality.

Regarding ICT, in this project they were mainly used to enhance community participation and project dissemination. To achieve this, the "Weerproof" website was developed, serving as a central reference point for Amsterdam's adaptation initiatives, projects, stories, and research (Amsterdam Weerproof., n.d.). The primary purpose of this website is to raise awareness and engage the local community regarding the importance of preparing the city for extreme weather conditions, adopting a diversified approach for residents and professionals. This is facilitated through a series of well-structured features and sections. The platform Weerproof acts as a catalyst for various facets of the project, fostering active participation and facilitating the sharing of experiences and knowledge among all stakeholders involved. Special attention has been given to the community, which has been engaged in various forms. These include participatory design of strategies through dedicated workshops, the creation of self-management areas such as Hasebroekstraat and Bellamystraat, where green areas are managed by residents themselves. Furthermore, community awareness has been promoted through the online dissemination of dedicated material, collaboration with schools and universities for courses and activities related to the topic, and the organization of thematic events in neighborhoods to showcase green solutions and success stories. Noteworthy is the carefully crafted communication strategy, tailored to use language aimed at making the population feel involved in climate change consequences and simultaneously co-owners of the adopted solutions.

2.2 Copenhagen "Climate Adaption Plan"

The flood experience of July 2011, which caused damages exceeding 1 billion dollars, led to a reconsideration of Copenhagen's resilience capacity (Madsen et al. 2019). However, this dramatic event marked a turning point, transforming the negative episode into an opportunity to improve the city. To this end, planning tools such as the "Copenhagen Climate Adaptation Plan" of 2011 (Copenhagen Municipality, 2011) and the "Cloudburst Management Plan" of 2012 (City of Copenhagen, 2012) were developed with the aim of reducing the city's vulnerability and exposure to extreme weather conditions. The Danish approach focuses on integrating infrastructure solutions with NBS. Projects like the redevelopment of Tåsinge Square in the St. Kjeld neighborhood represent a step forward in implementing these strategies. The process of revitalizing the square began in 2013, as part of a broader strategic plan devised by Tredje Natur, with the ambition to create the world's first resilient square (Tredje Natur, 2011). In the past, the area was primarily dedicated to vehicular traffic, with a central roundabout, occupying a flat area of 8.000 square meters (City of Copenhagen, 2016). The project altered vehicular circulation by reducing traffic flow to transform the intersection into a green park that provides both ecosystem services and recreational opportunities for residents simultaneously. The terrain was shaped by creating artificial hills, thus increasing the permeable surface area of the urban space by 206% (Mezzi, 2015). Besides the square, the entire St. Kjeld neighborhood has been rethought. Among the general objectives was to control the flow of stormwater by adopting solutions to prevent flooding through the use of vegetated surfaces and bike lanes serving as surface channels for drainage. Additionally, floodable squares, known as "water squares," were designed, which can be used as gathering places during pleasant days while serving as rainwater detention basins during heavy rainfall. In Copenhagen's adaptation plan, ICT was utilized not only for monitoring rainwater levels and predicting extreme weather events but also for disseminating strategies and implemented projects through the creation of a dedicated website (City of Copenhagen, n.d.). The main goal of the website is to inform residents about the intentions and progress of the adaptation strategy, fitting into a broader and articulated citizen engagement strategy (Engberg, 2017). The "Klimakvarter" website can be divided into two main categories: the first includes top-down projects, those initiated and carried out by the municipality independently or in partnership with private entities. This category is further divided into smaller thematic sections, referring to the type of spaces and the scale of interventions: urban spaces, courtyards, buildings, streets. Each section contains articles detailing the project from its inception to its current status. In the second category, found in the "your neighborhood" section, bottom-up projects are presented, showcasing initiatives that originated from residents themselves, who independently contributed to the cause by installing green roofs, neighborhood gardens, etc. A specific section of the site is dedicated to all documents produced during the various project phases, informational material, and residents' FAQs. Only in the Danish language version of the site—a limitiation for the international dissemination of the project—is it possible to access the "climate map," an interactive map where Copenhagen's climate change adaptation interventions are geolocalized. Using a specific function, users can select hot spots to create their own thematic "green" tour. The composed map can then be sent via email, with the algorithm automatically arranging the shortest route to reach all locations. On-site, by clicking on the various locations on the interactive map recieved, users gain access to an informative caption to learn about the project's characteristics. The city's renewal atmosphere, combined with a new sensitivity to environmental issues and the availability of regenerated spaces, contributed to the emergence of a revitalized social life in the Osterbro neighborhood. In a short time, residents organized themselves to create a committee to coordinate activities at Tåsinge Square, such as events, concerts, or markets, and some of them spontaneously started projects to transform their roofs into green roofs, inspiring more people to follow them (City of Copenhagen, 2015). To facilitate the dissemination of the project's values and raise awareness among the population, an animated short film titled "Kan du hjælpe?" (Can you help?) was produced, inspired by residents' actions, along with a musical tribute to Copenhagen and the democratic process that led to the development of a greener city.

2.3 Melbourne "Urban Forest Strategy"

The "Urban Forest Strategy" project covers an area of 17 square kilometers in the center of Melbourne.

Despite being part of a broader strategy to combat climate change, the project is focused on addressing the heat island effect as the primary risk factor. It focuses on the widespread use of NBS aimed at enhancing the provision of green infrastructure in most urban areas available—streets, parks, gardens, squares, riverbanks, railway corridors, balconies, rooftops—thus giving substance to the concept of urban forestation (Gulsrud et al., 2018). To achieve the set objectives, the city of Melbourne has heavily relied on NBS, which in this project include: the creation of green corridors to connect parks, forests, agriculture, and green architecture; the reduction of asphalted parking areas in favor of permeable green areas; the expansion of urban agriculture through urban gardens; the transformation of school, university, and hospital courtyards into green areas; the reduction of paved surfaces and an increase of green

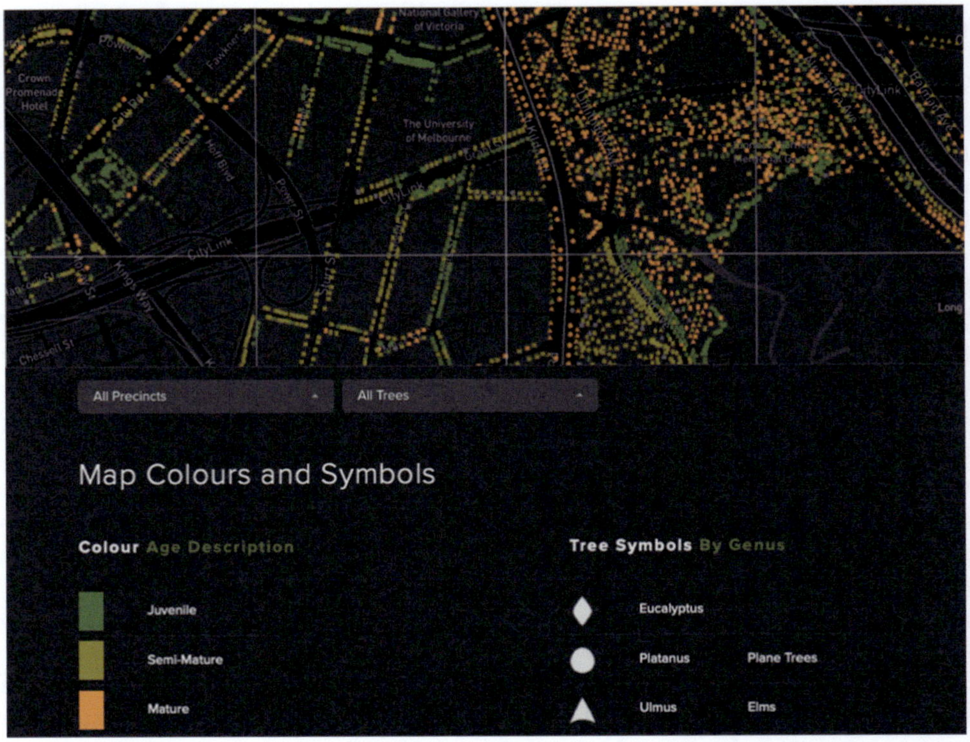

Fig. 3. Interactive map of urban tree distribution (Source: City of Melbourne)

surfaces on rooftops (Melbourne Municipality, 2012). Additionally, phytoremediation techniques have been applied, involving the use of trees and shrubs to reclaim disused and polluted soils. The entire urban/peri-urban green infrastructure system was designed considering adaptability to local conditions and biodiversity. In this regard, it was important to select plants suitable for Melbourne's urban context and climatic conditions (Melbourne Municipality, 2011). What makes the Melbourne case study particularly interesting from our perspective is the integration of NBS and ICT.

Technological innovations have been systematically integrated into NBS, both to monitor their effects and to support communication and population engagement. In this case, ICT was employed from the outset to support the planning and design phase. Through territorial data analysis and the use of advanced simulation models, it was possible to assess the expected effect of NBS on the urban environment. Detailed meteorological and climatic data, including temperatures, humidity, wind flows, and precipitation, were collected and analyzed to understand the area's specific conditions and identify zones at higher risk of the urban heat island effect. These data were integrated with information on existing vegetation and tree distribution, allowing the development of complex mathematical models. Considering various factors such as urban morphology, building layout, and soil characteristics, these models were used to simulate the impact of NBS on temperature and air quality. This approach allowed for evaluating the effectiveness of different strategies, such as tree planting along streets, the creation of urban parks, and the installation of green roofs on buildings, in reducing temperatures, mitigating air pollution, promoting biodiversity, and generating other environmental benefits from adopting NBS (Melbourne Municipality, 2011). The Urban Forest Strategy project saw significant involvement of ICT in various areas. In particular, ICT played a key role in cataloging trees and green areas, giving rise to an "Interactive Urban Forest Map" (Fig. 3). This interactive map not only allowed monitoring the health status of plants but also facilitated planning interventions such as pruning or replacement. ICT also played an essential role in communication and engaging local communities. Digital platforms, social media, and mobile applications were used to inform the population about the project's objectives and implementation phases, encouraging active participation and sharing results. These digital communication initiatives significantly contributed to raising awareness among citizens about the importance of urban forestation, promoting active involvement in caring for the city's environment. The participation process began from the early project stages. The Urban Forest Strategy, in fact, was developed in consultation with the community between November 2011 and April 2012. A broad community engagement process was initiated to offer all community members the opportunity to provide feedback through a variety of channels. Information was disseminated trough a wide range of media channels, including social media, to generate widespread community awareness of the strategy.

3 Discussion

After carefully examining the characteristics of the three analyzed case studies, this section initiates a discussion that, starting from them, expands to consider climate change adaptation projects more broadly. The goal is to identify effective approaches on which to base future adaptation strategies. This analysis has led to the identification of three categories of interventions, representing useful perspectives for understanding the strategies adopted in the case studies: NBS, ICT, and CE (Table 2). In the case studies we have analyzed, there are several examples of the use of NBS. Regardless of the specific purpose for which these solutions are employed, they have the advantage of providing ecosystem services that simultaneously address multiple risk factors. For example, transforming a traffic roundabout into a green square in the St. Kjeld neighborhood of Copenhagen not only increases ground-level permeability but also simultaneously mitigates the urban heat island effect generated by the asphalt that previously covered most of the area. In addition to the direct effects related to risk factor mitigation, NBS also provide indirect positive effects. One of these is the increase in biodiversity: using natural solutions compared to infrastructural ones has the indirect advantage of creating living spaces for flora and fauna, which become fundamental resources especially within the city. All three analyzed adaptation plans have contributed more or less directly to this purpose; furthermore, where planting has been linear and pervasive in the urban fabric, as in the case of Amsterdam, it has become a sort of "green infrastructure" for plant and animal species—such as birds and insects—facilitating their movements within the urban fabric (Bekar 2024). Another secondary positive effect is the increase in livability of spaces where nature is present: areas surrounded by nature produce greater physical and psychological well-being, making them more attractive than other places where nature is absent (Cox et al., 2017). This contributes to activating these spaces, fostering citizen participation and making them lively and frequented.

The second category we want to focus on in this study is ICT, which includes all hardware and software solutions that connect the physical world to the virtual world, leveraging the potential of digital tools. At first glance, ICT seems to be a solution at odds with NBS, as they direct intervention strategies toward more artificial orientations and therefore opposite to those of NBS. At first glance, it almost seems like a choice between these two approaches. However, ICT should not be considered substitute solutions for NBS; rather, they are most effective when they complement their artificial component by serving nature. This is the case of the Melbourne Urban Forest Strategy project, which, more than others, has been able to exploit the potential of digital technologies within a plan linked to the renaturalization of the city. Reviewing the applications described in the section dedicated to the Australian project, we can see how ICT is not configured as an independent solution but as support tool for other solutions. ICT can contribute in all phases of the adaptation strategy: in the initial phase of context analysis and screening, through surveys with precision sensors that assess the degree of risk and the most vulnerable areas of the territory; in the design phase, using parametric software to obtain an optimized spatial configuration based on a set of reference climatic indicators; in the community engagement and results dissemination phase, through applications that can stimulate interaction with citizens or make them aware of interventions in a more pervasive manner; and finally, in the monitoring phase, aimed at real-time control of the correct functioning or condition of the adopted solutions.

The third category of solutions related to climate change is "Community Engagement" (CE), which refers to a set of processes aimed at actively involving members of a local community in various decision-making and implementation phases. In the context of climate change adaptation projects,

Table 2. Main findings divided by categories (Source: By author)

Nature-Based Solutions (NBS)	The nature-based solutions adopted in the studied cities have proven effective in reducing climate risks, such as flooding and urban heat islands, while providing ecological and social benefits. Examples like Amsterdam's green-blue roofs and Copenhagen's floodable parks demonstrate how NBS not only address climate issues but also improve biodiversity and urban livability
Information and Communication Technologies (ICT)	The use of digital technologies in all three case studies has significantly improved environmental monitoring, urban design, and community engagement. In Melbourne, ICT integrated with real-time environmental data optimized NBS implementation
Community Engagement (CE)	Active citizen engagement proved crucial for the long-term success and sustainability of climate adaptation initiatives in all three cities. Community involvement was promoted through co-design workshops and self-management of green spaces, as seen in Amsterdam and Melbourne
Integrated Approach of NBS, ICT, and CE	The case studies highlight that the synergistic integration of NBS, ICT, and CE provides more comprehensive and sustainable results compared to strategies focusing on just one aspect. The combination of technological solutions with nature-based approaches and active community participation offers a replicable model for other cities developing climate adaptation strategies

community engagement becomes crucial. There are several possibilities for CE, which depend not only on the degree to which citizens are involved but especially on the phase in which they are called to participate. In the projects analyzed in the previous paragraph, we have seen strategies mainly focus on dissemination. This form of CE can be seen as the most moderate, yet it shouldn't be dismissed. The dissemination of a climate change adaptation project involves the spread and sharing of information, results, and practices developed within the project with a wide range of stakeholders, including local communities. Essentially, it aims to ensure that the knowledge and solutions developed within the project are accessible and used by as wide an audience as possible to raise awareness and inspire virtuous behaviors. This process can take place through various channels and strategies, many of which are found in the analyzed case studies. However, in dissemination, the communication channel remains one-way: citizens are informed about what is happening, but they cannot provide feedback or intervene. Other forms of CE, instead, are based on two-way communication, where citizens can actively participate in various project phases, directly influencing the choices of designers or policy makers. Depending on the phase in which citizens are called to participate, we can distinguish different forms of CE. If applied to the initial phase, we can speak of co-design, where citizens are invited to sit at the design table with a wide range of actors and stakeholders with the aim of integrating knowledge, experiences, and different perspectives to develop more effective strategies and interventions tailored to the specific needs of a particular intervention area. However, engagement can begin even earlier, in the phase of data collection aimed at analyzing the territory: "citizen science" is a collaborative approach that involves volunteers, even non-professionals, in the scientific investigation process. This involvement can take place in multiple ways, including the collection and analysis of data scattered throughout the territory using technologies embedded in smartphones to turn citizens into real-time data collection probes (Fraisl et al., 2022). Such an approach directly involves community members in contributing to obtaining more numerous and accurate data on a global, regional, or local scale, allowing technicians to have a more complete and detailed view of the phenomena studied and the territory. The same can happen at the end of the project, to collect data aimed at monitoring the status or progress of the project. Once the decision-making and implementation phases of climate change adaptation strategies are completed, some projects can be handed over to the community to involve them through self-management activities and bottom-up spontaneous projects. In the first case, the community itself is responsible for managing and organizing initiatives related to the improvement of the urban environment in which it lives. This involvement can take place at different levels, from design decisions to the implementation and daily management of the projects. On the other hand, bottom-up community projects refer to a participatory and inclusive approach where ideas, initiatives, and decisions start from the grassroots, from citizens rather than from the municipality. This type of approach values the experience, knowledge, and direct needs of the local community, allowing for the development of projects that are more in line with the real needs and aspirations of citizens.

4 Conclusions

The analysis conducted on international case studies of climate change adaptation in urban areas has provided a multifaceted overview of best practices and challenges associated with this field. Through the study of three intervention categories—Nature-Based Solutions (NBS), Information and Communication Technologies (ICT), and Community Engagement (CE)—we have highlighted some significant progress made in this area and, above all, the importance of creating synergy among these three aspects to achieve an effective adaptation strategy. If we were to hypothetically propose a strategy that completely excludes the use of NBS, emphasizing solely the construction of engineering infrastructures, this would entail a considerable increase in costs, especially in the initial phase. This approach would also have a significant impact on habitats, as it would require intensive use of materials and resources for their processing, consequently increasing greenhouse gas emissions and heavily affecting the overall ecological footprint. A strategy that does not employ ICT would miss a valuable opportunity to enhance its efficiency in climate change mitigation actions, foregoing the benefits associated with technological progress achieved in these years. It would lack the ability to collect and process a large amount of data (territorial, environmental, climatic, etc.), limiting the capacity to make informed decisions and predict potentially harmful phenomena in advance. Moreover, the absence of evaluation and monitoring systems for the impacts of adaptation actions would make it difficult to implement timely corrections. Additionally, the use of new technologies has become increasingly widespread even in communication channels, and foregoing the use of ICT would mean losing the primary means of dialogue with citizens, who are increasingly considered a fundamental resource by administrations. A strategy that overlooks CE would lead to the realization of solutions that lack consensus, which, at best, would be unwelcome to citizens, with negative consequences on the effectiveness of the action itself, if not outright obstructed in their implementation. Furthermore, without careful listening, it would be much more difficult to intercept solutions that reflect the real needs of local communities.

The selected case studies, therefore, have the merit of implementing actions that involve all three listed categories, using a holistic and multicriteria approach. However, it is important to emphasize that despite their multiple strengths, they are not without limitations, and there are areas for improvement on various fronts. None of the three case studies covers all aspects comprehensively, and only by observing them simultaneously can we form a comprehensive reference framework. As for future research, the realm of ICT undoubtedly represents the area with the widest development potential. The rapid evolution and diffusion of technologies open limitless research horizons, perhaps still difficult to imagine: consider the use of artificial intelligence and its ability to analyze meteorological, climatic, and environmental data to identify patterns, trends, and correlations that can be useful for predicting and managing the impacts of climate change (Luccioni et al., 2021); or the construction of digital twins, virtual replicas of a physical environment, such as a city or a natural area, to simulate scenarios and assess the impact of different adaptation strategies without directly intervening in the real environment (Henriksen et al., 2023); blockchain technology, which can be used to trace and ensure the sustainable origin of natural resources; or precision sensors, to collect high-resolution and real-time data on various environmental parameters (Fusero, Massimiano, 2019). Each of these solutions will accelerate the capacity to counter the effects of climate change, and it is up to us to find a way to derive the greatest benefit from these tools. In conclusion, the analysis of case studies has highlighted the importance of a holistic and collaborative approach to addressing climate change in cities. The synergistic integration of NBS, ICT, and CE can significantly contribute to the resilience and sustainability of urban areas facing rapidly evolving climate challenges, with particularly relevant implications when considered within the framework of the goals promoted by the European Green New Deal.

Attributions

Although this text is the result of collaborative research, paragraphs 1 & 2 can be attributed to Paolo Fusero; Sects. 2.1, 2.2 & 2.3 to Maura Mantelli; Sects. 3 & 4 to Lorenzo Massimiano.

All the tables and Fig. 1 are created by the authors.

Figure 2 was sourced from the website of the Dakdokters Studio https://dakdokters.nl/portfolio-items/dakpark-vivaldistraat/

Figure 3 was sourced from the website of the City of Melbourme http://melbourneurbanforestvisual.com.au.

References

Amsterdam Municipality. Amsterdam Rainproof: every drop counts (2023). https://weerproof.nl/app/uploads/2023/05/RAINPROOF-MAGAZINE-ENGELS_compressed.pdf. Accessed 10 May 2024

Amsterdam Weerproof. Samen maken we het verschil! https://weerproof.nl. Accessed 10 May 2024

Axelsson, C., Soriani, S., Culligan, P., Marcotullio, P.: Urban policy adaptation toward managing increasing pluvial flooding events under climate change. J. Environ. Planning Manage. **64**(8), 1408–1427 (2021). https://doi.org/10.1080/09640568.2020.1823346

Bekar, T.: Urban water and management in sustainable cities. Environ. Sci. Sustain. Dev. **9**(4), 1–18 (2024). https://doi.org/10.21625/essd.v9i4.1062

Butt, N., et al.: Opportunities for biodiversity conservation as cities adapt to climate change. Geo Geogr. Environ. **5**(1), e00052 (2018). https://doi.org/10.1002/geo2.52

C40. (2024). https://www.c40.org. Accessed 10 May 2024

City of Copenhagen. Copenhagen climate projects: annual report 2016 (2016). https://kk.sites.itera.dk›apps›kk_pub2›pdf. Accessed 10 May 2024

City of Copenhagen. Copenhagen climate projects: annual report 2016 (2016). https://kk.sites.itera.dk›apps›kk_pub2›pdf. Accessed 10 May 2024

City of Copenhagen. Copenhagen's First Climate resilient neighbourhood (2016). http://klimakvarter.dk/wp-content/uploads/2015/08/Copenhagens-first-climate-resilient-neighbourhood_WEB_low.pdf. Accessed 10 May 2024

Copenhagen Municipality. Copenhagen climate adaptation plan (2011). https://en.klimatilpasning.dk/media/568851/copenhagen_adaption_plan.pdf. Accessed 10 May 2024

Cox, D.T.C., Shanahan, D.F., Hudson, H.L., et al.: Doses of neighborhood nature: the benefits for mental health of living with nature. Bioscience **67**(2), 147–155 (2017). https://doi.org/10.1093/biosci/biw173

Dai, L., Wörner, R., van Rijswick, H.F.M.W.: Rainproof cities in the Netherlands: approaches in Dutch water governance to climate-adaptive urban planning. Int. J. Water Resour. Dev. **34**(4), 652–674 (2018). https://doi.org/10.1080/07900627.2017.1372273

Davis, M., Abhold, K., Mederake, L., Knoblauch, D.: Nature-based solutions in European and national policy frameworks. In: NATURVATION Project Deliverable 1.5. Horizon 2020 Grant Agreement No 730243, European Commission (2017)

Engberg, L.A.: Climate adaptation and citizens' participation in Denmark: experiences from Copenhagen. In: Hughes, S., Chu, E., Mason, S. (eds.) Climate Change in Cities. The Urban Book Series. Springer, Cham (2018). https://doi.org/10.1007/978-3-319-65003-6_8

Field, C.B., Barros, V.R., Stocker, T.F., Dahe, Q.: Managing the Risks of Extreme Events and Disasters to Advance Climate Change Adaptation: Special Report of the Intergovernmental Panel on Climate Change. Cambridge University Press (2012). https://doi.org/10.1017/CBO9781139177245

Fraisl, D., Hager, G., Bedessem, B., et al.: Citizen science in environmental and ecological sciences. Nat. Rev. Methods Primers **2**, 64 (2022). https://doi.org/10.1038/s43586-022-00144-4

Frantzeskaki, N., Castán Broto, V., Coenen, L., Loorbach, D.: Urban Sustainability Transitions: The Dynamics and Opportunities of Sustainability Transitions in Cities. Routledge (2017)

Fusero, P., Massimiano, L.: Precision environmental planning: state of the art and possible applications. In: Planning for Transition, Proceedings of the AESOP 2019 Conference (2019). ISBN 978-88-99243-93-7

Geneletti, D., Zardo, L.: Ecosystem-based adaptation in cities: an analysis of European urban climate adaptation plans. Land Use Policy **50**, 38–47 (2016). https://doi.org/10.1016/j.landusepol.2015.09.003

Gulsrud, N.M., Hertzog, K., Shears, I.: Innovative urban forestry governance in Melbourne? Investigating 'green placemaking' as a nature-based solution. Environ. Res. **161**, 158–167 (2018)

Hartigan, M., Fitzsimons, J., Grenfell, M., Kent, T.: Developing a metropolitan-wide urban forest strategy for a large, expanding and densifying capital city: lessons from Melbourne. Australia. Land **10**(8), 809 (2021). https://doi.org/10.3390/land10080809

Henriksen, H.J., Schneider, R., Koch, J., et al.: A new digital twin for climate change adaptation, water management, and disaster risk reduction (HIP digital twin). Water **15**(1), 25 (2023). https://doi.org/10.3390/w15010025

IPCC. Climate Change 2007: Synthesis Report. IPCC, Geneva, Switzerland (2007)

Kabisch, N., Korn, H., Stadler, J., Bonn, A.: Nature-Based Solutions to Climate Change Adaptation in Urban Areas: Linkages between Science, Policy and Practice. Springer (2017). https://doi.org/10.1007/978-3-319-56091-5

Luccioni, A., Schmidt, V., Vardanyan, V., Bengio, Y.: Using artificial intelligence to visualize the impacts of climate change. IEEE Comput. Graphics Appl. **41**(1), 8–14 (2021). https://doi.org/10.1109/MCG.2020.3025425

Madsen, H.M., Mikkelsen, P.S., Blok, A.: Framing professional climate risk knowledge: extreme weather events as drivers of adaptation innovation in Copenhagen, Denmark. Environ Sci Policy **98**, 30–38 (2019). https://doi.org/10.1016/j.envsci.2019.04.004

Melbourne Municipality. Urban Forest: diversity guidelines (2011). https://www.melbourne.vic.gov.au/SiteCollectionDocuments/urban-forest-diversity-guidelines.pdf. Accessed 10 May 2024

Melbourne Municipality. Urban Forest Strategy (2012). https://www.melbourne.vic.gov.au/SiteCollectionDocuments/urban-forest-strategy.pdf. Accessed 10 May 2024

Mezzi, P.: San Kjeld, Copenhagen: the first climate-change-adapted neighbourhood. In Abitare Magazine (2015). https://www.abitare.it/en/architecture/sustainable-elements/2015/03/01/san-kjeld-copenhagen-first-climate-change-adapted-neighbourhood/

Pettifor, A.: The case for the Green New Deal. Verso Books (2020)

Revi, A., et al.: Urban areas. In: Field, C.B., Barros, V.R., Dokken, D.J., et al. (eds.) Climate Change 2014: Impacts, Adaptation, and Vulnerability. Part A: Global and Sectoral Aspects. Cambridge University Press, pp. 535–612 (2014). https://www.ipcc.ch/report/ar5/wg2/

Seto, K.C., et al.: Human settlements, infrastructure and spatial planning. In: Edenhofer, O., Pichs-Madruga, R., Sokona, Y., et al. (eds.) Climate Change 2014: Mitigation of Climate Change. Contribution of Working Group III to the Fifth Assessment Report of the IPCC. Cambridge University Press (2014). https://www.ipcc.ch/report/ar5/wg3/

Sørensen, E., Torfing, J.: Co-creating ambitious climate change mitigation goals: the Copenhagen experience. Regul. Gov. **16**, 572–587 (2022). https://doi.org/10.1111/rego.12374

Siemens (2013). https://www.preventionweb.net/news/c40-siemens-inaugural-city-climate-leadership-awards-honour-ten-cities-excellence-urban. Accessed 10 May 2024

Siemens (2014). https://news.siemens.co.uk/news/c40-and-siemens-inaugural-city-climate-leadership-awards-honor-ten-cities-for-excellence-in-urban-sustainability. Accessed 10 May 2024

Tredje Natur. The first climate district (2011). https://www.tredjenatur.dk/en/portfolio/the-first-climate-district/

The Influence of Outdoor Space Geometry and Greenery Configuration on Students' Thermal Perception During Moderate Seasons in an Arid Climate

Hadeer Emad Sobhy, Hatem Mahmoud, and Omar Mohamed Galal

Abstract

Improving the outdoor thermal sensation helps improving the quality of outdoor spaces for pedestrians and users, consequently contributes to making cities more sustainable. Space geometry and space greenery are acknowledged for their impact on outdoor thermal comfort. However, limited experimental studies examine both strategies simultaneously. Two spaces at university campus at El-Giza, Egypt were selected as a case study. Space 1 had a favorable geometric configuration as it had 75% of its perimeter enclosed by 6 floors building and oriented towards the prevailing wind, however, it was completely paved. Space 2 had more than 60% of its surface covered by greenery with 50% of its perimeter enclosed by the same building and its western and southern sides were opened. The relevant climatic parameters were measured and surveys were distributed among 290 students over six days in spring and four days in fall during typical educational day hours in the two spaces. Throughout all measurements taken, Space 1 enjoyed cooler air temperature, Mean Radiant Temperature (MRT) and Physiological Equivalent Temperature (PET). Students with "Neutral" thermal conditions in Space 1 were twice the number of their peers in Space 2. For all the responses of the students, Thermal Sensation Vote (TSV) was found to have strong relation with PET, MRT and surface temperature. The results objectively (PET) and subjectively (self-reported survey) imply that variations in space geometry manipulation and space greenery can be associated with significant climatic and sensational variations for the students and how they use the outdoors.

Keywords

Outdoor thermal comfort · Physiological equivalent temperature (PET) · Thermal sensation vote (TSV) · University Campuses · Young Adults

H. E. Sobhy (✉) · O. M. Galal
Department of Architectural Engineering and Environmental Design, College of Engineering, Arab Academy for Science, Technology and Maritime Transport, Smart Village Campus, 6th October City, Giza, Egypt
e-mail: Hadeeremad@adj.aast.edu

O. M. Galal
e-mail: omar.galal@aast.edu

H. Mahmoud
Sustainable Architecture Program, Egypt-Japan University of Science and Technology E-JUST, New Borg El-Arab City, Alexandria 21934, Egypt

Department of Architecture, Faculty of Engineering, Aswan University, Aswan 81542, Egypt

H. Mahmoud
e-mail: hatem.mahmoud@ejust.edu.eg

1 Introduction and Background

Studies on outdoor thermal comfort are essential for comprehending how people react to their thermal surroundings. This knowledge can serve as a foundation for forecasting how urban outdoor space users will behave under different thermal conditions. It has become evident that creating comfortable outdoor urban spaces can enhance the quality of life. Given that over half of the world's population live in cities, outdoor recreational areas have a substantial impact on people's physical, social, and mental health (Aghamolaei et al., 2022; Baruti et al., 2019; "Population Reference Bureau," 2012). The improvement of the thermal conditions of outdoor spaces can result in energy savings in buildings as people will spend more time outdoors. Furthermore, the access restrictions have complicated the role that parks and outdoor areas play in

B. He et al. (eds.), *Towards the Framework of Livable and Resilient Cities* Advances in Science, Technology & Innovation, https://doi.org/10.1007/978-3-031-97849-4_20

promoting psychological and physiological recovery during the pandemic crisis (Rung et al., 2011; Samuelsson et al., 2020).

Outdoor thermal comfort is one of the most important indicators that affects the usability of outdoor urban spaces from the perspective of the residents of urban environments (Boukhelkhal & Bourbia, 2016). Some climatic variables, like air temperature, wind speed, relative humidity, and mean radiant temperature (MRT) can significantly affect outdoor thermal comfort. By understanding regional climatic changes and thermal comfort indicators at the micro-scale, improving the quality of outdoor spaces with all its economic benefits becomes possible. On the other hand, outdoor thermal comfort is not solely affected by environmental variables; there are additional parameters, that have a major impact on the thermal sensation of the space users such as activity level and clothing (ASHRAE 55, 2017; "ASHRAE Handbook," 2009). The characteristics of the built environment (space geometry, materials, and vegetation) affect thermal comfort.

In previous studies, personal factors such as age, gender, and body weight were reported to have a significant effect on human thermal comfort (Heidari et al., 2017; Krüger et al., 2010; Zhen et al., 2021). Many studies comprehensively choose the individuals formulating their samples without controlling for age variations (Elnabawi et al., 2016; Huang et al., 2015; Li et al., 2017; Lucchese et al., 2016). In the study at hand, through an experimental approach, we aim to explore the extent to which the built environment can affect the thermal sensation of outdoor space users. Outdoor thermal comfort was investigated in many previous studies with different climate classifications(Lin, 2009; Mahmoud, 2011; Sangkertadi, 2014). Many of the outdoor thermal comfort oriented studies in the MENA region examine the outdoor thermal comfort through simulation studies such as in (Fahmy & Sharples, 2009). Fewer studies mix simulations with field measurements (Elgheznawy & Eltarabily, 2021; Galal et al., 2020) or studies that solely depend on field measurements (Elnabawi et al., 2016; Salata et al., 2016). However, experimental studies focusing outdoor space users' thermal sensation in spaces with geometrical and vegitational diversity are limited.

The study at hand discusses the influence of building form, outdoor greenery and associated microclimatic conditions on the thermal sensation of outdoor space users performing different activities. The sample of individuals used within this study is formed from university students who are within the same age range (Young adults), having the same socio-economic background, and they exist in the same place and are exposed to the same local climatic conditions. The study was spatially performed within two university outdoor spaces that are significantly different in terms of their geometric configuration, orientation, and green cover. Nonetheless, It should be noted that shade was available for the students in the two spaces (the shadow of the building in space 1 and pergulas in space 2), yet, the overall shade area was bigger in space 1. The climatic measurements were simultaneously taken in the two spaces. Hence, it was possible to attribute any change in the thermal sensation of users solely to variations in the built environment. The study at hand has two main objectives; a) explore how space geometry, materials, and greenery influence the space user's outdoor thermal perception, b) investigate the relation between thermal sensation votes of the users using the outdoor spaces and the different climatic variables

Fig. 1. Workflow demonstrating the methodological steps conducted in the study (Source: By author).

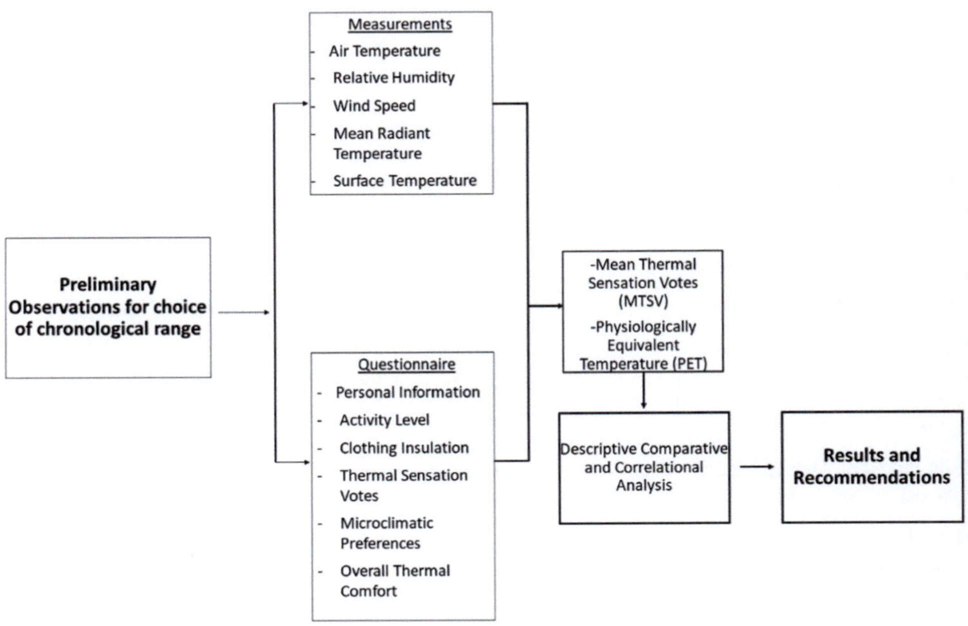

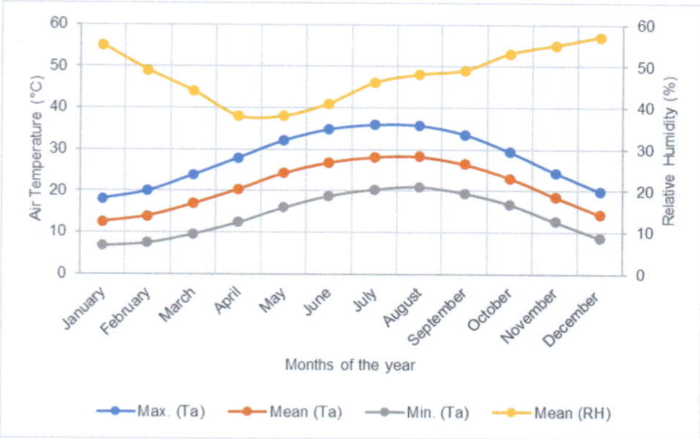

Fig. 2. Monthly mean, the maximum, minimum air temperature at the 6th of October, Giza, Egypt (Source by the author according to data from (6th of October Weather & Climate | Temperature & Weather by Month - Climate-Data.Org, 2022).

that they are subjected to, and c) How microclimatic variations are associated with changes in the way students use the space.

2 Data and Methods

The analysis was methodologically divided (as shown in Fig. 1) into the following primary categories; a) self-reported results that adhere to ASHRAE 55 and ISO standards, b) field climatic measurements.

2.1 Study Domain

The study was carried out at Smart Village in Giza, Egypt. According to Köppen's classification of climates (Peel et al., 2007), Smart Village has a BWh which is a dry and hot desert climate (Köppen, 1900). As a result, there is only an average of 4 mm of rainfall per year, mostly in winter during January. The minimum average temperature throughout winter is 12.5 °C (as shown in Fig. 2) which is considered warm. The summer is generally hot and dry. August is the hottest month and has an average temperature of 28.3 °C. May is regarded as the driest month of the year because it has 0mm of rainfall (*6th of October Weather & Climate*, 2022). Therefore, May have the lowest relative humidity, at 37.7%, while December has the greatest relative humidity of 56.6%.

Our study compared the climatic variables that were measured in two spaces and their usability to the users according to their different gender. The study was conducted in a university campus in Smart Village which is situated at (30.072°N, 31.018°E) and 75 m above sea level (*Smart Village Topographic Map*, 2022).

The study was conducted within a university campus but in two different spaces as shown in Table 1 (Fig. 3) during the same days. The first space (Space 1) is the primary outdoor public area (as shown in Fig. 5) which is oriented to the north with a slight inclination to the east and facing the direction of the prevailing wind in an unobstructed manner with an approximate area of 1148 m^2 and was surrounded by a glazed 6 floors building which partially shaded some parts of the space for long periods of the day. The majority of the surrounding building's facade is glazed. The flooring finish is made of granite and interlocking marble tiles. The second space (Space 2) (as shown in Fig. 8) is not as exposed to the prevailing winds and orientated to the south with a 1333 m^2 and was mainly designed as to be the entertainment space for the students with two garden areas of which a small area (approximately 150 m^2) is shaded by wooden pergolas and surrounded by palm trees. Except for minor portions that are partially shaded by nearby buildings and pergolas, the majority of Space 2 is exposed to solar radiation. 60% of the area of Space 2 is covered by greenery. The paved parts of the space are floored with interlocking concrete tiles. The remaining areas have grass as ground cover. The building façade overlooking Space 2 is also glazed.

2.2 The Climatic Field Measurements

The field measurements were taken in spring and fall. Among the limitations of this study is that we were not able to conduct the analysis during summer and winter as the number of students was significantly lower in summer. The measurements were carried out 3 times per day and for ten days during moderate seasons for the microclimatic parameters that affects outdoor thermal comfort based on ASHRAE Standard (ASHRAE 55, 2017). The measured parameters were air temperature (°C), wind speed (m/s), and relative humidity (%), in addition, to mean radiant temperature (MRT) (°C), and ground surface temperature (°C) in compliance with previous studies (Khalili et al., 2022;Elnabawi et al., 2016; Huang et al., 2015; Li et al., 2017; Lucchese et al., 2016). In this study, Physiological equivalent temperature (PET) was used as an indicator of outdoor thermal comfort. While newer thermal comfort indices were developed like Universal Thermal Climate Index (UTCI), PET was found to be highly correlated with UTCI under different climatic conditions (Gulyás et al., 2009; Johansson & Emmanuel, 2006; Matzarakis et al., 2007; Thorsson et al., 2006). For a human biometeorological examination within the urban climates, the PET validity has been proved in different climatic conditions such as hot arid and hot humid conditions, and was already included in the German VDI3787. At last, but not least, PET can be calculated using the open-access software Rayman.

The survey responses for the pedestrians' subjective thermal perceptions were gathered while the measurement

Table 1. In the first row there is a satellite image for AASTMT university buildings highlighting the two selected spaces. (Map data: Google Earth, Image @ 2022 Maxar Technologies). In the 2nd and 3rd rows, there is a mass model to indicate the locations of the vintage points from where the spaces were observed in addition to real photos for Space 1 and Space 2 respectively. (Source: By author)

<table>
<tr><td>Google Earth Image</td><td colspan="3">Fig. 20.3 A satellite image for AASTMT university buildings (Source: Google Earth, Image @2022 Maxar Technologies and edited by author).</td></tr>
<tr><td>Spaces</td><td>Observation Position</td><td>Real Space Picture</td><td>Measuring tool's location</td></tr>
<tr><td>Space 1</td><td>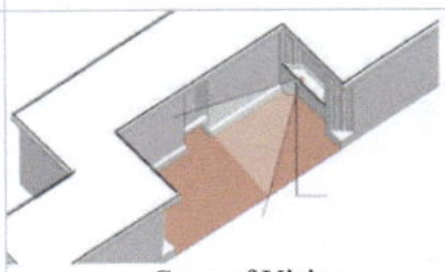Cone of Vision
The orange area is the first space, and the photos were taken from the second-floor terrace, as shown in the diagram.
Fig. 20.4 Mass Model for Space 1 (Source: by author).</td><td>Fig. 20.5 Picture for space 1 (Source: by author).</td><td>Fig. 20.6 Location of the tools in space 1 (Source: by author).</td></tr>
<tr><td>Space 2</td><td>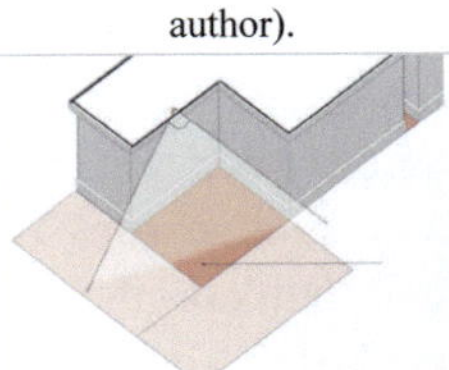Cone of Vision
The orange area is the second space and the photos were taken from the roof of the 6th floor of the building adjacent to the space
Fig. 20.7 Mass Model for Space 2 (Source: by author).</td><td>Fig. 20.8 Picture for space 2 (Source: by author).</td><td>Fig. 20.9 Location of the tools in space 2 (Source: by author).</td></tr>
</table>

tools were in place (as shown in Table 2). Two video cameras were placed throughout the day (as shown in Fig. 4 and Fig. 7) to determine the pattern of space use/ occupancy by the students in both spaces (Kim & Brown, 2022). The study survey was conducted in March, May, and November in 2022 with six days of measurements in spring and four days of measurements in fall.

During the monitoring campaign, each session lasted for approximately 25 min long as the globe thermometer required 20 min before giving accurate measurements for the globe thermometer. The sessions were repeated three times a day.

A portable Pasco wireless weather sensor with GPS (PS-3209) (as shown below in Fig. 10) was used to measure air temperature, wind speed and relative humidity. A thermal camera (testo 865) (as shown in Fig. 13) was used to measure the surface temperature; Globe temperature (Tg) was measured with a 150 mm diameter black matte globe thermometer (as shown in Fig. 11) and connected to a Testo 440 device (as shown in Fig. 12), placed at a height of 1.1 m at two places (as shown in Fig. 6 and Fig. 9). The following Eq. 1 was also used to calculate the mean radiant temperature (Tmrt) ("ASHRAE Standards 55," 2004; *ISO 7726*, 1998).

Table 2. Measuring tools specifications and features (Source: by author).

Parameter monitored	Sensor Brand and Model	Tool's Picture	Accuracy	Response/ Adjustment Time	Resolution	Measurement Range
Air Temperature	Pasco Wireless Weather Sensor with GPS (PS-3209)	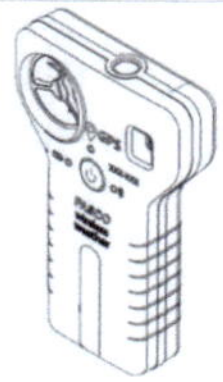 **Fig. 10.** Pasco weather sensor tool (Source: by author according to Pasco 013-15518A)	± 0.2 °C	5 min.	0.1 °C	- 40 to +125 °C
Relative Humidity	Pasco Wireless Weather Sensor with GPS (PS-3209)		± 2 %	5 min.	0.1 %	0 to 100 %
Wind Speed	Pasco Wireless Weather Sensor with GPS (PS-3209)		3 % of reading	5 min.	0.1 m/s	0.5 to 15 m/s (winds of up to ~ 33 mph)
Mean radiant temperature	Testo Globe Thermometer (D = 150mm)	 **Fig. 11.** Globe Thermometer (Source: by author according to test equipment depot)	At 22 °C ± 1 digit (type K thermocouple, class 1)	Approx. 30 min		0 to +120 °C
	Testo 440 device	 **Fig. 12.** Testo 440 device (Source: by author according to testo 440 0560 4401)	(± 1 digit) ± 0.4 °C (-40 to -25.1 °C) ± 0.3 °C (-25 to +74.9 °C) ± 0.4 °C (+75 to +99.9 °C) ± 0.5 % of m.v. (remaining measuring range)	5 min.	0.1 °C	- 40 to 150 °C
Thermal camera	Testo 865	 **Fig. 13.** Thermal camera (Source: by author according to testo 865 thermal imager)	± 2 °C, ± 2 % of measurement value (larger value applies)			- 20 to +280 °C

$$Tmrt = \left\{(Tg + 273)^4 + \left[\frac{(1.1x10^8 x Va^{0.6})}{(\varepsilon g\ x\ D^{0.4})}\right]x(Tg - Ta)\right\}^{\frac{1}{4}} - 273$$

where Tg is the globe temperature (°C), Va is the wind speed (m/s), Ta is the air temperature (°C), εg is the globe's emissivity (0.95 for the black globe), and D for the globe diameter (150 mm). The PET was then calculated by inputting the measured climatic parameters into the software Rayman model version 1.2 (Matzarakis, 2007). Rayman is a software that was developed in the Meteorological Institute, University of Freiburg, Germany, according to the German Engineering Society Guidelines (VDI, 1998). Rayman is regarded as one of the most rigorous models for bioclimatic calculations. The model requires values of air temperature, wind speed, relative humidity and mean radiant temperature as inputs in addition to some personal data as (Height, weight, age, gender, clothing insulation, activity level) to calculate PET as an output.

2.3 The Survey

A self-reporting survey was conducted along with the objective microclimatic monitoring, and the represented sample was comprised mainly of students with approximately equal ratio between males and females. In this study all the surveyed students were all within the same age range, and their positions were not further than 30 m from the measuring tools (Ruiz & Correa, 2015). The questionnaire had mainly three parts following ("ASHRAE Standards 55," 2004; *ISO 7730*, 2005, p. 773). The personal information was requested in the first section of the thermal comfort questionnaire**,** following (Nikolopoulou et al., 2003; Noémi & János, 2007) such as (e.g., age, gender, weight, height, activity, time spent in outdoor spaces, and motivation behind their choice to the outdoor space). In this part beside the personal information, we also ask the students about the reason for choosing this space, how long have they been in the space, how often do they use this space, and their current activity. The second section recorded, the clothing layers, along with the color of their clothing. Equivalent values for these data were then estimated Clo-units in addition to Metabolic rate (W/M2) based on the ("ASHRAE Handbook," 2009). All the aforementioned values were used as inputs in Rayman to calculate PET for each survey. The third and final section dealt with the subjective measurements such as the participants existing thermal perception and preferences for all climatic parameters, with the ASHRAE 7-point thermal sensation vote (TSV) scale (−3, cold; −2, cool; −1, slightly cool; 0, neutral; 1, slightly warm; 2, warm; 3 hot) Following Elnabawi et al. (2016) and Mahmoud (2011). While this is among the main limitations of the study at hand, in future research, we plan to use the more recent 9- point Thermal Sensation Vote (TSV) scale which will allow for more precise results. The McIntyre preference scale where each participant specifies if they want the climatic parameters to be ("cooler", "no change", or "warmer") was used. Finally, the overall thermal comfort (uncomfortable/ acceptable/ uncomfortable) was also reported. The outputs of the survey were used as inputs for the calculation of the Mean Thermal Sensation Vote (MTSV) in addition to documenting the user's personal characteristics. Two hundred and ninety participants were surveyed (of which 43% were females and 57% were males). The majority were between 18 and 23 years of age (85%) and 15% or the age between 23 to 30 years old. All the surveyed students used the outdoors of the university daily with a minimum duration of 20 min.

3 Results and Discussion

The microclimatic variables were measured and analyzed for both spaces (as shown below in Fig. 14 and Fig. 15.). The measurements were taken during the daily break times and for six days in spring (as shown in Fig. 14.) and four days in fall (as shown in Fig. 15). The geometric variations between the two spaces have led to miscellaneous differences in the microclimatic conditions as per demonstrated by the measurements. Due to its orientations, space 1 has consistently enjoyed lower MRT, air temperature, and average surface temperature values in comparison to space 2. This was reflected in the overall thermal sensation of the students. For example, during the hottest examined day, air temperature reached as high as 35 °C in space 2 while not exceeding 30 °C in space 1. At the same timing, the average PET values in space 1 were approximately 30 °C while exceeding 50 °C in space 2. Two additional points are worth noticing.

Firstly, it appears that the thermal performance of the two spaces is more sensitive to changes in space geometry during fall in comparison to spring. This can be noted as the differences between the calculated PET and MRT values are much higher in fall in comparison to spring. Secondly, despite not containing any vegetation or green ground cover, Space 1 enjoyed a consistently improved thermal environment in comparison to Space 2.

To further improve the understanding of the participants' thermal perception, the TSVs as per gathered from the surveys were compared as shown in Fig. 16. The results show that the female students tend to be more resilient to warmer conditions and more sensitive to cooler weather in comparison to the male students. In Space 1, a higher percentage of female with "neutral" vote was recorded (52.05%) verses (47.37%) for male. In Space 2, the highest percentages were for the TSV = +1 for the male and female with 39.6% and 40% respectively. Additionally, there is a significantly larger percentage of students voting "warm" and "hot" in Space 2 in comparison to Space 1. This indicates that MRT and air temperature are highly accompanied with change in the thermal sensation of the students. These findings are consistent with what was reported in previous studies (Butera, 2018; Nikolopoulou et al., 2001).

A correlation analysis was conducted to access the relationships between various climatic variables and the users' thermal sensation vote (TSV) in the two examined spaces to determine the Pearson correlation coefficient r as shown in Table 3. The results indicated that not all climatic variables had a similar influence on TSV under the moderate climate conditions. Some variables exhibited a stronger correlation with the thermal comfort perception of the users such as (PET, surface temperature, and MRT), while others showed a weaker or negligible effect as (relative humidity and wind speed).

Fig. 14. Comparison between the microclimatic conditions in the two spaces in spring seasons at the observed timings (Source: by author).

Table 3. The Pearson correlation coefficient (r) values for the relation between the different climatic variables and TSV (Source: by author).

Correlation Coefficient (R-value) Moderate Seasons	TSV
Air Temperature	0.58
Relative Humidity	−0.33
Wind Speed	−0.27
Mean Radiant Temperature	0.67
Surface Temperature	0.71
PET	0.74

P-value was lower than 0.05 for all the microclimatic variables examined. This variation suggests that different environmental factors contribute differently to thermal sensation in moderate climates, highlighting the complexity of thermal comfort assessments.

According to previous results, a further investigation was made by analyzing the participants' thermal preferences according to the different microclimatic parameters as shown in Fig. 17. This facilitated understanding how each climatic parameter affects the participant's overall thermal comfort. In Space 2, most participants preferred a cooler air temperature and mitigated solar radiation with a percentage of 87.3% and 78.5% respectively. In Space 1, "No change" represented the

Fig. 15. Comparison between the microclimatic conditions in the two spaces in fall season at the observed timings (Source: by author).

largest portion of the recorded votes across all the examined parameters. However, this was not the case in Space 2 as the largest portion preferred cooler air temperature and weaker solar radiation. However, no clear preference was noted for wind speed, relative humidity, the surface temperature.

Generally, in Space 2 50% felt that the climatic conditions were thermally "acceptable" while 43% felt "uncomfortable". Conversely, in Space 1, only 16% of the users felt uncomfortable (as shown in Fig. 18). All the aforementioned results confirm that space 1 has consistently enjoyed improved thermal conditions according to measurements and the users' perception. Space 1 had no vegetation or greenery within it, it was completely paved while space 2 had the majority of its ground covered with grass and contained a number of trees and palm trees. Hence, manipulating space geometry should be prioritized in comparison to utilization of landscape as it proved to have a significantly stronger influence on outdoor thermal comfort. These findings are in alignment with prior studies that have demonstrated the strong relationship between space geometry and PET (Adly et al., 2023; Galal et al., 2020). The study confirm that outdoor users showed difference in thermal sensation votes according to the space they used. The space geometry and the covered areas achieve stronger effect than the landscape zones. This

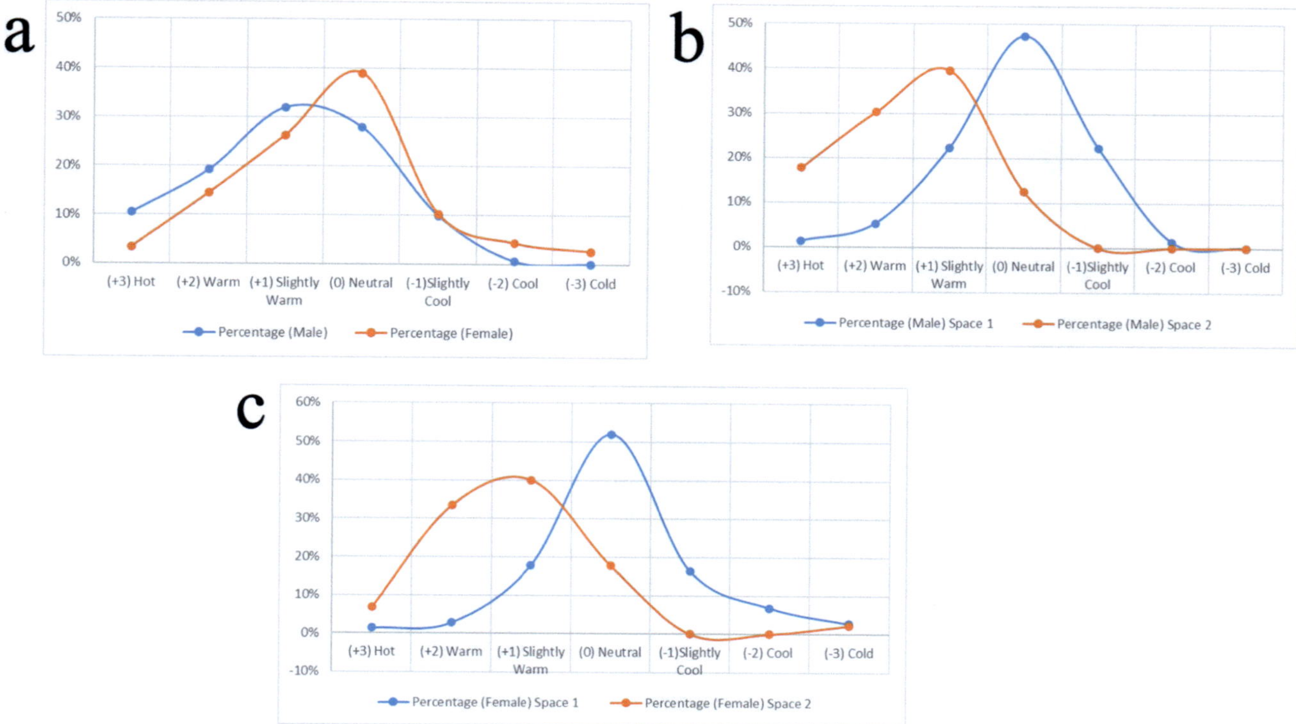

Fig. 16. a) Thermal sensation votes (TSVs) percentage distribution in the two spaces for all the students, b) TSV percentage distribution for males in the two spaces, and c) TSV percentage distribution for females in the two spaces (Source: by author).

Fig. 17. Thermal preferences vote during the moderate seasons in the two spaces, for all the measured climatic parameters: air temperature, wind speed, relative humidity, solar radiation, and surface temperature (Source: by author).

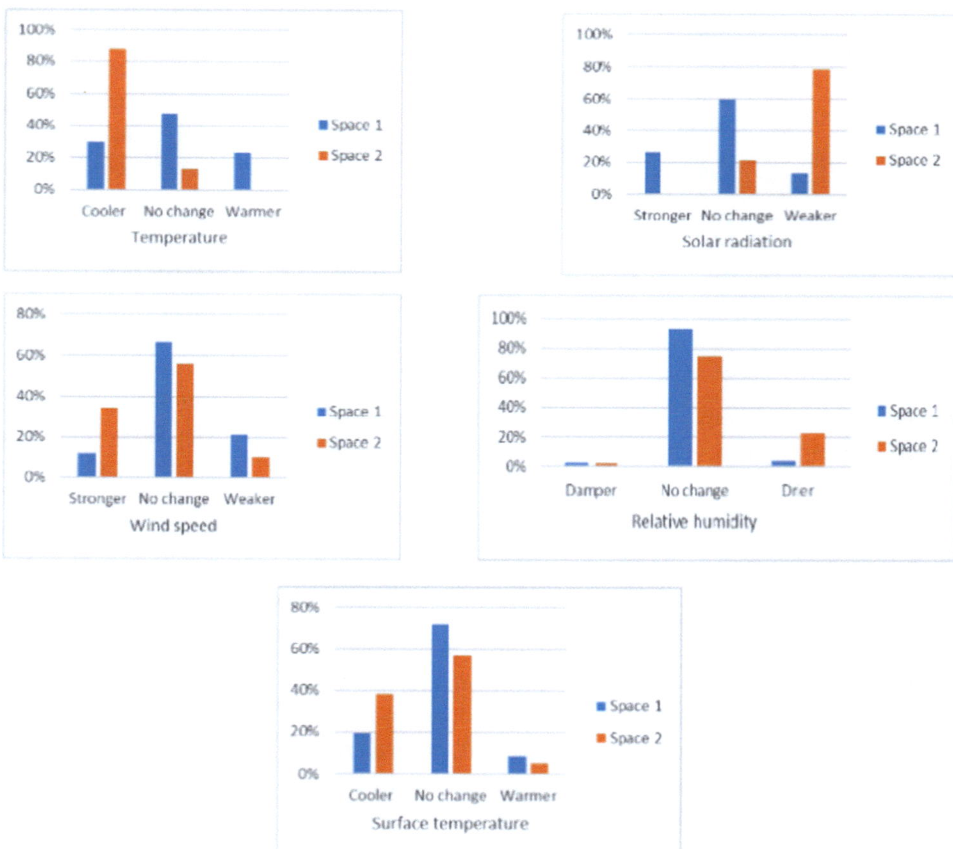

Day	Time	Photo	Thermal camera	Measurements				Analysis		Observation
				Air temp	Relative humidity	Wind speed	Mean radiant temp.	No. of users (Usability)	Activity level	Users choices for location
17/5	First Space at 12:00 pm	Sitting Standing Walking Playing 	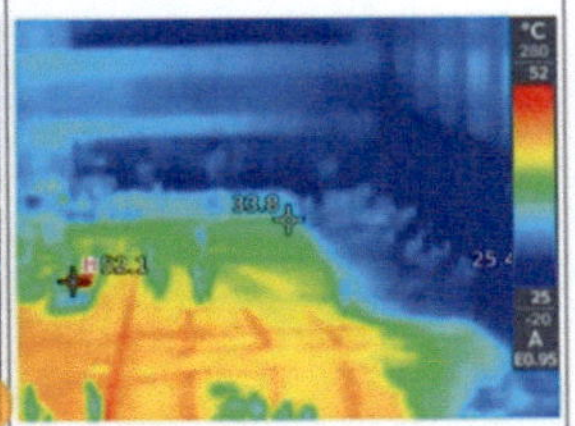	27	46.4	0.9	30.2	146 person (70 in sun and 76 in shade)	- 14 playing - 14 standing - 15 walk by - 103 sitting	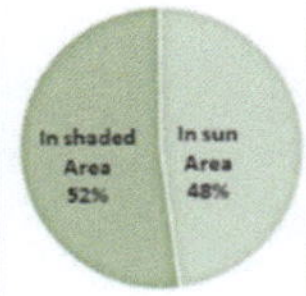
	Second Space at 12:00 pm	Sitting Standing Walking Playing 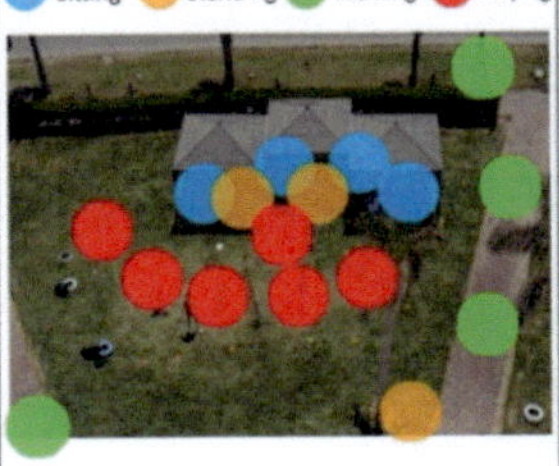	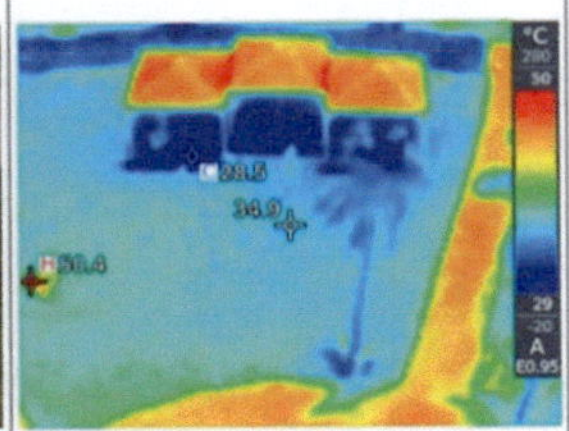	30.6	39.4	0.4	50.8	13 people (All in the sun except 5 sitting and 2 standing under wooden shading)	- 6 playing - 5 sitting - 2 standing	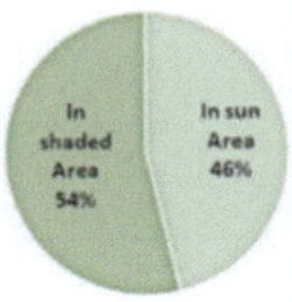

Fig. 19. Comparison of surface temperature and the distribution of the space users in both spaces on 17th of May 2022 at noon (12 pm) (Source: by author).

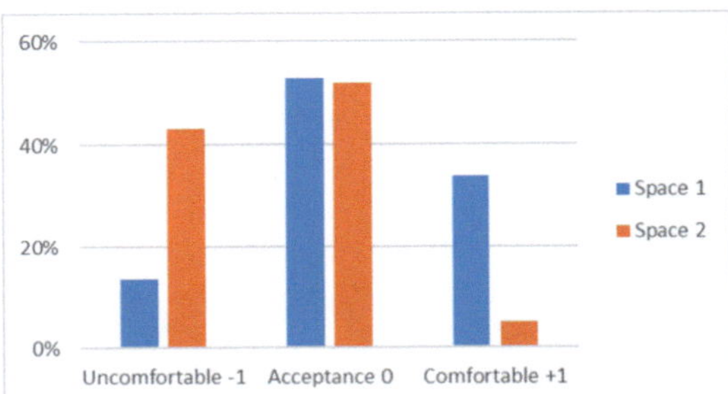

Fig. 18. Thermal preferences vote during the moderate seasons in the two spaces for the thermal comfort conditions in general (Source: by author).

could be achieved in other outdoor spaces that are exposed to sun by using light structure as pergolas, kiosks which is also mentioned before by (Mahmoud, 2011) in his study in outdoor park and he compared betweeen different zones with different in landscape and design.

Although space 1 and space 2 are used at the same time on the same day, it has been noticed that the distribution of space users in each differs due to the variations in the microclimatic conditions in each space as shown in Fig. 19. In space 1, availability of larger shaded areas (resulting from the building) provided the users with more comfort, particularly under similar climatic conditions, encouraging social interactions. Space 2, on the other hand, may have fewer seating options and limited shading, which leads users to pass through or use the space for the activities occurring there. Despite the spaces sharing the same external conditions, these internal differences result in distinct patterns of space usage. On the other hand, since the study was conducted at a university campus, we had to exclude the extreme seasons as the number of students present in campus was significantly lower at the summer and winter semesters. Hence, the behavior or the preference of the students to use the outdoor spaces may change under more extreme weather conditions.

4 Conclusion

In this experimental study, a comparative analysis was conducted to examine the degree to which outdoor space configuration impacts thermal perception of outdoor space users. The study took place during moderate seasons in 2022 in Giza, Egypt. Values of air temperature, relative humidity, wind speed, MRT, and surface temperature were simultaneously measured in two spaces.

Space 1 was well configured geometrically but contained no greenery. Space 2 was poorly oriented and less enclosed but 60% of its area was covered by greenery. The sample examined was formulated from 290 participants of students (young adults). Each participant was asked to fill out a questionnaire designed according to ASHRAE 7-point, and (ISO 10551, 1995). The climatic conditions and the personal qualities of the students (clothing and activity) were all used as inputs to calculate the PET using Rayman. PET calculation represented an objective tactic to estimate the thermal comfort levels in the two spaces, while the surveys was performed to induce the subjective thermal perception of the space users.

According to objective measurements, PET was found to be consistently lower in Space 1 throughout all the examined timings in spring and fall. The subjective evaluations demonstrated that Space 1 was also more thermally comfortable. "Neutral" thermal sensation was the most commonly voted with 50% and 49.15% in spring and fall at space 1 followed by slightly warm with 26.58%, however, in space 2 the slightly warm sensation was the most chosen vote with 26.58% and 44.26% followed by warm at 28% and 40% in spring and fall. The majority of the students in space 1 has consistently chosen "No change" for all the examined climatic parameters. In Space 2, the majority reported to prefer "Cooler" air temperature and more mitigated solar radiation. Hence, it becomes evident through objective and subjective analysis that space 1 was more thermally comfortable than space 2 despite the availability of shaded areas for the students in the two spaces (the shadow of the building in space 1 and pergolas in space 2). This may indicate that the abundance of shade beyond the exact sitting area has stronger impact on improving the thermal performance of outdoor spaces than implementing shade in the seating areas only even if the majority of the space was covered with greenery. This is further reinforced with the fact that surface temperature of grass covered and shaded surface in Space 2 was higher than a shaded paved surface in Space 1 despite the fact that opposite was true when the surfaces were exposed. Another aspect that may be associated with the students' discomfort is the facades. For the majority of day, the facades near Space 1 were completely shaded, while the facades near Space 2 had significantly higher surface temperature as it was also exposed to the sun. Among the other worth noting results is that the number of female students feeling "comfortable" was slightly higher than the male students despite their heavier clothing. This indicates that female students may be more resilient to warmer climatic conditions during moderate season in comparison to male students. Generally, variation in number of users in the two spaces or in each space under different climatic conditions indicates the possibility of the presence of a relationship between the climatic conditions in outdoor spaces and the number of users within them. While a significant body of literature was developed to discuss how the physical characteristics of space (like the uses overlooking it or the way it is furnished) affects the number of users within it, limited studies investigated how the thermal performance of outdoor spaces may affect the activities and the number of users within it. Future research may tackle this gap. On the other hand, it was found that PET, surface temperature, MRT and air temperature and the strongest influence on the users' outdoor thermal perception, while wind speed and relative humidity had an insignificant impact. Overall, the findings of this study may be useful for planners and architects who design new schemes or develop existing projects should they prioritize outdoor thermal comfort in their designs.

Acknowledgments There were no funding entities for this research.

References

6th of October Weather & Climate (2022). https://en.climate-data.org/africa/egypt/giza-governorate/6th-of-october-3677/

Adly, M.A., Mahmoud, H., Galal, O.M.: The impact of densification and orientation manipulation on outdoor thermal comfort at social housing in arid regions: a sensitivity analysis. HBRC J. **19**(1), 523–541 (2023)

Aghamolaei, R., Azizi, M.M., Aminzadeh, B., O Donnell, J.: A comprehensive review of outdoor thermal comfort in urban areas: effective parameters and approaches. Energy Environ. **34**, 0958305X2211161 (2022). https://doi.org/10.1177/0958305X221116176

ASHRAE Standard 55: ANSI/ASHRAE Standard 55. Thermal Environmental Conditions for Human Occupancy; American Society of Heating, Refrigerating and Air-Conditioning Engineers: Atlanta, GA, USA, 2017 (2017)

ASHRAE Handbook: American society of heating, refrigerating, and air-conditioning engineers (2009). https://www.academia.edu/39168321/Inch_Pound_Edition_2009_ASHRAE_HANDBOOK_FUNDAMENTALS

ASHRAE. ASHRAE standard 55: thermal environmental conditions for human occupancy. Atlanta: American Society of Heating, Refrigerating, and Air- Conditioning Engineers (2004)

Baruti, M., Johansson, E., Åstrand, J.: Review of studies on outdoor thermal comfort in warm humid climates: challenges of informal urban fabric. Int. J. Biometeorol. **63**, 1449–1462 (2019). https://doi.org/10.1007/s00484-019-01757-3

Boukhelkhal, I., Bourbia, F.: Thermal comfort conditions in outdoor urban spaces: hot dry climate -Ghardaia- Algeria. Procedia Eng. **169**, 207–215 (2016). https://doi.org/10.1016/j.proeng.2016.10.025

Butera, F.: Sustainable neighborhood design in tropical climates, pp. 51–73 (2018). https://doi.org/10.1016/B978-0-08-102074-6.00017-6

Elgheznawy, D., Eltarabily, S.: The impact of sun sail-shading strategy on the thermal comfort in school courtyards. Build. Environ. **202**(2021), 108046 (2021)

Elnabawi, M.H., Hamza, N., Dudek, S.: Thermal perception of outdoor urban spaces in the hot arid region of Cairo. Egypt. Sustain. Cities Soc. **22**, 136–145 (2016). https://doi.org/10.1016/j.scs.2016.02.005

Salata, F., Golasi, I., de Lieto Vollaro, R., de Lieto Vollaro, A.: Outdoor thermal comfort in the mediterranean area. a transversal study in Rome, Italy. Build. Environ. **96**(2016), 46–61 2016

Gulyás, Á., Matzarakis, A., Unger, J.: Differences in the thermal bioclimatic conditions on the urban and rural areas in a southern Hungarian city (Szeged) (2009)

Heidari, S., Amindeldar, S., Khalili, M.: The effect of personal and microclimatic variables on outdoor thermal comfort: a field study in Tehran in cold season. Sustain. Cities Soc. **32**, 153–159 (2017). https://doi.org/10.1016/j.scs.2017.03.024

Huang, K.-T., Lin, T.P., Lien, H.: Investigating thermal comfort and user behaviors in outdoor spaces: a seasonal and spatial perspective. Adv. Meteorol. **2015**, 4–14 (2015). https://doi.org/10.1155/2015/423508

ISO 7726:1998—Ergonomics of the thermal environment—Instruments for measuring physical quantities (1998). ITeh Standards Store. https://standards.iteh.ai/catalog/standards/iso/99f92eea-d1b3-48b4-8a3c-8e5a5112718a/iso-7726-1998

ISO 7730:2005- Ergonomics of the thermal environment—Analytical determination and interpretation of thermal comfort using calculation of the PMV and PPD indices and local thermal comfort criteria (2005). ISO. https://www.iso.org/cms/render/live/en/sites/isoorg/contents/data/standard/03/91/39155.html

ISO 10551: ISO (1995). https://www.iso.org/cms/render/live/en/sites/isoorg/contents/data/standard/01/86/18636.html

Johansson, E., Emmanuel, R.: The influence of urban design on outdoor thermal comfort in the hot, humid city of Colombo, Sri Lanka. Int. J. Biometeorol. **51**, 119–133 (2006). https://doi.org/10.1007/s00484-006-0047-6

Kim, S.W., Brown, R.D.: Pedestrians' behavior based on outdoor thermal comfort and micro-scale thermal environments, Austin TX. Sci. Total. Environ. **808**, 152143 (2022). https://doi.org/10.1016/j.scitotenv.2021.152143

Köppen: Wladimir Köppen and the Köppen Climate Classification System | SciHi Blog (1900). http://scihi.org/wladimir-koppen/

Krüger, E.L., Givoni, B., Rossi, F.A.: outdoor comfort study in Curitiba, Brazil: effects of gender, body weight and age on the thermal preference, vol. 12 (2010)

Li, L., Zhou, X., Yang, L.: The analysis of outdoor thermal comfort in Guangzhou during summer. Procedia Eng. **205**, 1996–2002 (2017). https://doi.org/10.1016/j.proeng.2017.10.070

Lin, T.-P.: Thermal perception, adaptation and attendance in a public square in hot and humid regions. Build. Environ. **44**(10), 2017–2026 (2009)

Lucchese, J., Mikuri, L., De Freitas, N., Andreasi, W.: Application of selected indices on outdoor thermal comfort assessment in Midwest Brazil. Int. J. Energy Environ. **7**, 291–302 (2016)

Mahmoud, A.H.A.: Analysis of the microclimatic and human comfort conditions in an urban park in hot and arid regions. Build. Environ. **46**(12), 2641–2656 (2011)

Matzarakis, A.: Climate, thermal comfort and tourism. In: Climate Change and Tourism-assessment and Coping Strategies, pp. 139–154 (2007)

Matzarakis, A., Rutz, F., Mayer, H.: Modelling radiation fluxes in simple and complex environments—Application of the RayMan model. Int. J. Biometeorol. **51**, 323–334 (2007). https://doi.org/10.1007/s00484-006-0061-8

Fahmy, M., Sharples, S.: On the development of an urban passive thermal comfort system in Cairo. Egypt. Buil. Environ. **44**(2009), 1907–1916 (2009)

Nikolopoulou, M., Baker, N., Steemers, K.: Thermal comfort in outdoor urban spaces: understanding the human parameter. Sol. Energy **70**(3), 227 (2001)

Nikolopoulou, M., Lykoudis, S., Kikira, M.: Thermal comfort in outdoor spaces: field studies in Greece. In: 5th International Conference on Urban Climate, IAUC-WMO (2003)

Noémi, K., János, U.: Human bioclimatological evaluation with objective and subjective approaches on the thermal conditions of a square in the centre of Szeged (2007). https://core.ac.uk/display/45379520?utm_source=pdf&utm_medium=banner&utm_campaign=pdf-decoration-v1

Galal, O.M., Sailor, D.J., Mahmoud, H.: The impact of urban form on outdoor thermal comfort in hot arid environments during daylight hours, case study: New Aswan. Build. Environ. **184**(2020), 107222 (2020)

Peel, M.C., Finlayson, B.L., Mcmahon, T.A.: Updated world map of the Köppen-Geiger climate classification. Hydrol. Earth Syst. Sci. Discuss. **11**(5), 1633–1644 (2007)

Population reference bureau: new security beat (2012). https://www.newsecuritybeat.org/2012/08/prbs-2012-world-population-data-sheet/

Ruiz, M.A., Correa, E.N.: Adaptive model for outdoor thermal comfort assessment in an oasis city of arid climate. Build. Environ. **85**, 40–51 (2015). https://doi.org/10.1016/j.buildenv.2014.11.018

Rung, A.L., Broyles, S.T., Mowen, A.J., Gustat, J., Sothern, M.S.: Escaping to and being active in neighbourhood parks: park use in a post-disaster setting. Disasters **35**(2), 383–403 (2011). https://doi.org/10.1111/j.1467-7717.2010.01217.x

Samuelsson, K., Barthel, S., Colding, J., Macassa, G., Giusti, M.: Urban nature as a source of resilience during social distancing amidst the coronavirus pandemic (2020). https://doi.org/10.31219/osf.io/3wx5a

Sangkertadi, S.R.: New equation for estimating outdoor thermal comfort in humid-tropical environment. Eur. J. Sustain. Dev. **3**(4), 43–52 (2014)

Smart Village topographic map. Topographic Maps (2022). https://en-gb.topographic-map.com/map-fzqjkl/Smart-Village/

Thorsson, S., Honjo, T., Lindberg, F., Eliasson, I., Lim, E.-M.: Thermal Comfort and Outdoor Activity in Japanese Urban Public Places. Environ. Behav. **39**, 660–684 (2006). https://doi.org/10.1177/0013916506294937

VDI (VDI 3787). Part I: Environmental meteorology, methods for the humanbiometeorological evaluation of climate and air quality for the urban and regionalplanning at regional level. Part I: Climate. VDI guideline 3787. Part 2. Berlin:Beuth (1998).

Zhen, M., Dong, Q., Chen, P., Ding, W., Zhou, D., Feng, W.: Urban outdoor thermal comfort in western China. J. Asian Archit. Build. Eng. **20**(2), 222–236 (2021). https://doi.org/10.1080/13467581.2020.1782210

Revitalizing Food Deserts Through Hydrophilic Architecture: A Case Study of Proposing Community Gardens in University of Arizona Dormitories

Yostina Yacoub, Khalid Dewidar, and Marianne Nabil

Abstract

Hydrophilic design interventions offer a promising solution to solving issues associated with water scarcity in arid regions, particularly when it comes to combating food deserts. Tucson, situated in the Sonoran Desert, faces these challenges, with significant portions of its population residing in food deserts due to limited access to fresh produce.

This paper explores the application of hydrophilic units on building facades, focusing on Maricopa Hall at the University of Arizona in Tucson as a case study.. These units, using desiccant materials like Calcium Chloride, extract water from ambient humidity and are particularly effective in Tucson's low-humidity environment. Through previous experimental testing, the units demonstrated a capacity to produce 1.8ml of water per day per unit. This paper aims to propose the use of these hydrophilic units as a case study and explores their application in redirecting the produced water for irrigating a variety of edible plants suitable for Tucson's climate, thus combating food deserts.

Through multiple qualitative, quantitative and numerical explorations it was found that if the units were arranged on Maricopa Hall's southern facade, they would be able to produce approximately 47 L/day. This yield would be able to support a large range of leafy greens, legumes, herbs, fruits and vegetables, thereby enhancing food security among students living in Maricopa Hall.

Results highlight that integrating hydrophilic units into building facades not only contributes to water sustainability efforts but also promotes local food production, aligning with the United Nations' Sustainable Development Goals (SDG). The paper addresses SDG 2 by enhancing food security through community gardens, SDG 6 by promoting sustainable water practices with hydrophilic architecture, SDG 11 by integrating sustainable design in urban dormitories, and SDG 12 by advocating for responsible consumption and production in food cultivation, and SDG13 by leveraging the adaptive capacity of buildings to optimize the use of prevailing weather conditions. This research underscores the transformative potential of hydrophilic design in addressing critical environmental and social challenges in arid, water-scarce regions, offering a pathway to sustainable urban interventions.

Keywords

Hydrophilic Design · Water Harvesting · Food Deserts · Resilient Architecture · Sustainable Facade · Alternative Water Resources · Double Skin Facade · Hydrophilic Architecture · Community Gardens · Sonoran Desert Plants

List of Abbreviations

1. SDG: Sustainable Development Goals
2. CaCl2: Calcium Chloride

Y. Yacoub (✉)
Tyler School of Art and Architecture - Temple University, Philadelphia, PA, USA
e-mail: yostina.yacoub@temple.edu

K. Dewidar · M. Nabil
Department of Architectural Engineering, The British University in Egypt, El-Sherouk City, Egypt
e-mail: khaled.Dewidar@bue.edu.eg

M. Nabil
e-mail: marianne.nabil@bue.edu.eg

B. He et al. (eds.), *Towards the Framework of Livable and Resilient Cities* Advances in Science, Technology & Innovation, https://doi.org/10.1007/978-3-031-97849-4_21

1 Introduction

The word "Hydrophilic" is derived from the word "Hydrophile", meaning a molecule or a molecular body that attracts water molecules and tends to be dissolved by water (Lag et al. 2008). There are many materials and compounds that exhibit hydrophilic properties, such as desiccant materials. The use of desiccants in water extraction is very beneficial as their chemical compositions allow them to absorb humidity even in the driest climates (Hao et al. 2017). Desiccant materials can absorb water vapour molecules at night when humidity is high, then when exposed to atmospheric heat in the daytime can desorb the vapour in the form of water (Roy et al. 2013).

The application of hydrophilic design in architecture and landscape design can come in many different forms. Hydrophilic textiles have been developed and used as a self-sufficient, and sustainable way of irrigating different types of vegetation. The hydrophilic textiles are wrapped around different forms of vegetation during night-time when humidity levels are high and temperatures are low. In the daytime, high temperatures release the water from the textile, irrigating the plants the textile is wrapped around (Sponsh 2018).

This paper explores the application of hydrophilic units, as designed by Yacoub (2024), on building facades in water-scarce regions to address food deserts. The study chooses Maricopa Hall, a dormitory at the University of Arizona in Tucson, Arizona as a case study. The study evaluates the efficacy of these units in generating water and compares the results with the water requirements of various edible plants suitable for Tucson's environment. This research contributes to and aligns with the United Nations' Sustainable Development Goals 2 (Zero Hunger), 6 (Clean Water and Sanitation), 11 (Sustainable Cities and Communities), and 12 (Responsible Consumption and Production), highlighting their potential impact on sustainable urban food production and water management.

The objectives of this paper will be achieved through a mixed-method approach utlizing both quantitative, and qualitative approaches. Firstly, the paper will examine the hydrophilic units designed by Yacoub et al. to develop an understanding of their design, mechanism, and water yield (2024). Secondly, a comparative analysis will be conducted between Cairo's climate, where the hydrophilic units were tested, and Tucson, Arizona's climate, where the application of these units is proposed. Thirdly, the study explores a case study to assess the feasibility of applying the hydrophilic block units in a water-scarce region suffering from a food desert. Afterwards, it will explore the Maricopa Hall dormitory at the University of Arizona as the chosen case study, including its significance within the campus and the student population it serves, to identify the most appropriate facade for the application of these hydrophilic units. Following this, the paper will analyze the types of plants suitable for growth in the proposed community gardens serving Maricopa Hall, focusing on low-water plants that thrive in Tucson's climate. Finally, the study will refer to multiple sources to quantitatively analyze the weekly water requirements of each plant species and compare these needs with the water yield from the hydrophilic units.

2 Methodology

To develop a comprehensive understanding of the hydrophilic units and its application on a case study, a series of diverse methodologies was employed.

A mixed-method approach, comprising qualitative and quantitative approach was utilized in this research. A comparative analysis of the environmental conditions and climate data in both Cairo, where the unit block was experimented, and Tucson, where the case study will be applied, was conducted, aiming to identify the differences and similarities in climate between both regions. This analysis extended to evaluation of potential plant species for community gardens, focusing on their adaptability to Tuscan's, climate, and their water requirements. Numerical methods were applied to estimate the water yield from the hydrophilic units, determining their effectiveness in producing sufficient water for irrigation and calculating the number of community garden plots that could be supported by the water produced from the units installed on Maricopa Hall. A preliminary framework for the application of these hydrophilic units is established in Fig. 1, with Maricopa Hall serving as a case study to document these procedural steps that allow the prototype to from the experimental phase to practical application (Groat and Wang 2013).

3 Water Scarcity and Food Deserts in Arid Regions

According to the United Nations, a water-scarce region is defined as a region where the demand for water exceeds the supply available (Hadi Dehghani et al. 2021). Water-scarcity can be further exacerbated in arid regions that typically have low precipitation rates. In turn, many of these water-scarce, arid regions become food desserts. Limited water availability can impact agricultural practices, contributing to food deserts where access to fresh and nutritious food is constrained.

The Sonoran Desert is one of the largest deserts in North America, covering approximately 260,000 km^2. The desert spans across parts of the United States and Mexico. In the U.S, it is primarily located in the states of Arizona and California. The Arizonian part of the Sonoran Desert is home to several cities, such as Tucson with a population of approximately

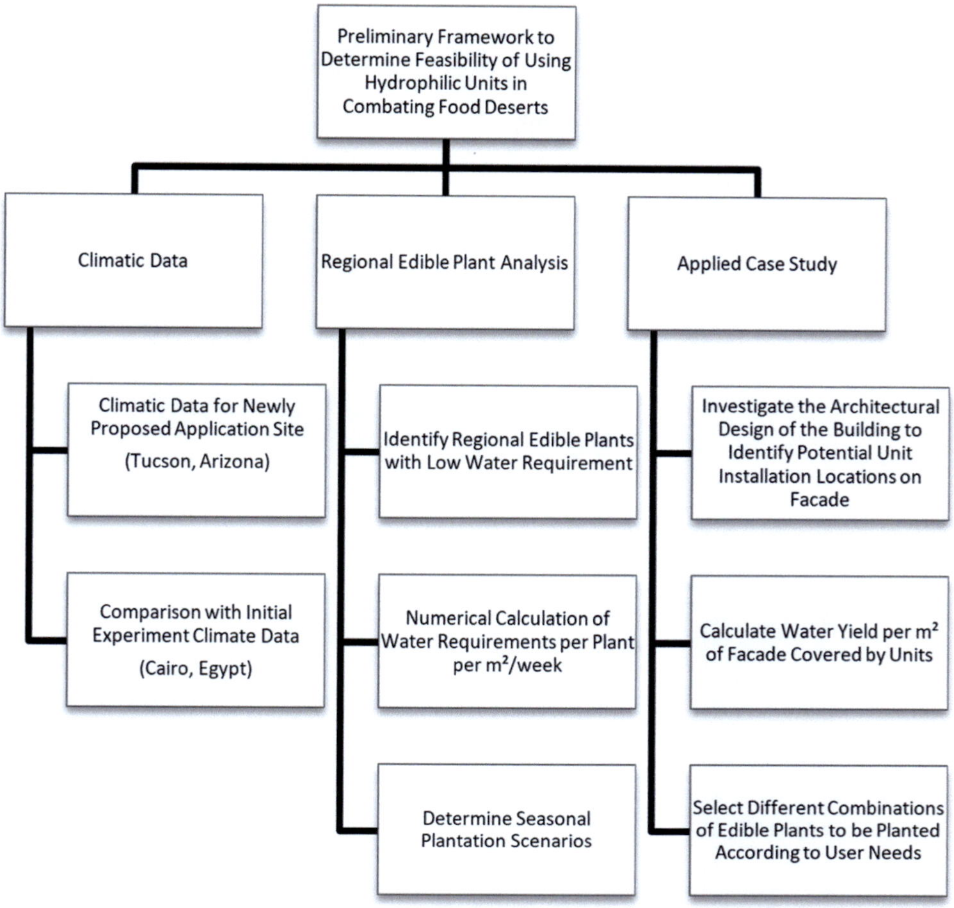

Fig. 1. Preliminary Framework to Determine Feasibility of Using Hydrophilic Units in Combating Food Deserts (Author 2024)

546,574 as of July 2022 respectively (U.S. Census Bureau 2024) Due to the arid climatic conditions and water scarcity of the Sonoran desert, cities in Arizona such as Tucson have become food deserts. Studies have shown a significant number of Tucson's inhabitants reside in areas classified as food deserts, situated more than 1.6 km away from access to fresh and nutritious food (Bennion 2022).

4 Hydrophilic Architectural Cladding Unit for Water Production

The units were designed to be vertically installed as cladding on a building's facade. They utilise the hydrophilic desiccant material Calcium Chloride (CaCl2) with a 30% concentration level to extract water from ambient humidity as demonstrated in comparable experiments conducted by Talaat et al. (2018). Desiccant materials have proven ability in absorbing ambient humidity in the atmosphere (Anokye-Bempah et al. 2023). As depicted in Fig. 2, the units are specifically engineered to enhance the absorption of humidity and its conversion into water. Each unit measures 20 cm × 20 cm × 7 cm and is constructed from plexiglass measuring 5 mm in thickness.

Their compact size allows rapid filling with humid air, optimising the efficiency of CaCl2 in moisture absorption. The plexiglass traps solar heat (Casati 2024), promoting temperature rise within the unit, which triggers vapour desorption (Yacoub et al. 2024).

Figure 2 illustrates that the prototype features small Ø2mm pores on its lower external side, enabling entry of humid air into the unit. As heat extracts vapour, excess moisture evaporates upwards and exits through Ø1mm pores positioned on the upper internal side. This placement of air entry and exit pores on opposite sides (Fig. 3) promotes cross ventilation, facilitating airflow within the unit to enhance CaCl2's absorption of humidity. The smaller Ø1mm pores at the top of the unit retain humid air longer, while allowing eventual release of excess air. Additionally, a Ø4mm pore was incorporated for inserting a sensor to monitor temperature and humidity levels inside the unit during experiments (Yacoub et al. 2024).

The units are angled at the top to channel condensed water downward from the interior surfaces of the plexiglass. At the bottom, they are sloped to serve as a reservoir for collected water. Figure 2 depicts these features. In Fig. 3, the CaCl2 is positioned on a detachable fixture resting at the unit's lower section. This fixture measures 20 cm × 6 cm and consists of

Fig. 2. Illustrations made on Rhinoceros 6 showing external and internal face of unit (Yacoub, in press)

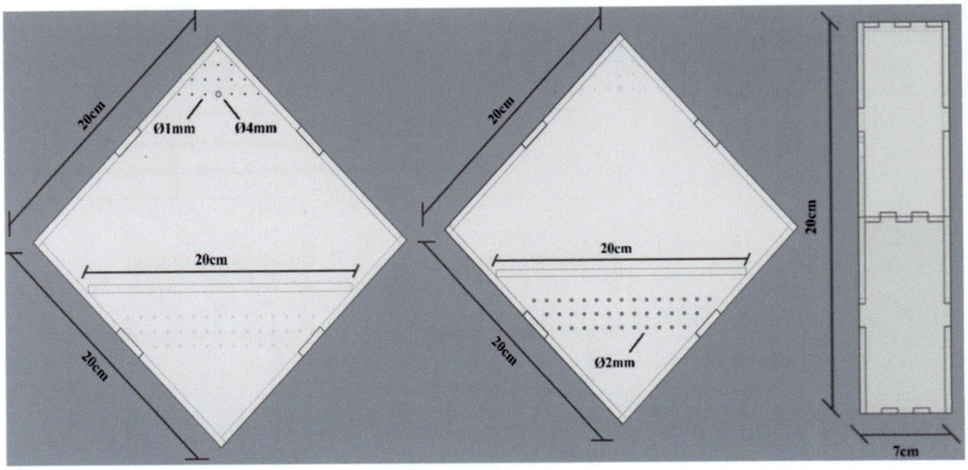

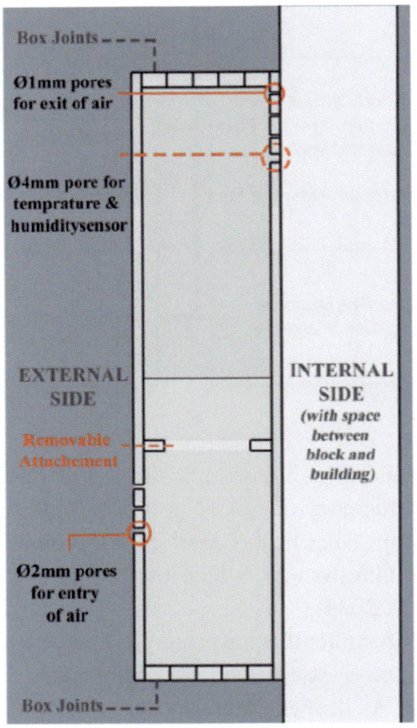

Fig. 3. Cross section in unit showing arrangement of pores for air entry and exit and connection of unit to building facade (Yacoub et al. 2024)

a thin plexiglass frame wrapped with black fabric interwoven around a metal wire to maximise surface area, as demonstrated in experiments by Talaat et al. (2018), giving the fabric a total surface area of 0.17 m^2 (Yacoub et al. 2024).

Figure 3 also highlights the box joint connections between unit components, facilitating easy assembly, disassembly, and replacement of desiccant materials or unit parts as necessary. The modular design of these units allows for versatility in shape and arrangement. Designed for prolonged operation with minimal maintenance, the units function autonomously over extended periods due to the cyclic nature of absorption and desorption processes, which rely on atmospheric humidity and heat triggers (Yacoub et al. 2024).

5 Testing Water Generation Under Controlled Environmental Conditions

The unit has undergone a rigorous series of experimentation. It was left overnight repeatedly in atmospheric humidity of approximately 50% and an average temperature of 40 °C, as a result the unit was able to generate 1.80 ml of water per day (Yacoub et al. 2024).

Yacoub et al. concludes that if a small facade measuring 100m^2 contained the hydrophilic units, it would produce approximately 4.5 L/day (2024).

The unit underwent testing in Cairo, Egypt, during the summer of 2020. Cairo experiences annual temperatures ranging from 10 °C to 40 °C, with an average annual relative humidity of 56%. In contrast, Tucson experiences annual temperatures ranging from 6 °C to 38 °C, with an average annual relative humidity of 25% (WeatherSpark 2024). Hydrophilic desiccant materials have demonstrated the capability to absorb humidity effectively even in very dry conditions, suggesting their suitability for use in Tucson's climate. The desorption process of these materials depends on atmospheric temperatures, which are comparable to those experienced annually in Egypt.

6 Maricopa Hall - Tucson, Arizona

The water produced from the hydrophilic unit has the potential to be redirected for various uses both inside and outside the building. Since this water is not fresh and will not be tested, it can be utilised for indoor cooling systems, fire protection, humidification appliances, car washing, and plant irrigation.

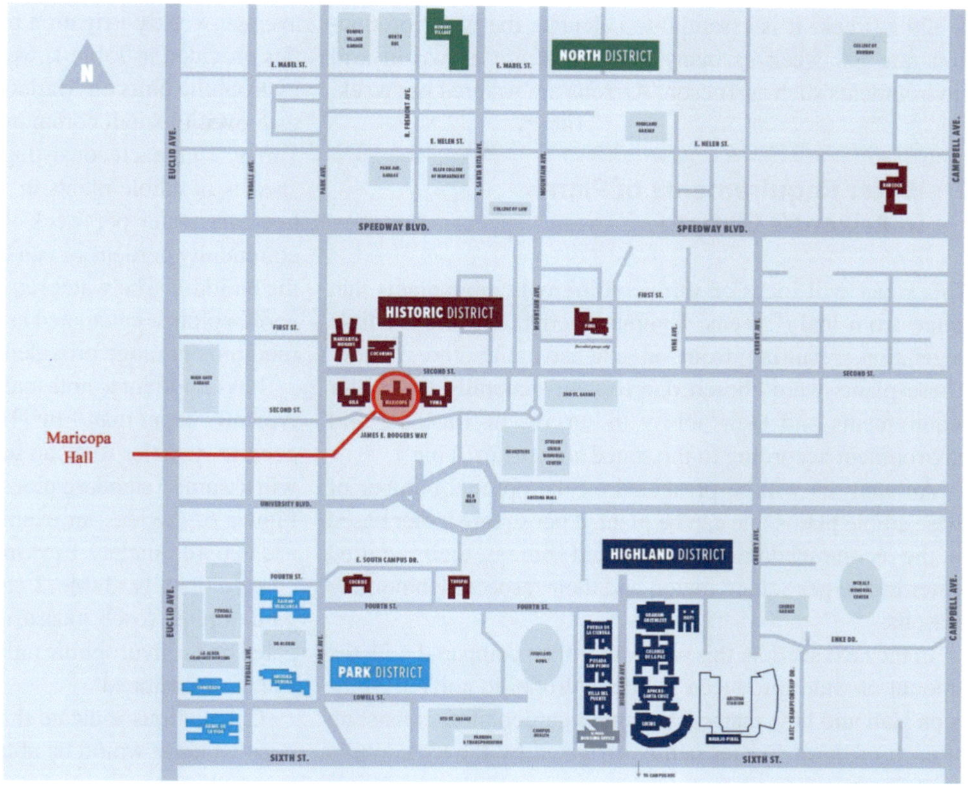

Fig. 4. University of Arizona. (n.d.). Dorm Map. University of Arizona Housing. Retrieved July 10, 2024, from https://housing.arizona.edu/dorms/dorm-map Edits made to the image via Photoshop by Yostina Yacoub.

This paper will focus on the feasibility of using this water for irrigating community gardens to address the issue of food scarcity, particularly in student dormitories in Tucson.

Maricopa Hall on the University of Arizona's campus, shown in Fig. 4, was selected for a number of reasons. The first is that it is occupied by over 100 students. It is surrounded by two other dormitories of similar capacity, thus the proposed community gardens would serve a reasonable amount of students. Introducing a community garden in a student dorm setting in a food scarce area not only strengthens community bonds but also promotes gardening skills, environmental awareness, and healthy eating habits. It lowers food costs, contributes to urban greening, enhances community empowerment, and fosters a sense of ownership among residents (Lampert et al. 2021).

This paper presumes that the hydrophilic units will attach as a double-skin facade and the water will be collected in storage tanks and redirected according to the required usage.

Since the efficiency of the facades depends greatly on atmospheric heat, it is logical to place them on the southernmost side of Maricopa Hall (Lionar 2024). Figure 5 presents an aerial view of Maricopa Hall's layout, illustrating the unblocked southern side where the units will receive maximum heat exposure. After calculating the square footage of Maricopa Hall's southern facade and subtracting the area of all windows, doors, and other openings, it was determined that approximately 1044 m^2 of wall space is available.

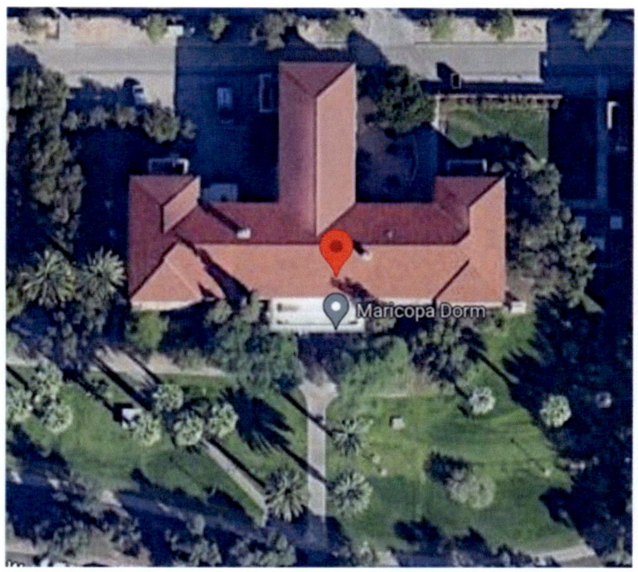

Fig. 5. Google LLC. (n.d.). Aerial image of *Maricopa Hall layout, University of Arizona campus.* Google Maps. Retrieved July 10, 2024, from [https://goo.gl/maps/2ez9q4MvA8dT1N4L6]

According to Yacoub et al. (2024), the hydrophilic units produce approximately 4.5 L/day for a facade measuring 100 m^2. For a facade measuring 1044m^2, this translates to about 47 L/day. Multiplying this by 7 gives the weekly water production by the hydrophilic units on Maricopa Hall, which is 329 L/week. It is essential to calculate the water production rate per week as many plants that are grown in arid environments such as Tucson, Arizona are watered per week.

7 Water Requirements of Plants in Arizona's Climate

This paper will focus on thirteen different edible plants that range from leafy greens, legumes, herbs, and other edible vegetation, resulting from an extensive literature review. These plants were chosen due to their generally low water requirements and their ability to survive in Tucson's arid environment according to the stated sources in Table 1.

An analysis will be presented on the optimal number of these edible plants that can be planted per square meter based on the recommendations in the stated sources, their required water intake per square meter, and their respective blooming seasons.

In the next section, this analysis will be compared with the amount of water produced by the hydrophilic units in Maricopa Hall and how many community garden plots consisting of each of these species can be irrigated by the hydrophilic units.

8 Findings and Results

Comparing the amount of water produced by the hydrophilic units on Maricopa's southern facade (329L/week) with the edible plants, their quantity per square metre, and their average weekly irrigation requirements based on the quantities provided in Table 1, we can conclude the following: The hydrophilic units on Maricopa's facade would be able to irrigate over 13 small community garden plots (each measuring 1 m by 1 m) each consisting of one species of the mentioned species of edible plants in Table 1 and have about 150 L of left over water per week that can be used to irrigate more community gardens or can be redirected for use elsewhere in the building. The water requirement of all the 13 community garden plots is calculated as a total of the average of the water requirement range provided in Table 1.

It is important to note that most community garden plots are typically larger than 1 m^2. However, for the ease of providing generic statistics that can be easily scaled to larger sizes, we will assume a standard plot size of 1 m by 1 m (Kordon 2024). Figure 6 provides an example of community garden plots, each approximately 1 m^2 in size, to aid in visualization.

Additionally, Table 2 shows the quantity of community garden plots (each measuring 1 m by 1 m) that can be irrigated by the hydrophilic units on Maricopa Hall for each plant species mentioned.

Calculations indicate that the hydrophilic units on Maricopa's facade would be able to irrigate a minimum 14 small community garden plots, specifically dedicated to growing malabar spinach, or alternatively, a maximum of 109 community garden plots, specifically dedicated to growing prickly pear cacti in Tucson's climate. Practically, if different kinds of plants are chosen from the table, then, the number of community garden plots will be in accordance with the irrigation needs and the proposed kinds to be within this range. These findings highlight the significant range in irrigation capacity

Table 1. Selected Edible Plants: Water Requirements, Planting Density, and Blooming Seasons in Tucson

Plant Type		Plants/m^2	Approx. Range of Water Requirement Per Week/m^2 (L)	Season	Source
Leafy Greens	Malabar Spinach	200	20–25 L	Late Spring - Early Fall	(Chaurasiya et al. 2021)
	Amaranth	8	20–25 L	Late Spring - Early Fall	(Mukuwapasi et al 2024)
	Purslane	23	15–20 L	Late Spring - Early Fall	(Wood 2014)
Legumes	Tepary Beans	25	15–20 L	Late Spring - Early Fall	(Molosiwa et al. 2022)
	Lentils	50	10–15 L	Late Fall - Early Spring	(Ness 2015)
Herbs	Rosemary	5–11	5–10 L	Perennial	(Al-Fraihat et al. 2023)
	Lavender	4–5	5–10 L	Perennial	(Paraskevopoulou et al. 2020)
	Thyme	20–30	5–10 L	Perennial	(Iftikhar et al. 2023)
	Oregano	14	5–10 L	Perennial	(Hancioglu et al. 2020)
Other	Prickly Pear Cactus	1	0–5 L	Perennial	(Wilson et al. 2019)
	Chiltepin Peppers	2–3	10–15 L	Late Spring - Early Fall	(Grant 2020)
	Turnips	9	15–20 L	Late Fall - Early Spring	(Schuh 2022)
	Tomatoes	3	15–20 L	Late Spring - Early Fall	(Ruelas-Islas et al. 2022)

Fig. 6. Liverpool City Council. (n.d.). *Community Gardens.* Liverpool City Council. Retrieved July 10, 2024, from https://www.liverpool.nsw.gov.au/environment/community-gardens

Table 2. Irrigation Capacity of Hydrophilic Units on Maricopa Hall for Selected Plant Species in Tucson

Plant Type		Plants/ m^2	Average Water Requirement Per Week/m^2 (L)	Season	Number of Community Garden Plots Irrigated by Hydrophilic Units on Maricopa Hall per Plant Species
Leafy Greens	Malabar Spinach	200	23 L	Late Spring - Early Fall	14
	Amaranth	8	23 L	Late Spring - Early Fall	14
	Purslane	23	18 L	Late Spring - Early Fall	18
Legumes	Tepary Beans	25	18 L	Late Spring - Early Fall	18
	Lentils	50	13 L	Late Fall - Early Spring	25
Herbs	Rosemary	5–11	8 L	Perennial	41
	Lavender	4–5	8 L	Perennial	41
	Thyme	20–30	8 L	Perennial	41
	Oregano	14	8 L	Perennial	41
Other	Prickly Pear Cactus	1	3 L	Perennial	109
	Chiltepín Peppers	2–3	13 L	Late Spring - Early Fall	25
	Turnips	9	18 L	Late Fall - Early Spring	18
	Tomatoes	3	18 L	Late Spring - Early Fall	18

provided by the hydrophilic units, depending on the plant species being cultivated.

9 Conclusion

Hydrophilic design offers a promising solution to water scarcity and food insecurity in arid regions, as demonstrated by its application at Maricopa Hall at the University of Arizona in Tucson. This study evaluated the effectiveness of hydrophilic units in irrigating community garden plots. These units were previously shown to produce significant amounts of water, essential for growing various edible plants suited to Tucson's climate.

Tucson faces significant challenges in ensuring access to fresh produce, particularly for populations in arid zones, due to food deserts exacerbated by water scarcity. By integrating hydrophilic units on Maricopa Hall, this research explored their potential to support community gardens and enhance food security, especially for students. The study employed a mix of quantitative, and qualitative methods, along with developing a preliminary framework, to comprehensively evaluate the units' performance and determine the preliminary methodological framework for application.

The results underscored the feasibility of using water from hydrophilic units to sustainably cultivate edible plants. Findings indicated that the units could irrigate 14 plots of malabar spinach or 109 plots of prickly pear cacti weekly, highlighting their versatility in Tucson's climate. This research underscores the technological innovation of hydrophilic design in architecture and its significant impact on social and environmental sustainability, promoting local food production and resilience against food deserts in Tucson.

10 Recommendations

Future studies should explore broader deployment opportunities in urban settings by integrating hydrophilic units with existing building infrastructure to maximise water production efficiency. It is also recommended to conduct experiments in Tucson's climate to refine findings specific to Arizona's climate zone, enhancing the precision of hydrophilic unit efficiency assessments.

Additionally, extended studies should be undertaken to monitor hydrophilic units' performance across various seasons and conditions, providing valuable insights into their durability and efficiency.

Furthermore, collaboration with researchers from the University of Arizona's School of Agriculture could further enhance this research, leveraging their agricultural expertise to optimise the efficiency of the hydrophilic units on community gardens in student dormitories.

Finally, advocating for policies that facilitate the adoption of hydrophilic design in building codes and urban planning processes will be crucial for streamlining implementation and promoting sustainable water practices in arid regions.

11 Relation to Previous Work

This research builds on our prior work in hydrophilic facade technologies, pushing the boundaries of sustainable design by integrating advanced material science with architectural innovation. Our earlier studies, set for publication in Springer, introduced the concept of hydrophilic architectural cladding units, that can function as a water-harvesting facade unit that passively extracts moisture from the the ambient humidiy in the atmosphere. While those findings established the feasibility of the system, this paper advances the conversation by expanding the applications of hydrophilic cladding units beyond conventional architectural cladding.

By bridging the gap between theoretical research and real-world application, this study marks a crucial step toward revolutionizing how buildings interact with their environments. Unlike existing passive water-harvesting solutions, this approach is scalable, architecturally integrative, and tailored for regions facing severe water scarcity. The findings presented here not only refine the prior work but also lay the groundwork for future iterations, further cementing the role of architecture as an active participant in environmental resilience.

12 Declaration

The authors declare that they have no conflicts of interest related to this research. No funding was received for this study. This research did not involve human participants or animal subjects.

Regarding data availability, all data generated or analyzed during this study are included in this published article.

Dedication:

I dedicate this work to God, whose guidance and blessings have been my constant source of strength. To my family and friends, for their unwavering support and encouragement. To the scholars across the globe who have contributed to this research with their inspiring insights, collaborative efforts, and invaluable feedback. To everyone who sees the potential of design in addressing global challenges, thank you.

I am profoundly grateful for my mentors, Professor Khaled Dewidar and Associate Professor Marianne Nabil, for their invaluable guidance and contributions throughout this journey. Their expertise has been instrumental in refining both the methodology, the broader vision of this study and laying the foundation for this work.

Special thanks also go to Temple University and the Tyler School of Art and Architecture for their intellectual support and funding, which made this research possible. Their unwavering commitment to fostering innovative and impactful scholarship has allowed this research to continue to flourish.

References

Al-Fraihat, A.H., Al-Dalain, S.Y., Zatimeh, A.A., Haddad, M.A.: Enhancing rosemary (rosmarinus officinalis, L.) growth and volatile oil constituents grown under soil salinity stress by some amino acids. Horticulturae **9**(2), 252 (2023). https://doi.org/10.3390/horticulturae9020252

Anokye-Bempah, L., Han, J., Kornbluth, K., Ristenpart, W., Donis-González, I.R.: The use of desiccants for proper moisture preservation in green coffee during storage and Transportation. J. Agric. Food Res. **11**, 100478 (2023). https://doi.org/10.1016/j.jafr.2022.100478

Bennion, N., Redelfs, A.H., Spruance, L., Benally, S., Sloan-Aagard, C.: Driving distance and food accessibility: a geospatial analysis of the food environment in the Navajo Nation and border towns. Front. Nutr. **9**, 904119 (2022). https://doi.org/10.3389/fnut.2022.904119

Casati, E., Allgoewer, L., Steinfeld, A.: Solar thermal trapping at 1,000 °C and above. Device **2**(7), 100399 (2024). https://doi.org/10.1016/j.device.2024.100399

Chaurasiya, A., Pal, R.K., Verma, P.K., Katiyar, A.R., Kumar, N.: An updated review on Malabar Spinach (basella Alba and basella rubra) and their importance. J. Pharmacognosy Phytochem. **10**(2), 1201–1207 (2021). https://doi.org/10.22271/phyto.2021.v10.i2p.13974

Hancioglu, N.E., Kurunc, A., Tontul, I., Topuz, A.: Growth, water use, yield and quality parameters in oregano affected by reduced irrigation regimes. J. Sci. Food Agric. **101**(3), 952–959 (2020). https://doi.org/10.1002/jsfa.10703

Iftikhar, T., Majeed, H., Zahra, S.S., Waheed, M., Niaz, M., Bano, N.: Thyme. In: Essentials of Medicinal and Aromatic Crops, pp. 399–429 (2023). https://doi.org/10.1007/978-3-031-35403-8_16

Lampert, T., Costa, J., Santos, O., Sousa, J., Ribeiro, T., Freire, E.: Evidence on the contribution of community gardens to promote physical and mental health and well-being of non-institutionalized individuals: a systematic review. PLOS ONE **16**(8), e0255621 (2021). https://doi.org/10.1371/journal.pone.0255621

Lionar, R., Kroll, D., Soebarto, V., Sharifi, E., Aburas, M.: A review of research on self-shading façades in warm climates. Energy Build. **314**, 114203 (2024). https://doi.org/10.1016/j.enbuild.2024.114203

Molosiwa, O.O., Baleseng, L.B., Chirwa, R.: Effect of plant spacing on agronomic performance and fodder quality of four tepary bean (phaseolus acutifolius A. Gray) cultivars. Int. J. Agron. **2022**, 1–10 (2022). https://doi.org/10.1155/2022/6590608

Mukuwapasi, B., Mavengahama, S., Gerrano, A.S.: Grain amaranth: a versatile untapped climate-smart crop for enhancing food and nutritional security. Disc. Agric. **2**, 44 (2024). https://doi.org/10.1007/s44279-024-00057-8

Paraskevopoulou, A.T., Tsarouchas, P., Londra, P.A., Kamoutsis, A.P.: The effect of irrigation treatment on the growth of lavender species in an extensive green roof system. Water **12**(3), 863 (2020). https://doi.org/10.3390/w12030863

Ruelas-Islas, J. del, Rubiños-Panta, J. E., Peinado-Fuentes, L. A., Mendoza-Pérez, C., Martínez-Ruiz, A., Escobosa-García, I.: Tomato water consumption in the greenhouse according to the number of stems. Revista Mexicana de Ciencias Agrícolas, (28), 137–147 (2022). https://doi.org/10.29312/remexca.v13i28.3269

Hadi Dehghani, M., Karri, R.R., Lima, E.: Green Technologies for the Defluoridation of Water. Elsevier, Amsterdam (2021)

Kordon, S.: Cultivating benefits: a review of the types of community gardens enhancing urban life. In: Climate Change, Carbon Management and Green Space Systems in Architectural Sciences (2024). https://doi.org/10.5281/zenodo.13882440

Roy, W.R., Helfferich, F.G., Petruzzelli, D.: Migration and Fate of Pollutants in Soils and Subsoils, p. 169. Springer, Berlin (2013)

Wang, D., Groat, L.N.: Architectural Research Methods. Wiley, Hoboken (2013)

Yacoub, Y., Dewidar, K., Nabil, M.: Harvesting hope: hydrophilic architectural cladding unit for combating water scarcity in hot-arid regions. In: Advances in Science, Technology and Innovation (ASTI). Springer, Heidelberg (in press)

Hao, X., Geng, S., Yuan, L., Luo, B.: Study of composite scheme of absorption/desorption method and condensation method for extracting water from air. In: Procedia Engineering, pp. 2069–2075. Elsevier Ltd. (2017). https://doi.org/10.1016/j.proeng.2017.10.096

Talaat, M.A., Awad, M.M., Zeidan, E.B., Hamed, A.M.: Solar-powered portable apparatus for extracting water from air using desiccant solution. Renew. Energy **119**, 662–674 (2018). https://doi.org/10.1016/j.renene.2017.12.050

U.S. Census Bureau. U.S. Census Bureau quickfacts: Arizona; Tucson City, Arizona (2024). https://www.census.gov/quickfacts/fact/table/AZ,tucsoncityarizona/PST045222

Weather Spark. (n.d.). Compare the climate and weather in Cairo and Tucson. https://weatherspark.com/compare/y/96939~2857/Comparison-of-the-Average-Weather-in-Cairo-and-Tucson

Sponsh. (n.d.). Technology - Sponsh. Accessed 11 Dec 2019. https://sponsh.co/technology/

Grant, A.: Uses for Chiltepin Peppers: How To Grow chiltepin chili peppers (2020). https://www.gardeningknowhow.com/edible/vegetables/pepper/how-to-grow-chiltepin-peppers.htm

Ness, C.: Lentils, a mighty force for improving the food system. University of California (2015). https://www.universityofcalifornia.edu/news/lentils-mighty-force-improving-food-system

Schuh, A. M. (2022). Growing turnips and rutabagas in Home Gardens. UMN Extension. https://extension.umn.edu/vegetables/growing-turnips-and-rutabagas#watering-794810

University of California Agriculture and Natural Resources. (n.d.). Common purslane (Portulaca oleracea). Weed Gallery: Common purslane--UC IPM. https://ipm.ucanr.edu/PMG/WEEDS/purslane.html

Wilson, H., Wyatt, M.A., Zilliox, T.: Prickly pear cactus: Food of the Desert. Cooperative Extension | The University of Arizona (2019). https://extension.arizona.edu/pubs/prickly-pear-cactus-food-desert

Wood, S.: Purslane, weed it or eat it. UC Master Gardener Program of Alameda County (2014). https://acmg.ucanr.edu/Over_the_Fence/Purslane_Weed_It_or_Eat_It/

University of Arizona. (n.d.). Dorm Map. University of Arizona Housing. Accessed 10 July 2024. https://housing.arizona.edu/dorms/dorm-map. Edits made to the image via Photoshop by Yostina Yacoub

Google LLC. (n.d.). Aerial image of Maricopa Hall layout, University of Arizona campus. Google Maps. Accessed 10 July 2024. https://goo.gl/maps/2ez9q4MvA8dT1N4L6

Google LLC. (n.d.). Southern facade of Maricopa Hall, University of Arizona campus. Google Maps. Accessed 10 July 2024. https://goo.gl/s2Suy9L6R6E5Pa5V6

Liverpool City Council. (n.d.). Community Gardens. Liverpool City Council. Accessed 10 July 2024. https://www.liverpool.nsw.gov.au/environment/community-gardens

Wind Sheltering Techniques for Reducing High Temperature Air Penetration in Spaces with Outdoor Acclimatization Strategies

Francisco Ruda Sarria, Paulo García-Melgar, Cristian Romero García, and Servando Álvarez Domínguez

Abstract

The urban climate is a widely studied topic by the scientific community due to its impact on the global population. The general increase in temperatures in urban environments, caused by the Urban Heat Island effect, leads to heat accumulation problems in streets and public spaces, degrading people's outdoor experiences. Therefore, it is essential to implement effective outdoor ventilation strategies to dissipate accumulated heat and improve thermal comfort in public spaces. When designing an outdoor acclimatization strategy, it is crucial to study the inflow of external air. If the air inside the target zone is constantly being renewed, no acclimatization approach will significantly improve thermal comfort. For that reason, in this paper different wind sheltering strategies for outdoor open spaces are proposed and studied with simplified CFD models in order to measure the reduction of air renewal within a desired zone where acclimatization techniques are used. Two different approaches are being tested: the first one is based on urban canyon-like designs where the target zone is located underneath the floor level of the surroundings buildings whereas the second approach is by means of using windbreaks structures to disrupt the flow creating an isolated recirculation zone immediately after the obstacle. After defining both approaches, a 2D CFD model is created and different conditions are tested to compare which approaches reduces the incoming air, studying the streamlines pattern and the air renewal rate. After the study, it has been stated that the suggested techniques effectively reduces the air income within the desired zones, manifesting the value of this research. Finally, the main objective is to suggest methods to improve the quality of outdoor acclimatization strategies in order to recover the use of open spaces in a world climatically changing.

Keywords

UHI · Outdoor acclimatization · Wind shelters · Thermal comfort · CFD

1 Introduction

Effects related to climate change are in the spotlight of the scientific community due to their impact on Urban climate and people's quality of life. Firstly, the overheating occurring in the cities has garnered attention for quite some time (Nazarian et al., 2022; Santamouris & Kolokotsa, 2015). This phenomenon causes a local temperature increment within urban landscapes and streets which can affect negatively the life of every citizen (Pisello et al., 2018). Furthermore, heatwaves have a huge impact on people's health (Gasparrini & Armstrong, 2011; Smoyer-Tomic et al., 2003). This phenomenon is evolving, increasing not only the number of times a given city is affected by a heatwave in a year but also the maximum temperature reached and the time it lasts (Habeeb et al., 2015; Kuglitsch et al., 2010). In the face of these challenges, the pursuit of effective strategies to mitigate the negative impacts on the thermal quality of open spaces becomes essential (Fernandez Milan & Creutzig, 2015; Shahmohamadi et al., 2010). In this context, outdoor acclimation

F. R. Sarria (✉) · P. García-Melgar · C. R. García
Department of Energy Engineering, University of Seville, Seville, Spain
e-mail: fruda@us.es

P. García-Melgar
e-mail: pgarcia21@us.es

C. R. García
e-mail: cromero3@us.es

S. Álvarez Domínguez
Higher Technical School of Engineering, University of Seville, Seville, Spain
e-mail: salvarez@us.es

B. He et al. (eds.), *Towards the Framework of Livable and Resilient Cities* Advances in Science, Technology & Innovation, https://doi.org/10.1007/978-3-031-97849-4_22

emerges as a crucial approach, where the implementation of bioclimatic designs and wind control strategies can play a pivotal role in enhancing thermal comfort in public areas, thereby contributing to the creation of more sustainable and liveable urban environments.

1.1 Outdoor Acclimatization

Outdoor acclimation of open spaces focuses on creating bioclimatic designs to acclimate public areas using natural cooling mechanisms to improve thermal comfort (Wilkinson, n.d.). However, this has not been deeply studied as it is not as easy to reach good thermal comfort as indoors due to technological and economic limitations, although some authors have tried to give some information about its benefits (Krüger et al., 2017; Lam & Lau, 2018). For this purpose, some researchers point out (Kang et al., 2013; Stathopoulos, 2006) that wind is important in outdoor human thermal comfort. When using active cooling mechanisms in space it is necessary to control the unexpected airflow inlet because if not, the amount of air renewal would be excessive and thermal control would be unfeasible (Dec et al., n.d.). Because of this, wind control strategies are needed to create quasi-isothermal air conditions in a given open space whether or not using active cooling features. Although reducing air ventilation might seem undesirable for improving air quality, in this application, the implications in air quality are not considered because these strategies are considered to be applied in open spaces where usually no pollution issues are usually found, because there is zero or low traffic and very few pollutant sources (squares, pedestrian streets, city parks, etc.). Therefore, if it is wanted to improve thermal comfort in these zones, some mechanism are needed to prevent the high temperature air form outside that zone penetrating into the target zone.

The sheltering strategy (Gandemer, 1979) is the one proposed in this paper. Sheltering can be defined as an urban landscape modification to change and control airflow patterns in the main wind direction to help reach good thermal comfort in a given open space. This can be accomplished via two different types of modifications: setting windbreak-like structures surrounding the area of interest or at least in the principal wind direction or creating a street canyon-like morphology where the target space is placed below floor level. Both strategies are very valuable when the temperature of the surroundings of the target zone (for example, a public square) is considerably high. With this mechanisms, it is seek to reduce the high temperature air penetration in the zone, which combined with other active cooling methods lead to the reduction of air temperature inside a desired zone therefore achieving good thermal confort.

The first approach has been studied deeply with other purposes not directly related to thermal comfort (Borges & Viegas, 1988; Cleugh, 1998). The goal of the windbreak is to disturb the airflow which in its absence would break straight into the space making active cooling mechanisms useless. With it, the flow is forced to change its direction creating a recirculation flow just after the windbreak (Heisler & Dewalle, 1988; Vigiak et al., 2003). This is a brand new method for outdoor acclimatization, thus there is no much information about the results of air penetration and the applicability to this field in the literature, so this motivates this study to mitigate this. On the other hand, the second approach's applicability to open space acclimation has not been studied as much as the previous one. However, it isolates the area from outside conditions (Karimimoshaver et al., 2021), setting perfect conditions to apply active cooling strategies as it forms indoor-like conditions. This approach has been applied successfully in real-life projects (Medina et al., 2022).

1.2 Objectives

The main objective is the characterization of the implications and effects on airflow due to the sheltering mechanism. This way a database is provided where qualitative and quantitative information about the airflow pattern and air penetration is given for each sheltering configuration. With this, full information about the penetration of air with high temperature is provided, and then it can be discussed if the suggested strategy or configuration is appropriated for the specific case where it is applied.

This has been carried out by CFD bidimensional simplified where different scenarios and configurations of the obstacles are tested. The results focus not only on the flow distortion caused by the sheltering but also on the numerical penetration results in terms of airflow rate into the recirculation zone (which is not desirable as the external air high temperature difficulty cooling the space), which is expected to be low to achieve good sheltering efficiency. This also mitigates the lack of information about these types of results, because it is an aspect not researched in depth. All of this is presented as charts or equations with enough information on the main results for it to be used in the preliminary design of public open spaces with natural climate control methods in real-life projects (Fig. 1).

2 Methodology

In this study, a CFD simulation of a wide variety of cases is done for each sheltering strategy. For that purpose, here is provided the explanation of each technique in depth as well as the configuration of the CFD software used (Fluent) and the list of the cases studied to obtain representative air flow and air renewal results.

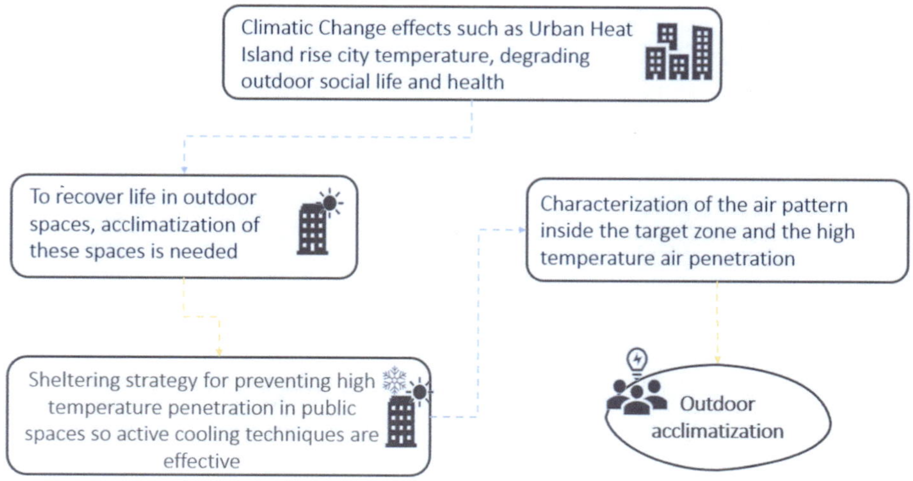

Fig. 1. Scope of the study (Source: by authors)

2.1 Urban Canyon-Like Sheltering

This sheltering approach puts the open space into an urban canyon-like configuration, where the open space of interest is below floor level, creating a depression on the ground and thus preventing the external air from entering the zone easily, highly reducing air renewal (Nosek et al., 2016; Salizzoni et al., 2009). This causes the isolation of the open space, where different natural ventilation methods can be employed to reach the desired thermal comfort, as shown in some projects (Medina et al., 2022).

The main parameter in this type of sheltering is the height-to-width aspect ratio (AR), obtained as $AR = \frac{H}{W}$, where H is the reference height of the interest area and W is its width. This mainly determines the apparition and position of recirculation's inside the space. In this study, only AR < 0.5 is considered as it is usually the range of public open spaces designed for social life such as avenues or squares. In this paper, the reference height is fixed at H = 10m, and the range of AR is from 0.1 to 0.5, increasing in steps of value 0.05, so at first 9 study cases are done.

For evaluating the airflow pattern and the renewal rate of this technique, some indicators are used. The first indicator extracted from all the simulations is the flow streamlines within the area of interest. These describe the air as it moves inside it. In addition, they are useful to easily visualize air recirculation and where its origin is located. Because of this, a subsequent deep study of the number and position of them is done.

The other indicator common to all of the cases studied is the air mass flow rate or air renewal (ACH) in units of kg/s. This indicator is an illustrative measure of the amount of air which enters the target area and consequently the air renewal inside it. Because of the bi-dimensional discrete nature of the CFD study, the expression for ACH is in Eq. 1. This air income is the name given to the high temperature external air which penetrates into the desired zone where the temperature reduction is needed.

$$ACH = 1[m] \cdot \rho \sum_{0}^{k=N} \Delta\overline{w}_{k_{roof}} \cdot \Delta x_k \tag{1}$$

where ρ is the density of the fluid, Δx_k is the size of the k cell located in the entrance line of the zone (this means the distance between two adjacent nodes) and $\Delta\overline{w}_{k_{roof}}$ is the mean vertical velocity between the two nodes from one cell. The last indicator of interest is the equivalent air mass flow rate (ACH_{eq}) which provides an illustrative measure of the portion of the air inflow which comes from outside the sheltered zone and not from the vortex itself, which can be obtained by an energy balance. ACH_{eq} is defined as the flow rate at the same temperature as the undisturbed flow (T_{ext}) obtained from Eq. 2, where T_k is the mean temperature of the air in cell k and T_{area} is the mean temperature in the area. This is a quality indicator because it measures the effective net air renewal.

$$ACH_{eq}(kg/s) = 1[m] \cdot \rho \frac{\sum_{0}^{k=N} \Delta\overline{w}_{k_{roof}} \cdot \Delta x_k \cdot \left(T_k - T_{area}\right)}{T_{ext} - T_{area}} \tag{2}$$

In order to perform the simulation, the numerical domain must be defined. In this case, the numerical domain used is a simplification of an urban canyon seen in Fig. 2, as well as its boundary conditions. In addition, there are two zones where the grid is different, in the outer zone the mesh (H/20), whereas in the inner zone the mesh is finer (H/(100*AR)) because this is where the disturbance of the flow affects more to the result.

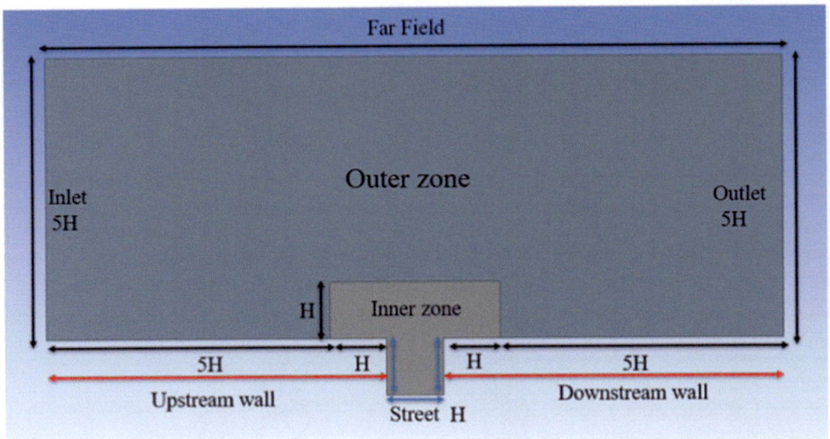

Zone	Boundary Condition
Inlet	Velocity Inlet. x-Velocity $U = V_{ref}\left(\frac{y}{H}\right)^{0.16}$
Outlet	Pressure Outlet
Far Field	Symmetry
Upstream	Wall. Roughness: C_s=1; z_0=0.1m
Downstream	Wall, no slip condition
Street	Wall, no slip condition

Fig. 2. Numerical domain and boundary conditions for street canyon-like sheltering (Source: by authors)

Fig. 3. Parts of the stream separation across a windbreak (Raine J.K. & Stevenson, 1977)

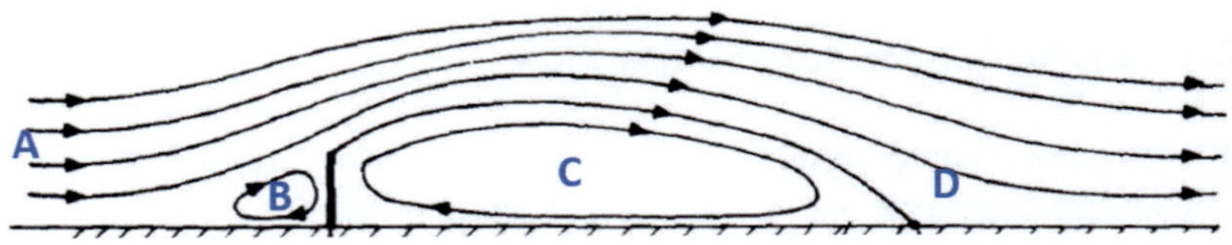

2.2 Windbreak Sheltering

The use of barriers to shield from the wind is an old method well developed. Sheltering windbreaks-like structures can be trees, stone walls, embankments or anything that can deviate air (Heisler & Dewalle, 1988). When airflow encounters a tall obstacle, part of it does not pass and recirculates in front of the obstacle, another part ascends it and then descends progressively until reaching the floor, and finally the rest of the air rises above the obstacle and forms a recirculation right after it (Fig. 3). It can be found four zones well segregated:

- Zone A: Non-disturbed flow before and after the obstacle.
- Zone B: Flow disturbed right before the obstacle.
- Zone C: Vortex right after the obstacle.
- Zone D: After the vortex.

Different impermeable obstacles are tested to deeply analyse the dynamics of the airflow and the features of outdoor sheltering with CFD software. The chosen geometries include the most common configurations found. Changing key parameters (such as obstacle height or inclination) will enable the extrapolation to different cases not directly studied. For this aim, three geometry configurations have been developed: double-sided step, double-sided slope and finally slope-step setup. Then the height is modified between 1, 1.5 and 2 m. Besides that, the slope is also changed to see the influence of the inclination in the flow disturbance, so two values are tested, one 30º and the other 45º. Furthermore, the external conditions are changed by testing three different reference wind speeds: 1, 2 and 4 m/s.

Table 1. Cases studied for windbreak sheltering (Source: by authors)

Obstacle type	Wind speed (m/s)	Obstacle height (m)	Inclination (º)
Double-sided step	1,2,4	1	-
	1,2,4	1.5	-
	1,2,4	2	-
Double-sided slope	1,2,4	1	45
	1,2,4	1	30
	1,2,4	1.5	45
	1,2,4	1.5	30
	1,2,4	2	45
	1,2,4	2	30
Slope-step	1,2,4	1	45
	1,2,4	1	30
	1,2,4	1.5	45
	1,2,4	1.5	30
	1,2,4	2	45
	1,2,4	2	30

The total amount of cases studied is gathered in Table 1.

For these cases, a simplified numerical domain is defined similar to the previous sheltering technique, but with a simple tetrahedral grid with a cell size of 0.04H which has been proved to provide fine enough results within a relatively short time, because reducing its size does not change the solution.

Fig. 4. Real life application of a windbreak obstacle for sheltering (Source: by authors)

The indicators used in this case are the flow pattern and ACH described previously. However, here the ACH definition is slightly different (Eq. 3), as the entrance line is a vertical line located in the centre of the recirculation, where, Δy_k is the size of the k cell located in the vertical line and $\Delta \overline{v}_{k_{line}}$ is the mean horizontal velocity between the two nodes from one cell.

$$ACH = 1[m] \cdot \rho \sum_{0}^{k=N} \Delta \overline{v}_{k_{line}} \cdot \Delta y_k \quad (3)$$

This type of sheltering have been used in real life applications combined with several other cooling techniques, for example in Seville, where a public square in a street used this sheltering approach for reducing the external air high temperature penetration (Palomo Amores, 2024). The barrier used in this case was a fountain, which can be seen in Fig. 4.

3 Results

3.1 Results for Urban Canyon-Like Sheltering

Firstly, the airflow pattern shows different behaviour depending on the aspect ratio (Fig. 5). Although only one recirculation is found in every case tested, the position of it changes. For very low AR, the vortex does not occupy all the desired zones, being close to the left wall. However, as AR increases, the vortex changes its position, approaching the opposite wall for AR = 0.3. Finally, between AR = 0.3 and AR = 0.5, the recirculation tends to settle on the centre of the area.

Secondly, attending the air renewal results (Fig. 6), its value sharply descends from AR = 0.1 until AR = 0.25, where it reaches its minimum and then comes to a hand still. The value of it is approximately 2 kg/s whereas it descends to approximately 0.25 kg/s On the other hand, the new indicator ACH_{eq} exhibits another important behaviour. The air entering the target zone of interest coming from the exterior is only approximately half of the total for low AR, but it increases up to 75% for higher AR, which implies that the air renewal is more effective.

3.2 Results for Windbreak Sheltering

The airflow pattern results for this sheltering mechanism are very similar for each type of obstacle tested (Fig. 7). A unique vortex is formed downstream whose centre is located at the same height approximately as the obstacle and a distance of approximately 6*H m from the end of the obstacle for the first type, and slightly inferior for the other two. It means that the height of the obstacle directly affects the position on the vortex and the space where the air penetration can be controlled. In general, the stepped one provide more space for cooling as in is less thick. The air that has been cooled by different active cooling methods inside the recirculation zone, will be kept within it, with very few high temperature air penetration as the obstacle prevents the external air from entering directly into the cooled space.

Finally, the results of air renewal or air penetration are displayed. It shows similar trends for all the cases studied, as the dependence of ACH for both obstacle height and wind speed are linear. The air penetration for the step obstacle is higher than the other, so that means that isolation effectiveness

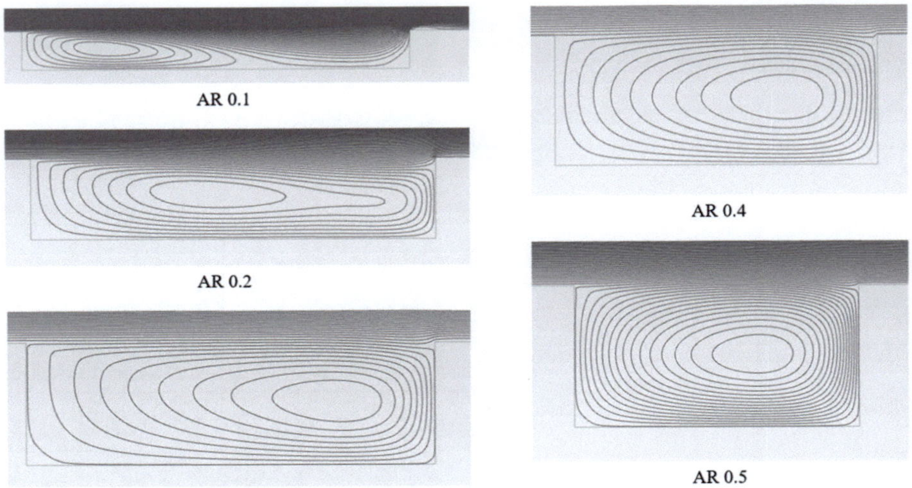

Fig. 5. Airflow pattern results for street canyon-like sheltering (Source: by authors)

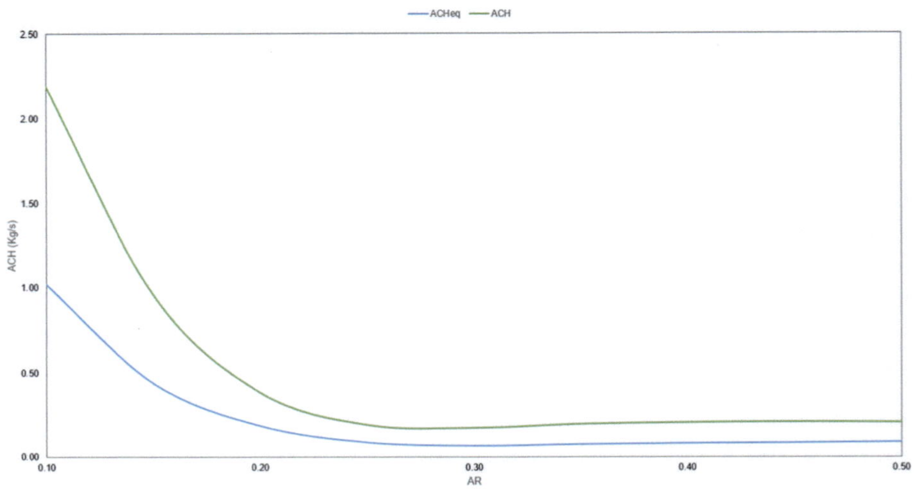

Fig. 6. Air renovation for street canyon-like sheltering (Source: by authors)

Fig. 7. Airflow pattern for windbreak sheltering (Source: by authors)

is lower (Fig. 8), reaching the value of approximately 1.40 kg/s for the case of H = 1 m and V_{ref} = 4 m/s. When comparing the three morphologies tested, In the most beneficial case, the windspeed is the lowest (1 m/s) and with a height of 1m is enough to shelter the desired space, with the step-type the air penetration is ~0.12 kg/s, for the slope-type is ~0.10 kg/s and for the mixed step-slope type is ~0.15 kg/s. It can be seen that the slope-type obstacle has the smallest air penetration when compared with the other, which means that it is the most efficient type for isolating a space for applying active

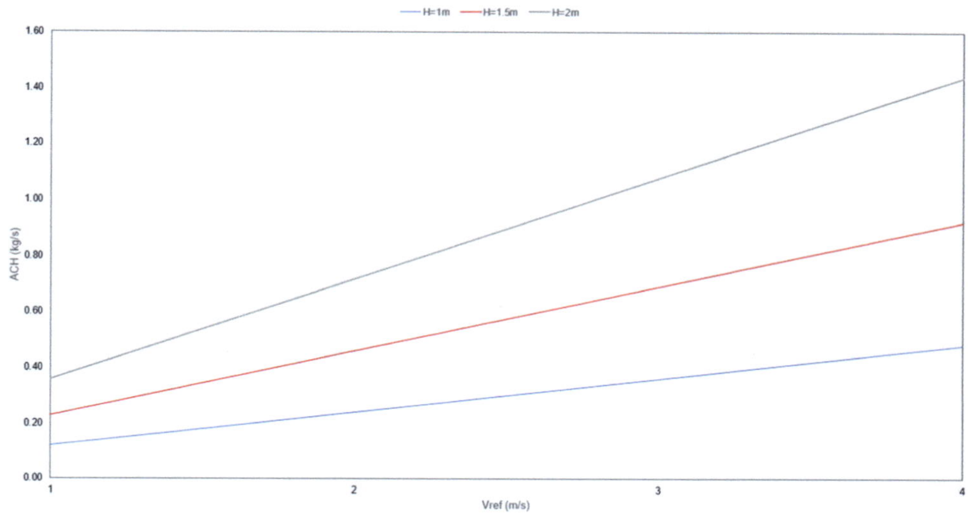

Fig. 8. Air renovation results for step-type obstacle (Source: by authors)

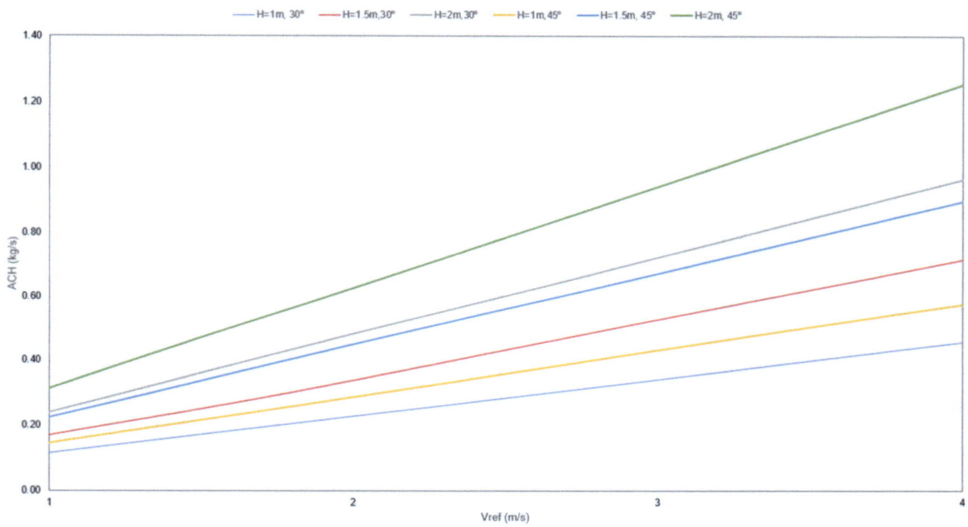

Fig. 9. Air renovation results for slope-type obstacle (Source: by authors)

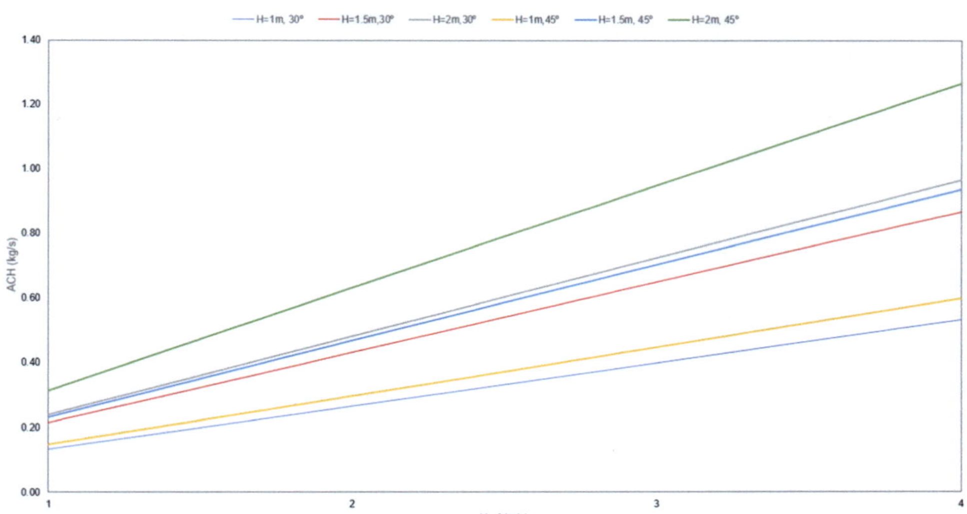

Fig. 10. Air renovation results for step-slope type obstacle (Source: by authors)

cooling techniques. This coincides with the example of real application of this sheltering strategy in Fig. 4, which is a slope-type obstacle and it is a small obstacle.

On the other hand, the results for the slope obstacle (Fig. 9) show also a linear relationship between V_{ref} and ACH, with a maximal value of ACH = 1.20 kg/s for the case of a obstacle with height 2m, inclination 45° and a external airspeed of 4 m/s. In addition, it can be seen the effect of the slope inclination. The ACH value for the case of 45° is nearly 30% higher than the case of 30° for each windspeed tested. This means that it is more valuable to have a flatter obstacle as the air penetration significantly with the reduction of the inclination. However, this can be also prejudicial, as a flatter obstacle occupies more space, thus creating a smaller isolated zone for applying active cooling method. This way, a balance between both effects have to be done. This agrees with the solution for the step type obstacle, because the more inclination the slope has, the more it resembles a step type obstacle, which have found to have the less isolation potential.

Finally, for the mixed slope-step case (Fig. 10) the effect of the slope inclination is not that relevant, as for the cases of H = 1m and H = 1.5m the ACH results are very similar for both inclination values tested. However, when the external airspeed is high enough, the effect of the inclination becomes noticeable. In general, this mixed morphology offers less isolation potential than the other morphologies.

4 Conclusion

The results previously presented compose a high-quality database to use in the initial stages of designing an outdoor acclimatization mechanism because it provides information about the airflow pattern and air renovation for different sheltering techniques. This way, different approaches for isolating a desired zone are presented, with quantitative information about the air flow rate affecting the area.

The airflow pattern for the street canyon-like sheltering shows a different behaviour depending on the AR value. For very low AR, the vortex is displaced towards the left wall, which implies that the air renewal is considerably higher than the rest of the cases.

Then, for the windbreak sheltering, the airflow patterns are almost identical in each case tested. However, the air penetration results show that for the step type, the values are higher than the others. In addition, a linear dependence of the air renovation with both V_{ref} and obstacle height has been proved.

With this information, urban city planners are able to try different city layouts for enhancing thermal comfort outdoor in places where temperatures can be extremely high, making impossible social life and degrading people's quality of life. Depending on the space and resources available, they can opt for one of the strategies presented here to mitigate this effect, without having to do preliminary studies about the effectivity of them. Some examples of the real life applications of this strategies have been showed, highlighting the applicability of this study.

To summarize, this study has allowed the creation of a sort of catalogue where a characterization of the sheltering strategy for outdoor acclimatization is done. These results can be used in the initial stages when designing strategies for the adaptation of cities to climate change, as many of the indicators provided (for instance, the air renewal) are usually inputs.

4.1 Limitations

The main limitation of this study is the bi-dimensional nature of the simulations conducted. Although this can be useful for reduce computational cost and time in the simulations, the results do not represent the reality exactly as three-dimensional effects are not considered. However, this can provide illustrative results that can be helpful in the initial steps of designing acclimatization methods to discuss which of the strategies apply with a specific configuration. Then, once having decided it, further three-dimensional simulations can be done to enrich the acknowledgement of the air penetration for the specific case.

Acknowledgments This study has been funded by the projects "CONSTANCY - Resilient urbanisation methodologies and natural conditioning using imaginative nature-based solutions and cultural heritage to recover the street life" (Grant Agreement PID2020-118972RB-I00) and the project "NATURBEAM - Lighting the way to a greener future to restore urban habitability through nature-based solutions" (Grant Agreement TED2021-130416B-I00) by the Spanish Ministry of Science, Innovation and Universities.

The first author is supported through the grant agreement C23.I1.P03.S01.01 ANDALUCÍA from "Programa Investigo" funded by the Recovery, Transformation and Resilience Plan of Spanish Government.

References

Borges, A.R., Viegas, D.X.: Shelter effect on a row of coal piles to prevent wind erosion. J. Wind Eng. Ind. Aerodyn. **29** (1988)

Cleugh, H.A. (1998). Effects of windbreaks on airflow, microclimates and crop yields

Dec, E., Babiarz, B., Sekret, R. (eds.): Analysis of temperature, air humidity and wind conditions for the needs of outdoor thermal comfort. https://doi.org/10.1051/e3sconf/20184400028

Fernandez Milan, B., Creutzig, F.: Reducing urban heat wave risk in the 21st century. Curr. Opinion Environ. Sustain. **14**, 221–231 (2015). https://doi.org/10.1016/j.cosust.2015.08.002

Gandemer, J. (1979). Wind shelters. J. Ind. Aerodyn. 4

Gasparrini, A., Armstrong, B.: The impact of heat waves on mortality. Epidemiology **22**(1), 68–73 (2011). https://doi.org/10.1097/EDE.0b013e3181fdcd99

Habeeb, D., Vargo, J., Stone, B.: Rising heat wave trends in large US cities. Nat. Hazards **76**(3), 1651–1665 (2015). https://doi.org/10.1007/s11069-014-1563-z

Heisler, G. M., Dewalle, D.R.: Effects of windbreak structure on wind flow. Ecosyst. Environ. **22**(23), 41 (1988)

Kang, K.N., Song, D., Schiavon, S.: Correlations in thermal comfort and natural wind. J. Thermal Biol. **38**(7), 419–426 (2013). https://doi.org/10.1016/J.JTHERBIO.2013.06.001

Karimimoshaver, M., Khalvandi, R., Khalvandi, M.: The effect of urban morphology on heat accumulation in urban street canyons and mitigation approach (2021).https://doi.org/10.1016/j.scs.2021.103127

Krüger, E.L., Tamura, C.A., Bröde, P., Schweiker, M., Wagner, A.: Short- and long-term acclimatization in outdoor spaces: exposure time, seasonal and heatwave adaptation effects. Build. Environ. **116**, 17–29 (2017). https://doi.org/10.1016/J.BUILDENV.2017.02.001

Kuglitsch, F.G., et al.: Heat wave changes in the Eastern Mediterranean since 1960. Geophys. Res. Lett. **37**(4) (2010). https://doi.org/10.1029/2009GL041841

Lam, C.K.C., Lau, K.K.L.: Effect of long-term acclimatization on summer thermal comfort in outdoor spaces: a comparative study between Melbourne and Hong Kong. Int. J. Biometeorol. **62**(7), 1311–1324 (2018). https://doi.org/10.1007/s00484-018-1535-1

Medina, D.C., et al.: Climatic control of urban spaces using natural cooling techniques to achieve outdoor thermal comfort. Sustain. Switz. **14**(21) (2022). https://doi.org/10.3390/su142114173

Nazarian, N.: Integrated assessment of urban overheating impacts on human life. Earth's Fut. **10**(8), e2022EF002682 (2022). https://doi.org/10.1029/2022EF002682

Nosek, Š, Kukačka, L., Kellnerová, R., Jurčáková, K., Jaňour, Z.: Ventilation processes in a three-dimensional street canyon. Boundary-Layer Meteorol. **159**(2), 259–284 (2016). https://doi.org/10.1007/S10546-016-0132-2

Palomo Amores, T.R.: Enfoque avanzado para la mejora del confort en espacios exteriores: cuantificación de la resiliencia, catálogo de componentes y metodología de prediseño de estancias mediante técnicas de control climático. 1. (2024). https://dialnet.unirioja.es/servlet/tesis?codigo=332176&info=resumen&idioma=ENG

Pisello, A.L., Saliari, M., Vasilakopoulou, K., Hadad, S., Santamouris, M.: Facing the urban overheating: recent developments. Mitigation potential and sensitivity of the main technologies. Wiley Interdiscip. Rev. Energy Environ. **7**(4), e294 (2018). https://doi.org/10.1002/WENE.294

Raine, J.K., Stevenson, D.C.: Wind protection by model fences in a simulated atmospheric boundary layer. J. Ind. Aerodyn. **2**, 159–180 (1977)

Salizzoni, P., Soulhac, L., Mejean, P.: Street canyon ventilation and atmospheric turbulence. Atmos. Environ. **43**(32), 5056–5067 (2009). https://doi.org/10.1016/j.atmosenv.2009.06.045

Santamouris, M., Kolokotsa, D.: On the impact of urban overheating and extreme climatic conditions on housing, energy, comfort and environmental quality of vulnerable population in Europe. Energy Build. **98**, 125–133 (2015). https://doi.org/10.1016/J.ENBUILD.2014.08.050

Shahmohamadi, P., Che-Ani, A.I., Ramly, A., Maulud, K.N.A., Mohd-Nor, M.F.I.: Reducing urban heat island effects: a systematic review to achieve energy consumption balance. Int. J. Phys. Sci. **5**(6), 626–636 (2010). http://www.academicjournals.org/IJPS

Smoyer-Tomic, K.E., Kuhn, R., Hudson, A.: Heat wave hazards: an overview of heat wave impacts in Canada. In: Natural Hazards, vol. 28 (2003)

Stathopoulos, T.: Pedestrian level winds and outdoor human comfort. J. Wind Eng. Ind. Aerodyn. **94**(11), 769–780 (2006). https://doi.org/10.1016/j.jweia.2006.06.011

Vigiak, O., Sterk, G., Warren, A., Hagen, L.J.: Spatial modeling of wind speed around windbreaks. CATENA **52**(3–4), 273–288 (2003). https://doi.org/10.1016/S0341-8162(03)00018-3

Wilkinson, M., (ed.): Outdoor thermal comfort 1553 Figure 1. The Olgyay bioclimatic chart (11) (adapted with permission by)

Importance of Design of Air Emitters in the Healthcare Facilities: Computational Fluid Dynamic Maps for Air Distributions Under Mediterranean Climate

Rosa Francesca De Masi, Nicoletta Del Regno, Antonio Gigante, Silvia Ruggiero, Francesco Tariello, and Giuseppe Peter Vanoli

Abstract

Indoor environmental quality is a key objective in the building performance improvement. After the COVID-19 emergency, a lot of healthcare facilities are working to improve the design of air emitter and their management. However, in the existing structure, the scientific literature underlines the lack of methodologies for the evaluation of the state of fact in term of air quality and influence of position or adoption of heating, cooling and ventilation systems mainly in the patient rooms. Thus, the paper, by means of a case study, is aimed to highlight the potentialities of computational fluid dynamic based map for studying the air distribution and the incidence of external and internal forcing in the patient rooms focusing on the positive role that the mechanical ventilation could have in the reduction of infections and in the improvement of environmental conditions. Starting form monitored data in a patient's room of "A. Cardarelli" hospital of Campobasso (city of central Italy) a numerical model was calibrated through the Design Builder graphic interface of Energy Plus. According to the results of the study, the type of diffusers and the location of the exhaust fans influence the airflow patterns. Since there is a wide variety of parameters that define each room, among these also the outdoor conditions, it is not possible to reach a general conclusion on the optimal room design and supply and exhaust locations.

Keywords

air-emitters · healthcare facilities · CFD · air distribution · Mediterranean climate

R. F. De Masi · A. Gigante · S. Ruggiero
DING-Department of Engineering Piazza Roma, University of Sannio, 21, 82100 Benevento, Italy
e-mail: rfdemasi@unisannio.it

A. Gigante
e-mail: gigante@unisannio.it

S. Ruggiero
e-mail: sruggiero@unisannio.it

N. Del Regno (✉) · G. P. Vanoli
Department of Medicine and Health Sciences, University of Molise, Vincenzo Tiberio, Via G. Paolo II Contrada "Tappino", 86100 Campobasso, Italy
e-mail: n.delregno@studenti.unimol.it

G. P. Vanoli
e-mail: giuseppe.vanoli@unimol.it

F. Tariello
Department of Agricultural, Environmental and Food Sciences, University of Molise, Via F. De Sanctis, 86100 Campobasso, Italy
e-mail: francesco.tariello@unimol.it

1 Introduction

The indoor comfort, prerequisite of the indoor environmental quality, is described as a condition of well-being of the individual within the environment [1]. Healthcare structures are particularly complex environments since people with different comfort needs coexist: patients, due to their limited mobility, need lower metabolic rates while health personnel need higher metabolic rates [2]. Moreover, good air quality inside hospital rooms favours on the one hand a faster recovery of the patient and on the other a comfortable working environment for the personnel. Furthermore, good ventilation reduces the spread of diseases in surgery blocks and operating theatres. Lu et al., by means of numerical approach, have studied the distribution of contaminants within a hospital room using different types of ventilation (layered, mixed, downward and displacement) and demonstrated that layered ventilation minimizes the risk of contamination of personnel

B. He et al. (eds.), *Towards the Framework of Livable and Resilient Cities* Advances in Science, Technology & Innovation, https://doi.org/10.1007/978-3-031-97849-4_23

[3]. Sheng et al., with the aim to improve the thermal conditions have proposed a new type of vertical induction ventilation [4]. Instead, some studies found that the positions of the inlet and outlet vents significantly affect the pathogen removal efficiency [5] and also the thermal comfort [6]. Briefly, within healthcare structure, good ventilation should be on of main aspects to take into consideration during the design phase or the refurbishment of existing buildings [7] for both assuring low risk of contamination and for reaching well-being. There is an obvious need for interdisciplinary research in the field to address the pressing challenges [8]. A multi-disciplinary study combining advanced computer modeling, experimental testing, and on-site experimental designs could provide evidence to identify optimum ranges for temperature, humidity and ACH along with appropriate ventilation design strategies [6]. In this frame, the paper's aim is the introduction of a method for evaluating possible design strategies for the heating, cooling and ventilation systems that allow minimizing infections and making healthier hospital environment. The potentiality of computational fluid dynamics simulation is investigated in order to explore the influence of different moving paths and emitters position as well as ventilation efficiency under different air supply configurations.

1.1 Review of Literature

A careful analysis of the literature has shown that the study of air distribution was conducted mainly following the spread of the 2020 SARS COVID-2 virus in order to reduce the presence of contaminating particles within hospital departments. From the study by Dao and Kim [9], to reduce the concentration of droplets from sneezing and coughing, through a CFD analysis carried out to lower the concentration of contagious particles in an isolation room, it was found that there is a greater efficiency of removal of droplets when the air exhaust grille is above the patient's head. According to Wang et al. [10], to reduce the risk of cross-infection, it is necessary to use a bottom-in and top-out air distribution to expel contaminant particles from the ceiling. Alternatively, two air exhaust grilles can be used on the lower wall behind the patient's bed [11].

Islam et al. [12] used a CFD analysis to demonstrate that increasing the air velocity inside the room improves the extraction of air from the room.

2 Proposed Study General Description

Hospitals are very complex structures, so their fundamental requirement is to have an environment as "clean" as possible from the point of view of contaminants. Following the Covid-19 pandemic, the prerequisites for reaching healthy objectives have become even more stringent. There are several scientific studies that compare the ventilation strategies but there is not a standardized method or the analysis of the influence of external and internal parameters based on real monitored data in patient rooms. For these reasons, the work focuses on the ventilation strategy adopted in a real hospital room of the surgery department of the "Cardarelli" hospital in Campobasso (Italy). Starting from the validation of a numerical model by means of measured carbon dioxide concentration, the natural ventilation, actually present in the room, is compared with the mechanical one with two different emitters as possible refurbishment configuration. More in detail, it is considered:

- installation of a circular diffuser in the center of the ceiling and extraction grille on the door;
- installation of a wall vent and extraction grille on the door.

In order to create distribution maps of indoor quality, the calibrated model was simulated in term of energy needs and indoor parameters then, the module for the computational fluid dynamics (CFD) of DesignBuilder [13] is used. CFD applied to buildings provides the designer with information on probable air velocities, pressures and temperatures occurring in the room with specified boundary conditions including climate, internal heat gains and active systems.

DesignBuilder CFD is based on a method known as the finite volume that involves the solution of a set of partial differential equations describing the transport of momentum, energy and turbulence quantities. Partial differential equations are converted into a set of simultaneous algebraic equations. More in detail, the room under analysis is divided into a set of non-overlapping adjoining rectilinear cells (finite volume grid). Algebraic equations are set up for each grid cell and the whole set of equations solved using a numerical method. In the two refurbishment selected configurations, the boundary conditions on the incoming air temperature as well as its flowrate are updated according to the type of system and the technical datasheet of emitters. The methodological approach is summarized in Fig. 1.

2.1 Experimental Set-Up

In order to calibrate a numerical model, a measurement campaign was carried out during ten days of summertime in a multiple room (Fig. 2) of the surgery department, with

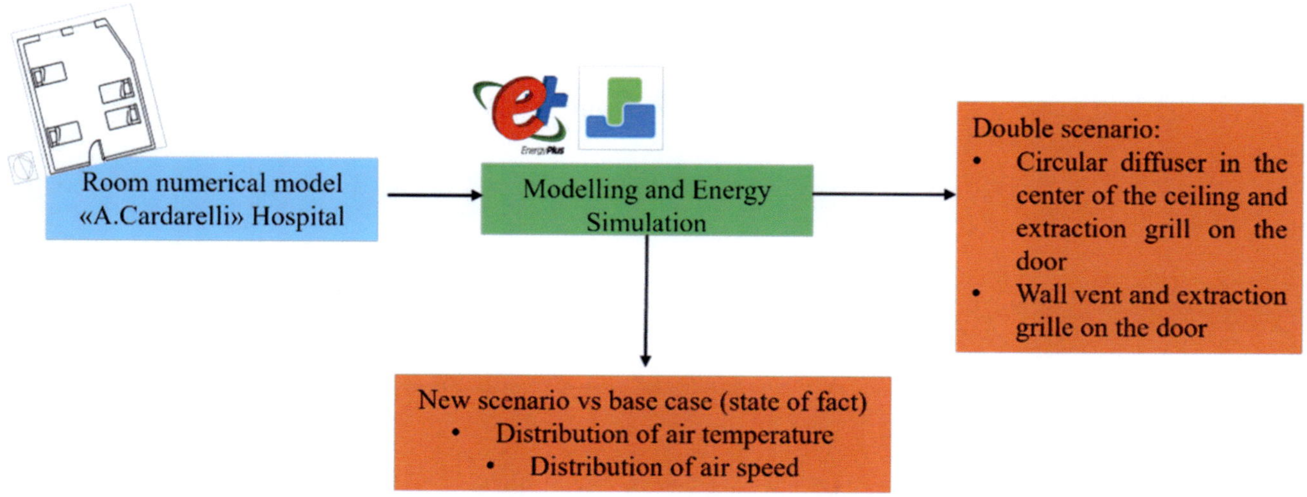

Fig. 1. Methodological approach (Created by the authors)

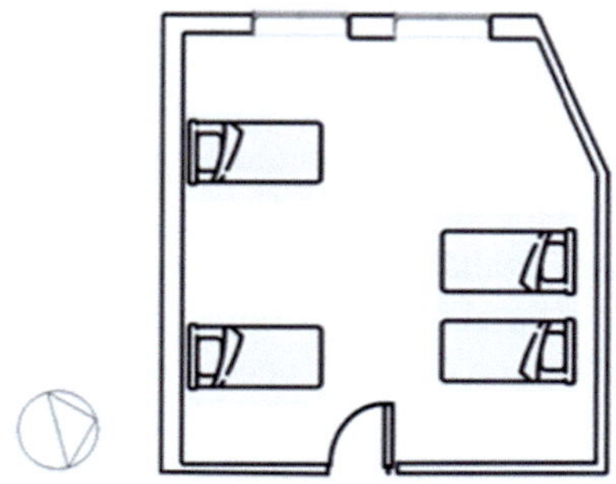

Fig. 2. Multiple room (Created by the authors)

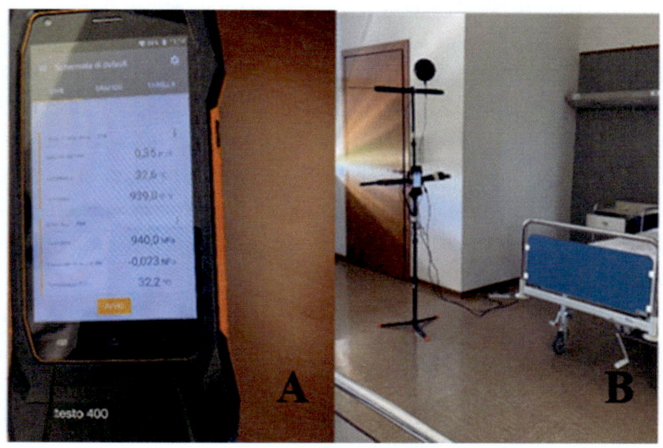

Fig. 3. Data-logger (A); (B) stand-alone station installed in the multiple room (Created by the authors)

four beds, (35 m^2) characterized by one external wall and two windows west exposing. According to Gargaro et al.[14], the external walls, composed by hollow bricks, are characterized by a thickness of 30 cm and a thermal transmittance of 1.0 W/m^2K. The glazed area is 4.9 m^2 and it is characterized by aluminium frame with single clear glass (thermal transmittance 5.7 W/m^2K) and mobile external screens.

"Testo 400 IAQ and comfort kit" standalone monitoring station was installed from 25 July to 3 August 2022, in order to evaluate the main comfort parameters, such as the average radiant temperature, air temperature, air velocity and humidity. Relative. This station consists of a data logger and various probes connected by cable or via Bluetooth (Fig. 3). The main characteristic of each sensor are showed in Table 1.

Table 1. Characteristics of each sensor (Table created by the authors)

Parameter	Range	Accuracy
Air temperature	0 ÷ 50 °C	± 0.5 °C
Relative Humidity	5 ÷ 95%	± 3% (10 ÷ 35% RH) ± 2% (35 ÷ 65%RH) ± 3% (65 ÷ 90%RH)
Air speed	0 ÷ 5 m/s	± (0.03 m/s + 4% of m.v.)
Globe temperature	0 ÷ 120 °C	Tolerances Class 1, EN 60584–2

2.2 Mechanical Ventilation Design

The room is not equipped by an air-condition system; the radiators are installed and connected to a central basement boiler for the heating request. Natural ventilation is permitted through the windows meanwhile there is not centralized ventilation. The maintenance office of the hospital, also considering the indications of patients, is working on the refurbishment of the active system and mong the interventions there is the installation of a centralized heating, ventilation and air-conditioning system. The analysis contributes to the design

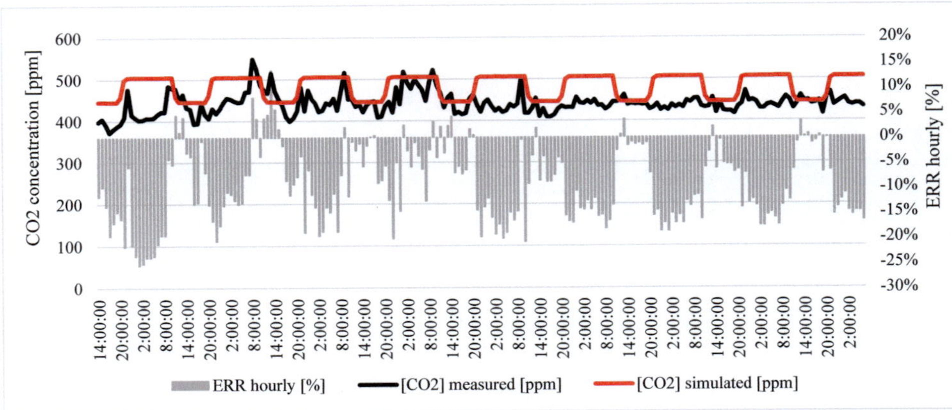

Fig. 4. Comparison and simulated and measured for carbon dioxide concentration (Created by the authors)

by comparing two different terminals: a circular diffuser positioned in the center of the ceiling and a wall inlet vent. As regards the diffuser, the air car's thermostatic variable geometry helical diffuser was considered, with a nominal diameter of 250 mm. Needing to condition the room, a launch degree of 67.5° was chosen. Furthermore, a diffuser was chosen with an air flow rate of 521 m^3/h and an air intake speed of 3 m/s [15]. For the inlet vent, double row of individually adjustable fins from the "tecno-ventil" is considered. In particular, a vent with an air flow rate of approximately 500 m^3/h was chosen [16]. The extraction grill, in both cases, was positioned on the door.

3 Results

3.1 Calibration Results

The numerical model of the room was simulated by means of DesignBuilder, graphical interface of Energy Plus [17] with a time step of 6. The numerical model was calibrated following the indications of M&V Guidelines [18] when hourly data are available. Figure 4 shows the hourly error (ERR_{hourly}) for the carbon dioxide concentration (gray bars), the measured carbon dioxide concentration trend (in black), and the simulated carbon dioxide concentration trend (red). From the Figure, we can see that the hourly error always remains within the ± 30% range, as required by the guidelines for calibrating a model. Moreover, the Mean Bias Error (MBE) is -9% and the average annual error $ERR_{average\ hourly}$ is -9% and thus lower than threshold that is for both indicators

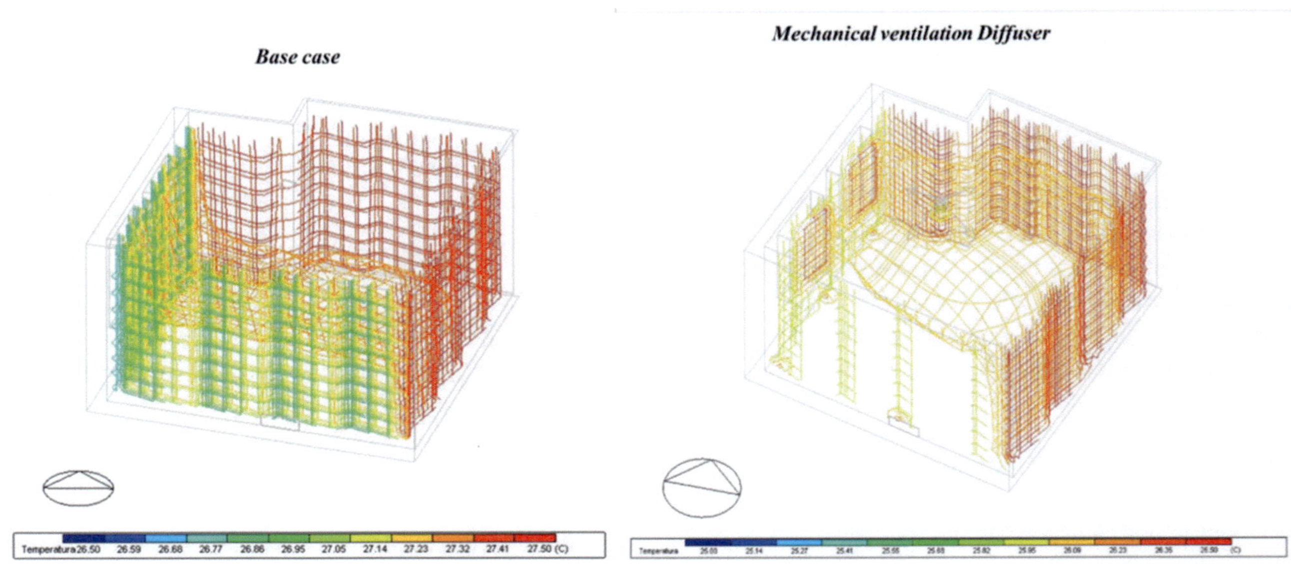

Fig. 5. Temperature distribution in the base case and with a diffuser in the center of the ceiling (Created by the authors)

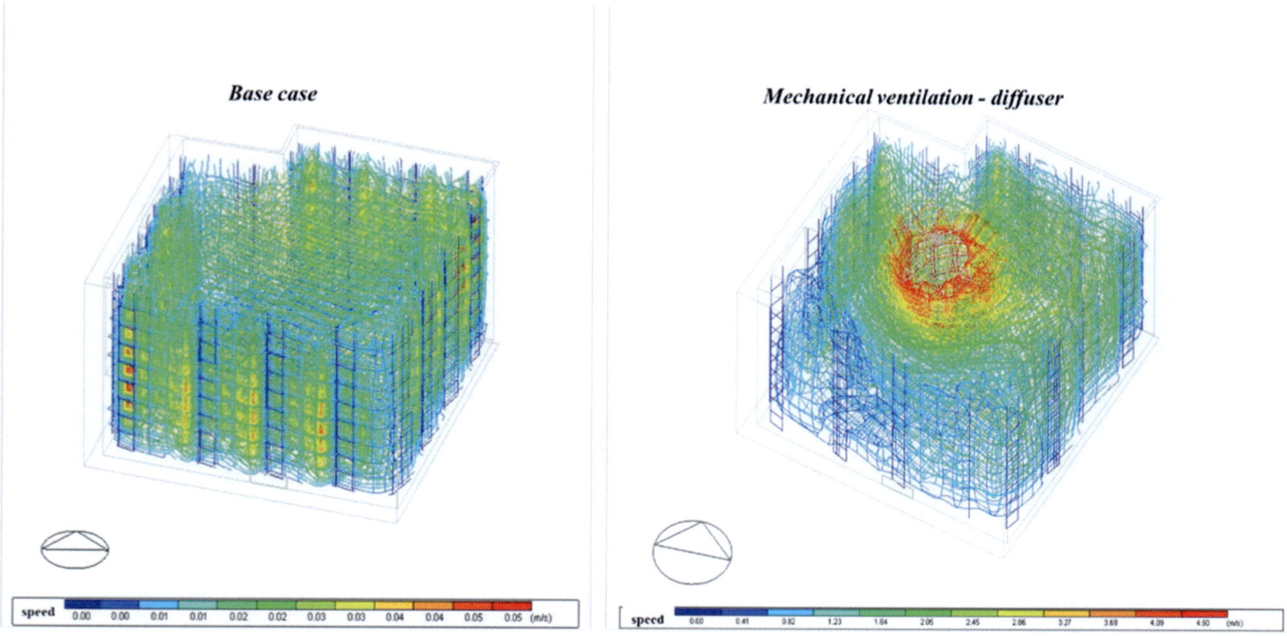

Fig. 6. Air speed distribution in the base case and with a diffuser in the center of the ceiling (Created by the authors)

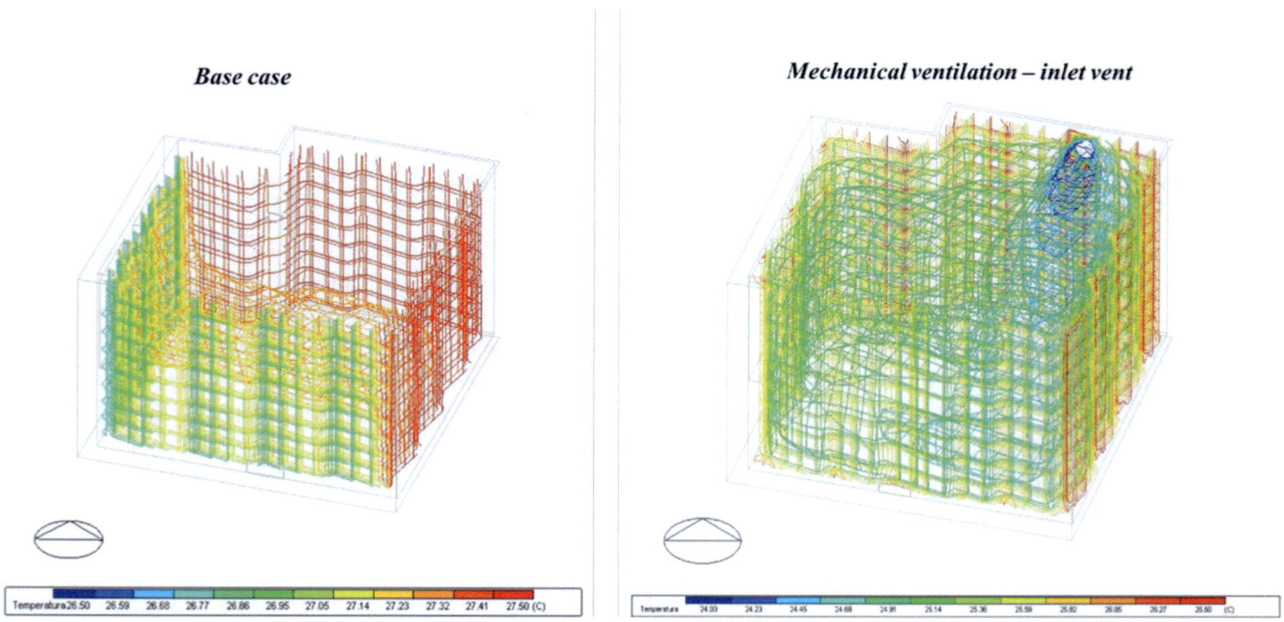

Fig. 7. Temperature distribution in the base case and with inlet grid installed on the wall (Created by the authors)

± 10%. Therefore, the numerical model can be considered calibrated and consistent with real behaviour.

3.2 CDF Results

3.2.1 Circular Diffuser

Figure 5 shows the temperature distribution in the base case and considering the diffuser that leads into the room fresh air in neutral condition (26 °C) mainly for the renovation and pollutant removal. The ventilation uniforms the thermal field with a maximum temperature within the patient position near 26 °C meanwhile it is 27 °C when the natural ventilation is considered. Due to t Coanda effect, the air sticks to the ceiling and then falls on the walls, uniforming the temperatures and reducing the residence time of the air by 70%. As shown in Fig. 6, the diffuser assures a rapid equalization of the speed and therefore the area occupied by the patient is not affected by the jet.

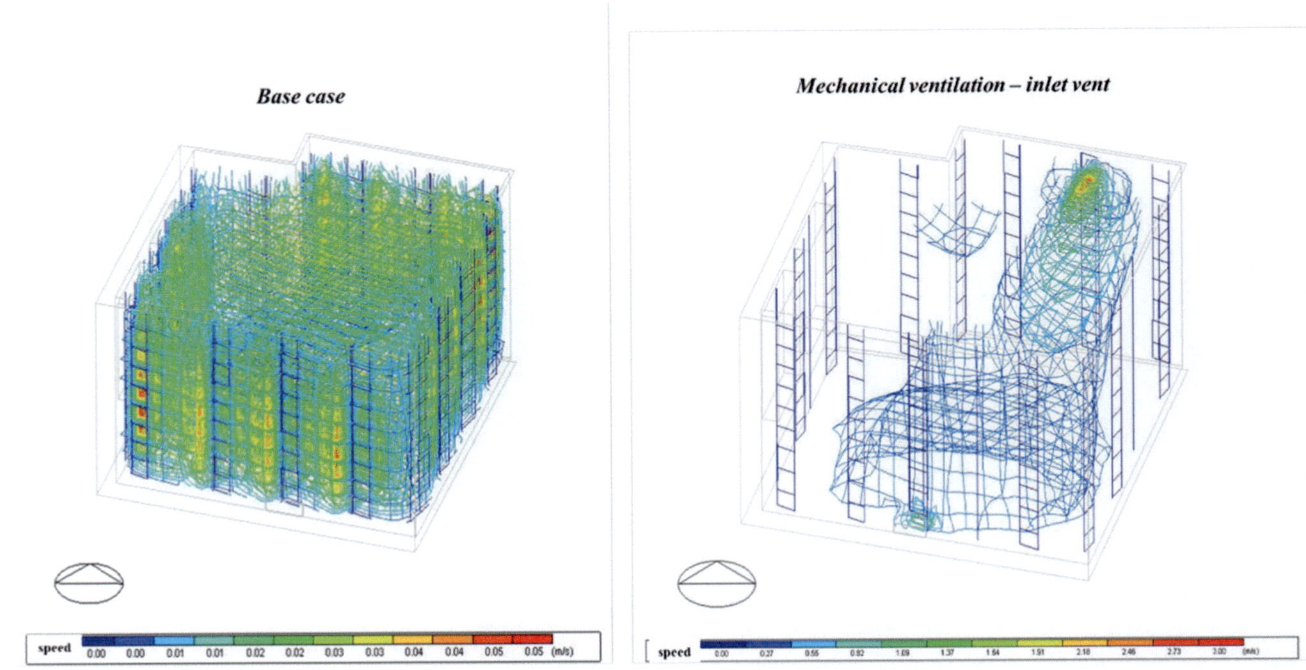

Fig. 8. Air speed distribution in the base case and with inlet grid installed on the wall (Created by the authors)

Figure 7 shows the temperature distribution when the inlet vent is used and in this case the maximum value decrease for effect of higher induction assured by the emitter and it reaches approximately 25 °C in the patient position. Also in this case the best condition is recorded in the presence of ventilation since in proximity to the patient the air speed is around 0.25 m/s (Fig. 8), the value required by the regulations for comfort. The residence time is reduced of 80% and this assures the better behaviour of this type of emitter.

Results indicate that using CFD provides a complete spatial-temporal vision for analyzing the indoor conditions that can be combined with the reconstruction of the potential airborne pathogen dispersion induced by hospital; this could facilitate the interdisciplinary study of nosocomial comfort and security among building engineers, medical staff, infection control specialists and epidemiological experts.

4 Conclusions

Maintaining a low infection risk and a healthy environment in a naturally ventilated patient room should require high airflow and stringent design parameters that avoid deviations in the direction of airflow between zones with different types of patients. In addition, overheating is a significant problem due to climate change. Given these complexities and challenges, it is crucial to approach ventilation design in healthcare settings with a multidisciplinary and evidence-based perspective. A multiple room, for four people, in the surgery department of the "A. Cardarelli" Hospital in Campobasso (Italy) was considered as a real case study. By replacing natural with mechanical ventilation it was found:

- the thermal field becomes more uniform with a maximum temperature near the patient of about 26 °C;
- with the central diffuser the air residence time is reduced by 70% and by 80% with the vent, compared to natural ventilation;
- for both vent and diffuser, the area occupied by the patient is not affected by the air jet;
- in the case of the vent placed on the wall the temperature reached inside the room is about 25 °C near the patient and the air speed is about 0.25 m/s;

Therefore, for the case study, the best solution seems to be the adoption of a wall air intake, but more generally, the design of the best solution should take into account internal environmental factors, uncertainty in pathogen behaviour, human preference and compliance with protocols and insufficient or obsolete data.

References

1. Yuan, F., et al.: Thermal comfort in hospital buildings – a literature review. J. Build. Eng. **45**, 103463 (2022)

2. Khalid, W., Zaki, S.A., Rijal, H.B., Yakub, F.: Investigation of comfort temperature and thermal adaptation for patients and visitors in Malaysian hospitals. Energy Build. **183**, 484–499 (2019)
3. Lu, Y., Oladokun, M., Lin, Z.: Reducing the exposure risk in hospital ward by applying stratum ventilation systems. Build. Environ. **183**, 107204 (2020)
4. Sheng, S., Yamanaka, T., Kobayashi, T., Choi, N., Chou, N.: Experimental study and CFD modelling of four-bed hospital ward with all-air wall induction unit for air-conditioning. Build. Environ. **222**, 109388 (2022)
5. King, M.F., Noakes, C.J., Sleigh, P.A.: Modelling environmental contamination in hospital single-and four-bed rooms. Indoor Air **25**(6), 694–707 (2015). https://doi.org/10.1111/ina.12186
6. Pereira, P.F.D.C., Broday, E.E., Xavier, A.A.D.P.: Thermal comfort applied in hospital environments: a literature review. Appl. Sci. **10**(20), 7030 (2020). https://doi.org/10.3390/app10207030
7. Rahman, N.M.A., Haw, L.C., Fazlizan, A., Hussin, A., Imran, M.S.: Thermal comfort assessment of naturally ventilated public hospital wards in the tropics. Build. Environ. **207**, 108480 (2022)
8. Tsang, T., Mui, K., Wong, L.: Computational fluid dynamics (CFD) studies on airborne transmission in hospitals: a review on the research approaches and the challenges. J. Build. Eng. Part A **63**, 105533 (2023)
9. Dao, H.T., Kim, K.: Behavior of cough droplets emitted from COVID-19 patient in hospital isolation room with different ventilation configurations. Build. Environ. **209**, 108649 (2022)
10. Wang, J., Cao, X., Chen, Y.: An air distribution optimization of hospital wards for minimizing cross-infection. J. Clean. Prod. **279**, 123431 (2021)
11. Cho, J.: Investigation on the contaminant distribution with improved ventilation system in hospital isolation rooms: effect of supply and exhaust air diffuser configurations. Appl. Therm. Eng. **148**, 208–218 (2019)
12. Islam, M.T., Chen, Y., Seong, D., Verhougstraete, M., Son, Y.: Effects of recirculation and air change per hour on COVID-19 transmission in indoor settings: a CFD study with varying HVAC parameters. Heliyon **10**, e35092 (2024)
13. Design Builder v.4.7 (2016). http://www.designbuilder.co.uk/
14. Gargaro, D., Rainieri, C., Fabbrocino, G.: Structural and seismic monitoring of the "Cardarelli" Hospital in Campobasso. In: X International Conference on Structural Dynamica, EURODYN 2017, pp. 936–941 (2017)
15. AirCar. Technical Datasheet. DGVT.pdf (aircar.it)
16. TecnoVentil. Technical Datasheet. da.pdf (tecnoventil.it)
17. U.S. Department of Energy, Energy Plus simulation software, Version 8.1.0
18. U.S Department of Energy Federal Energy Management Program November 2015, M&V Guidelines: Measurement and Verification for Performance-Based Contracts Version 4.0

Potential Heat Island Phenomenon in Florence: Microclimatic Assessment Through Digital Urban Modelling

Neri Banti, Cecilia Ciacci, Frida Bazzocchi, and Vincenzo Di Naso

Abstract

The Urban Heat Island (UHI) phenomenon presents significant challenges for urban environments, with thermal stress leading to discomfort and potential health issues for residents. Italian urban centres are particularly exposed to outdoor overheating conditions and summer heat waves, also due to the ongoing increase in external temperatures. The available mappings of UHIs are based on zoning from satellite data with broad meshes that do not allow detailed analysis at the local scale. Building upon these, the research aims to evaluate microclimatic conditions within an urban district in Florence, to precisely identify areas susceptible to discomfort. Using an urban district modelling workflow based on Grasshopper plugins for environmental analysis (Ladybug, Honeybee, Dragonfly), the study accurately replicates the volumetric and physical characteristics of buildings within the area of interest and evaluates their impact on the local microclimate through dedicated simulations. Climate data available at the urban scale are processed in relation to the specific characteristics of the site under examination to obtain data relevant to local conditions. The research delves into two distinct sections of the district under study, including a road intersection area and a continuous road segment. The results highlight the presence of areas where individuals may be exposed to thermal stress conditions ranging from moderate to severe, particularly significant in unshaded portions where the mitigating action of winds is insufficient. By pinpointing in detail, the areas most susceptible to overheating phenomena, the results allow for the precise identification of interventions aimed at alleviating discomfort and improving urban liveability. In this perspective, the study represents the initial phase of a broader research endeavour, laying the groundwork for the development of mitigation strategies tailored to the specific needs of the urban environment considered.

Keywords

Urban heat island · urban district modelling · discomfort · thermal stress · climate resilience

1 Introduction

The ongoing climate change is leading to a significant and widespread rise in external air temperatures. At the European level, EUROSTAT data indicate that during the last year, the number of cooling degree days (CDD) [1] increased by about 25% compared to data from the 1980s, with projections estimating a potential rise of around 100% by 2050 [2]. The Mediterranean area has proven to be particularly exposed to climate alterations and their negative consequences. According to the ISTAT report [3], in 2020, the average outdoor air temperature in Italy rose by approximately 1.2 °C in regional capitals, with a forecasted increase of 2 °C by 2050 [4].

Adapting to this changing climatic context is particularly challenging in urban areas, due to their low resilience and adaptability that make them vulnerable in case of extreme meteorological events [5]. Urban-built environments are also subject to the Urban Heat Island (UHI) phenomenon,

N. Banti (✉) · C. Ciacci · F. Bazzocchi · V. Di Naso
Department of Civil and Environmental Engineering, University of Florence, Santa Marta Street, 50139 Florence, Italy
e-mail: neri.banti@unifi.it

C. Ciacci
e-mail: cecilia.ciacci@unifi.it

F. Bazzocchi
e-mail: frida.bazzocchi@unifi.it

V. Di Naso
e-mail: vincenzo.dinaso@unifi.it

B. He et al. (eds.), *Towards the Framework of Livable and Resilient Cities* Advances in Science, Technology & Innovation, https://doi.org/10.1007/978-3-031-97849-4_24

which leads to higher outdoor temperatures compared to surrounding rural areas. These conditions are exacerbated by increasingly frequent and prolonged heat waves during the summer months. Such heat stress conditions pose heightened mortality risks to the global population and negatively impact the quality of life and well-being in urban spaces [6].

Despite these issues, more than half of the global population currently lives in urban agglomerations, a proportion projected to increase up to 70% by 2050. This highlights the urgent need for sustainable urban development, as emphasized by Goal 11 of the Sustainable Development Goals (SDGs) promoted by the United Nations, which aims to make cities resilient and environmentally friendly.

Given this context, the growing interest in this topic led to an important proliferation of scientific research about the theme. A consistent number of scientific publications can be retrieved in the literature dealing with urban environments and UHI at different levels and with different purposes [7].

Studies have focused on identifying potential hotspot areas at district, urban, or territorial levels and establishing simulation and assessment frameworks. Analyses of the climate conditions have been performed in the past in the urban area of Florence to assess their variability and derive valuable insights by installing an extensive sensor network [8]. More recently, studies have utilized Land Surface Temperature (LST) data from satellite observations to identify hot and cool spots in Italian cities. Morabito et al. conducted an extensive research on 10 different large cities in Italy [9] and, with a similar methodology, the Tuscany area comprised of the provinces of Florence, Prato and Pistoia was analysed in detail to provide a summer hotspot classification [10]. Geographical Information System (GIS) have been successfully adopted in this kind of analysis: Ellena et al. investigated micro-scale UHI risks within the urban context of the city of Turin [11] while other authors, coupling the potential of a GIS-based methodology with the evidence of real climate data collected by sparse weather stations, highlighted the correlation between urban fabric features and UHI intensity in the metropolitan area of Rome [12].

Along with environmental and microclimate simulation software, alternative evaluation tools recently emerged and became progressively widely adopted for these analyses, especially when considering the evaluation of urban planning strategies and amelioration interventions to contrast UHI. Grasshopper, a visual programming platform integrated into the 3D modelling software Rhinoceros, offers different options for research thanks to its flexibility, the wide range of plugins, and the possibility to integrate different simulation engines [13].

This methodological approach was followed by Bajšanski et al. when evaluating the possible improvements in outdoor comfort following strategic urban planning decisions in Novi Sad (Serbia) [14]. Focusing on the Mediterranean climate, Evola et al. evaluated the climate peculiarities of the built environment considering an urban canyon located in Catania (southern Italy) [15].

Going beyond the assessment of current conditions and criticalities in urban environments, several research were performed considering different intervention strategies and solutions to contrast UHI. Notably, the adoption of greening and vegetation emerged as effective ameliorating solutions widely tested and discussed in the literature considering both green facades and green roofs and trees planting [16] as well as the adoption of cool materials in urban spaces [17], with urban planning becoming a topic of increasing interest (Reitberger et al. 2024). Similar research were conducted also considering the Florence context, trying to evaluate the influence that different design options for public squares can have on UHI reduction [19, 20].

The study presented in the paper focuses on evaluating microclimate conditions in a designated area of Florence focusing on human comfort metrics such as the Universal Thermal Climate Index (UTCI) to evaluate potential discomfort for the population. The research produced a 3D model of the area of interest using Rhinoceros and Ladybug tools, overcoming the limitations of site-specific weather data. The primary goal is to evaluate the microclimate conditions of the case study area, particularly focusing on urban canyons vulnerable to summer heat waves. By pinpointing possible site-specific criticalities, the study aims to identify locations suitable for implementing mitigation strategies to enhance human comfort in the outdoor environment. Recent and updated studies focusing on the microclimate characterization of the metropolitan area of Florence are available [21] as well as site-specific building-specific assessments [22], while contributions about the hot-spot definition and retrofit solutions are primarily related to industrial sites [23]. Environmental simulations for larger parts of residential districts are still missing. The ones retrievable in the literature refer to urban squares [19, 20], with microclimate conditions that can substantially differ in terms of local conditions and applicable redevelopment solutions from narrower streets. In addition, the detection of potential urban heat islands is conducted in the literature recurring to Sat-EO data, providing insights primarily related to the external temperatures of built surfaces, and not guaranteeing adequate data granularity for urban planning and the implementation of mitigation strategies. The 3D environmental simulations foreseen in the study were meant to cover these knowledge gaps and provide indications replicable also in other districts with similar characteristics and climate conditions in the same city or other regions.

2 Case Study Area

The urban district analysed in the research is adjacency to the historical centre of Florence, as highlighted in Fig. 1. This portion of the city developed starting approximately in

Fig. 1. Geographical overview of the analysis area and an aerial view of the portion under consideration. Source: By author.

1860, following the abatement of the ancient city walls and the expansion in areas previously scarcely inhabited and mainly devoted to agricultural activities. According to the chronological classification of buildings provided by Tuscany Region according to the sources available, the complete urbanization of the area was reached by 1956. For this reason, most of the buildings included in the district show features of traditional local architecture typical of the end of the XIX century and the beginning of XX. Most buildings are 2 or 3-storey facilities, realized with masonry, simply plastered, and painted for external finishing. Wooden roofs are double-sloped and covered with traditional tiles. Only a few reinforced concrete buildings can be retraced. These features are sufficiently representative of similar conditions found in other parts of Florence, suggesting that the findings could be applicable to comparable districts.

The post-unitary expansion of the city centre outside the ancient city walls led to the adoption of solutions analogue to the ones retrievable in the case study area selected also in other zones.

Via Gioberti is the main artery of the neighbourhood and, like other streets in the district, accommodates a single asphalted lane for one-way vehicular traffic with parking stalls and sidewalks on both sides. For the scope of the analysis conducted in the research, a portion of about 0.06 km^2 was considered, corresponding to the central section of Via Gioberti, for a total length of approximately 360 m.

This district was selected based on the evidence from local thermal maps [8] and the results of previous research conducted by Guerri et al. that assessed the presence of a potential summer hotspot in this portion of the city [10]. In addition to the hazardous conditions mentioned above, the case study area was selected given the high number of individuals potentially exposed to UHI-related issues, such as heat strokes and thermal discomfort. The district has a significant population density of about 15800 ab/km^2, for a total of approximately 1700 people living in the area analysed, according to data provided by the online regional databases [24] and their elaborations in the QGIS environment [25]. Alongside the resident population, the presence of commercial activities also attracts a considerable number of occasional visitors. In this regard, implementing UHI mitigation strategies could not only reduce thermal discomfort but also improve the quality of public spaces promoting socialization, safety, and sustainability.

3 Methodology

The method followed in the research is synthesized in the diagram illustrated in Fig. 2. The workflow is based on the integration and combination of Grasshopper plugins for energy and environmental analysis: Dragonfly, Urban Weather Generator (UWG), Butterfly, and Ladybug. The different components of each plugin and the purpose of their application will be detailed in the following paragraphs. The approach adopted was meant to overcome the current imitation of on-field data to produce detailed environmental simulations of the district of interest. The modelling procedure of the area of interest allowed to faithfully reproduce the geometry of the urban layout and buildings, to investigate local effects and site-specific conditions. In addition, the same Visual Programming Language script developed in this case can be applied for new simulation tasks, by simply applying it to a new geometrical model and changing the values of the variables for climate and surface characteristics.

Globally, the research workflow can be divided into two different main phases.

The first one revolves around the global setting of the research, including urban district modelling and climate data acquisition. At first, data about both the urban layout of the addressed area and the climate conditions were acquired. For the former, provided by Tuscany Region's online repositories [24] were exploited and.dxf format files

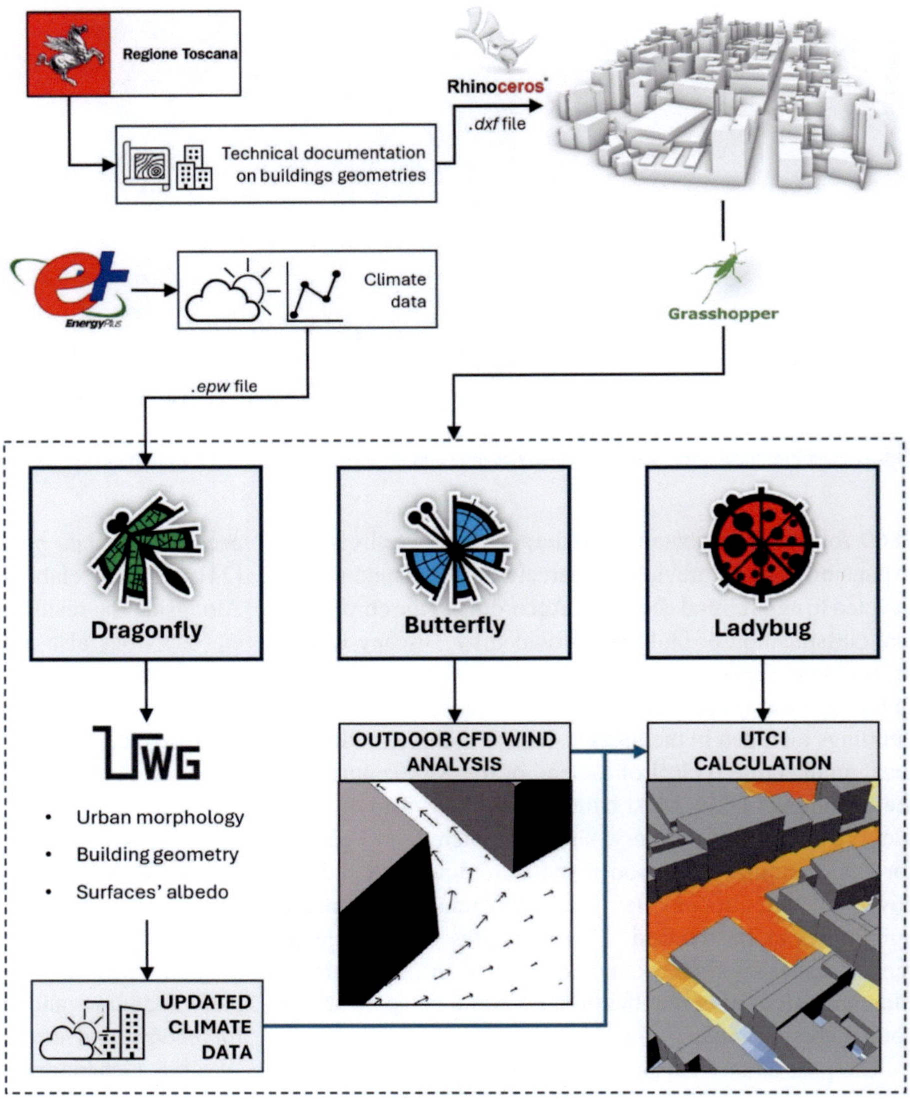

Fig. 2. Research methodology workflow. Source: By author.

containing buildings' geometry and district layout were downloaded. The raw data obtained were then processed using Rhinoceros, a graphical 3D modelling software chosen because of its flexibility and interoperability with different plugins. Starting from the geometries provided by the downloaded files, the 3D model of the urban area of interest was subsequently refined and adjustments were made to remove overlapping or duplicate surfaces and prevent possible issues during the following simulation stages. In addition, all non-essential graphic and informational elements have been removed by deleting the layers in which they were contained, reducing the file size, and preventing potential unwanted interferences for modelling and subsequent simulations. In the model, also buildings surrounding the area of interest were included to account for their solar and wind-shielding effect in environmental simulation.

As for climate data, the.epw file referring to the weather station installed in Firenze Peretola was obtained from the freely accessible online EnergyPlus database [26]. A comparison with real data obtained following an on-site experimental campaign was not possible given the initial stage of the research. These current limitations will be addressed in the following stages of the research. The UWG tool, applied in the research to overcome this limitation, has been validated by different authors in the literature considering both real measured weather data [15] and ENVI-met simulations [27]. To constantly enhance its reliability, researchers proposed also possible improvements to the UWG tool, refining its calculations related to radiation, the presence of vegetation, and convective heat transfer [28].

The second part of the methodology is dedicated to simulation tasks, through an algorithm defined in the Grasshopper environment. The visual appearance of the original algorithm

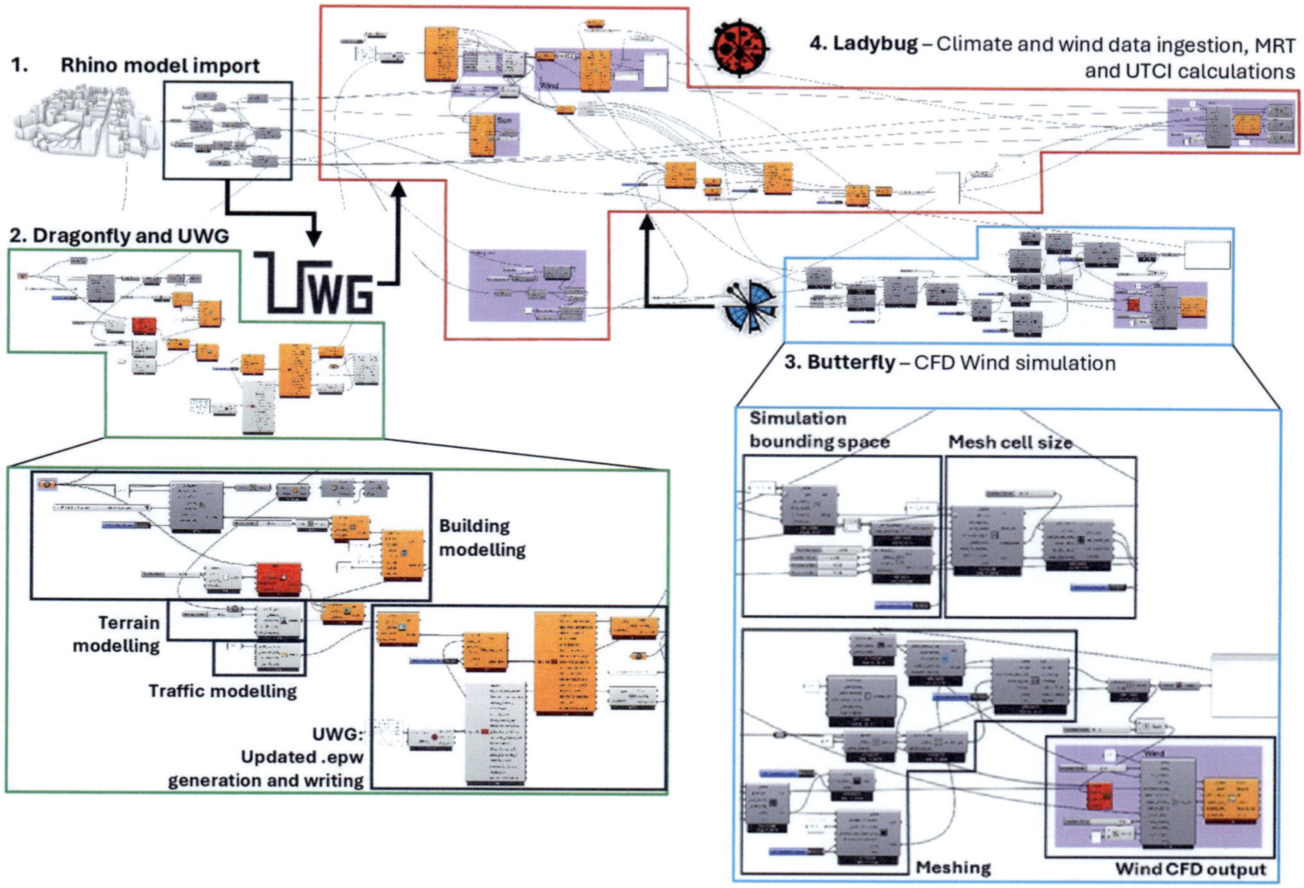

Fig. 3. Grasshopper script for simulation purposes. Source: By author.

drafted by the authors is presented in Fig. 4, where the components that refer to the different authoring plugins are mapped. The Grasshopper script set out by the authors is made up of 4 different main sections that will be detailed referring to the numbering adopted in Fig. 3.

1. Rhino model import

This part of the algorithm was set to simply get the geometries of the buildings created within the Rhinoceros modelling environment to later use them for the simulations to be performed. Given the practically plain morphology of the site, all the buildings were assigned the same ground level eliminating the negligible height differences present in the original.dxf model.

2. Dragonfly and UWG

UWG component in the Dragonfly plugin was used in this section to update the original Firenze Peretola.epw file and produce a new one accounting for site-specific conditions induced by building geometries and surfaces albedo. At first, an average for floor-to-floor height equal to 3 m was imposed to vertically fraction the building geometries and a window-to-wall ratio equal to 30% was set considering the characteristics of buildings observed through on-site surveys. Similarly, roof and wall albedo were set in this phase. For the former, a traditional roof covering made up of terracotta roof tiles and ridge tiles was generally registered in the buildings encompassed in the area of interest, as confirmed by the aerial views available. Following the evidence of research available in the literature on traditional roofing systems in central Italy [29], an albedo equal to 0.4 was assumed. As for the walls facing the streets, most of the buildings feature traditional facades plastered and painted with light or neutral colours. For this reason, a reflectance index of 0.6 has been applied for the terrain element an average albedo equal to 0.3 was calculated considering that the road surfaces are covered with bitumen asphalt and both sidewalks and private courtyards, except for rare exceptions, are paved with stones. The UWG tool also offers the opportunity to set vehicular traffic conditions and a medium traffic setup compatible with urban areas was imposed. New.epw file modified to account for the site-specific characteristics was finally produced.

3. Butterfly – CFD wind simulations

Outdoor airflow conditions were simulated recurring to the Butterfly plugin, an object-oriented Python library for computational fluid dynamics (CFD) simulations utilizing OpenFOAM as a calculation engine. A reference cell size of 1 m was assigned to adequately mesh the model of the district. Input wind conditions were derived from the Firenze Peretola.epw file in terms of prevailing direction and average velocity. The CFD outputs let to assess the actual wind intensity and direction in the district considering the inferences of existing buildings and the urban layout of the main roads in the district.

4. Ladybug – Climate and wind data ingestion, MRT and UTCI calculations

Various components of the Ladybug plugin were used to finally produce comfort analyses, importing the UWG modified.epw file and the outcomes of the Butterfly CFD for wind settings. A grid analysis 3 m spaced and located at a height of 1.5 m over the ground level was set to reproduce human bodies' positions within the area of interest, excluding all the points included within the building's geometry. The input parameters necessary for The UTCI calculation were retraceable from the cited.epw source or derived using intermediate components: air temperature, relative humidity, wind velocity, and mean radiant temperature (MRT). The latter was calculated through a dedicated component by setting data sourced from the UWG modified climate file. The albedo of asphalt roads was set equal to 0.15 based on the evidence available in the literature to account for the ageing of the surfaces [30]. UTCI index was hence calculated for each point of the analysis grid and plotted in the model to produce effective visual outputs for the overall research. All the analyses were conducted considering the 21st of June as the reference day, and simulations were performed considering in particular the central hours of the day for a total of 4 h between 10 a.m. and 2 p.m.

4 Results and Discussion

The simulations carried out through the UWG component let to assess how the characteristics of the built environment considered can influence microclimate conditions. Figure 4 provides a comparison between the climate conditions registered at the Firenze Peretola weather station and the simulated ones. The district considered is characterized by warmer conditions with respect to the airport site, where the weather station is installed. These findings are aligned with the literature on the theme [31].

Considering the specific conditions of the analysed district, the increase in the outdoor temperature is more significant during summer months and can be quantified around 3 °C. Notably, the trends of the two different datasets are comparable from a qualitative point of view. The monthly per-hour data graph collects the average value registered at each hour in every month. As appreciable from the graph, the UWG simulated temperatures are constantly higher especially at night, when a maximum shift of about 6 °C can be observed during July. However, a possible overestimation of external temperature during nighttime was registered by other authors in the literature [32]. The presence of paved and built surfaces, the shielding effect of buildings on cooling winds, and the lack of green areas are the main influential factors causing this temperature rise, as recognized also by other authors for the Florentine context [33].

The CFD analysis carried out allowed to evaluate wind distribution and air flows within the district. An average wind speed was set based on the real climate data available (2.63 m/s) and, similarly, the prevailing wind direction was imposed. Winds are blowing from N-W. The velocity and direction obtained following the wind analysis let to assess a general predominance of winds aligned with the main W-E streets and some recirculation effects in correspondence of the crossroads.

Wind analysis outputs were included in the UTCI index calculation that was conducted to highlight possible issues and heat-related criticalities in the urban district analysed. Figure 5 provides a graphical representation of the outcomes. North-South-oriented streets are more subject to overheating conditions as they cannot benefit from the shadowing effect produced by the surrounding buildings. Average UTCI values in these roads range between 29.7 °C and 31.4 °C resulting in moderate to strong heat stress conditions for pedestrians, according to the evaluation scale proposed by the calculation method [34]. On the other hand, as briefly anticipated above, Via Gioberti Street canyons portions are characterized by lower UTCI because the surrounding buildings, arranged along an East-West direction, prevent both direct solar radiation and mitigate the impact of warm winds coming from the South-West. As recognized in the literature, orientation and width-to-height ratio in urban street canyons are crucial to determining their behaviour towards the UHI effect [35] as well as the shading effect of the buildings flanking the street [36].

According to the results obtained, crossroad sections are the most vulnerable areas within the districts and were hence analysed in detail. In these portions, the wind's mitigating action is not sufficient to counteract the direct incident solar radiation and a general increase in the UTCI value is registered. As represented in Fig. 6, the first crossroad considered

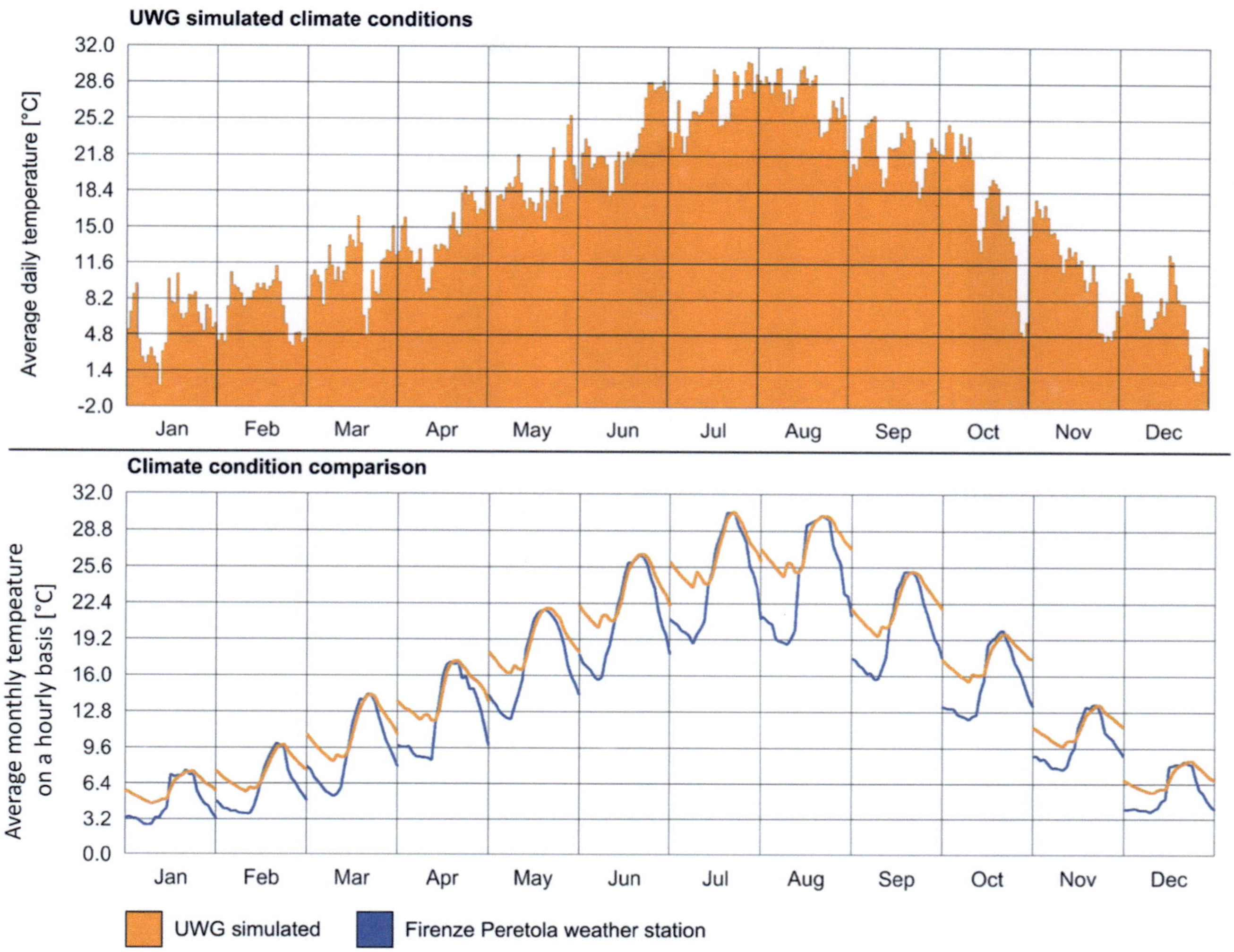

Fig. 4. Comparison between UWG simulated and real weather data collected at the Firenze Peretola weather station. Source: By author.

(a in the Figure) shows a configuration commonly retraceable in the city both in terms of urban layout and buildings' characteristics. Given the limited space available in similar intersections, acting on the facades of the buildings prospecting them could represent a possible and viable mitigation strategy to be implemented. The installation of vertical greenery could enhance their thermal performances, improve the aesthetic quality [37] and mitigate outdoor temperature. In addition, as crossroads sections may be more subjected to higher concentrations of pollutants concentration, the dust and particle removal potential of vegetation could provide another possible positive contribution to enhancing life quality in urban areas.

On the other hand, the second crossroad considered (b in the Figure) is characterized by the presence of a more recent facility hosting a retail activity. The latter, built using cast-in-place reinforced concrete technology, is a single-story structure lower than the surrounding ones. The plain roof is used as a parking space, accessible to a side ramp and completely covered with asphalt. Similarly, at the ground level, both sidewalks and traffic roads are totally paved, without any vegetation or flowerbeds installed. Wong et al. remarked that the lack of urban green infrastructures (UGI) and related evapotranspiration processes in built environments contributes to exacerbating the UHI phenomenon in dense settlements [38]. Concurrently, the low albedo of the built surfaces in this specific area represents a detrimental factor for thermal heat stress [39]. For this portion, a set of different mitigation strategies could be implemented. The wide sidewalks could be modified and rearranged to host trees capable of lowering pollution levels and casting shadows during the hottest days. Similarly, the parking area on the roof of the commercial facility could be dismantled to replace the actual covering layer with a reflecting coating or with the installation of a green roof solution. This second option could also allow to obtain a green area within a densely built district that is currently suffering from the lack of a public relational and leisure area, resulting in a beneficial social impact as underlined by other authors in the literature [40].

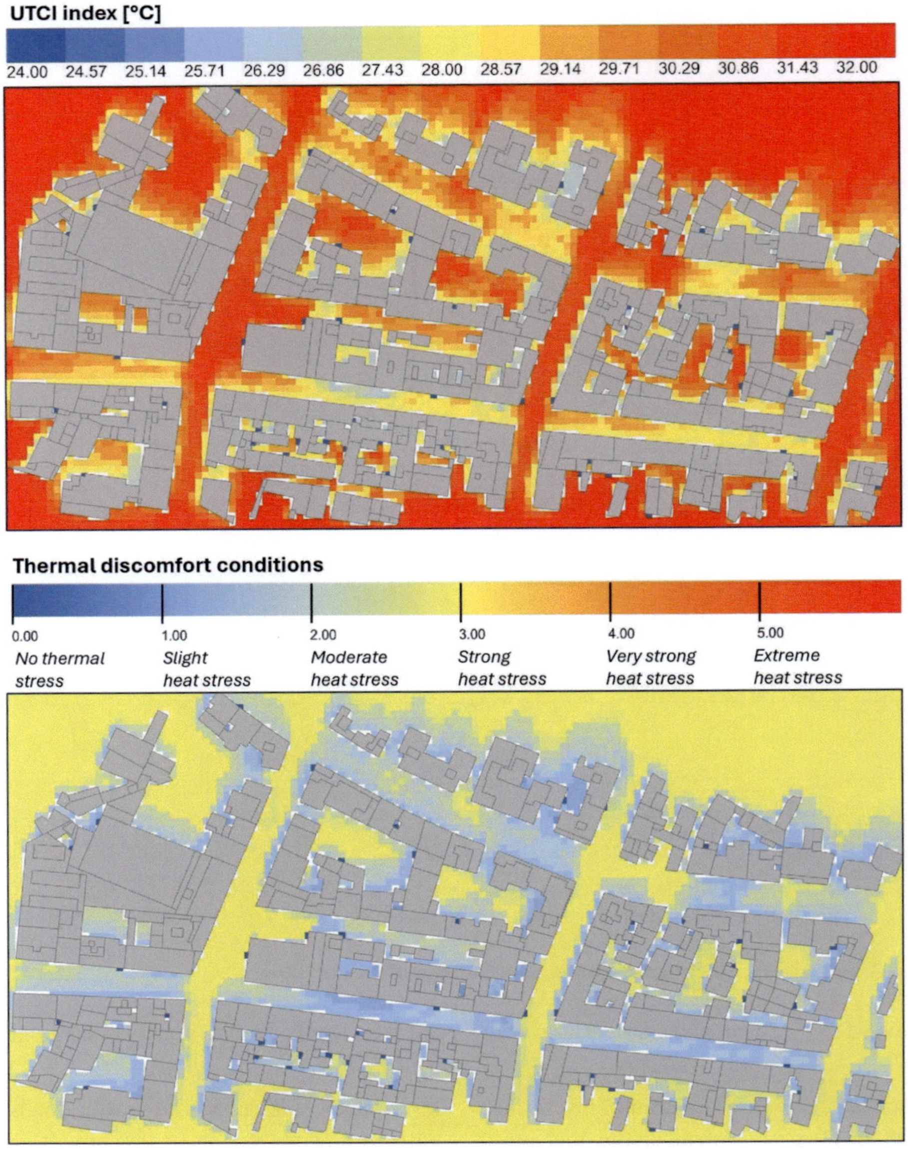

Fig. 5. UTCI index and thermal discomfort assessment in the district under consideration – 21st of June – 1 pm. Source: By author.

In addition, the implementation of cool pavement finishing can be evaluated, at least in crossroads, to lower the surface temperatures in road sections and mitigate UHI conditions, as well as the installation of shade fabrics across the streets between buildings to provide beneficial sheltering of solar radiation.

5 Conclusions

The study presented enabled the definition of the microclimate characteristics of a specific urban area located in Florence that is potentially exposed to heat-stress conditions for occupants in case of summer heat waves. The UTCI index was the benchmark parameter used to assess the comfort of pedestrians in public open-air spaces. Environmental simulations were performed using Ladybug plugins and the Urban Weather Generator tool in the Grasshopper environment to evaluate the microclimate features of the designated urban space.

Grasshopper and Ladybug proved to be valuable tools easing 3D modelling procedures compared to conventional software and enabling deeper analysis thanks to the flexibility of the algorithms and the integration of multidisciplinary plugins. Through the UWG tool, the original weather data available for the city of Florence were modified to reflect the ones retrievable in the urban district considered. The simulated external air temperature mainly rises during summer months by about 3 °C, with a peak shift during night hours that reaches a maximum of about 6 °C during the summer months.

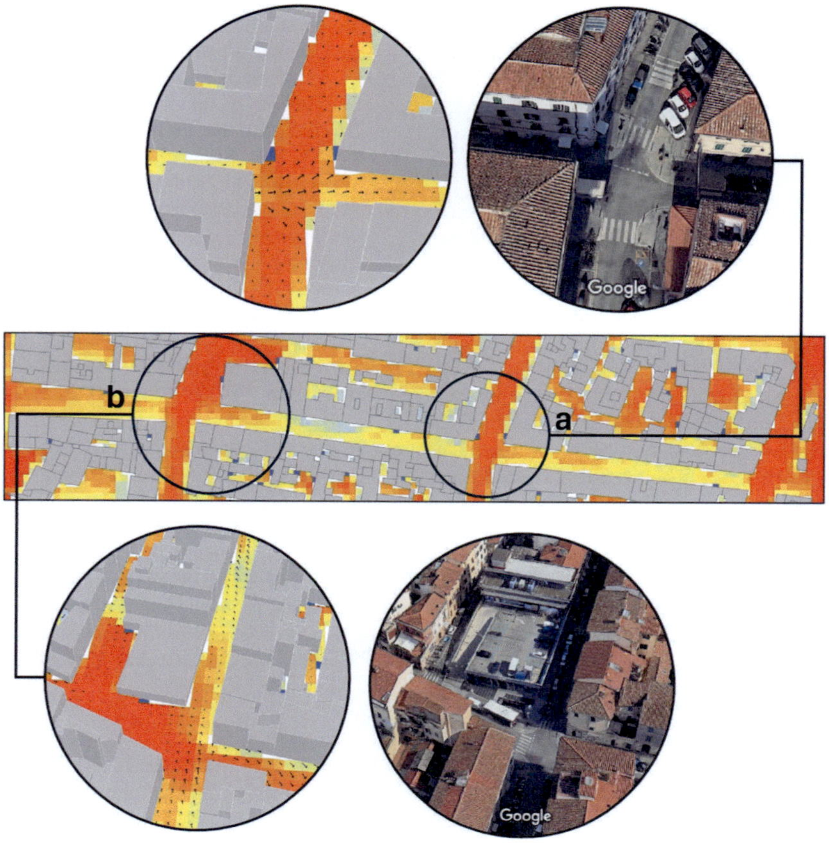

Fig. 6. Detailed analysis of crossroad sections. Source: By author.

Regarding the impact of UHI on pedestrian comfort, UTCI values ranging from 29.7 °C (moderate heat stress) to 31.4 °C (strong heat stress) were registered in N-S oriented roads. W-E-oriented street canyons are characterized by generally light heat stress levels, mainly due to the shadowing and shielding effects of the flanking buildings. Crossroads sections emerged as particularly vulnerable and exposed to UHI risks. A detailed examination of the two main intersections led to the identification of the most significant criticalities and suggested intervention measures. The low albedo of finishing materials, the lack of green areas, and the geometry of the urban layout in this densely built urban area emerge as critical factors to address. Furthermore, the possible installation of vertical and/or horizontal greenery on existing buildings as mitigation measures should be carefully addressed from constructive, technological, and structural perspectives.

The research is conceived as a preliminary stage toward the definition of possible retrofit solutions tailored to the urban context in the city of Florence to tackle UHI impact. In this initial phase, some approximations were introduced, such as assuming average reflectivity levels for the built surfaces, using a wide mesh to encompass the area of interest, and considering a limited time interval as well as limitations such as the lack of real on-site monitoring data. However, the study led to assessing the microclimate conditions in the district with a precision not achievable through currently available satellite data elaborations and to define some possible intervention areas for UHI mitigation and the related solutions. Future developments of the research will involve a general refinement of the results presented and the evaluation of redevelopment measures to mitigate the value of the UTCI parameter and enhance outdoor human comfort through dedicated modelling and simulation tasks.

Conflict of Interest The authors have no conflicts of interest to declare that are relevant to the content of this chapter.

References

1. Eurostat. Heating and cooling degree days - statistics. https://ec.europa.eu/eurostat/statistics-explained/index.php?title=Heating_and_cooling_degree_days_-_statistics
2. Faggian, P., Riva, G.M., Bonanno, R.: Stima dei gradi-giorno per la valutazione del fabbisogno energetico nazionale nei prossimi decenni. Ing dell'Ambiente, **5**, 273–287 (2018)
3. Ferruzza. A., Vignani, D.: I Cambiamenti Climatici: misure statistiche (2020)

4. Tagliacozzo, G., Baldoni, T., Del Bufalo, E., Zannoni, S.: Rapporto SDGs 2023. Informazioni statistiche per l'Agenda 2030 in Italia (2022)
5. Datola, G.: Implementing urban resilience in urban planning: a comprehensive framework for urban resilience evaluation. Sustain. Cities Soc. **98**, 104821 (2023). https://doi.org/10.1016/j.scs.2023.104821
6. European Environment Agency: Climate change as a threat to health and well-being in Europe: focus on heat and infectious diseases (2022)
7. Cheval, S., Amihăesei, V.A., Chitu, Z., et al.: A systematic review of urban heat island and heat waves research (1991–2022). Clim. Risk Manag. **44**, 100603 (2024). https://doi.org/10.1016/j.crm.2024.100603
8. Petralli, M., Massetti, L., Brandani, G., et al.: Le mappe termiche della città di firenze: la ricerca alla base delle strategie di adattamento e mitigazione degli estremi termici in città – thermal maps of florence: research supports urban planners and policymakers to take action towards urban tempera. In: il clima cambia le città Conferenza sull'adattamento climatico in ambito urbano. Venice, pp. 23–24 (2013)
9. Morabito, M., Crisci, A., Guerri, G., et al.: Surface urban heat islands in Italian metropolitan cities: tree cover and impervious surface influences. Sci. Total. Environ. **751**, 142334 (2021). https://doi.org/10.1016/j.scitotenv.2020.142334
10. Guerri, G., Crisci, A., Messeri, A., et al.: Thermal summer diurnal hot-spot analysis: the role of local urban features layers. Remote Sens. **13**, 1–29 (2021). https://doi.org/10.3390/rs13030538
11. Ellena, M., Melis, G., Zengarini, N., et al.: Micro-scale UHI risk assessment on the heat-health nexus within cities by looking at socio-economic factors and built environment characteristics: the Turin case study (Italy). Urban Clim **49**, 101514 (2023). https://doi.org/10.1016/j.uclim.2023.101514
12. Battista, G., Evangelisti, L., Guattari, C., et al.: Space-time estimation of the urban heat island in Rome (Italy): overall assessment and effects on the energy performance of buildings. Build. Environ. (2023). https://doi.org/10.1016/j.buildenv.2022.109878
13. Perini, K., Ottelé, M., Giulini, S., et al.: Quantification of fine dust deposition on different plant species in a vertical greening system. Ecol. Eng. **100**, 268–276 (2017). https://doi.org/10.1016/j.ecoleng.2016.12.032
14. Bajšanski, I.V., Milošević, D.D., Savić, S.M.: Evaluation and improvement of outdoor thermal comfort in urban areas on extreme temperature days: Applications of automatic algorithms. Build. Environ. **94**, 632–643 (2015). https://doi.org/10.1016/j.buildenv.2015.10.019
15. Evola, G., Costanzo, V., Magrì, C., et al.: A novel comprehensive workflow for modelling outdoor thermal comfort and energy demand in urban canyons: results and critical issues. Energy Build **216**, 109946 (2020). https://doi.org/10.1016/j.enbuild.2020.109946
16. Kim, J., Khouakhi, A., Corstanje, R., Johnston, A.S.A.: Greater local cooling effects of trees across globally distributed urban green spaces. Sci. Total. Environ. **911**, 168494 (2024). https://doi.org/10.1016/j.scitotenv.2023.168494
17. Morales-Inzunza, S., Gonzalez-Trevizo, M.E., Martínez-Torres, K.E., et al.: On the potential of cool materials in the urban heat island context: scalability challenges and technological setbacks towards building decarbonization. Energy Build., 113330 (2023). https://doi.org/10.1016/j.enbuild.2023.113330
18. Reitberger, R., Palm, N., Palm, H., Lang, W.: Urban systems exploration: a generic process for multi-objective urban planning to support decision making in early design phases. Build. Environ. **254**, 111360 (2024). https://doi.org/10.1016/j.buildenv.2024.111360
19. Romano, R., Gallo, P., Donato, A.: Evaluation of mitigation strategies of the urban heat island effect in Mediterranean area. The case study of largo Annigoni in Florence (Italy). In: Littlewood, J., Howlett, R.J., Capozzoli, A.J.L. (eds.) Urban and Transit Planning. Towards liveable communities: urban places and design spaces, pp. 531–541. Springer Nature Singapore, Singapore (2020). https://doi.org/10.1007/978-3-030-97046-8_6
20. Romano, R., Bologna, R., Hasanaj, G., Arnetoli, M.V.: Adaptive design to mitigate the effects of UHI: the case study of Piazza Togliatti in the municipality of scandicci. Smart Innov Syst Technol **163**, 531–541 (2020). https://doi.org/10.1007/978-981-32-9868-2_45
21. Cureau, R.J., Balocco, C., Pigliautile, I., et al.: On urban microclimate spatial-temporal dynamics: evidence from the integration of fixed and wearable sensing and mapping techniques. Environ. Res. **262**, 119795 (2024). https://doi.org/10.1016/j.envres.2024.119795
22. Balocco, C., Pierucci, G., De Lucia, M.: An experimental method for building energy need evaluation at real operative conditions. A case study validation. Energy Build. **266**, 112114 (2022). https://doi.org/10.1016/j.enbuild.2022.112114
23. Guerri, G., Crisci, A., Morabito, M.: Urban microclimate simulations based on GIS data to mitigate thermal hot-spots: tree design scenarios in an industrial area of Florence. Build. Environ. **245**, 110854 (2023). https://doi.org/10.1016/j.buildenv.2023.110854
24. Geoscopio - Cartoteca.https://www.regione.toscana.it/-/geoscopio. Accessed 26 Oct 2022
25. QGIS
26. Energy Plus
27. Detommaso, M., Costanzo, V., Nocera, F.: Application of weather data morphing for calibration of urban ENVI-met microclimate models. Results and critical issues. Urban Clim. **38**, 100895 (2021). https://doi.org/10.1016/j.uclim.2021.100895
28. Xu, G., Li, J., Shi, Y., et al.: Improvements, extensions, and validation of the urban weather generator (UWG) for performance-oriented neighborhood planning. Urban Clim. **45**, 101247 (2022). https://doi.org/10.1016/j.uclim.2022.101247
29. Di Giuseppe, E., D'Orazio, M., Di Perna, C., et al.: Riflettanza e solar reflectance index (SRI) di manti per coperture in laterizio. In: Laterservice (ed.) (2017)
30. Sen, S., Roesler, J.: Aging albedo model for asphalt pavement surfaces. J. Clean. Prod. **117**, 169–175 (2016). https://doi.org/10.1016/j.jclepro.2016.01.019
31. Arnfield, A.J.: Two decades of urban climate research: a review of turbulence, exchanges of energy and water, and the urban heat island. Int. J. Climatol. **23**, 1–26 (2003). https://doi.org/10.1002/joc.859
32. Boccalatte, A., Fossa, M., Thebault, M., et al.: Mapping the urban heat Island at the territory scale : an unsupervised learning approach for urban planning applied to the Canton of Geneva. **96** (2023) https://doi.org/10.1016/j.scs.2023.104677
33. Massetti, L., Petralli, M., Orlandini, S.: Paper 586: Effect of green areas on summer air temperatures in Florence. Weather Forecast. **1**, 10–13 (2012)
34. International Union of Physiological Sciences-Thermal Commission: Universal Thermal Climate Index (UTCI) assessment scale: UTCI categorized in terms of thermal stress. J. Therm. Biol., 75–106 (2003)
35. Ibrahim, Y., Kershaw, T., Shepherd, P., Elwy, I.: A parametric optimisation study of urban geometry design to assess outdoor thermal comfort. Sustain. Cities Soc. **75**, 103352 (2021). https://doi.org/10.1016/j.scs.2021.103352
36. Deng, J.Y., Wong, N.H.: Impact of urban canyon geometries on outdoor thermal comfort in central business districts. Sustain. Cities Soc. **53**, 101966 (2020). https://doi.org/10.1016/j.scs.2019.101966

37. Magliocco, A., Perini, K.: The perception of green integrated into architecture: installation of a green facade in Genoa, Italy. AIMS Environ. Sci. **2**, 899–909 (2015). https://doi.org/10.3934/environsci.2015.4.899
38. Wong, N.H., Kwang Tan, A.Y., Chen, Y., et al.: Thermal evaluation of vertical greenery systems for building walls. Build. Environ. (2010). https://doi.org/10.1016/j.buildenv.2009.08.005
39. Xi, Y., Wang, S., Zou, Y., et al.: Seasonal surface urban heat island analysis based on local climate zones. Ecol. Indic. **159**, 111669 (2024). https://doi.org/10.1016/j.ecolind.2024.111669
40. Berto, R., Stival, C.A., Rosato, P.: Enhancing the environmental performance of industrial settlements: an economic evaluation of extensive green roof competitiveness. Build. Environ. **127**, 58–68 (2018). https://doi.org/10.1016/j.buildenv.2017.10.032

Evaluation of the Air Quality Inside a Hindu Temple: A Case of Mahakaleshwar Temple Ujjain

Apurva Sharma and Anupama Sharma

Abstract

Temples are a significant part of India's cultural legacy, spanning from the northern to the southern regions, and harboring an extensive assemblage of deities. Many temples, dating back hundreds of years, remain well-preserved and serve as focal points for cultural activities. Ujjain, a historic city located on a plateau in the Malwa region of the Vindhyachal mountains in Madhya Pradesh, is home to the Mahakaleshwar Temple, a prominent religious site. However, the increasing number of worshipers has led to a decline in the temple's indoor environmental quality (IEQ). This study aims to evaluate the indoor air quality (IAQ) and infection risk in the Mahakaleshwar Temple by analyzing ventilation rates and their impact on airborne transmission. Carbon dioxide was used as a tracer gas to calculate the air changes per hour (ACH), and the Wells-Riley model was employed to assess the relationship between ventilation rates and infection probabilities. The results highlight that insufficient natural ventilation and poor air circulation have led to discomfort among devotees, with reported issues such as suffocation, irritability, and nasal congestion. Elevated CO_2 levels, reduced oxygen levels, and increased temperature and humidity have further worsened the indoor environment. The findings underscore the critical need for improved ventilation systems in high-occupancy religious buildings to mitigate health risks and enhance comfort for devotees.

Keywords

Air quality · Ujjain temple · Ventilation · CO_2 · Air changes per hour

A. Sharma (✉) · A. Sharma
Department of Architecture and Planning, MANIT, Bhopal, India
e-mail: apurva_2007_sharma@yahoo.com

List of Abbreviations

IEQ:	Indoor Environment Quality
IAQ:	Indoor Air Quality
VOCs:	Volatile organic compounds
POPs:	Persistent organic pollutants
PMx:	Nanoparticles
NOx:	Nitrogen oxides
SOx:	Sulphur oxides
CO_2:	Carbon dioxide
CO:	Carbon monoxide
MVE:	Mechanical Ventilation System
ACH:	Air Changes per Hour
ppm:	Parts per million
RH:	Relative humidity

1 Introduction

There are some distinguishing characteristics that distinguish holy temples from other kinds of constructions. In most cases, they have a substantial volume and are occupied at particular times in order to carry out operations that are clearly defined. The temple has the capacity to accommodate a large number of people in a relatively short amount of time. In spite of the fact that the weather circumstances may be excellent, this ultimately leads to crowded enclosed areas that cause discomfort due to the temperature. In addition, factors such as inadequate ventilation, the growth of microorganisms in carpets, on walls, or the components of heating, air-conditioning systems, and ventilation are the causes of poor indoor air quality, which may even pose a hazard to human health. The intention of the numerous research that have been carried out on the subject of IEQ in temples was to investigate this particular aspect. The thermal comfort and the quality of the air inside the building

B. He et al. (eds.), *Towards the Framework of Livable and Resilient Cities* Advances in Science, Technology & Innovation, https://doi.org/10.1007/978-3-031-97849-4_25

are two of the most important factors IEQ that are essential for the members of the congregation to feel calm and to achieve a sense of serenity and tranquilly. (Abdullah et al., 2016).

The Mahakal temple Situated on the bank of the holy river Shipra, Ujjain is one of the most important religious cities of India since time memorial. There are many famous temples like Mahakaleshwar Mandir, Harsiddhi Mandir, Gopal Mandir, Mahakali Mandir, Navagarh Mandir etc. Mahakaleshwar Mandir is among one of the largest temples in India, situated 28.71 mt. Above the mean sea level and measuring 10.77 mt. by 10.77 mt. It is one of the twelve Jyotirlingas. Ujjain is one of the four places in India where the famous Kumbh Mela is held. It is known as Simhastha (Benjamin et al., 2014).

1.1 Indoor Air Quality (IAQ)

According to Abdul–Wahab et al. (2015), the term IAQ is utilised to describe the quality of air that is contained within a confined space. As a result of the fact that the majority of human activities take place in confined spaces and the pollutant concentrations inside are frequently higher than those outside (ANSI/ASHRAE Standard, 2017), it is equally important to be exposed to airborne pollutants in the interior environment as it is to be exposed to them in the outdoor air (Seppänen et al., 1999). It is of the utmost importance to make certain that the air inside the building does not pose any health risks, in addition to ensuring that the temperature provided is comfortable (Edwards et al., 2004). Bioaerosols, radon, asbestos, volatile organic compounds (VOCs), persistent organic pollutants (POPs), particulate matters of various sizes including nanoparticles (PM_x), nitrogen oxides (NOx), sulphur oxides (SO_X), carbon dioxide (CO_2), microorganisms, carbon monoxide (CO), and allergens are some examples of the types of biological and chemical pollutants that can cause chronic or acute symptoms when inhaled in contaminated air (Standard, 1992). These pollutants can cause symptoms such as nausea, headaches, and breathe shortness. [(Zabiegała et al., (2017), Gao et al., (2014), Wargocki and Wyon, (2007), Seppanen and Fisk, (2002), Mendell, (1999) Zhang et al., 2017, O. Seppanen, 2002)].

1.2 Transferring Moisture Versus Above-Ground Spaces

A high level of dampness, particularly during the summer months, is one of the characteristics that set underground places apart. Humid environments in underground settings have notable impacts on human comfort, health, operational safety, and equipment longevity. Microbial development can result from dampness in subsurface areas, leading to corrosion and damage to building equipment and surfaces. In light of this, it is essential to exercise control over the moisture, which mostly originates from the dampness of the walls as well as the hot and humid air. The following factors are responsible for the wetness that is present in the wall: (Kosonen et al., 2019).

- Evaporation of leftover water from building
- Groundwater penetration through the coating layer
- External humid air penetrates into the interior

There are a number of physiological and psychological effects that are frequently associated with the environment of underground space, including feelings of boredom, psychological despair, and a sense of anxiety. Factors such as lack of visibility, sunlight, proximity, excessive humidity, and poor air quality contribute to this issue.

1.3 Mechanical Ventilation System (MVE)

The MVE ventilation system utilizes a fan to supply or extract air from a building, employing exhaust air terminal devices, ducts, and roof/wall outlets. In single exhaust mechanical systems, air enters the house through externally placed air transfer devices, windows, and leakages (CEN, 2007). The Mahakal Temple is mechanically ventilated.

1.4 Airborne Disease

Airborne diseases are our oldest known enemies. Airborne diseases spread through, as the name suggests, the airborne route. This means when a person coughs, sneezes, talks he is releasing certain droplet nuclei which are nothing but small drops of body fluid which contain the disease-causing pathogen. This release of bodily fluid, as stated before happens at a very high rate and contains the disease-causing pathogens from one per droplet nuclei to thousands per nuclei. On release from a person these droplets areas are lies that means they are released into the air in very small units getting the pathogens and they tend to linger into the air for long periods of time depending on various factors until they have a chance to enter into the breath of another and infected person and cause infection. In this scenario we have to understand that buildings can play a key role and in buildings it is ventilation that is critical (Rudnick et al., 2003).

The aim of this study is to examine the Mahakal temple's indoor environmental quality, which has declined due to increased worship attendance. Due to the lack of natural ventilation in the Mahakal temple, devotees have experienced suffocation, irritation, and nasal congestion. The indoor atmosphere has deteriorated due to poor air circulation and ventilation. CO_2 has increased while O_2 has decreased. The above-normal temperature and humidity have also caused discomfort.

The rest of the paper is structured as follows: Sect. 2 provides an overview of the existing research and identifies the gaps in the literature. Section 3 outlines the objectives of the proposed work. Section 4 details the methods employed in the proposed work. Section 5 presents the simulation results. Finally, Sect. 6 concludes the paper and discusses potential future directions.

2 Related Work

The review of existing literature underscores the importance of understanding the transmission of airborne infections in high-density indoor environments, particularly in religious spaces like temples. Birnbach et al. (2010) highlight that evaluations of healthcare facility designs often focus on post-construction critiques, which can reveal preventable flaws. They suggest pre-planning ventilation systems during the design phase to enhance efficiency and safety, reducing the need for costly retrofitting. Similarly, Joseph and Rashid (2007) emphasize the role of indoor air quality (IAQ) in creating safe and effective environments, noting its significance alongside noise and lighting in healthcare and other densely occupied spaces.

Sachdeva et al. (2018) identify the need to mitigate occupationally acquired tuberculosis risks, noting that current airborne infection control guidelines lack effective implementation. This finding underscores the broader necessity for robust infection control strategies in high-density spaces. Asif et al. (2018) further explore how modern architectural trends, particularly energy-efficient building designs, can inadvertently compromise IAQ, including in hospitals, by limiting ventilation and exacerbating health risks.

Xu et al. (2022) focus on the role of ventilation in epidemic control, showing how poor airflow management allows droplet nuclei, particularly those smaller than 10 microns, to remain airborne for extended periods. This prolonged suspension increases the likelihood of airborne transmission, especially in confined spaces with inadequate ventilation. Similarly, Dai and Zhao (2020) highlight the importance of maintaining sufficient ventilation rates in confined environments to minimize infection risks. They demonstrate how recirculation-based air conditioning systems can worsen pathogen propagation and calculate infection probabilities using the Wells-Riley model, estimating COVID-19 quanta emission rates between 14 and 48 per hour.

A review of ventilation standards and guidelines reveals significant gaps. Most guidelines focus on general thermal comfort and IAQ but fail to account for user-generated pollutants or the unique challenges of religious spaces like temples. Notably, none explicitly identify temples as a distinct building type. Furthermore, while numerous studies examine infection risks in residential and commercial spaces, there is a dearth of research focusing on religious buildings in the Indian context.

Key factors influencing indoor airborne infections include ventilation rates, user activity levels (respiratory rates), duration of occupancy, space volume, and virus decay influenced by temperature, relative humidity, and solar radiation. These findings emphasize the critical need to address ventilation challenges in religious buildings, where traditional architectural features and crowd density create unique risks. This study aims to bridge these gaps by focusing on temples and providing tailored ventilation solutions to mitigate airborne infection risks.

3 Research Objectives

1. To find out probability of infection risk based on on-site ventilation rate of a Hindu temple.
2. For ventilation rates of selected case use simulation software to project the observed value.
3. To identify relation between ventilation rate and risk of infection through airborne pathogens.
4. Suggest scope of further study to develop some benchmark for ventilation.

4 Proposed Work

The indoor environmental state of the temple is investigated in this study. The condition of the temple has deteriorated as a result of an increase in the number of devotees attending services there recently. Several devotees have reported experiencing suffocation, exhaustion, dizziness, irritability, and nasal congestion due to the absence of vents and natural ventilation in the Mahakal temple. Since there is insufficient air circulation and natural ventilation, the condition of the interior environment has declined. It has been observed that CO_2 levels have increased while O_2 levels have decreased. In addition, temperature and humidity levels have risen above normal levels, causing discomfort.

The Wells-Riley equation, employed for assessing the likelihood of airborne transmission of infectious diseases like tuberculosis within enclosed spaces, can be problematic due to its assumption of steady-state conditions and the need for accurate measurement of outdoor air supply rates. These rates are often challenging to measure and tend to fluctuate over time.

Generally, the air quality at Mahakaleshwar Temple starts deteriorating in late October. The winters are the worst-hit season in terms of air pollution.

Using an Extech SD800 CO_2, Humidity and Temperature Datalogger, we gathered information on the temperature within the nandi hall of the Mahakal Temple on September 31, 2021. This information included the temperature inside and outside, the relative humidity, the concentration of carbon dioxide, and the air pressure.

We utilized subscale to model the airflow within an indoor space in order to estimate the Air Changes per Hour (ACH) value, which was determined to be 0.3 VENTILATION RATE PER PERSON (L/S/PER). Then we used ACH value for calculating infection risk in mahakaal temple using wells riley model.

4.1 Problem Identification

In order to provide devotees with medical assistance in the event of an emergency, two medical camps are established annually during the Navratri festival season. A significant number of individuals have reported experiencing symptoms such as shortness of breath, fainting, giddiness, pain, and nasal conjunctivitis. The vast majority of the children shed tears as they make their way into the temple, which is packed with followers of prayer.

4.2 Wells–Riley Equation

For the purpose of estimating the probability of airborne transmission of an infectious agent indoors, Riley et al. (1978) developed what is now called the Wells–Riley equation

$$Prob = \frac{N}{Sp} = 1 - \exp\left(-\frac{I_{bgt}}{O}\right) \tag{1}$$

The following variables are used in this equation: Prob, which represents the probability of infection for susceptible, N, which stands for the number of disease cases; Sp, for susceptible; I, for infectors; b, for breathing rate per person (m^3/s); g, for quantum generation rate by an infected person (quanta/s); t, for total exposure time (s); and O, for outdoor air supply rate (m^3/s).

The frequently misunderstanding parameter q, which represents the rate at which an infected person generates infectious quanta, is crucial to comprehending Eq. 1. The concept of a quantum dosage, sometimes known as an infectious dose, was proposed by Wells (1955) to explain the unpredictable nature of airborne infection. Contrary to what some writers have said, just because there could be a quantum of infection involving a lot of organisms doesn't imply that more than one organism starts the infection (Nicas, 1996). Even when a quantal dose is significantly larger than one organism, Wells noted (p. 126), "infection bears a Poisson relation to this dose…" To the host, it seems as though a solitary invading organism has found a weak spot where it can multiply and infect. Therefore, the average chance of becoming infected is 63% $(1 - e^{-1})$, which is essentially an infectious dose of 63% (ID_{63}), when exposed to one quantum of infection. When it comes to certain creatures, the odds of any one individual discovering and exploiting a susceptible locus are really slim.

Assumptions:

1. It has been commonly used to predict the concentration of contaminants indoors, based on the premise that the airspace is well-mixed.
2. The susceptible person or population is present in the premises starting from t = 0–25 min.
3. The quanta emission rate is equivalent to Quanta emission rate of SARS-CoV-2.

4.3 CO_2-Based Risk Equation

The assumption of well-mixed airspace has been utilised in predicting indoor pollutant concentrations, such as in the development of the Wells-Riley equation, although its accuracy has not been extensively verified. Furthermore, we presupposed that ventilation much outweighs other methods of infectious particle clearance, such as viability loss, filtration, diffusional deposition and gravity on surfaces, and others (Hao et al., 2014). See below for more on this assumption.

These assumptions are based on the idea that the rate of quantum generation (g), the rate of breathing (b), the number of infectors (I), and the number of people in the ventilated space (P_v) all stay the same. The quantum concentration in the ventilated space (C) = concentration of quanta in the exhaled breath of infectors (g/b) times the volume fraction of air in the space that was exhaled by infectors ($E_{in}I/P_v$):.

$$C = \frac{E_{in} I g}{P_v b} \tag{2}$$

This means that the avg. Quantum concentration $\hat{C}$ for the entire exposure time (t) can be calculated directly from the avg. Proportion of interior air that is exhaled $\hat{E}_{in}$.

$$\hat{C} = \frac{\hat{E}_{in} I g}{P_v b} \tag{3}$$

The average number of quanta breathed by a susceptible person $\hat{\alpha}$ can be calculated from the following equation:

$$\hat{\alpha} = bt\hat{C} \tag{4}$$

Infected individuals' numbers follow a Poisson distribution, as pointed out by Wells (1955). The likelihood that a sensitive individual does not contract an infection is thus equal to $e^{-\hat{\alpha}}$, and the converse is true for the probability of infection (Prob):

$$\text{Prob} = 1 - e^{-\hat{\alpha}} \tag{5}$$

substitute $\hat{C}$ from Eq. 3 into Eq. 4 and using the resulting expression for $\hat{\alpha}$ into Eq. 5 gives

$$Prob = \frac{N}{Sp} = 1 - \exp\left(-\frac{\hat{E}_{in}lgt}{P_v}\right) \tag{6}$$

Equation 6 is applicable in a wide variety of situations, including steady-state and non-steady-state settings, as well as situations where the rate of indoor air supply changes over time.

5 Simulation Results

The infection rate is estimated based on the fluctuation in CO2 content, which is influenced by the number of visitors in the temple. We have gathered the primary data for the proposed project during peak hours in the morning and evening, when there is a strong influx of devotees in the months of October. Sample data is shown in Fig. 4. The gathered for the stated parameters and the CO_2 concentration during a period of 15 days. The main dataset consists of the following variables: date, time, outdoor temperature (TEMP_OUT), indoor temperature (TEMP_IN), outdoor relative humidity (RH_OUT), indoor relative humidity (RH_IN), number of persons, CO2 level, atmospheric pressure, and area (m^2). The suggested study is simulated using simscale to quantify the risk of infection.

The Figs. 1, 2 and 3 depict several perspectives of the darshan hall in the Mahakal Temple.

5.1 Factors Related to the Environment Considered for This Research Work

The volume is 3265 cubic metres and the atmospheric pressure is measured in atmospheres i.e. 1, the average daily temperature in degrees Celsius is 27.5. The relative humidity is 73%. The duration of the event is measured in minutes i.e. the number of repetitions in a day is 25. The ventilation rate measured in outside air per hour is 45. The CO_2 concentration at its highest point is 1050 parts per million (ppm). Time at which the degradation reaches 63% of its initial value is 3. The decay of virus per hour is 2.664. The ventilation rate per person is expressed in litres per second per person is 0.3.

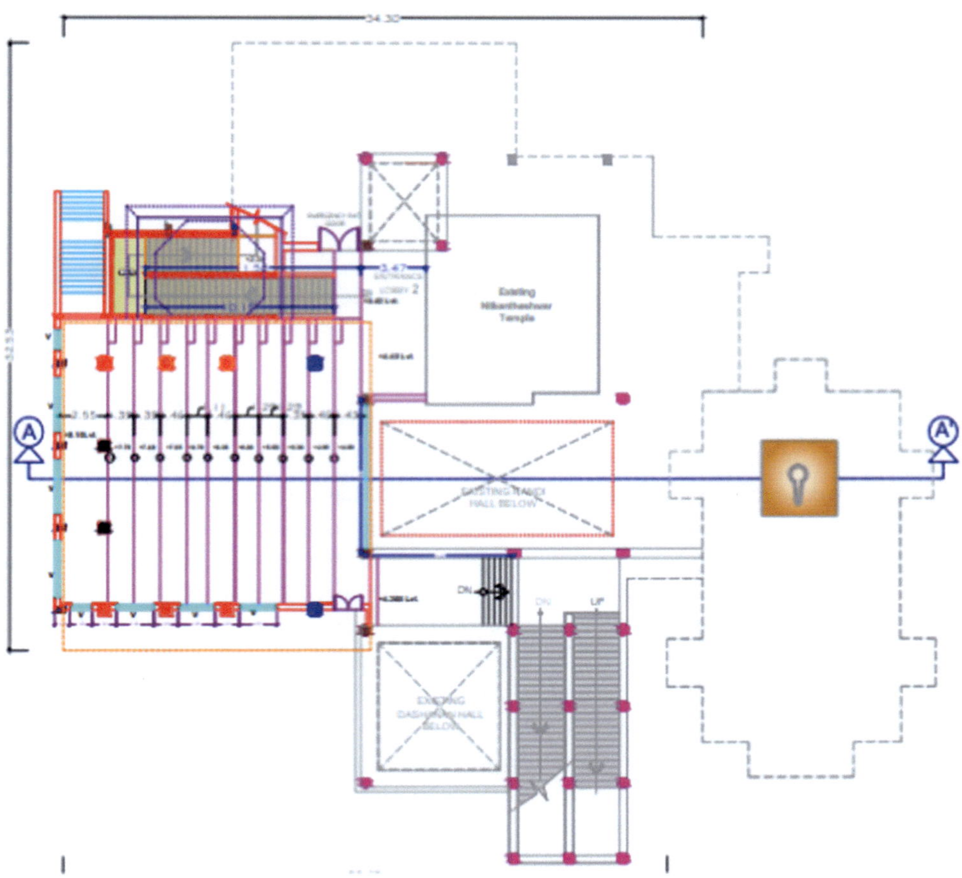

Fig. 1. Upper-level plan of existing Darshan Hall of Mahakaal temple (Source: Author)

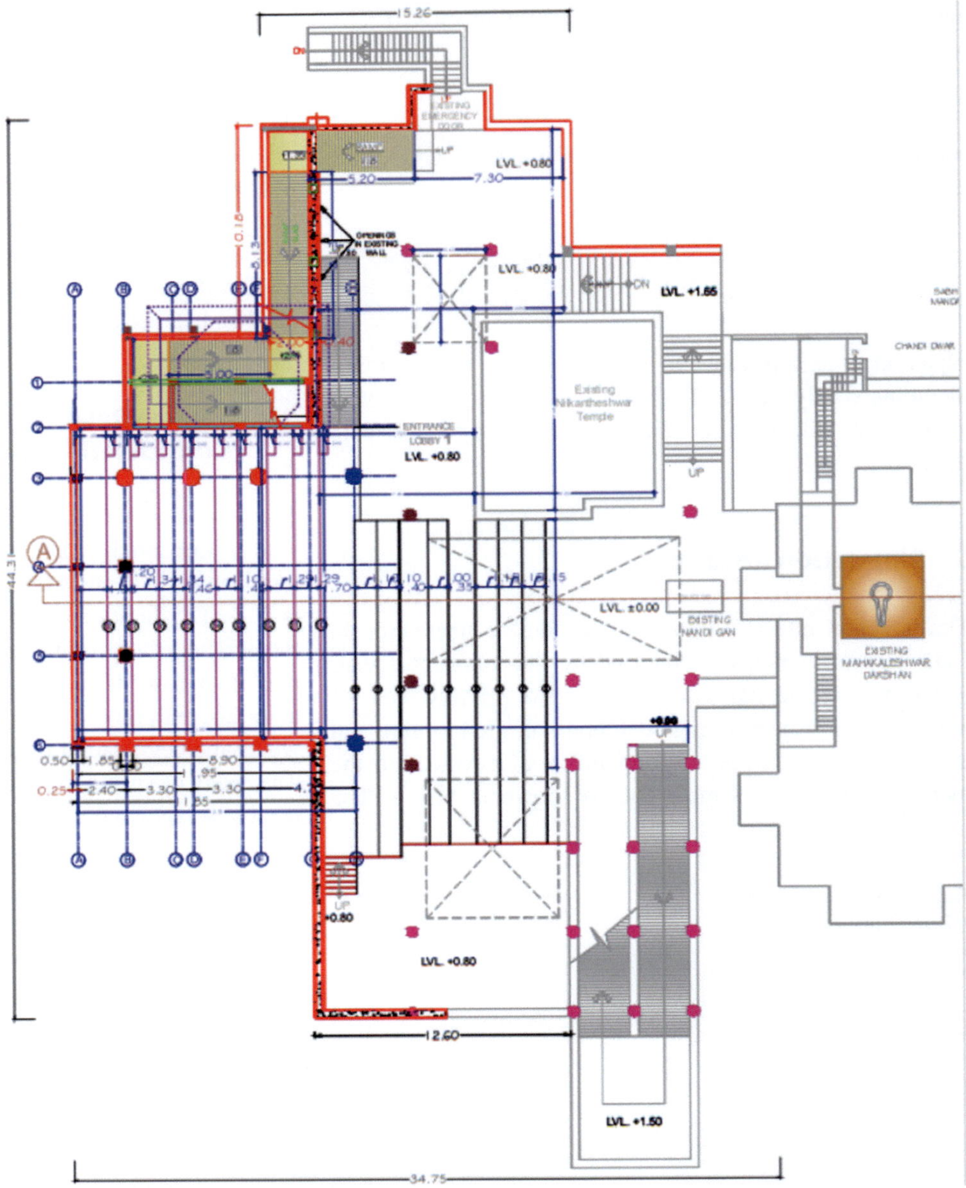

Fig. 2. Upper-level plan of existing Darshan Hall of Mahakal temple (Source: Author)

5.2 Variables Pertaining to Individuals and Actions Within the Room

Maximum occupancy of 1000 and individuals who are contagious is 1, Percentage of the population that is immune is 0, Individuals who are vulnerable to a particular condition or disease is 999. The density is calculated by dividing the area by the number of people, expressed in square feet per person is 7.3, the density of individuals per square meter is 1.5. The density measured in cubic meters per person, is 3.3, Rate of respiration (cubic meters per hour) is 5.16 Quanta Increase in performance resulting from variations. Rate of exhalation of quanta (infected) is 970. Conditional outcomes for an INDIVIDUAL and a SINGLE EVENT (assuming a positive number of infected individuals, usually 1).

Utilizing the Wells-Riley infection model to quantify the quantity of infectious doses breathed. The equation labelled as Eq. (1) in the research paper authored by Miller et al. in 2020. Individual infection probability is 2.6 percent. Conditional outcomes for all participants and a single event (assuming a given number of infected individuals is 1). The infection risk parameter (H) is equal to h2 persons per cubic meter (m3) is 4.99. The Relative Infarction Risk parameter (H_r) is 0.005. CO_2 is not necessary for estimating infection and can be disregarded for simplicity when assessing risk. The average carbon dioxide (CO_2) mixing ratio is 2614 parts per million (ppm).

CO_2 results are not necessary for estimating infection risk and can be disregarded for the sake of simplicity. The average carbon dioxide (CO_2) mixing ratio is 2614 parts per million (ppm).

Fig. 3. Longitudinal section showing vision angle of existing darshan hall. (Source: Author)

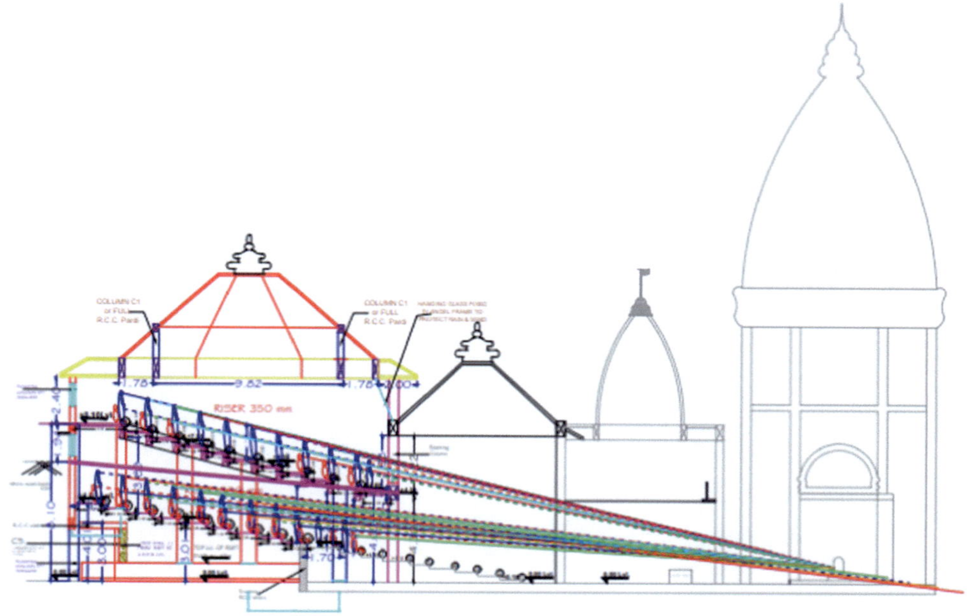

Concentration of exhaled carbon dioxide that is subsequently inhaled by an individual (measured in parts per million) is 916.15 The amount of carbon dioxide (CO_2) that has to be inhaled for one hour to cause a 1% infection rate is measured in parts per million (ppm) is 702.

5.3 The Simulation Process Comprises the Following Steps

1. Create a model design using Rhino 7.
2. Upload the pre-existing model to Simscale for examination.
3. Choose carbon dioxide as the material and specify the initial circumstances.
4. Specify the boundary conditions, such as pressure at the input and exit sides, as well as any walls.
5. Log of mesh generation.
6. Results of simulations (plots showing convergence).

The graph depicted in Figs. 5 and 6 illustrates four distinct lines, for different time period, each representing a different variable (humidity rate, CO_2 level, temperature, and chance of infection). These lines are drawn on the same time scale along the x-axis. The Garb Gufa's enclosed structure and limited air circulation contribute to the likelihood of it having the highest temperature (Kalogirou et al., 2003). Entry and exit points may exhibit reduced temperatures as a result of their direct exposure to the external environment. Humidity in the Garb Gufa may reach its peak due to limited evaporation or inadequate ventilation, similar to temperature. Entry and exit locations may experience reduced humidity as a result of the introduction of fresh air. The concentration of people and limited air movement in the Garb Gufa may result in the highest levels of CO_2. Entry and exit locations may exhibit reduced levels of CO_2 as a result of improved ventilation. Elevated temperatures are likely to correlate with an increased likelihood of infection. This phenomenon may be attributed to heightened viral activity or a compromised immune response under elevated temperatures. The temperature and CO_2 concentrations gradually rise within the Garb Grah, but they remain relatively constant at the Entry and Exit sites. Similarly, the likelihood of infection is constantly greater in the Garb Gufa in comparison to the other sites. Elevated temperatures may

date time	date	time	TEMP_OUT	TEMP_IN	RH_OUT	RH_IN	no of ppl	CO2	atm pressure	Area (m2)
30-09-2021 12:15	30-Sep	12:15	31.0	30.5	55	61.1	100	441	0.99	675
30-09-2021 13:15	30-Sep	13:15	31.0	29.2	55	65.4	125	487	0.99	675
30-09-2021 14:15	30-Sep	14:15	32.0	29.3	54	65	120	436	0.99	675
30-09-2021 16:35	30-Sep	16:35	32.0	31.0	54	62.5	90	400	0.99	675
30-09-2021 17:35	30-Sep	17:35	31.0	29.5	60	66.7	100	445	0.99	675
30-09-2021 18:35	30-Sep	18:35	28.0	28.8	70	69.5	100	411	0.99	675
30-09-2021 19:30	30-Sep	19:30	26.0	28.8	83	71	150	501	0.99	675

Fig. 4. Sample dataset (Source: Author)

Fig. 5. Simulations results (convergence plots) of temperature, CO_2, RH% inside grabhagrah during morning (Source: Author)

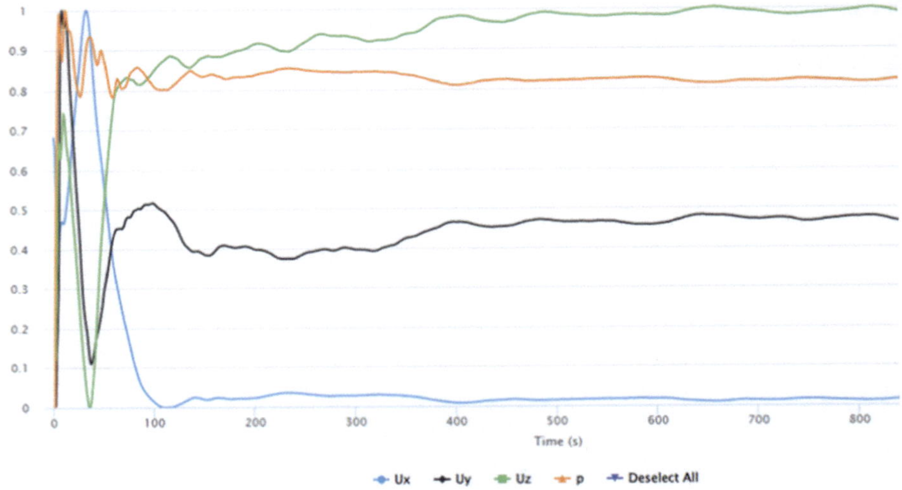

Fig. 6. Simulations results (convergence plots) of temperature, CO_2, RH% inside graph during evening (Source: Author)

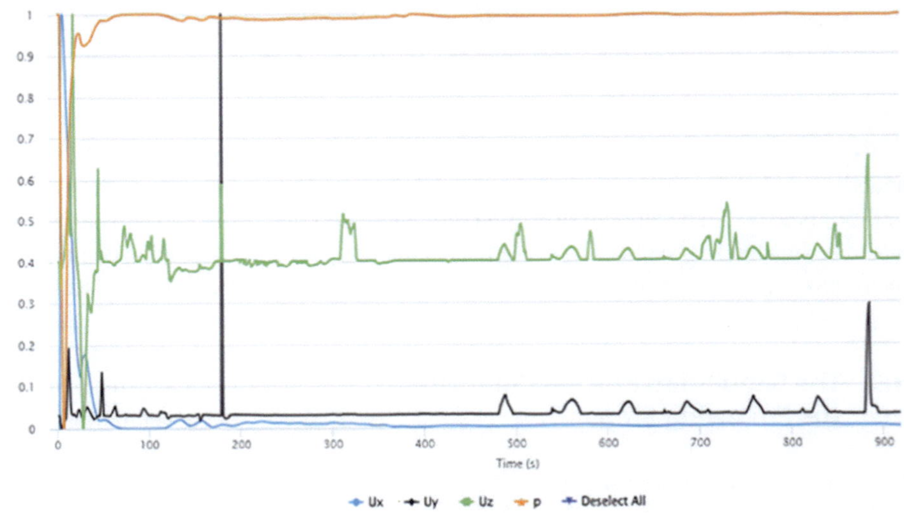

Fig. 7. Simulations results (convergence plots) of humidity inside the temple (Source: Author)

be correlated with increased amounts of CO2 and a heightened likelihood of illness. In the same vein, elevated humidity is associated with an increased likelihood of infection as it promotes the spread of viruses.

The rate of humidity inside the Mahakaal temple is depicted in Figs. 7, and 8 exhibit the solution elements of the ventilation system in the Mahakaal temple. The Fig. 9 presents the projected probabilities of infection, offering valuable

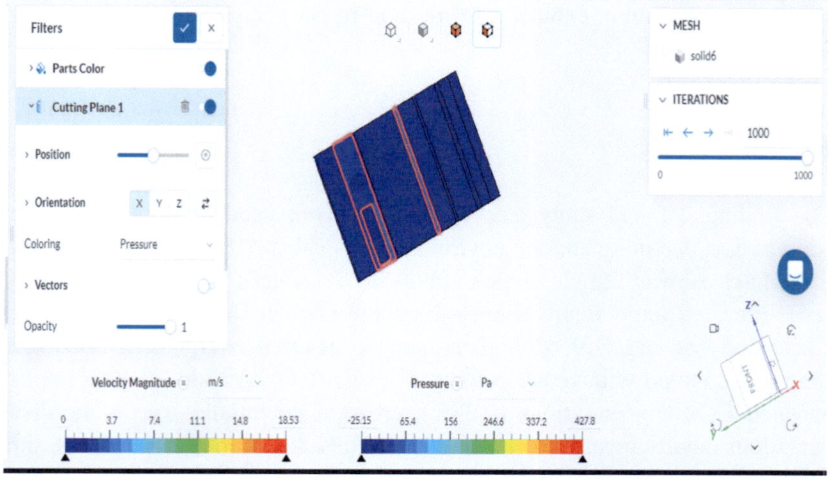

Fig. 8. Simulations results of solution fields (Source: Author)

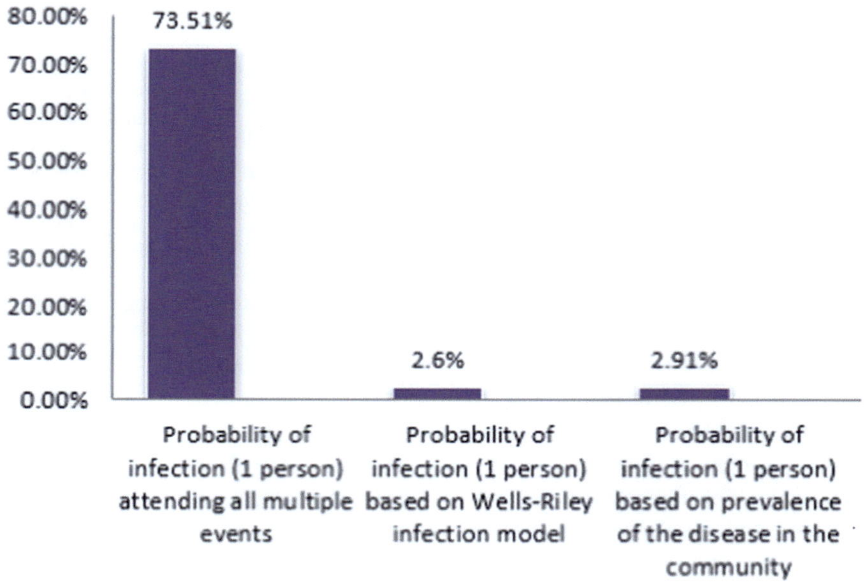

Fig. 9. Probability of infection for different case in Mahakaal temple (Source: Author)

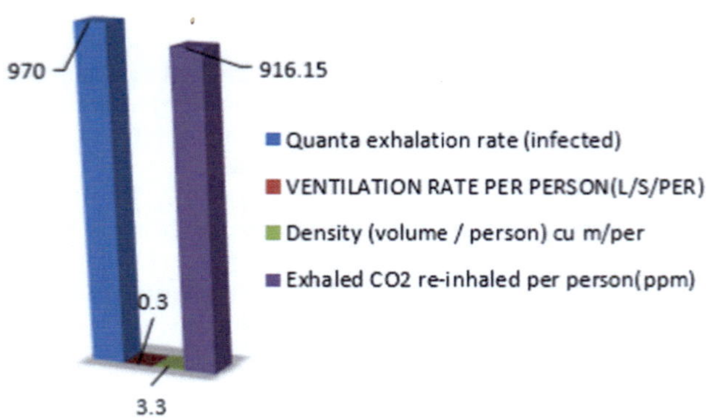

Fig. 10. Quanta Emission rate and Ventilation rate (Source: Author)

insights into the potential dangers and factors that affect infection rates within the temple. The graph illustrates the influence of variables such as event attendance, the Wells-Riley model, and community prevalence on the probability of infection. By comparing the various probabilities, one can identify specific regions or activities in the temple that present a greater danger of infection. This information can be used to develop effective methods to reduce the risk. The Mahakaal Temple in Ujjain's ventilation system shows a 2.6% infection risk for a devotee who spends 25 min in the Nandi Hall and Grabhgrah. This computation uses the Wells Riley model, which assumes one sick individual. The Wells Riley model assumes one sick person per event, hence a single staff member at Nandi Hall or Grabhgrah for a day has a 73.51% chance of infection. The temple administration reported 2000 visitors every hour at Mahakaal Temple, therefore minimising this risk is crucial to preventing future pandemics. Research is needed to standardise ventilation parameters since Air Changes per Hour (ACH) considerably reduces indoor airborne illness risk.

The Fig. 10 illustrates that the Quanta exhalation rate for infected individuals is 970, the ventilation rate per person is 0.3 L/S/PER, the density (volume per person) is 3.3 cum/

per, and the amount of exhaled CO_2 re-inhaled per person is 916.15 ppm.

6 Discussion

The findings of this study emphasize the urgent need to address the declining indoor environmental quality (IEQ) in Mahakaleshwar Temple, which affects both devotees and staff. Elevated temperature levels within the Garbha-Grah, consistently at least 1.9 °C higher than the external environment, coupled with reduced oxygen levels (0.66%) and elevated CO2 concentrations (0.25%), create a potentially hazardous environment Luo et al., (2007). These conditions are further exacerbated by inadequate ventilation, lack of natural air circulation, and high occupancy levels.

The observed infection risks, as calculated using the Wells-Riley model, indicate a 2.6% chance of infection for devotees during short visits and a staggering 73.51% for staff exposed for an entire day (Seppanen and Fisk, 2004). This highlights the role of ventilation in controlling airborne infections and underscores the necessity for improving air circulation, particularly in spaces like the Garbha-Grah and Nandi Hall, which experience the highest congregation density.

The current ventilation system does not comply with recommended standards for high-occupancy spaces. Low air changes per hour (ACH) contribute to the persistence of pollutants and increased humidity, leading to discomfort. Additionally, noise levels of 100 dB and the mixed odours from incense, flowers, and perspiration detract from the spiritual and sensory experience expected in such religious spaces.

These findings align with existing literature on the impact of poor ventilation in high-density indoor environments. However, this study addresses a unique gap by focusing on temples, which are not typically considered in standard ventilation guidelines. This highlights the need for research and policies specific to such cultural and religious settings, considering their distinct architectural designs and usage patterns.

Practical implications of this study include the development of customized ventilation systems for temples that ensure adequate air circulation while respecting the cultural and architectural integrity of these spaces. Implementing mechanical ventilation systems with higher ACH, optimizing air flow, and incorporating natural ventilation where feasible can significantly mitigate health risks and enhance the comfort of devotees.

Future research should explore cost-effective solutions, the role of advanced filtration systems, and the integration of renewable energy technologies in temple ventilation designs. Moreover, longitudinal studies could evaluate the long-term effects of improved ventilation on health outcomes and visitor satisfaction in religious settings.

7 Conclusion

The study reveals critical findings about the indoor environmental quality (IEQ) of Mahakaleshwar Temple, highlighting significant challenges. The temperature in the Garbha-Grah is consistently higher than outside, CO_2 levels surpass permissible limits, and oxygen levels drop by 0.66%, contributing to discomfort and health risks for devotees. The infection risk is estimated at 2.6% for a devotee spending 25 min in confined areas and 73.51% for staff exposed throughout the day, emphasizing the urgent need for ventilation improvements.

Key implications include the necessity for establishing ventilation benchmarks tailored to high-occupancy religious spaces to mitigate airborne infection risks and enhance visitor comfort. Further research should focus on culturally appropriate and cost-effective ventilation strategies for such environments.

Authors Contributions All authors contributed to the study conception and design. Material preparation, data collection and analysis were performed by Apurva Sharma. The first draft of the manuscript was written by apurva Sharma and all authors commented on previous versions of the manuscript. Dr. Anupama Sharma has supervised and approved the final manuscript. All authors read and approved the final manuscript."

Funding The authors did not receive support from any organization for the submitted work.

Ethical Approval This study does not contain any studies with human or animal subjects performed by any of the authors.

Consent to Participate Not applicable.

Consent to Publish Manuscript has been read and approved for submission by all the named author.

Competing Interests The authors have no competing interests to declare that are relevant to the content of this article. All authors certify that they have no affiliations with or involvement in any organization or entity with any financial interest or non-financial interest in the subject matter or materials discussed in this manuscript.

Data Availability Statement Data will be made available on request.

Disclosure of Interests The authors have no competing interests to declare that are relevant to the content of this article.

References

Abdullah, F.H., Majid, N.H.A., Othman, R.: Defining issue of thermal comfort control through urban mosque façade design. Procedia-Soc. Behav. Sci. **234**, 416–423 (2016)

Abdul-Wahab, S.A., En, S.C.F., Elkamel, A., Ahmadi, L., Yetilmezsoy, K.: A review of standards and guidelines set by international bodies for the parameters of indoor air quality. Atmos. Pollut. Res. **6**(5), 751–767 (2015)

Asif, A., Zeeshan, M., Hashmi, I., Zahid, U., Bhatti, M.F.: Microbial quality assessment of indoor air in a large hospital building during winter and spring seasons. Build. Environ. **135**, 68–73 (2018)

Birnbach, D.J., Nevo, I., Scheinman, S.R., Fitzpatrick, M., Shekhter, I., Lombard, J.L.: Patient safety begins with proper planning: a quantitative method to improve hospital design. BMJ Qual. Saf. **19**(5), 462–465 (2010)

Edwards, D.A., et al.: Inhaling to mitigate exhaled bioaerosols. Proc. Nat. Acad. Sci. **101**(50), 17383–17388 (2004)

Gao, J., Wargocki, P., Wang, Y.: Ventilation system type, classroom environmental quality and pupils' perceptions and symptoms. Build. Environ. **75**, 46–57 (2014)

Hao, H., Lin, Y., Kowli, A.S., Barooah, P., Meyn, S.: Ancillary service to the grid through control of fans in commercial building HVAC systems. IEEE Trans. Smart Grid **5**(4), 2066–2074 (2014)

Joseph, A., Rashid, M.: The architecture of safety: hospital design. Curr. Opin. Crit. Care **13**(6), 714–719 (2007)

Kalogirou, S., Eftekhari, M., Marjanovic, L.: Predicting the pressure coefficients in a naturally ventilated test room using artificial neural networks. Build. Environ. **38**(3), 399–407 (2003)

Lal Benjamin, N., Sharma, S., Pendharker, U., Shrivastava, J.K.: Air quality prediction using artificial neural network. IJCS **2**(4), 07–09 (2014)

Luo, Z., Zhao, J., Gao, J., He, L.: Estimating natural-ventilation potential considering both thermal comfort and IAQ issues. Build. Environ. **42**(6), 2289–2298 (2007)

Nicas, M.: Refining a risk model for occupational tuberculosis transmission. Am. Ind. Hygiene Assoc. J. **57**(1), 16–22 (1996)

Riley, E.C., Murphy, G., Riley, R.L.: Airborne spread of measles in a suburban elementary school. Am. J. Epidemiol. **107**(5), 421–432 (1978)

Rudnick, S.N., Milton, D.K.: Risk of indoor airborne infection transmission estimated from carbon dioxide concentration. Indoor Air **13**(3), 237–245 (2003)

Sachdeva, K.S., et al.: Tuberculosis infection control measures at health care facilities offering HIV and tuberculosis services in India: a baseline assessment. Indian J. Tuberc. **65**(4), 280–284 (2018)

Seppänen, O.A., Fisk, W.J., Mendell, M.J.: Association of ventilation rates and CO2 concentrations with health and other responses in commercial and institutional buildings. Indoor Air **9**(4), 226–252 (1999)

Singh, A., Pandey, Y., Kumar, A., Singh, M.K., Kumar, A., Mukhopadhyay, S.C.: Ventilation monitoring and control system for high rise historical buildings. IEEE Sens. J. **17**(22), 7533–7541 (2017)

Śmiełowska, M., Marć, M., Zabiegała, B.: Indoor air quality in public utility environments—a review. Environ. Sci. Pollut. Res. **24**, 11166–11176 (2017)

Standard, A.S.H.R.A.E.: Thermal environmental conditions for human occupancy. ANSI/ASHRAE **55**, 5 (1992)

Wargocki, P., et al.: Ventilation and health in non-industrial indoor environments: report from a European multidisciplinary scientific consensus meeting (EUROVEN). Indoor Air **12**(2), 113–128 (2002)

Wargocki, P., Wyon, D.P.: The effects of moderately raised classroom temperatures and classroom ventilation rate on the performance of schoolwork by children (RP-1257). Hvac&R Res. **13**(2), 193–220 (2007)

Xu, C., Liu, W., Luo, X., Huang, X., Nielsen, P.V.: Prediction and control of aerosol transmission of SARS-CoV-2 in ventilated context: from source to receptor. Sustain. Cities Soc. **76**, 103416 (2022)

Zhang, X., Wargocki, P., Lian, Z., Thyregod, C.: Effects of exposure to carbon dioxide and bioeffluents on perceived air quality, self-assessed acute health symptoms, and cognitive performance. Indoor Air **27**(1), 47–64 (2017)

CEN, E.: 15242 Ventilation for buildings-calculation methods for the determination of air flow rates in buildings including infiltration. European Committee for Standardization (CEN), Brussels, Belgium (2007)

Dai, H., Zhao, B.: Association of the infection probability of COVID-19 with ventilation rates in confined spaces. In: Building simulation, vol. 13, pp. 1321–1327. Tsinghua University Press (2020)

Li, A., et al.: Ventilation and environmental control of underground spaces: a short review. In: E3S Web of Conferences, vol. 111, p. 01039. EDP Sciences (2019)

Mendell, M.J.: Indoor environments and student performance. Indoor Air **9**(4), 227–252 (1999)

Seppanen, O.A., Fisk, W.J.: Summary of human responses to ventilation (2004)

Wells, W.F.: Airborne contagion and air hygiene: an ecological study of droplet infections. Commonwealth Fund (1955)

Study on Circular Building Design Criteria of Human Settlements

An-Pi Chang and Yan-Hong Xiong

Abstract

In 2013 and 2021, China announced the "Circular Economy Development Strategy and Near-Term Action Plan" and the "Action Plan for Carbon Dioxide Peak Before 2030," respectively. Both are pivotal strategies for advancing sustainable development. Against this policy backdrop, the concept of a circular economy is progressively being integrated into the sustainable development strategies of various industries, giving rise to residential circular buildings. This study aims to establish a multi-criteria decision-making evaluation framework for residential circular building projects, which will aid stakeholders in selecting the optimal plan during the pre-project planning phase. In the preliminary stages of the research, mind mapping was employed to integrate project operations, and evaluation criteria were established through group decision-making by project stakeholders. The Analytic Hierarchy Process (AHP) and Eigenvector method were applied to calculate weights, resulting in an evaluation framework comprising 4 dimensions and 24 criteria. The outcomes and contributions of this study is envisaged to serve as a reference for stakeholders in residential circular building projects during the pre-project planning phase, aiding them in selecting the optimal plan.

Keywords

Circular economy strategy · Sustainable development · Circular Building · Pre-project planning · Analytic hierarchy process

A.-P. Chang (✉) · Y.-H. Xiong
Research Institute of Architecture and Human Settlements, Fujian University of Technology, Fuzhou, China
e-mail: chang.anpi@gmail.com

1 Introduction

In 2013, the Ellen MacArthur Foundation published Towards the Circular Economy, which shifted the paradigm of resource use by promoting a circular economy model. This model integrates industrial and biological resource systems through rethinking, redesigning, and remanufacturing processes, moving away from the traditional linear economic model (Ellen MacArthur Foundation 2013). In the same year, Chian introduced the Circular Economy Development Strategy and Near-Term Action Plan. In October 2021, China also released a guiding document on the country's work to achieve carbon peaking and carbon neutrality goals under the new development philosophy. This document outlines five major tasks, including creating a green, low-carbon and circular economy, improving energy efficiency, increasing the share of non-fossil energy consumption, lowering CO_2 emissions and boosting the carbon sink capacity of ecosystems. In order to implement the circular strategy, China Resources Recycling Group was established on October 18, 2024 to completely, accurately and comprehensively implement the new development concept, deeply implement the comprehensive conservation strategy, adhere to recycling, turn waste into treasure, adhere to innovation-driven, Open cooperation, strive to smooth the resource circular supply chain, create a national and functional resource circular platform, and build a multi-level, three-dimensional, and efficient resource ciraular system. Given this context, developing circular building assessment criteria to accelerate the integration of the building industry into the circular economy is an important area for continued research.

Circular Economy (CE) drives sustainable innovation by addressing resource shortages and high energy consumption issues through the integration of the concept into the building industry. Circular Building (CB) is an extension of CE principles and is categorized under the domain of Green

B. He et al. (eds.), *Towards the Framework of Livable and Resilient Cities* Advances in Science, Technology & Innovation, https://doi.org/10.1007/978-3-031-97849-4_26

Fig. 1. Research Process. (cited by author)

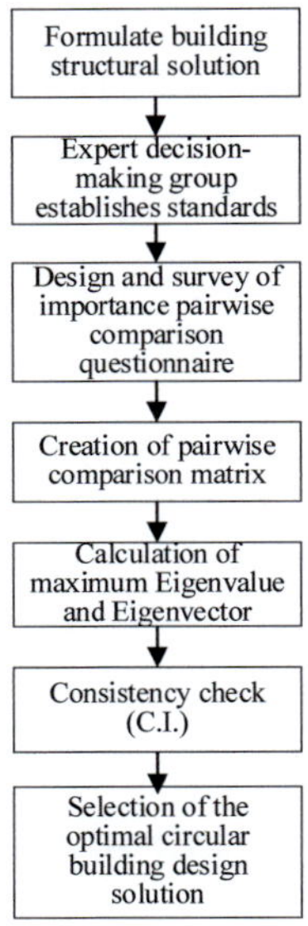

Economy (GE). The concept of CE is widely recognized as a critical turning point towards sustainable development. Presently, many countries have adopted CE as part of their national development strategies, expanding its application across various industries and progressively implementing CE legislation and action plans. This serves as a national strategy to achieve "Net Zero Emissions" and "carbon peaking and carbon neutrality" goals. The essence of a Circular Economy is to reduce the exploitation of resources and the emission of carbon dioxide (Chang and Chang 2020). Specifically, in 2023, the building sector accounted for 37% of total industrial greenhouse gas emissions, 40% of resource consumption, and 40% of waste generation (Bergmans et al. 2023). To overcome the challenges faced by the building industry, we can employ various strategies, such as integrating circular design principles that consider factors affecting the lifespan and future reuse of materials and structures throughout each phase of a project (Graaf et al. 2022).

2 Research Process and Methods

2.1 Research Process

As illustrated in Fig. 1, the research process is divided into three main parts. The first part involves selecting the structural type of the building. The project team will evaluate various building structures and choose three that feature modularity, lightweight materials, and disassemblability. These choices will align with the requirements for circular buildings and serve as the basis for selecting the best solution for residential circular buildings. The preliminary planning and design of the building project will significantly impact the overall effectiveness of resource generation, reuse, and circularity of the project.

The second part focuses on establishing evaluation criteria for residential circular building projects. The criteria will be based on literature and standard practices in building projects. They will integrate the essence of circular building principles throughout different stages of the lifecycle. This includes selecting or introducing new evaluation criteria related to building regulations, structural components, recyclable resources, and monitoring systems for both hardware and software.

The third part involves the AHP questionnaire and computational analysis. Based on the AHP analysis process, an importance pairwise comparison scale questionnaire will be developed to create a pairwise comparison matrix. This matrix will be used to determine the maximum Eigen Value (λmax), and the consistency index (C.I.). Finally, overall computations will be performed to identify the optimal solution for residential circular buildings.

2.2 Research Methods

2.2.1 Analytic Hierarchy Process (AHP)

AHP is chosen for its simplicity, clarity, and ease of use, which significantly influences the project's overall operations and analysis outcomes. The Analytic Hierarchy Process (AHP) is commonly applied in multi-criteria decision-making tasks. This study uses AHP's hierarchical independence to perform importance pairwise comparisons of project alternatives, aiming to select the optimal solution for residential circular buildings. As depicted in Fig. 2, the hierarchical structure consists of four levels, with the top level being the Goal, the second level being the Dimensions, the third level being the criteria (which are not singular but multiple criteria within the second-level dimensions) and the bottom level, which would be the Alternatives to be evaluated and selected as the best solution for the study.

2.2.2 Semi-structured Expert Interviews

Semi-structured interviews are particularly well-suited for project engineering research. Besides covering general aspects of building projects, face-to-face interviews and experience sharing can effectively capture respondents' experiences and their genuine thoughts on residential circular building research. This approach helps to refine project evaluation processes. The project team will develop interview content and corresponding importance scale questionnaires based on the four stages of the building life cycle. The results of the interviews will be reviewed by the project decision-making group to determine whether they should be included as evaluation criteria. The interviews will be

Fig. 2. Hierarchical Structure of AHP for Residential Circular Building. (cited by author)

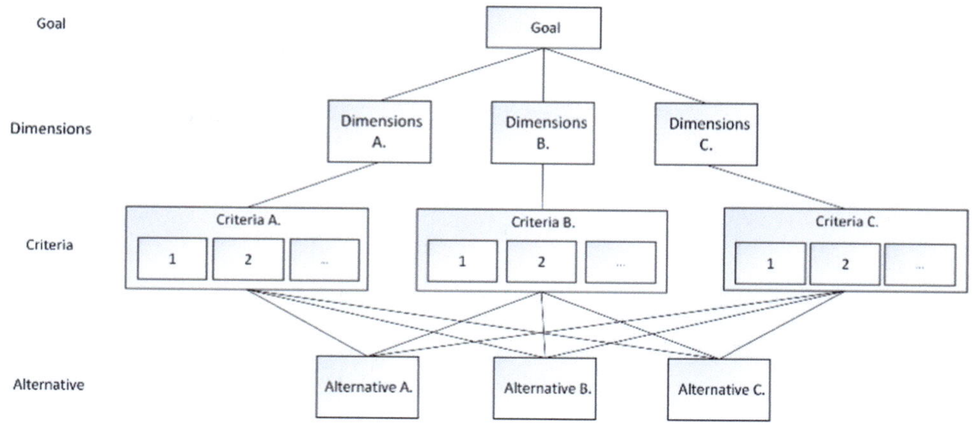

conducted in a named format, allowing for repeated confirmation of consistency with respondents if significant differences in opinions arise. Given the project's independence, limited resources, and specialized nature, and the high level of interdisciplinary collaboration required in circular building projects, the decision-making group will consist of key stakeholders involved in the project. This setup is beneficial for building team consensus and enhancing team project execution. Group decision making and the integration of opinions among experts are an integral process. This study imposed stringent requirements on the practical experience and professional backgrounds of the experts. 10 individuals took part in the research interviews, while 15 individuals contributed to the scale questionnaire. The distribution of expertise among the participants is depicted in Fig. 3.

3 Establishing AHP Evaluation Criteria Framework

3.1 Development of Circular Building Structural Solutions

Durability and resilience are critical features to consider when selecting materials and components for buildings, ensuring that quality or value is maintained over time and use (Rahla et al. 2021). Since the main structural components of a building account for over 60% of the total structure, they are crucial in evaluating the best project solution. Traditional Reinforced Concrete (RC) structures have issues such as heat retention, slow heat dissipation, high weight, and significant resource consumption. Additionally, the process of manufacturing, transportation, and recycling are not environmentally friendly, and the separation technology is underdeveloped, making RC structures unsuiT-able 26. as a green circular building core. Commonly used main structural materials in construction projects include Steel Reinforced Concrete (SRC), Precast Concrete (PC), and Steel Structure (SS). For the circular building project, the decision-making group selected three structural types that offer modularity, lightweight properties, and disassemblability, aligning with the principle of retaining the maximum value of building materials. These characteristics make them appropriate for evaluating the best circular building solution.

3.2 Establishment of Evaluation Criteria

The lifecycle of a construction project involves complex and detailed tasks, but these tasks follow clear processes and requirements. The selection of evaluation criteria is based on the seven essential aspects of circularity: circular supply chain, resource minimization, resource regeneration and reuse, avoidance of secondary manufacturing pollution, resource sharing mechanisms, operational and maintenance management systems, and reverse logistics (Chang et al. 2017), serving as the foundation for selecting evaluation criteria.

Based on the seven essences of circular, in circular building projects, the building is treated as an environmental system with a focus on "human-centered" design elements. The criteria for evaluation must not only include conventional project tasks but also incorporate design elements specific to green circular buildings, as well as monitoring and control

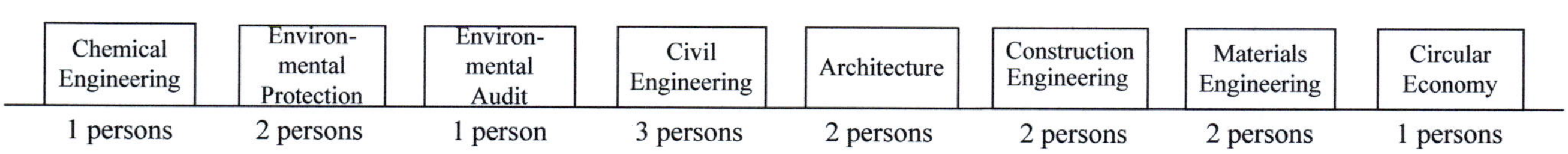

Fig. 3. Distribution of Expertise Fields in the Decision-Making Group. (cited by author)

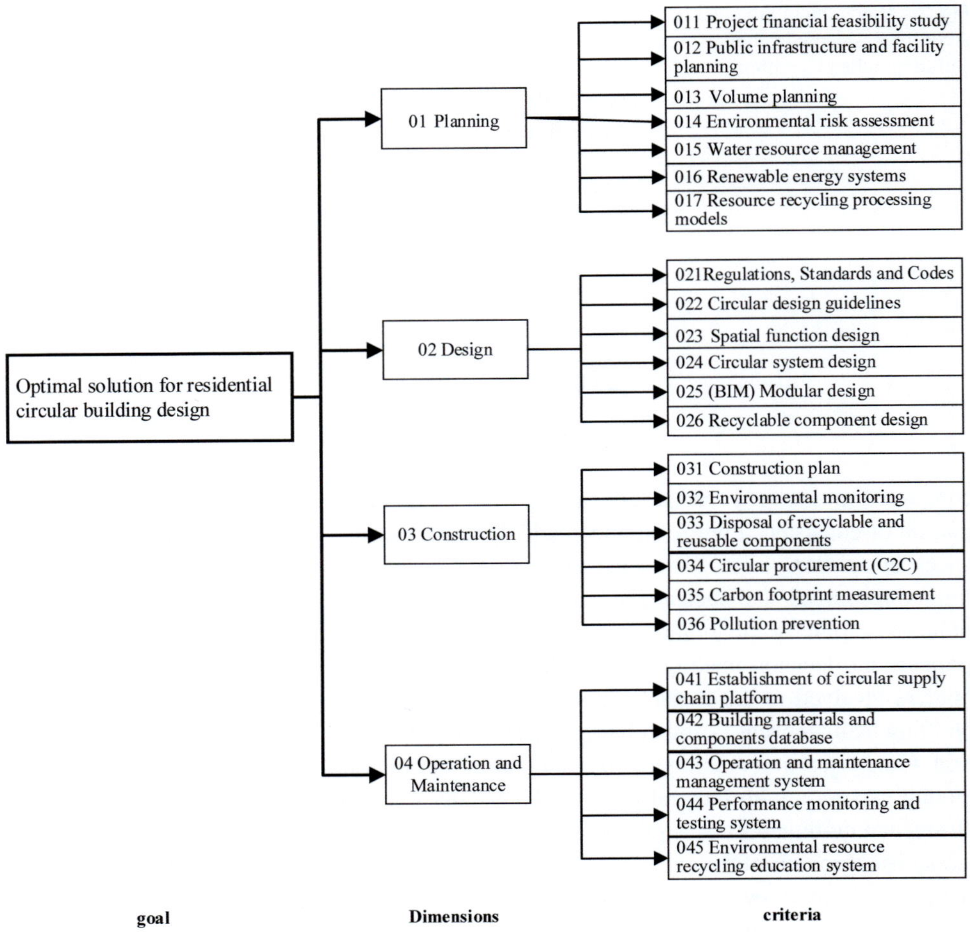

Fig. 4. Hierarchical Structure of Circular Building Projects. (cited by author)

systems. Additionally, the design should account for natural environmental resources such as land, water, sunlight, and wind (China's relationship with natural geographical environment elements), to optimize both external and indoor space environments. Figure 4. In order to meet the goals of circular buildings, in the phase of planning and design of the project, circular operations at each stage of the life cycle must be considered, and establish an evaluation structure with 4 dimensions and 24 criteria.

Table 1. Pairwise Comparison Matrix for Criteria under Planning Dimension

01_Criteria	011	012	013	014	015	016	017	υ
011	1	1	7	5	3	9	7	0.352
012	1	1	5	3	5	5	5	0.292
013	1/7	1/5	1	1/3	1/5	1/5	1	0.039
014	1/5	1/3	3	1	1	1	1/3	0.076
015	1/3	1/5	5	1	1	1	1/3	0.089
016	1/9	1/5	1/5	1	1	1	1/3	0.046
017	1/7	1/7	1/5	3	3	3	1	0.106

λmax=7.544 C.I.=0.09 <0.1

4 Computational and Analysis Results

The case study involves collaboration with a social housing development project in New Taipei City, Taiwan. The project decision-making team comprises 10 members with over 10 years of professional experience in various fields including municipal government administration, project management, architectural design, environmental engineering, scientific research, and academia.

Table 1, 2, 3, 4 and 5 present the hierarchical computation results of this study. The pairwise comparison matrix calculations involve normalizing the row vectors to form a positive reciprocal matrix. In this matrix, all elements are positive, and the relationships are defined by the function $a_{ij} = 1/a_{ji}$, $a_{ii} = 1, \forall i, j$, where a_{ij} represents the cross-comparison values between factors i and j as determined by the decision-makers, indicating their level of importance. During this process, individual expert preferences are integrated beforehand using the geometric mean method. The pairwise comparison matrices must be checked for consistency (Consistency Index ≤ 0) before integration. This involves consolidating the pairwise comparison matrices from each expert into a single matrix (Chang et al. 2013).

Table 2. Pairwise Comparison Matrix for Criteria under Planning Dimension

02_Criteria	021	022	023	024	025	026	υ
021	1	1	7	5	9	5	0.395
022	1	1	5	1	7	3	0.259
023	1/7	1/5	1	1	5	1	0.089
024	1/5	1	1	1	5	3	0.154
025	1/9	1/7	1/5	1/5	1	1/3	0.029
026	1/5	1/3	1	1/3	3	1	0.074

λmax=6.406 C.I.=0.08 <0.1

Table 3. Pairwise Comparison Matrix for Criteria under Construction Dimension

03_Criteria	031	032	033	034	035	036	υ
031	1	3	1/3	1/5	1/5	1/3	0.070
032	1/3	1	1/3	1/5	1/5	1	0.055
033	3	3	1	1/3	1/5	1	0.117
034	5	5	3	1	1	3	0.299
035	5	5	5	1	1	5	0.360
036	3	1	1	1/3	1/5	1	0.098

λmax=6.4574 C.I.=0.09 <0.1

Table 4. Pairwise Comparison Matrix for Criteria under Operation and Maintenance Dimension

04_Criteria	041	042	043	044	045	υ
041	1	1	1	1/3	5	0.192
042	1	1	1	1/3	3	0.169
043	1	1	1	1	3	0.215
044	3	3	1	1	5	0.366
045	1/5	1/3	1/3	1/5	1	0.058

λmax=5.193 C.I.=0.05 <0.1

The final analysis results indicate that the weighted ranking of the alternatives is as follows: Alternative C (Steel Structure) scored 65.1% > Alternative B (Precast Concrete) scored 25% > and Alternative A (Steel Reinforced Concrete) scored 9.8%. This result aligns with expectations. From the perspective of circularity and reusability, the Steel Structure has advantages such as complete recyclability, lightweight, high resilience, durability, seismic performance, and ease of construction. Compared to Alternatives A and B, the Steel Structure allows for more systematic assembly on-site, with easier project timeline management. In terms of maintenance, the costs are similar across the three options, but Alternative C (Steel Structure) offers the highest recovery value in the long term. Overall, the Steel Structure (SS) best exemplifies the benefits of circular architecture.

5 Conclusions

This study collaborated on a public housing development project, which is significant in the context of residential building environments. The circular economy is guiding the construction industry towards sustainable development, inspiring innovations in green circular building concepts. The primary goal is to reduce resource extraction, decrease high energy consumption in buildings, and curb CO_2 emissions. The materials used in the case study must be renewable resources to embody the essence of the circular economy – "starting with the end in mind". In the early stages of the project, resource and facility-sharing mechanisms were considered, along with the disassembly and reassembly of building materials' life cycles, to extend the building's lifespan and maximize the value retention of various parts.

Table 5. Summary of Pairwise Comparison Matrices at All Levels

01_Planning

0.335						
011	012	013	014	015	016	017
0.118	0.098	0.013	0.025	0.030	0.016	0.035
Alternative A			Alternative B		Alternative C	
0.002			0.003		0.009	

03_Construction

0.488					
031	032	033	034	035	036
0.009	0.007	0.015	0.038	0.045	0.012
Alternative A		Alternative B		Alternative C	
0.000		0.002		0.003	

02_Design

0.488					
021	022	023	024	025	026
0.193	0.126	0.043	0.075	0.014	0.036
Alternative A		Alternative B		Alternative C	
0.002		0.005		0.014	

04_ Operation and Maintenance

0.488				
041	042	043	044	045
0.010	0.009	0.011	0.019	0.003
Alternative A	Alternative B		Alternative C	
0.000	0.001		0.001	

Regarding the cost and benefits of circular construction, the total cost is approximately 10% higher compared to traditional construction projects. However, adopting modular, lightweight, and detachable circular building designs, with the entire building considered as a system for environmental design, achieved an overall reuse rate of 85% and a circularity rate of 65%. Considering the limited resources and inflationary benefits, the long-term benefits of circular construction are estimated to increase by 15%. The building's insulation and heat dissipation systems are key to its energy-saving performance. By leveraging local building advantages and incorporating passive building design features, the environmental monitoring of the building envelope effectively reduced air conditioning usage by approximately 48%.

This study assessed the project lifecycle phases, including general construction practices and the essence of circular construction, to ensure that the evaluation system meets the needs of circular building projects. The primary contribution of this study is that the proposed model, validated through practical case studies, can assist project decision-makers in selecting the optimal execution plan based on actual environmental resource conditions and can serve as a reference during the initial planning phase of residential circular building projects.

References

Chang, A.P., Lin, J.D., Chou, C.C.: Analytic Network Process (ANP)-selection of the best alternative in the promotion of participation in infrastructure projects. Int. J. Pavement Res. Technol. **6**(5), 612–619 (2013). https://doi.org/10.6135/ijprt.org.tw/2013.6(5).612

Chang, A.P., Yu, C.H., Lin, J.D.: Quality definition and essential consideration of circular economy. Qual. J. **53**(5), 38–42 (2017). https://www.airitilibrary.com/Article/Detail?DocID=10173692-201705-201706070019-201706070019-38-42

Chang, A.P., Chang, S.L.: Build up circular city planning index with EAM. Earth Environ. Sci. **531**, 012016 (2020). https://doi.org/10.1088/1755-1315/531/1/012016

Bergmans, I., Bhochhibhoya, S., van Oorschot, J.: Assessing the circular re-design of prefabricated building envelope elements for carbon neutral renovation. J. Facade Des. Eng. **11**(2), 169–196 (2023). https://doi.org/10.47982/jfde.2023.2.A4

Ellen MacArthur Foundation. Towards the Circular Economy. Founding Partners of the Ellen MacArthur Foundation (2013). https://www.ellenmacarthurfoundation.org/towards-the-circular-economy-vol-1-an-economic-and-business-rationale-for-an

Graaf, D.D., Hamada, K., Gruis, V.: Circular Buildings constructing a sustainable future (2022). Accessed 5 Feb 2024. https://circulareconomy.europa.eu/platform/sites/default/files/nl-branding-circular-buildings.-f.pdf

Rahla, K.M., Mateus, R., Bragança, L.: Implementing circular economy strategies in buildings—from theory to practice. Appl. Syst. Innov. **4**(2), 26 (2021). https://doi.org/10.3390/asi4020026

Behavioral and Policy Approaches to Climate Change Mitigation

Sustainability Science: A Guide to Dealing with Problems in Climate Change and Solutions for Policy Practice

Ariel Shu Xian Toh

Abstract

Research on sustainability science has gathered diverse perspectives in terms of thematic definitions, principles, and theories or practices. This article provides an extensive review and an updated research agenda for the sustainability science field. This paper addresses knowledge gaps in advancing science and solutions for sustainable development. The author explains how sustainability science is an important field as an education mechanism in managing the problems of climate change, and to enable effective policy practices. This research outlines the nature of sustainability science, including a discussion about its role in addressing climate change and sustainability agendas. The research shows that the scope of sustainability science has broadened with connections to multidisciplinary scientific disciplines. At the same time, the author acknowledges that the grand complexity of climate change in the present time remains a concerning practice in the policy field. One key practice is advancing experiments with science, technology, and innovation (STI). STI is observed to be an integrated knowledge domain that is much required to function as a transformative model of sustainability science.

Keywords

Climate Change · Complexity · Education · STI · Sustainability Science

A. S. X. Toh (✉)
Marcharh Sustainability Science Excellence Centre, One Canada Square, London E14 5AA, UK
e-mail: arieltoh@nachaltigkeit.com

1 Introduction

1.1 Reflections on "Sustainability" and "Sustainability Science"

On May 8th, in the year 2024 at Malaysia; a professional sustainability workshop for companies was held at one of the Big Five Tech offices. During the workshop, one of the participants shared with an expert in sustainability about his concerns. The participant had perplexed introspection about defining sustainability. He (*the participant*) questioned how he and his children interpreted sustainability. He confidently shared that as he imagined young children playing on a wooden-made boat (commonly known as "sampan" in Malaysia); – that this perfect scenario can be define as sustainability for his interpretation. To him, he defined sustainability as the case that the future generation is at risk of not enjoying traditional cultures or contemporary approaches of transportation. He genuinely had a clear idea that sustainability would require future generations to imagined what life was like in the historical periods where technology was playing a less prominent role in everyday lives. As he shared this scenario and definition of sustainability with the expert, the sustainability expert was trying to understand whether this individual man understood truly why issues surrounding sustainability existed. The sustainability expert was benumbed that during the conversation with the participant who shared his view, he showed zero interest or compassion about environment or climate change. The men strongly defined sustainability as when human cultures and traditional consumption activities will cease to exist for future generations. This definition was out of the norm in the position of the sustainability expert. The shared interpretation about sustainability was an unconventional theory and practices related to sustainability.

Based on these real accounts, even in professional settings which are advocating for "Sustainability" objectives; every

B. He et al. (eds.), *Towards the Framework of Livable and Resilient Cities* Advances in Science, Technology & Innovation, https://doi.org/10.1007/978-3-031-97849-4_27

individual and society has a different interpretation of "Sustainability". However, if a sustainability practitioner takes a closer look, the underlying ethics of sustainability is inherently linked to a collective value concerning conservation, preservation, resiliency, quality of life, or well-being. All these values are more often prioritized as actions towards climate action, sustainability practices, and sustainability decisions. Nevertheless, the subject matter of sustainability has become more complex with the division of human ethics, disharmony of practitioners' values, or the lack of understanding about the historical accounts that led to the establishment of global policies to ensure sustainability practices become a regime. Sustainability therefore entails a body of knowledge, whereby this knowledge comprises a scientific comprehension about every concept, theory, and hypothesis concerning the world of sustainability. The world of sustainability is continuously evolving with multiple events such as climate change, to biodiversity crisis, to health risk, and societal challenges. All these events backed up by the idea of pursuing sustainability-oriented actions, and urgently demanding new solutions and societal interventions. Thus, sustainability science (SS) has become a pre-eminent field by the year 2020 to provide relevant knowledge and effective solutions addressing sustainability problems. The report from UNESCO (2017) advocated that sustainability science (SS) represents an important knowledge field to fulfill the 17 Sustainable Development Goals (SDGs). Following to this emerging field, scholars have launched initiatives to promote research and publication on the theme of Sustainability Science (SS). One of the oldest journal series was established since the year 2006: "*Sustainability Science*" journal published by *Springer*. This journal defined the scope of Sustainability Science as trans-disciplinary in an academic discipline, which addresses a holistic knowledge of global natural systems, social systems, and human systems. A similar perspective on sustainability science has been understood by another publisher: *PNAS*; which highlights that the theme of sustainability science investigates challenges in meeting the needs of the present and future generations. In another example of a journal called "*Sustainability Science and Technology*" published by *IOP Publishing*; sustainability science (SS) is dedicated to advances in science, technology, and engineering to contribute to planet sustainability and the UN SDGs. Reflecting on the conceptualization of SS based on different scientific publishing content, SS is clearly understood to be the following:

1. *Sustainability science (SS) aims to improve scientific knowledge and scientific theories, which leads toward efficient solutions for addressing sustainability outcomes and transforming the world.*
2. *Discussion focusing on all aspects of "Sustainability", are the foundations and core principles that shape the curriculum, practice, and applications of sustainability science (SS).*
3. *Experiments and education in the sustainability science (ss) field present information and the discovery of new knowledge concerning a worldview approach.*
4. *Sustainability science (SS) seeks all sciences and challenges the intersecting field of life sciences, humanities, and physical sciences to tackle planetary issues and global challenges.*

Sahle et al. (2024) research identified that to advance discovery in SS field, new research should explore the intersections of the theme around SS, cross-cutting issues, and innovations in sustainability education. In line with this recommendation, this chapter incorporates an all-inclusive review research technique that examines both theory and practices of SS. The all-inclusive review technique covers an exhaustive literature review process which includes an acumen analysis from the feedback provided by a panel of scientific experts.

1.2 Advancing Research and Practices on Sustainability Science (SS)

Scholars globally are determined to align with the mission of several calls for publishing and to uptake scientific pursuits, and thus; staging breakthrough innovations in sustainability. The field of SS has become more dominant post-year 2020; setting out interdisciplinary scientific methodologies and embedding systematic thinking approaches to tackle the world's pressing challenges. To some extent, the understanding of SS also covers subject matter in law, politics, and policy (Heinrichs et al., 2016). In the study of Tahvilzadeh et al. (2017), the authors identified the existing status quo in enabling sustainable development interventions. Reflecting at the introduction section, about the diverse interpretations of "sustainability" by society; every individual has a different depth of comprehension and depth of responsibility in sustainability issues. On the other hand, sustainability practices are also dependent on external factors. Amongst the factors include resources available to mobilize key actions, to increasing multiple events concerning society, health, world peace, or war; which can steer away societal focus to participate more actively in issues around climate change or sustainable development. In everyday situations, society therefore depend upon stimuli that can drive them to consistently persuade sustainability-related causes, while having an in-depth understanding of sustainability. The stimuli can be postulated to be steered by policies, or by normalizing the education of SS. On the other hand, designing effective policies will require a deeper understanding of different narratives about sustainability and climate change. Discourses in sustainability decisions are often due to contradicting perspectives of decision-makers. In this sense, the common "Stakeholder Theory" can be least important when relooking at the application of SS. The application of SS would more often challenge and investigate societal ethics, values, and belief systems as emphasized in the "Value Theory" by Joseph Raz. Therefore, establishing efficient policies and rational regulatory interventions are also required to form synergies in societal action.

Society's debate about climate change has a potential to shape future-centric theories and hypotheses which eventually unfolds to become a grounded approach to form solutions that are widely preached. Climate change therefore is knowledge-driven, depending on both reliable information and ethical practices of society. The implication is that the SS field is comparable to the cliché of factors of climate change and when understanding stimuli of efficient societal practices. SS is perceived to be a pivotal scientific discipline that can shape future decisions and effective policies to manage complex problems such as climate change. This study therefore investigates the background of the field of SS; intending to understand how advances in science, technology, and innovation (STI) are inter-related. This chapter includes discussions drawn from literature reviews that examine the underlying approaches of SS. The findings make connections to how SS is an important subject matter in the problems of climate change. Amongst the research questions that are addressed in this study are the following:

i) What is sustainability science (SS)?
ii) How can sustainability science (SS) play a role in climate change?
iii) Which transformative models enable sustainability science (SS)?
iv) What are examples of sustainability science (SS) approaches that are relevant to policy implementation?

The outcome of this research frames a brief guide that can be useful for academics, scientists, practitioners, decision-makers, educators, and any interested party who is searching for a general reference text that summarizes anterior principles and practices of SS. The analysis in this article is descriptive; incorporating a qualitative research approach. The content analysis performed in this research aims to capture a prudent view of the topic in an unbiased manner.

2 Methodology

The research began by selecting keyword-specific journal articles containing the terms - "sustainability", "sustainability science", and "climate change". These three keywords are prioritized because it is most relevant to answering the research questions. Additionally, the search results will provide a larger sample of literature focused on the focused topic. As there are also limited sources of Scopus journal articles intently containing detailed assessments on SS; the literature review intently referenced journal articles as far as two decades in history (E.g. year 2010). The rationale reason of doing so is due to the case that these historical journal articles provided the empirical scientific theory and history of SS. Additionally, some back-dated articles provided an evolutionary perspective on the subject matter. Bearing in mind that SS continues to be an emerging scientific discipline and a transforming academic field, this article undertakes a wide literature review matrix. The literature review entails analyzing 26 judicious articles that provided insights about SS in section three of this article. Apart from the literature review from Scopus-indexed articles, the study captured feedback and research findings from an international conference discussion between 18th to 20th September 2023, on the scope of engaging science policy to address the SDGs. The research findings shared in this conference provided observational data, where participants and the scientific committee verified the described interpretations of the principles of SS. The stakeholder experts' views at the conference also provided scientific recommendations mentioned in the work of ASX Toh (2023). These recommendations are then analyzed in the format of "reflective learning" to further synthesize SS-related educational experiences in both theory and practice.

The methodological framework of this research is shown in Fig. 1.

3 Understanding Connections Between Sustainability Science (SS) to Climate Change and Beyond

Over the three decades, scholars have actively researched the challenges and solutions concerning the world's environment, natural resources, and society's well-being. Reflecting on the urgent matter of climate change, there are advantages to learning better about how specific scientific disciplines or scientific practices are linked to socio-environmental processes. This includes the influence of policy, which requires contribution from an interdisciplinary or transdisciplinary science method. The table in this section provides a literature review matrix to analyze the underlying connections between SS to climate change, including policy dimensions. This literature review provided two main assessments:

1. Signals and indications where evolving scientific approaches such as SS are relevant to problems of climate change and policies.

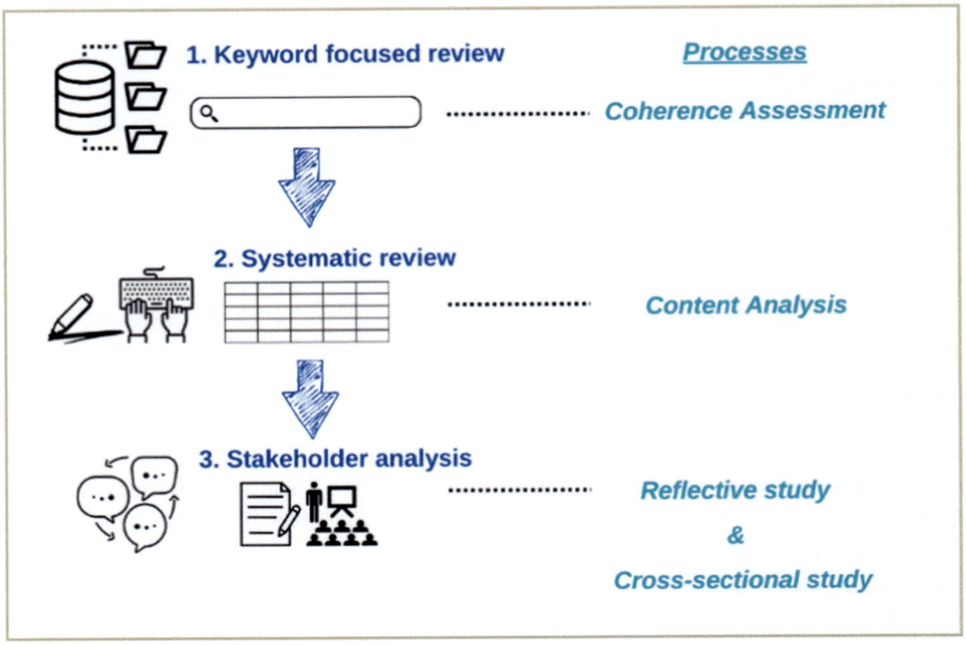

Fig. 1. Methodological framework of research (By author)

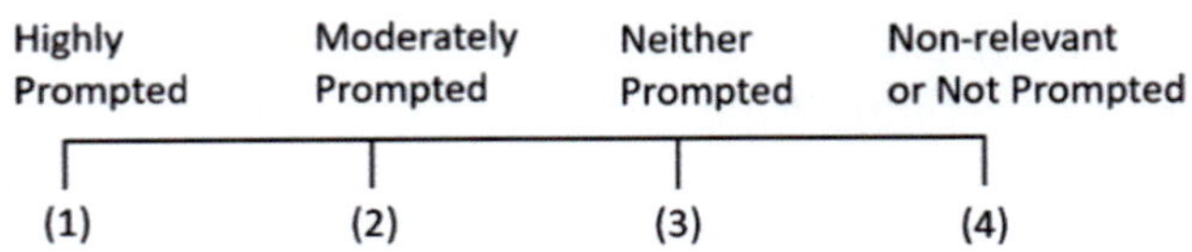

Fig. 2. Interval scale to measure the level of emphasis that was mentioned in the literature article on the field or subject matter of SS. (By author)

2. Across all types of world crises, a closer insight into how climate change subject matter is a complex theme. Thus, the analysis indicated evidence that sustainability is intertwined with a paradigm of the perceived skills, knowledge, approaches, or technical understanding of society. The last column of the literature review matrix includes key statements in the reviewed articles. The statements or concluding remarks are further assessed through a reflective learning process. In this process, the content is investigated in-depth to capture any indications whether *1. Is there a need for SS?;* and *2. Is SS a future step when addressing climate change issues?.*

An interval scale is also used to measure the approximation of the article prompting to what extent is the application of the field of SS, in the research paper studied. The interval scale is evaluated as shown in Fig. 2.

(1) "Highly prompted" means the literature is highly inclined to hint or prompt about the study of SS or indicated detailed context related to SS.
(2) "Moderately prompted" means the literature has hinted or prompted about the study of SS or context related to SS in some sections of the literature paper.
(3) "Neither prompted" means the literature has not inherently mentioned the study or context about SS, especially either using the term "SS". However, the literature may have subtly indicated other areas which can correlates to SS.
(4) "Non-relevant" or "Not prompted" means that the literature has not focused on any study of SS, or mentioned a focused context of SS.

The second last column in Table 1 which refers to "Accompanying Academic Discipline", highlights the academic discipline that is primarily existing to be applicable or relevant to the entire research article. This part of the assessment identifies which fundamental academic disciplines that are relatable to the domain of SS based on each literature source. Figure 3 shows the methodological framework to conduct the literature review matrix.

4 Discussion

The literature review matrix and stakeholder analysis of the research article discussed by ASX Toh (2023), provided updated perspectives about the following:

- The evolving global challenges concerning climate change (CC) and sustainable development (SD) have widely been studied. These studies shed discovery of information, which communicates scientific data advocating for sustainable solutions. These transitions are repeatedly based on the policies and framework of the Brundtland Commission and the 17 Sustainable Development Goals (SDGs).

Table 1. Literature Review Matrix in Content Analysis of Climate Change as a Foreground of Sustainability Science (By author)

Reference	Report on Challenges of Climate Change (CC)	Report on Solutions Related to Climate Change (CC)	Research Observations	Accompanying Academic Discipline	Approximation to Sustainability Science (SS) Author "Key statements" or reflections mentioned
Wittmann & Meissner (2024)	CC is ranked as the most critical global risk. This risk requires future-oriented knowledge Solutions to global problems are linked to policy transitions and state-level commitments Transition to Sustainable Development (SD) requires an immediate transformation in university education to enable transdisciplinary skills	A systemic approach in university scientific governance and education Cross-cutting fields such as science diplomacy with integrated subjects "Science diplomacy" as a solution to environmental challenges	"Global challenges ahead of humanity can only be tackled by trans-disciplinarity & open-mindedness of science." "Global sustainability considers the understanding of the role of science with a scientific approach that develops concepts or opens discourses" Measures in the field of STI to be coordinated with policy agents; & thus, to create trans-bordering communication and collaboration platforms	Global Studies Global Policies World Development Innovation & Technology	2 - [Moderately Prompted to SS] Study reflected on the growing existence of the SS curriculum & indicated the need for a "Global SS" approach
Hsien & Yeh (2024)	CC entails challenges in the domain of politics, economics, social, & cultural barriers, & inequality Integrated issues are linked to both CC & SD such as water-energy-food (nexus). These nexuses are also connected to the SDG framework	Formulate SDG strategies & embed climate risks into risk management frameworks Adequate education, good policy guidance, & consistent public participation are required to achieve SD	Trending topics in CC covers climate mitigation, climate adaptation, health, environmental ecology, & global security Research articles published during 2018–2020 showed an inclination in cross-disciplinary disciplines in addressing CC subject matter Trending solutions to CC entail impact assessment, policy, adaptation strategies, education, & synergies with SDG strategies	Environmental Education Environmental Policy	3 - [Neither Prompted to SS] "The study contributes to the development of interdisciplinary research to understand complex interconnections between CC" "SDG requires integration of knowledge from multiple fields & disciplines"

(continued)

Table 1. (continued)

Reference	Report on Challenges of Climate Change (CC)	Report on Solutions Related to Climate Change (CC)	Research Observations	Accompanying Academic Discipline	Approximation to Sustainability Science (SS) Author "Key statements" or reflections mentioned
Ellis et al. (2024)	Environmentalism, sustainability principles, & conservation practices must be culturally embedded in social identities or political decisions Multiple factors of equity and social inequality challenges exist in CC issues which requires frameworks to foster inclusivity	Nature-based Climate Solutions (NbCS) include practices in engineered ecosystems Nature-based Solutions (NBS) involve broad practices addressing societal pillars Natural Climate Solutions (NCS) emphasizes climate change mitigation practices that involve changes in human stewardship of ecosystems	Future CC solutions are holistic in principles, which ensures to also sustain global natural biodiversity & deal with complex social equity challenges Future solutions in CC are measured by science-based & climate-driven methodologies. e.g., ensuring climate scenarios are assessed	Applied Science Environmental Science Biology Engineering	3 - [Neither Prompted to SS] The study reflected that advances in science & technology measurement systems can establish better environmental monitoring methodologies or CC solutions The study proposes a holistic CC solution with the integration of applied science approach
Stefanis et al. (2024)	Challenges of CC & environmental sustainability must be addressed through integrated methodologies encompassing technology, innovation, & policy Transformative interventions in socio-technical systems are required to address CC problems Global disruptions such as health pandemic crisis are linked to CC factors	Transition to clean energy solutions requires consideration of multiple factors & dimensions to fully achieve the SDGs Ecological footprint compliance described in standards or policies solves complex environmental sustainability problems A multifaceted policy implementation fosters effective SD strategies	Research since the year 2020 focuses in the subject matter of sustainability, circular economy, CC, bioeconomy, decarbonization, & impact assessment Bioeconomy & biological processes in economic systems are intertwined with policy innovation International policy packages enables sustainability practices & establishment of a blueprint for CC solutions	Environmental Policy Innovation & Technology	4 - [Not Prompted to SS] Study reflected that innovation & policy are critical pillars to establish effective sustainability initiatives or framework Study reflected that academic themes are evolving in broader policy-oriented goals such as the SDGs, to circular economy, technological innovation, & sustainable resource management
Zare et al. (2024)	Traditional problem-solving approach or stand-alone natural science methods to address wicked problems such as CC is ineffective Knowledge gaps exist between the theory & practice of science, whereby the knowledge transfer & translation of skills are ineffective	Future research skills & models to tackle real-world challenges need to adopt practical knowledge & innovations from different disciplines Design science as a scientific discipline, can address ill-structured problems, & bridge knowledge gaps between science & practice	Innovation in socio-environmental systems improves knowledge transfer in science Depending on pure science or practical research as solutions of CC is inadequate SS gaps exist when knowledge domains of applied science and interdisciplinary science are lacking	Business Education	3 - [Neither Prompted to SS] Study reflected on solution-seeking approaches such as design science to address socio-environmental challenges

(continued)

Table 1. (continued)

Reference	Report on Challenges of Climate Change (CC)	Report on Solutions Related to Climate Change (CC)	Research Observations	Accompanying Academic Discipline	Approximation to Sustainability Science (SS) Author "Key statements" or reflections mentioned
Pickett et al. (2024)	Social & biophysical factors affects transformation of urban systems e.g., acceleration of social inequality or risk of environmental regeneration Urban world & global diversity are affected by new adaptation patterns of global environment systems	Establish a synthesis approach in science to develop understanding of urban science Solutions to CC requires transdisciplinary, with interlinkages to social-ecological-technological systems (SETS)	Understanding the subject matter & practical skills of urban planning or urban ecology addresses wide CC challenges Urban ecology studies enables knowledge transfer and effective practices across science, policy, and sustainability practices	Education Ecosystem studies Built environment Ecology	3 - [Neither Prompted to SS] Study reflected the need to employ "pluralistic & transformative mode for science to deal with CC"
Poliseli & Caniglia (2024)	Challenges of CC & environmental sustainability must be addressed through integrated methodologies encompassing technology, innovation, & policy Lack of attention in promoting collaboration across arts & physical sciences	Complex socio-environmental problems require multidimensional approaches. e.g., connecting knowledge from arts discipline to other scientific disciplines to tackle CC Inclusion of arts sciences & humanities disciplines with other scientific practices drives a system thinking process	Multidisciplinary discipline develops complementary solutions rather than a linear solution Using just one scientific method will be a weaker approach to address challenges of CC Philosophy of science studies enriches the understanding of multiple relationships across scientific theories, scientific practices, & global climate knowledge	Arts & Humanities Climate & Earth Systems Science	2 - [Moderately Prompted to SS] Study provided evidence of how CC issues are interconnected to the combination of climate science & multidisciplinary scientific disciplines
Trott et al. (2023)	Most research on impacts of CC focused on varied negative impacts. There are opportunities to consider how CC impacts can affect human development & social resiliency Perceived lack of scientific literature evidence about understanding how climate educational intervention can significantly promote positive human development	Youth participation in CC solutions is more active in community-centric initiatives Introducing environmental education at a community level, promotes greater human development in the specialization of CC issues Civic engagement through civic science enhances pro-environmental behaviours	CC issues involves a multidimensional process in the factors of social, political, & economics Informal education program aimed to upskill a community scientific knowledge & technology skill can promote climate resilience Building interconnectedness between technical scientific knowledge & knowledge of the natural environment systems permeates effective environmental remediation	Social science Citizen science	3 - [Neither Prompted to SS] Study reflected the role of civic science & STEM (Science, technology, engineering, and math) programs in addressing CC

(continued)

Table 1. (continued)

Reference	Report on Challenges of Climate Change (CC)	Report on Solutions Related to Climate Change (CC)	Research Observations	Accompanying Academic Discipline	Approximation to Sustainability Science (SS) Author "Key statements" or reflections mentioned
Keys et al. (2023)	Gaps in the field of Anthropocene science to predict future solutions & to address critical ecosystem services Conflict in Earth system boundaries & models for SD are affecting informed policy decisions	Practices of human-earth systems science require a transformation in practices Anthropocene science produces knowledge about the changing planet ecosystems & informs social-economic narratives in conservation	Anthropocene Science is an updated scientific discipline embedding ethics, values, behavioral changes, & planetary stewardship – all of which are human-environmental oriented Epistemological dimensions of existing scientific disciplines should be better understood	Systems science Complexity science Earth science Ecology	4 - [Not prompted to SS] Study reflects how the importance of a climate scenario projections approach is a better approach at addressing the broad dimensions of the geophysical, social, & technological world
Riggs et al. (2023)	Policy-level decision-makers often misunderstand how environmental systems are complex dynamic systems Gaps still exist in the methods of how research is conducted & how collaboration is engaged. More experiential learning & practice of knowledge is required in society	Transdisciplinary research & learning approaches address complex challenges of sustainability issues Importance to widened environmental education to embed soft-skills competencies such as communication, social science, & collaborative problem-solving knowledge Theory of Change (ToC) can aid organizations & agents clarify goals, making decisions, defining strategies, & monitor progress in the working scope of sustainability	Learning pathways can be emphasized to explore research questions & to innovate for new scientific frontiers Transdisciplinary science connects the nexus between science, policy, and innovation Transdisciplinary science deals with uncertainty, disruption, & transition processes of sustainability landscapes	Environmental science Social science	2 - [Moderately Prompted to SS] Study reflected the important role of embedding transdisciplinary systems in understanding environmental science Study referenced learning framework & learning pathways of SS
Song et al. (2023)	Integrated approaches are a necessity to solve sustainability challenges. This includes the way data-driven approaches provide climate information Management of big data & multi-faceted digital systems to address sustainability information is complex	Future cyber infrastructure systems to support big data in climate will entail advance analytics, artificial intelligence (AI), & open science Interoperability of data is part of a SS information system so that interdisciplinary or multidisciplinary information & models can operate	The majority of data for climate information stemmed from multiple scientific disciplines Future cyberinfrastructure systems must integrate an SS-driven framework The growing need for digital solutions for sustainability is driven by the contribution of multiple scientific disciplines (e.g. Ecology, engineering, social science, geosciences, health)	Computer science Data science Innovation & Technology	2 - [Moderately Prompted to SS] Study reflected that big data & cyber ecosystem for data management advances the field of SS

(continued)

Table 1. (continued)

Reference	Report on Challenges of Climate Change (CC)	Report on Solutions Related to Climate Change (CC)	Research Observations	Accompanying Academic Discipline	Approximation to Sustainability Science (SS) Author "Key statements" or reflections mentioned
Deng et al. (2023)	Addressing grand challenges related to sustainability & CC entails interdisciplinary solutions	Incorporation of environmental science knowledge provides an understanding of anthropogenic processes & socio-biological ecosystems	Each independent scientific discipline whether natural science or physical sciences, has the mutual ability to address the topic of CC & sustainability systems New scientific measurements, improved predictions, & spatial-temporal assessments can be made if interdisciplinary approaches are practiced	Environmental science Engineering Geology Education	1 - [Highly prompted to SS] Study reflected that combining interdisciplinary scientific disciplines such as environmental science, engineering, & geosciences can enable better scientific study in sustainability themes
Malt & Majid (2023)	Human behaviour & human perception of the natural environment are primary drivers to address CC Public policy has enormous power to influence increasing scale of anthropogenic activities Actions which causes CC are mainly driven by mankind intellect, emotions, & practices	Cognitive science plays a critical role to sustainability interventions. e.g., psychology discipline relates to motivational theories & social drivers to influence pro-environmental behaviours Understanding a sound knowledge of the earth & human challenges promotes effective reasoning when making sustainability decisions	Knowledge is translated via arts, humanities, & interacts through the theory of social sciences Decisions & solutions on sustainability can be affected by conceptualization of belief systems Forming "causality" is critical to establish connections of cause & effects of CC Knowledge systems (e.g., mental model) provide information on complex sustainability subject matter	Psychology Social science Information science	2 - [Moderately Prompted to SS] Study reflected on the use of anthropology, linguistic, philosophy, social science, & cognitive science perspective to establish Study reflected that investigating conceptual foundations of sustainability should explore beyond basic science
Kovacic, Z. & Marcos-Valls, A. (2023)	The understanding & practice of interdisciplinarity varies amongst different individuals Misconceptions in science perspectives & scientific communication exist	Enhancing communication & cultural exchange among scientists prevents scientific conflicts Science & technology studies (STS) embodies the qualities of interdisciplinary research & education	Recent European university funding research is planned in interdisciplinary settings to address CC CC is a common theme for all post-grad programs (i.e. PhD), with a focus for multiple disciplinary, trans-academic, & interdisciplinarity	Environmental science Education	4 - [Not prompted to SS] Study reflected on the challenges & opportunities of incorporating different scientific disciplines to respond to multi-dimensional issues such as CC

(continued)

Table 1. (continued)

Reference	Report on Challenges of Climate Change (CC)	Report on Solutions Related to Climate Change (CC)	Research Observations	Accompanying Academic Discipline	Approximation to Sustainability Science (SS) Author "Key statements" or reflections mentioned
Tambe et al. (2023)	In "moderately structured problems", science is debated by two parties with intent to strengthen perspectives or scientific arguments In "unstructured problems", there is no mutual agreement amongst decision makers Sustainability challenges intersects with societal behavior & affects across scientific knowledge, policy, scientific practice, & politics	Knowledge coalitions can be used as a strategic form to solve scientific arguments Different scientists play different defined roles to ensure that specific information, knowledge, or solutions are applicable to policy agents	Research acknowledged scientific evidence connecting knowledge & practice across science & policymaking; creating the term: "science policy practice interface' (SPPI) Construction of solutions or policy decisions are dependent on the three factors: *1. The way knowledge gaps are managed, 2. How coalition in decisions are formed, & 3. How competition in science is eliminated*	Policy Decision science	4 - [Non relevant to SS] Research reflected on how wicked problems & conflict in science practice can be intervened through strategic science Study reflected on how in sustainability scenarios; science & policy both can influence social agents' decisions. e.g., policy can divert social agents to over-ride scientific facts
Kuo, S. et al. (2022)	Global environmental problems spur new integrative & interdisciplinary studies to influence policy decisions When dealing with SS studies, the sustainability issues accounts complexity, deep uncertainty, incorporates temporal scale of factors & range of spatial data There are gaps in linking science & society to policy measures	SS approach drives co-production of knowledge, to which knowledge also leads to effective action Pathways of SS incorporates both science & technology in decision making for sustainability	SS embodies the study of science & society Socio-technical systems are linked to the sustainable trajectories in dealing with CC issues Research in SS makes it possible to construct systematic information (e.g., CC data)	Environmental science Earth science Physics Information science	1 - [Highly Prompted to SS] Study reflected on the historical conceptualization of SS from the year 2001 to 2014, "SS embeds co-production of knowledge in an interdisciplinary manner, cross-scale dimensions, & work at the interface of research & practice"

(continued)

Table 1. (continued)

Reference	Report on Challenges of Climate Change (CC)	Report on Solutions Related to Climate Change (CC)	Research Observations	Accompanying Academic Discipline	Approximation to Sustainability Science (SS) Author "Key statements" or reflections mentioned
Neta et al. (2022)	Robust evidence-based interventions are increasingly required to address climate health challenges There is a need to strengthen climate adaptation strategies & establish more capacity building to manage repetitive CC occurrences	Field of "Implementation science" bridges gaps in research & scientific practice, scaling up more effective climate health solutions Evidence-based health interventions will include tools, programs, & policies in scientific practice	A consolidated framework of "Implementation science" approach is discussed to conduct new climate health research or to deploy solutions The "Implementation science" framework considers multisectoral & multilevel sectors of decision makers The framework allows policy practice & solutions which consider inner settings (e.g., culture, governance) & outer settings (e.g., incentives, policy factors)	Health & medical science	3 - [Neither Prompted to SS] Study reflected on the establishment of a scientific model or scientific model to address resurging CC-related events Through the framework of a scientific model, establishing capacity building in the knowledge of "climate science" is suggested
Shrivastava et al. (2020)	CC drives global-change oriented research & solutions, requiring sustainability scholarship, innovation, & collaboration Business as usual approaches & economic growth challenges the carrying capacity of environmental systems. In this scenario, planetary boundaries are breached	SS must embed transdisciplinary settings in enterprises SS integrates social sciences, arts, humanities, & natural sciences SS is problem driven, with active social engagement to scale up global change	Science as a field, has inequalities in global communities. Access to Science is still limited at areas where CC is concerned Funding for Science initiatives are lacking Current SS approaches lack attention to social & human dimensions. Requires holistic integrated disciplines	Environmental science Education	1 - [Highly Prompted to SS] "SS emerged from recognizing the need to increase global-change research & solutions, which requires interdisciplinary & multidisciplinary approaches" "Global SS contributed to global agreements to address challenges of CC"
Chiu et al. (2020)	State-of-the-art scientific knowledge is required in today's urgent direction to establish rational & evidence-based policies	To address socio-economic impact, natural sciences should be integrated along with technology systems, management practices, & public policy	Nuances still exist in science-policy interface. e.g., geographic, temporal, or sociological limitations Replication of scientific methodologies leads to unreliable or invalidity in solutions where sub-problems are not effectively addressed	Education Pedagogy	2 - [Moderately Prompted to SS] SS embodies multi-disciplinary & transdisciplinary approaches to address wicked problems

(continued)

Table 1. (continued)

Reference	Report on Challenges of Climate Change (CC)	Report on Solutions Related to Climate Change (CC)	Research Observations	Accompanying Academic Discipline	Approximation to Sustainability Science (SS) Author "Key statements" or reflections mentioned
Ojha, H. R. (2020)	There are real accounts when investigating CC events, but social & policy dimensions are lacking Lack of interdisciplinary risk assessments in CC events leads to narrowly construed science-based recommendations New CC crisis is evolving while knowledge about environmental systems is also growing. Yet, more scientific work & policy solutions are required to intervene in CC	Different system thinking & systemic methods are required in scientific research to deal with large body of CC information, & to understand CC complex challenges Predictions, risk assessments, & projections about CC must link with the entirety of science, & to link both practice-based knowledge & policy communication	Lack of interdisciplinary approach to research & scientific solutions leads to incomplete results, maladaptation, or worsening CC events Standalone scientific discipline: e.g,. Environmental science, may gather information that are too narrowed to the theme. Environmental science studies must incorporate other techno-scientific disciplines SS is a multi-scalar problem driven field, guiding sustainability transitions through adaptive & action-oriented research & practices	Environmental science Education	2 - [Moderately Prompted to SS] Study reflected the current example in environmental science, where science as a research practice or knowledge seeking should engage in a SS setting Study proposed "Himalayan sustainability science", as a transdisciplinary & critical action research approach
Wamsler, C (2020)	Though sustainability scholarship & education increased analytical advancements; there is a lack of effective sustainability changes in the present world ecosystems Ethics & virtues of society need to be intervened through non-mainstream educational approaches	Societal inward dimensions & humanities pedagogies are increasingly needed to drive SD Societal motivation & personal practices to adopt sustainability stewardship require SS	The concept of cognitive science & theory of mindfulness can contribute to scientific transformation in society Experimenting with new teaching approaches can promote sustainability practices. e.g., Reflexive teaching Human-sustainability connections are enhanced through the activation of intrinsic transformation	Humanities Pedagogy Education	2 - [Moderately Prompted to SS] Study reflected the use of SS approach to promoting cognitive, emotional, & humanities-related transformation

(continued)

Table 1. (continued)

Reference	Report on Challenges of Climate Change (CC)	Report on Solutions Related to Climate Change (CC)	Research Observations	Accompanying Academic Discipline	Approximation to Sustainability Science (SS) Author "Key statements" or reflections mentioned
Mach & Kraan (2020)	Assessment of "uncertainties" in CC is a difficult process, as there are cascading impacts interlinked to several sources of knowledge CC impacts & responses are understood & valued differently, which causes inequitable decisions in SD CC is interconnected to socio-economic & environmental conflict	Science-policy interactions can aid in establishing effective practices & knowledge relevant to CC conflicts An integrated scientific methodology in CC studies, CC assessments, & CC solutions deals well with the multifaceted nature of CC issues	Science-society dynamics are important in CC research Diverse scientific evidence grounded by multiple disciplines is crucial to creating effective policy practices A combination of social dimensions & policy sciences leads to more effective climate adaptation solutions	Education Social science Policy	3 - [Neither Prompted to SS] Study reflected on the combinatorial science approach's role in addressing uncertainties & conflict in CC issues
McGreavy, B., & Hart, D. (2017)	Problems of CC are wicked, have tipping points, thresholds, & social conflicts CC influence other several socio-environmental dimensions, which will require a broader study approach Presence of "boundary work" to permit effective knowledge transfer & translation of information are challenging	SS is a promising conceptual & analytic framework to ensure SD Communication field drives socio-technical transitions & policy outcomes in SS	SS shapes collaborative research & socio-technical or policy-oriented outcomes Environmental communication approaches establish climate adaptation action SS & environmental communication are both contributing to policy outcomes	Environmental science Arts & Humanities	1 - [Highly prompted to SS] Study reflected the important emerging theme of SS & how communication shapes pathways to SS Study boundaries of the SS disciplines
Nelson, M. P. & Vucetich, J. A. (2012)	Role of science addressing global sustainability challenges is still unclear Ethical dimensions in sustainability causes market failures	Both technological & political knowledge must be better understood to form effective SS solutions Transdisciplinary approaches in SS can manifest holistic frameworks capable to deal with CC issues & beyond these issues	Ethics dimension is intertwined within the field of SS SS term can be redundant when if it is just focused on a scientific endeavor, along with the absence of ethics practice	Environmental science Education Innovation & Technology	1 - [Highly prompted to SS] Study reflected the critical issues around ethics to be included in the dimensions of SS

(continued)

Table 1. (continued)

Reference	Report on Challenges of Climate Change (CC)	Report on Solutions Related to Climate Change (CC)	Research Observations	Accompanying Academic Discipline	Approximation to Sustainability Science (SS) Author "Key statements" or reflections mentioned
Spangenberg (2011)	Changes in the system of scientific conduct are required to effectively address multiple dimensions of sustainability Advances in science & technology on its own (as silo pathways) are insufficient to fully achieve SD	Advances in basic scientific knowledge are required to establish effective real-world solutions SD solution embeds a systems science approach synchronized in a metasystem ecosystem to deal with complex & evolving systems of the world	Systems science approaches develop coherent information about sustainability Applied research or applied science can be integrated into SS SS must be interdisciplinary at minimum	Environmental science Pedagogy	1 - [Highly prompted to SS] Study reflected on the evolving concept of SS & how the worldview in sustainability necessitates SS as a fitted approach Study reflected how science, is an instrument for sustainability transformations
Jerneck et al. (2010)	Organization of scientific knowledge & scientific practices requires changes Traditional approaches to sciences operate in a paradigm which focuses on scientific objectivity	Interdisciplinary should be embedded in sustainability academics & practices Problem-solving approaches & critical research should be embedded as the cross-cutting framework in SS	SS bridges natural science & social science to address complex challenges SS is a field of knowledge & practice that actively responds to societal dimensions CC is a critical scientific study & knowledge domain to conceptualize SS	Environmental science Education	1 - [Highly promptedss to SS] Study reflected on the concepts of SS, approaches of SS, & the sustainability dimensions of SS

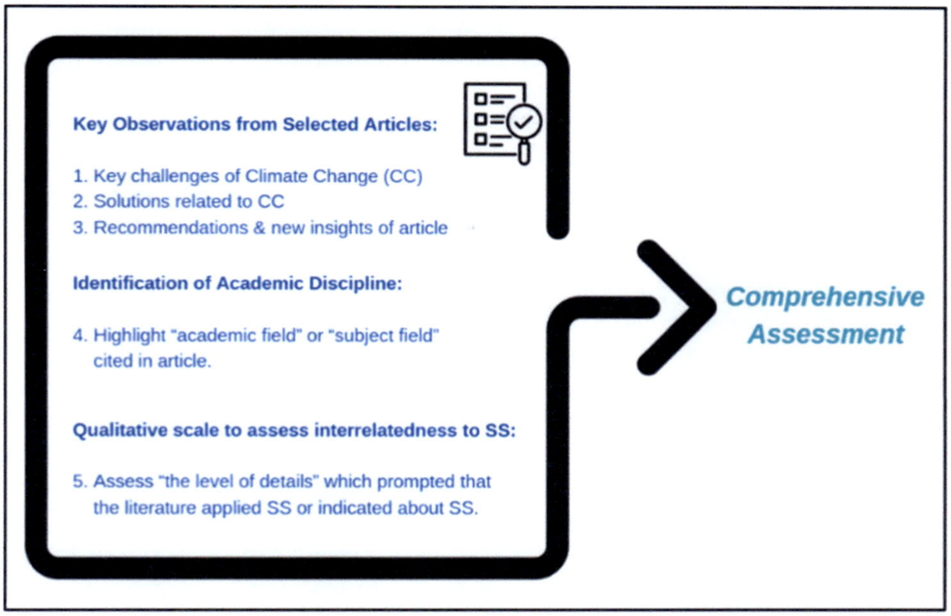

Fig. 3. Methodological framework of literature review matrix at Table 1. (By author)

- There are gaps in knowledge transfer and scientific practice to mobilize solutions or policies towards SD, including addressing climate change. These gaps require a new way of organizing knowledge and an evolutionary scientific approach.
- Both academic discipline and approaches to science education require a rethink of design, rethink of practices, rethink of approaches, along with integration into the present world issues such as climate change. Academic discipline and approaches to science education continue to evolve over time. Both theory and practices of scientific knowledge will continue to adapt to socio-environmental needs.
- Natural sciences or independent scientific approaches are old norms of achieving scientific objectivity, which do not lead to transformational changes that are required in the sustainability paradigm. Co-production of knowledge is necessary in science, which implies combining applied sciences and applied research.
- Sustainability Science (SS) has been an evolving scientific methodology or curriculum since cited in the literature from the year 2001. This field was acknowledged with a steady volume of publications focusing on the conceptualization of SS up to the year 2014. Since the year 2020s, there have been gaps in research about SS. There is a perceived need to investigate more about SS relevance and to define SS approaches. Some literature from the year 2020 hinted SS at a moderate level, with an emphasis on understanding the relevance of implementing multiple scientific disciplines. This implementation is often discussed through interdisciplinary, multidisciplinary, or transdisciplinary.

4.1 Main Findings

In the present time where wicked problems of sustainability exist, a state-of-the-art science is needed to create adaptive solutions which are precise and productive. Sustainability Science (SS) as a state-of-the-art science, responds to the multi-scalar dimensions of sustainability. Based on the content analysis from Table 1, 40% of the assessed 26 literature articles have identified similar challenges concerning global sustainability issues including climate change. For one primary challenge, climate change topics account for information from a variety of scientific disciplines. When processing this information, knowledge of sustainability is often connected to characteristics of complexity, to deep uncertainty, and deals with spatial context. Secondly, the climate change field demands a broader approach to science. This is because sustainability discussions are often assessed based on the theory of limits of growth, theory of change, theory of social sciences, cognitive sciences, Anthropocene theory, sustainability thresholds, complexity theory, systems theory, and socio-environmental conflict.

Based on these common challenges identified, historically; scientific research has prompted that the most common approach for solutions to climate change or sustainability is to go beyond the norm of sciences. This mean not ignoring that there are gaps between theory and practice of different knowledge domains. Yet, every sustainability stakeholder or "change agent" has different adaptation response to knowledges and theories. At the same time, at some point of preaching a knowledge or practicing a sustainable solution, there are challenges in terms of "work boundaries". For instance, a boundary can exist when one implementation of a solution or knowledge practice, cannot collaborate or harmonize well with another solution or knowledge practice. In a simpler scenario, a work boundary can exist when an individual is unable to learn, practice, or adapt multiple academic scientific disciplines. The underlying literature has also investigated that currently both societal impacts and policy impacts are lacking effective decisions due to the lack of best knowledge. On the other hand, climate change and sustainability problems have transboundary processes whereby multiple simultaneous events of climate change or socio-environmental problems will continue to challenge the most effective interventions for this chaos. For instance, while a European country is confronting flash floods; another Middle Eastern country is confronting extreme drought at the same time. In another example, while a least developed country needs to sacrifice its land-use for agriculture economy, a developed country needs to conserve its land-use for biodiversity wildlife. Due to this, the "best knowledge" about dealing with climate change is not always guaranteed by arguments of scientific theories. Scientific knowledge and scientific practices are prone to multiple perspectives which leads to conflict in perspectives, practices, and conflict of ethics. This nature is observed in the work of McGreavy, B., & Hart, D. (2017), Mach & Kraan (2020), Keys et al. (2023) and Kovacic, Z. & Marcos-Valls, A. (2023).

One way to address the scope where "best knowledge" can be obtained, and to mitigate conflict is to support the co-production of knowledge by bridging theory, practice, and sciences. Several literatures have indicated the role of "systems science" or "systematic approach", as a key methodology to rearrange education curriculum and to create a science-policy interface (SPI). The evidence of embedding a systems approach has been mentioned in the work of Ojha, H. R. (2020), Spangenberg (2011), Keys et al. (2023), Wittmann & Meissner (2024) and Poliseli & Caniglia (2024).

When considering the establishment of new science such as SS; there are several constructive models to support knowledge transfer to deal with multifaceted subjects. The following are three typical **approaches to constructing SS:**

i) Interdisciplinary model

Interdisciplinary comprise integrated studies that typically consider several theories or practices from a set or sub-sets of scientific disciplines. Interdisciplinary engages in the gaps in sciences, which creates novel insights or hybrid solutions through the intersection of each distinct scientific disciplines.

The interdisciplinary approach has been strongly suggested to redesign the scientific curriculum since the year 2010 (Jerneck et al., 2010). Several scholars - Spangenberg (2011), Ojha, H. R. (2020), Shrivastava et al. (2020), Kuo, S. et al. (2022), Kovacic, Z. and Marcos-Valls, A. (2023), Deng et al. (2023), Song et al. (2023), Hsien and Yeh (2024), and Zare et al. (2024), discussed the effectiveness of interdisciplinary approach.

ii) Transdisciplinary model

Transdisciplinary unites theory, practices, and knowledge of several disciplines for a common outcome or purpose. A collective approach process combines scientific disciplines, whereas synergies of sciences help to solve paradigms of sciences or knowledge in a synergetic way. Transdisciplinary seeks knowledge by transcending the horizon of the disciplines, in opposite perspective of depending on the strength of a solo discipline. Transdisciplinary creates complementary solutions to specific problems, where unconventional theory or practice can exist.

The transdisciplinary approach has been discussed since the year 1970s. Narratives on transdisciplinary to bridge knowledge and practice can be found in the work of Nelson, M. P. and Vucetich, J. A. (2012), Ojha, H. R. (2020), Chiu et al. (2020), Shrivastava et al. (2020) Riggs et al. (2023), Wittmann & Meissner (2024), and Pickett et al. (2024).

iii) Multidisciplinary model

Multidisciplinary often refers to the combination of multiple scientific disciplines (typically two or more); whereby these disciplines cross boundaries in practice. By understanding the strengths and insights from each scientific discipline, the scientific observations and solutions are complementary. For instance, one scientific discipline can provide a matching theory or provide supplementary practices to another scientific discipline. Through this model, uncovering identical features can lead to a more resilient process of translating knowledge into an outcome. On the other hand, this model often uses a blended research, teaching and learning processes to optimize the best information or practice from each scientific discipline. The optimization creates a harmonize effect of learning. Multidisciplinary model leads towards a more contextual or technical approaches, as the multiple discipline creates a complex and elaborate knowledge.

Both literature - Song et al. (2023), and Poliseli & Caniglia (2024) highlighted the role of multidisciplinary.

These three models are translated into a visual illustration depicted in Fig. 4.

Figure 4 showcase how the three different models are theorized, practiced, and operated. A legend is provided to understand the visual illustration. The changing width of vertical lines and changing width size of yellow dotted lines define

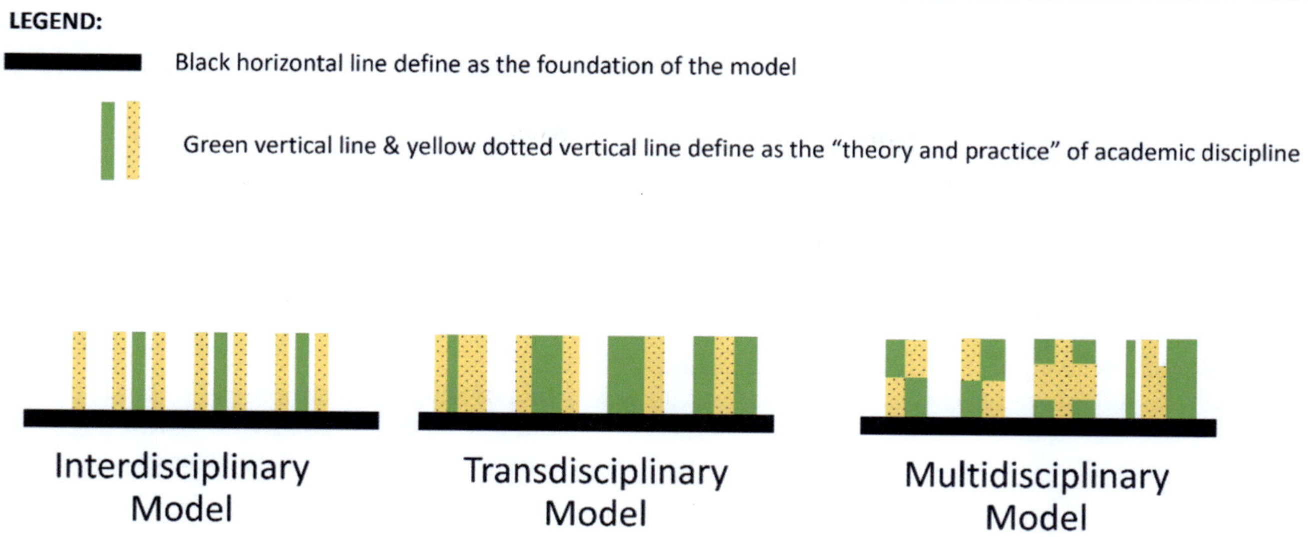

Fig. 4. Visual illustration describing the operations and definitions of three models to construct knowledge transfer in sustainability science. (By author)

the complexity of a subject matter. This means the academic discipline and subject matter have a higher density of theories and knowledge practices. These three models embed the co-production of knowledge, advancing scientific theories, and the study of cross-scale dimensions in a scientific discipline. In essential, to facilitate teaching and learning for sustainability subjects, as well as to facilitate research activities; these three models operate as an interface for transferring knowledge. Additionally, the three models strengthen skills which enhances precision in scientific practice.

4.2 Looking at Policy Practice Through the Lens of Sustainability Science

Both work of Chiu et al. (2020) and Tambe et al. (2023) have mentioned that a science-policy interface (SPI) is critical for the pressing global sustainability challenges. Despite the existence of international environmental policies and climate change regulations, there is a lack of active practice in society answering the urgent call to pro-actively engage in scientific solutions. Sustainability habits or sustainability related behaviors are much more engaged through knowledge channels that are easy for society to grasp, and through influence of society culture. Hence, when concerning any subject matter or technical context through "Science", "Science" may not be a primary source of how society will immediately engage or respond to climate change issues. Trott et al. (2023) have highlighted the role of "Citizen Science" and introduced sustainability education through a community-centric approach to shape more effective pro-environmental responses. Thus, today's present policies and regulatory interventions would require knowledge contribution involving civil society, along with multidisciplinary practitioners' knowledge which can provide a balanced view.

At the same time, conflict of individual virtues and ethics co-exist in the dynamics of socio-environmental decisions. The divergence of society perspectives, values, and commitment to sustainability leads to weak agreements that may not fully capture each complex dimension of sustainability. In this case, the dynamics of policy setting should consider a holistic view of the impacts of climate change (CC); and the interlinkages of social, environment, economics, politics, and science. To some extent, several literatures in the literature review matrix have discussed the impact of inadequate understanding of the complexity of both CC subjects and sustainability subjects. In a scenario when sustainability practitioners and policy decision makers practices deficient knowledge of a particular subject; this scenario can lead to flawed policy outcomes. If policy setting and regulation are backed up by weak grounds of knowledge; then both CC problems and sustainability challenges cannot be fully solved successfully. Moreover, the lack of understanding about the realms of "sustainability" and how "sustainability goals" are prioritized by every decision maker often challenges the status-quo of which solution is the fittest for a particular scenario. The problems of CC and all matters concerning sustainability require a post-norm science. A post-norm science would then mean a curriculum or approach that goes beyond the objectivity of science. Yet, going beyond the objectivity of science should not ignore the concepts of rationality, information reliability, ethics, safety, integrity, and quality. Thus, through the lens of SS, multiple disciplines of sciences can offer resiliency, balanced decisions, and rational policy practices that prevent behavioral prejudices or societal biases.

The present policies for climate change interventions such as environmental policies, climate policies, and energy policies; are often established through rigorous evidence-based assessments. For instance, a professional scientific panel will formally assess the emerging climate trends, potential risks, and future climate scenarios. This assessment will then contribute to measurable indicators that can be incorporated into climate policies.

Specific policies have predefined criteria for tracking the progress of climate change interventions. Both environmental and CC policies set realistic timelines to achieve sustainable development (SD) milestones by the year 2030. While adaptation measures are performed because of policy practice, periodic reviews are performed by policymakers and scientific panels to assess and discuss new advancements. This means that while current policy practices are aiming to mobilize relevant sustainability solutions based on the 17 SDGs for instance; there is a continuous need for science as a field to simultaneously expand and adapt. For example, SS is one discipline that matures during policy implementation processes. During policy practice, challenges in practices are better known, while opportunities to refine scientific approaches develop. Looking into the future horizon, such as post-year 2030 of the SD transitions; both policy systems and policy science will arrive at an essential period where SS is a requisite. As new policy measures are considered for the future of global issues, SS is both a methodology and scientific discipline that will be urgently called for to deal with any shortfall of predefined policy practices. In this sense, the concept of sustainability science (SS) is a discipline capable of these functions:

- SS creates a larger body of knowledge and precision in scientific implementation, which influences policy design and practice effectiveness.
- SS is a medium that extends beyond traditional scientific views, where the complexity and challenges of sustainability are thoroughly studied and understood; thus, leading toward more ethical policy decisions.
- SS is connected to the outcomes and impacts of policy interventions, policy recommendations, and policy-based practices. It identifies both the weaknesses and strengths of predefined criteria and regimes. It identifies the excluded variables which have influenced the outcomes of specific policy measures.
- SS conceives "functional" knowledge and a priori policy recommendations which also embed the dynamics of policy-based factors. To which, policy science is simultaneously integrated into SS research, education, or practices.

Based on the literature review matrix, policy science remains an essential research field and scientific discipline. At the same time, the implementation of policies and policy practices requires the presence of other interfaces of knowledge and discipline (i.e., SS).

The research method to prompt about the future scenarios of SS based on Fig. 2, provided a clear indication of the complexity of both the current scientific field and academic practice to implement a complete knowledge of SS.

4.3 Role of Science, Technology and Innovation (STI) as Transformative Models for Sustainability Science

Amongst several literatures, Wittmann and Meissner (2024) highlighted the research agenda for science, technology, and innovation (STI) to foster communication platforms relevant for policy practices. At least 30% of the content analysis from all 26 literature sources has a background of putting forth the importance of "science". To which discovering new methodologies, practices, and implementation of sciences are part of the natural process in establishing a more multidisciplinary model such as SS. A set of literature has also discussed the relevance of "innovation and technology (IT)" as a complementary field and subject matter in sustainability studies and research. To name a few, focusing on digitalization, emerging engineering disciplines, to artificial intelligence (AI); are some of the IT advancement which can be embedded in the study and knowledge management of SS field. IT systems and the digital world establish robust communication and collaboration platforms relevant to deal with large complex knowledge systems such as sustainability. Poliseli and Caniglia (2024) work provided evidence that both STI and policy science are examples of integrated methodologies to consider for the future of sustainability. Song et al. (2023) identified a simple example of the present era, where big data and varied information sources associated to climate change or sustainability would depend on the efficiency of STI systems. At the same time, the use of big data and the use of advance technology raises ethical concerns in society, and raises issue around control of intellectual property; as well as data privacy challenges. This section of the chapter arrives at a point of logical comprehension that the perspectives pointed out by Sahle et al. (2024) are indeed aligned to the research findings of this paper. For one logical observation, Sahle et al. (2024) observed that technology is an autonomous transformative environment that the present world adapts to. For the purpose of setting guidelines and good practices in research and development (R&D) activities, policies are thus critically required. This observation complements the main findings of this research in a way that both STI and policy science, are two essential knowledge and field practices to deal with the subject matter concerning sustainability.

STI requires specialization in a broad science ranging from science, technology, engineering, arts, mathematics, to humanities. The focused in science, technology, and innovation can translate to comprehensive and highly technical solutions that deal with complexity. Reflecting to the research of Malt and Majid (2023), effective sustainability practices and knowledge is conceptualized by the knowledge systems of society. If the present and future society are more tuned to the interactions of STI to sustainability subject matter (including climate change); then through SS, society can form a better understanding of the cause and effects of sustainability. This means having improved understanding of the "causality" factors of climate change events, or "causality" to climate change solutions.

5 Summary on the Principles of Sustainability Science (SS)

In addition to the earlier sections of the findings of the literature review and themes discussed, this chapter has met the research objective to highlight the identified principles of sustainability science (SS). Figure 5 showcases the principles of SS which are supplemented by the results of study and stakeholder analysis advancing from the work of ASX Toh (2023).

Exploring these SS principles adds further methodological approaches to carrying out SS academic curriculum, or to embed SS in any field. As SS remains as an emerging academic curriculum and growing field of knowledge, this research recognizes that innovation in SS must persist.

The identified principles highlight a research agenda to rethink the conventional theory of SS, and to redesign models of SS capable of addressing dynamic subject matter over an infinite time. The impressions intersecting with the principles is formed through the sum of several interrelated impressions of scientific study around the topic of SS, based on *Gestalt* theory (Desolneux et al., 2008). Learning from impressions involves repeated reflective learning to understand the complex and interconnected knowledge areas about SS. These impressions prompt knowledge areas that can be further studied to test models that can transform and support SS implementation.

6 Conclusion

Although the world is actively experimenting with new approaches to fulfill the 17 SDGs or climate-related goals; the current trajectory of climate crisis events demands the adoption of efficient and resilient models. Thus, SS is proposed as a field or subject matter that encompasses knowledge and scientific practices that solves the world's pressing issues. The literature review matrix and content analysis provided initial context about how both CC and policy science have a background in the field of SS. According to the different approaches to SS, there are three types of common educational models to support the practice of SS. When concerning

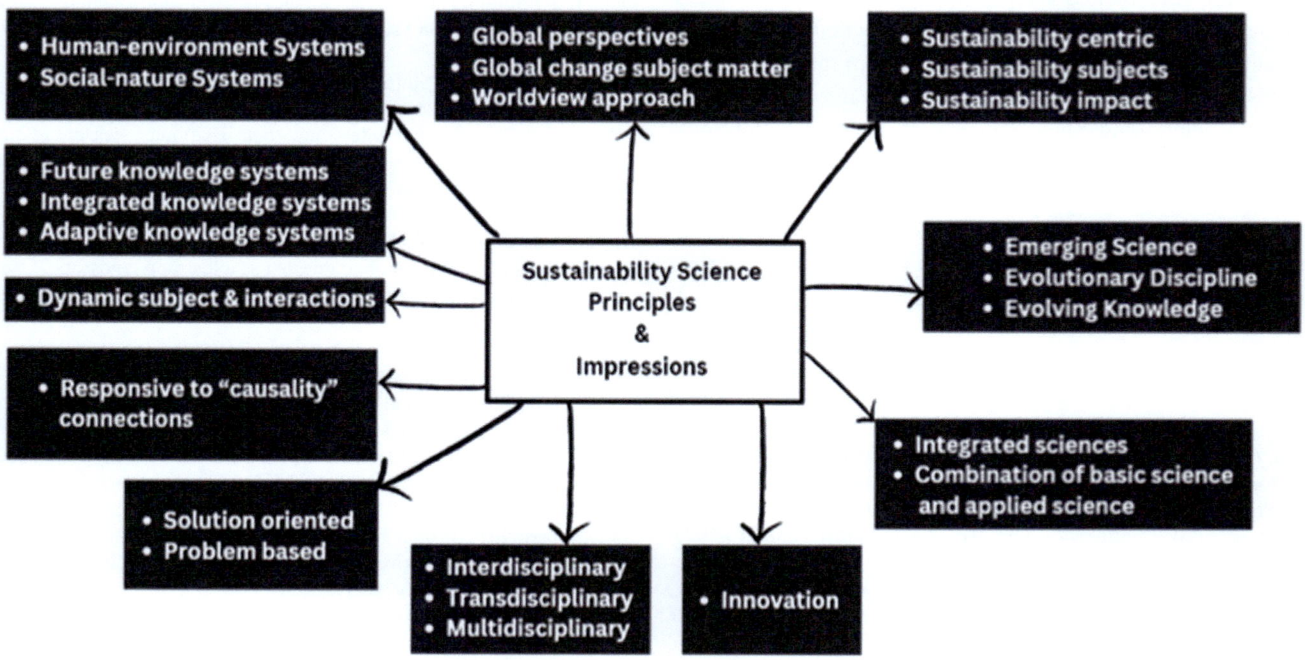

Fig. 5. Principles of Sustainability Science (SS) (By author)

global change problems such as CC, such complex study requires a comprehensive and integrated model such as STI to be embedded in SS. Effective policy practice for sustainable development (SD) can be achieved more successfully, whereby the concept of SS becomes a knowledge field to facilitate effective policy planning. Global progress on CC, is always concerned with the nexuses of the science-policy interface. This chapter has provided up-to-date scientific literature concerning the field of SS. The research findings have identified a current research gap about the linkages across STI, policy science, and SS. This guide includes models to enable the practice of SS, or to support any type of activity which embeds SS. This work also gathered expert review and recommendations of a scientific panel which raised mutual concerns around the academic practice and research practices. It is observed that in the present society setting, there are diverging specialization of knowledge and skills. This means SS can be seen as a possible remediation to conflicting practices, as SS can bridge gaps of knowledge and bridges scientific practices to create better synergies in solutions. SS also permits scientific practitioners to explore beyond the boundaries of their conventional scientific background.

Additionally, this chapter provided suggestions of how both STI, and policy science can create a more equitable society, by means society can collaborate better, communication is enhanced, and transfer of knowledge devises intelligent solutions capable of solving global challenges. At the same time, by reflecting on the three types of models to promote knowledge transfer in SS, society can upskill their knowledge by increasing the depth of expertise in a scientific field. This guide acknowledges that crafting an environment where SS as a field can be taught, embedded, or practiced would demand a dynamic socio-ecosystem that can engage in SS. Further investigation of the interface between STI, policy science, and SS is required at a larger institutional level, by means of involving sample studies from a variety of research institutions or higher education representatives. In this sense, an active engagement in the topic of these nexuses will cultivate the optimal framework and methodologies for SS.

References

Toh, A.S.X.: Sustainability Science Excellence Centre (SSEC): working together with higher education towards a sustainable future. In: Proceedings of International Conference on Sustainable Development (2023). https://ic-sd.org/wp-content/uploads/2023/10/2023-submission_243.pdf

Chiu, A., Aviso, K., Tan, R.: On general principles at the sustainability science-policy interface. Resour. Conserv. Recycl. **158**, 104828 (2020). https://doi.org/10.1016/j.resconrec.2020.104828

Deng, H., Giammar, D., Li, W., Vengosh, A.: Embracing the intersections of environmental science, engineering, and geosciences to solve grand challenges of the 21st century. Environ. Sci. Technol. **57**(30), 10907–10910 (2023). https://doi.org/10.1021/acs.est.3c04795

Desolneux, A., Moisan, L., Morel, J.-M.: From Gestalt Theory to Image Analysis. Springer, Heidelberg (2008)

Ellis, P.W., et al.: The principles of natural climate solutions. Nat. Commun. **15**(1) (2024). https://doi.org/10.1038/s41467-023-44425-2

Heinrichs, H., Martens, P., Michelsen, G., Wiek, A.:. Sustainability science. Springer eBooks (2016). https://doi.org/10.1007/978-94-017-7242-6

Hsieh, Y., Yeh, S.: The trends of major issues connecting climate change and the sustainable development goals. Disc. Sustainabil. **5**(1), 31 (2024). https://doi.org/10.1007/s43621-024-00183-9

Jerneck, A., et al.: Structuring sustainability science. Sustain. Sci. **6**(1), 69–82 (2010). https://doi.org/10.1007/s11625-010-0117-x

Keys, P.W., Badia, L., Warrier, R.: The future in anthropocene science. Earth's Future **12**(1) (2023). https://doi.org/10.1029/2023ef003820

Kovacic, Z., Marcos-Valls, A.: Institutionalising interdisciplinarity in PhD training: challenging and redefining expertise in problem-oriented research. Environ. Educ. Res. **29**(3), 473–488 (2023). https://doi.org/10.1080/13504622.2023.2174252

Kuo, S., Tang, T., Hsu, H., Lin, L., Chen, Y.: Challenges to govern a global sustainability science problem: lessons from a domestic climate change research project. Environ. Dev. Sustain. **25**(3), 2885–2904 (2022). https://doi.org/10.1007/s10668-021-02067-8

Mach, K.J., Kraan, C.M.: Science–policy dimensions of research on climate change and conflict. J. Peace Res. **58**(1), 168–176 (2020). https://doi.org/10.1177/0022343320966774

Malt, B.C., Majid, A.: Conceptual foundations of sustainability. Top. Cogn. Sci. **15**(3), 334–356 (2023). https://doi.org/10.1111/tops.12678

McGreavy, B., Hart, D.: Sustainability Science and Climate Change Communication. Oxford University Press eBooks (2017). https://doi.org/10.1093/acrefore/9780190228620.013.563

Nelson, M.P., Vucetich, J.A.: Sustainability Science: ethical foundations and emerging challenges. Nat. Educ. Knowl. **3**(10), 12 (2012)

Neta, G., et al.: Advancing climate change health adaptation through implementation science. Lancet. Planetary Health **6**(11), e909–e918 (2022). https://doi.org/10.1016/s2542-5196(22)00199-1

Ojha, H.R.: Building an engaged Himalayan sustainability science. One Earth **3**(5), 534–538 (2020). https://doi.org/10.1016/j.oneear.2020.10.009

Pickett, S.T.A., et al.: Shifting forward: urban ecology in perspective. Ambio (2024). https://doi.org/10.1007/s13280-024-02007-6

Poliseli, L., Caniglia, G.: Inter- and transdisciplinary reasoning for action: the case of an arts–sciences–humanities intervention on climate change. Sustain. Sci. (2024). https://doi.org/10.1007/s11625-024-01469-7

Riggs, R.A., et al.: Transdisciplinary science for strengthening forest systems in British Columbia: quesnel as a learning landscape. Can. J. For. Res. **53**(12), 1006–1016 (2023). https://doi.org/10.1139/cjfr-2023-0057

Sahle, M., Lahoti, S.A., Lee, S.Y., et al.: Revisiting the sustainability science research agenda. Sustain. Sci. (2024). https://doi.org/10.1007/s11625-024-01586-3

Shrivastava, P., Smith, M.S., O'Brien, K., Zsolnai, L.: Transforming sustainability science to generate positive social and environmental change globally. One Earth **2**(4), 329–340 (2020). https://doi.org/10.1016/j.oneear.2020.04.010

Spangenberg, J.H.: Sustainability science: a review, an analysis and some empirical lessons. Environ. Conserv. **38**(3), 275–287 (2011). https://doi.org/10.1017/s0376892911000270

Song, C.X., et al.: Cyberinfrastructure for sustainability sciences. Environ. Res. Lett. **18**(7), 075002 (2023). https://doi.org/10.1088/1748-9326/acd9dd

Stefanis, C., Stavropoulos, A., Stavropoulou, E., Tsigalou, C., Constantinidis, T.C., Bezirtzoglou, E.: A spotlight on environmental sustainability in view of the European Green Deal. Sustainability **16**(11), 4654 (2024). https://doi.org/10.3390/su16114654

Tahvilzadeh, N., Montin, S., Cullberg, M.: Functions of sustainability: exploring what urban sustainability policy discourse "does" in the Gothenburg Metropolitan Area. Local Environ. **22**(1), 66–85 (2017). https://doi.org/10.1080/13549839.2017.1320538

Tambe, S., et al.: Bridging science, policy and practice for sustainability: towards a conceptual framework. Environ. Sci. Policy **145**, 208–216 (2023). https://doi.org/10.1016/j.envsci.2023.04.007

Trott, C.D., Weinberg, A.E., Frame, S.M., Jean-Pierre, P., Even, T.L.: Civic science education for youth-driven water security: a behavioral development approach to strengthening climate resilience. Int. J. Behav. Dev. **48**(2), 145–155 (2023). https://doi.org/10.1177/01650254231188661

UNESCO. Guidelines on sustainability science in research and education (2017). https://unesdoc.unesco.org/ark:/48223/pf0000260600. Accessed 8 Aug 2024

Wamsler, C.: Education for sustainability. Int. J. Sustain. High. Educ. **21**(1), 112–130 (2020). https://doi.org/10.1108/ijshe-04-2019-0152

Wittmann, V., Meissner, D.: The nexus of Global Challenges and Global Studies: a trans-disciplinary global sustainability Science Curriculum. Disc. Sustainabil. **5**(1) (2024). https://doi.org/10.1007/s43621-024-00220-7

Zare, F., Jakeman, A.J., Elsawah, S., Guillaume, J.H.: Bridging practice and science in socio-environmental systems research and modelling: a design science approach. Ecol. Model. **492**, 1107919 (2024). https://doi.org/10.1016/j.ecolmodel.2024.110719

Climate Finance for Resilience Future: Pioneering Steps of the United Arab Emirates (UAE) in the Gulf Region

Fatima zahra Moussaid, Nor Razinah Mohd. Zain, Oumaima Tounchibine, and See Teong Ng

Abstract

Climate change presents a main challenge worldwide which requires urgent actions from different stakeholders. The integration of Islamic finance into discussions surrounding environmental sustainability emerges as a promising avenue for effective actions. In the Gulf region, known for its dominance in oil and gas production, countries play a dual role as both major contributors to climate challenges and key participants in global climate action efforts. Looking forward to 2025, the United Arab Emirates (UAE) emerges as a leader in climate action, demonstrating a notable emphasis on climate financing initiatives. This study employs a qualitative research methodology, utilizing a combination of library-based approaches and documentary analysis, to explore the complex intersections between climate finance, Islamic finance, and environmental sustainability within the UAE. Through an analysis of strategic approaches to climate financing in the country, incorporating Islamic finance, the study aims to elucidate the significance of environmental sustainability and its impact on climate mitigation practices adopted in the UAE. The findings highlight the importance of strategic approaches to climate finance among Islamic financial services providers in enhancing climate mitigation practices within the UAE. To this end, tailored Shariah-compliant instruments must be developed, while simultaneously fostering awareness, building capacity, and promoting strategic partnerships between relevant government entities and private sectors. By doing so, the best practices observed in the UAE demonstrate the potential for Islamic finance to emerge as a leading provider of climate finance, thereby advancing environmental sustainability goals.

Keywords

Islamic finance · Climate finance · Environmental sustainability · Climate mitigation · United Arab Emirates (UAE)

1 Introduction

Climate change has emerged as one of the most pressing global challenges of our time, demanding urgent and coordinated action from stakeholders across the world. The Intergovernmental Panel on Climate Change (IPCC) Sixth Assessment Report (IPCC, 2023) emphasizes the increasing urgency of the climate crisis, while the United Nations Environment Programme's Emissions Gap Report 2023 (UNEP, 2023) highlights that current policies are insufficient to meet the Paris Agreement goals. As countries grapple with the complexities of mitigating and adapting to climate change, innovative financing mechanisms have become increasingly crucial.

In this context, Islamic finance has gained attention for its potential to contribute to environmental sustainability efforts. The Islamic Finance Council UK and Global Ethical Finance Initiative's report in 2021(UKIFC et al., 2021) highlights the synergies between Islamic financial principles and sustainability objectives. The emergence of "green sukuk" as a tool for funding environmentally friendly projects, as noted in

F. Moussaid (✉)
National School of Business and Management, Ibn Tofail University, Kenitra, Morocco
e-mail: moussaid.fatimazahra@uit.ac.ma

N. R. Mohd. Zain · O. Tounchibine
IIUM Institute of Islamic Banking and Finance, International Islamic University Malaysia (IIUM), Kuala Lumpur, Malaysia
e-mail: mumtaz_razi@iium.edu.my

O. Tounchibine
Ibn Tofail University, Kenitra, Morocco

S. T. Ng
IIUM Institute of Islamic Banking and Finance, International Islamic University, Kuala Lumpur, Malaysia

B. He et al. (eds.), *Towards the Framework of Livable and Resilient Cities* Advances in Science, Technology & Innovation, https://doi.org/10.1007/978-3-031-97849-4_28

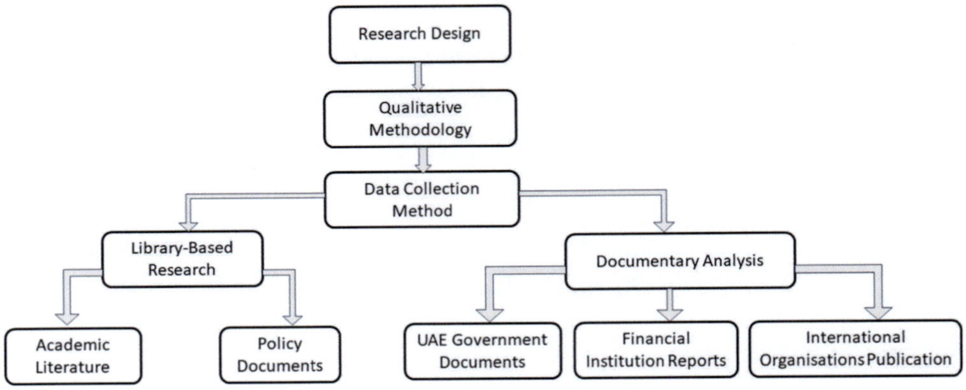

Fig. 1. Methodological Framework for UAE Climate Finance Study Source: by authors

the United Nations Environment Programme Finance Initiative's "Mobilising Islamic Banking for Climate Action" report (UNEP FI et al., 2023), demonstrates the growing alignment of Islamic finance with climate action.

The Gulf region, home to some of the world's largest oil and gas producers, occupies a unique position in the climate change narrative. These countries face the dual challenge of being significant contributors to greenhouse gas emissions, while also being called upon to play a pivotal role in global climate action. This dichotomy presents both challenges and opportunities for the region to leverage its financial resources and influences towards sustainable development.

Among the Gulf nations, the United Arab Emirates (UAE) has distinguished itself as a leader in climate action, particularly in the realm of climate financing initiatives. As the country hosted COP28 in 2023, its commitment to environmental sustainability and climate mitigation has become increasingly evident. The UAE's efforts to integrate Islamic finance principles with climate action strategies offer a compelling case study for examining the potentials of faith-based financial systems in addressing environmental challenges.

This research aims to explore the intersection of climate finance, Islamic finance, and environmental sustainability within the context of the UAE. Specifically, the study seeks to answer the following questions:

- How are Islamic financial institutions in the UAE incorporating climate finance into their strategies to enhance climate mitigation practices?
- What are the key challenges and opportunities in developing tailored Shariah-compliant instruments for climate finance in the UAE?
- How can the UAE's approach to integrating Islamic finance and climate action, including strategic partnerships and capacity building, serve as a model for other Gulf countries and the broader Muslim world?

To address these questions, this study employs a qualitative research methodology. The research design incorporates a library-based approach, drawing on existing literature and reports, as well as documentary analysis of policy documents, financial statements, and strategic plans of relevant institutions in the UAE (Fig. 1). This approach allows for a comprehensive examination of the strategic approaches to climate financing in the country, with a particular focus on the role of Islamic finance in enhancing climate mitigation practices.

By analyzing the UAE's experience, this research aims to contribute to the growing body of knowledge on sustainable finance in Islamic contexts and provide insights into the potentials of faith-based financial systems to address global environmental challenges. This study is particularly timely given the recent findings from the Standing Committee on Finance's fifth Biennial Assessment and Overview of Climate Finance Flows (UNFCCC, 2022) and the Global Landscape of Climate Finance 2023 report (CPI, 2023), which highlight the critical need for increased financial mobilization to meet international climate goals.

Building on this context, this paper explores the intersection of climate finance and Islamic finance, focusing specifically on the UAE. It begins with a literature review of climate finance's role in climate change mitigation and adaptation, alongside an overview of Islamic finance principles. The paper then examines the UAE's climate strategies, initiatives, and policies. It analyzes the role of Islamic finance in the UAE, highlighting Shariah-compliant products, case studies, and integration challenges. The discussion evaluates the UAE's strategic approaches and their impact on climate goals, offering lessons for other regions. The paper concludes with a summary of findings, recommendations for advancing Islamic climate finance, and future research suggestions.

2 The Convergence of Climate Finance and Islamic Finance: A Comprehensive Literature Review

This literature review explores the convergence of climate finance and Islamic finance, examining their potential synergies in addressing global climate challenges. It investigates climate finance's critical roles in supporting mitigation and adaptation efforts, provides an overview of Islamic finance principles and instruments, and explores the nexus between these two financial domains.

2.1 Climate Finance and its Roles in Climate Change Mitigation and Adaptation

Climate finance plays a pivotal role in both mitigation and adaptation strategies to address climate change globally (Haites, 2011; Hong et al., 2020). In terms of mitigation, it provides crucial fundings for sustainable development related projects that reduce greenhouse gas emissions, such as renewable energy installations, energy efficiency improvements, and sustainable transportation systems. These investments help countries transition to low-carbon economies and contribute to global emission reduction targets. For the adaptation, climate finance supports initiatives that build resilience against climate impacts, including flood protection infrastructure, drought-resistant agriculture, and early warning systems for extreme weather events. By funding these efforts, climate finance enables vulnerable communities (Moussaid et al., 2023) and ecosystems to adapt to the changing climate, reducing potential losses and damages.

The effectiveness of climate finance in driving mitigation and adaptation efforts is evident in various sustainable development related projects worldwide. For instance, the Green Climate Fund has supported large-scale solar power plants in developing countries, contributing to significant reductions in carbon emissions (GCF, 2019). Similarly, adaptation projects like mangrove restoration in coastal areas have enhanced natural barriers against sea-level rise and storm surges, protecting local communities and biodiversity. However, challenges persist in scaling up climate finance to meet the growing needs, particularly in developing countries that are most at risk from climate change impacts.

As we consider the landscape of climate finance and its crucial roles in combating climate change, it is important to recognize that diverse financial systems and principles that can contribute to this global effort. One of such system that has gained attention for its potentials in supporting sustainable development and climate action is Islamic finance. The principles and instruments of Islamic finance offer unique perspectives on ethical investing and risk-sharing that could complement and enhance climate finance initiatives. Understanding these principles can provide valuable insights into alternative approaches to funding climate mitigation and adaptation projects, and potentially broadening the pool of resources available for addressing this global challenge.

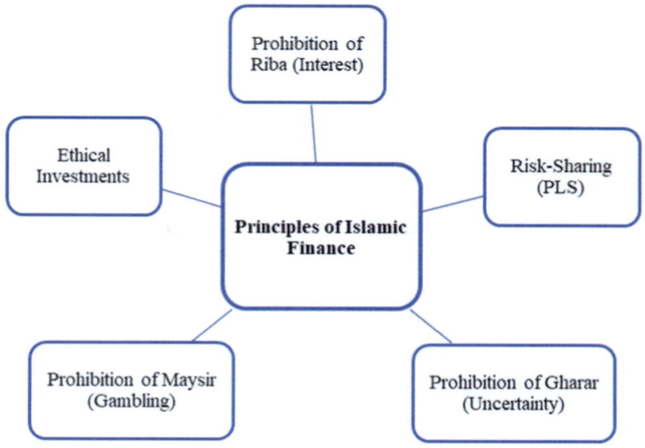

Fig. 2. Core Principles of Islamic Finance Source: by authors

2.2 Overview of Islamic Finance Principles and Instruments

Islamic finance is built upon a set of core principles derived from divinely revealed Shariah which can be traced back to its main sources, i.e. Al-Qur'an and Sunnah of Prophet Muhammad (bless and peace be upon him). These principles of Shariah aim to promote ethical, sustainable, and socially responsible financial practices (Ayub, 2009; Iqbal & Mirakhor, 2011). The key principles include (Fig. 2):

- Prohibition of riba (interest): any predetermined return on capital is forbidden.
- Risk-sharing: both parties in a financial transaction should share the risks and profits (profit-loss sharing or PLS).
- Prohibition of gharar (uncertainty): contracts must be clear, transparent, and avoiding excessive uncertainty.
- Prohibition of maysir (gambling): speculative activities are not allowed.
- Ethical investments: investments in Shariah non-compliant businesses, such as those that are deemed unethical (e.g., alcohol, gambling), are strictly prohibited.

To implement these principles in practice, Islamic finance has developed a range of financial instruments that comply with Shariah principles. These instruments serve various purposes, from basic banking services to complex investment structures. Table 1 illustrates a comparison of some of the most common instruments.

The diverse range of Shariah-compliant financial instruments, as illustrated in Table 1, has been instrumental in driving the robust growth and resilience of the Islamic Financial Services Industry (IFSI). These instruments, tailored to meet various financial needs while adhering to Islamic

Table 1. Comparison of Islamic Financial Instruments Source: by authors

Instrument	Detail	Risk Sharing	Asset-Backed	Profit Distribution	References
Murabahah (Cost-Plus Financing)	A sale where the seller discloses the cost and profit margin	Low	Yes	Fixed	(Hanif, 2014)
Ijarah (Leasing)	Similar to conventional leasing, but with specific Shariah-compliant conditions	Medium	Yes	Fixed/Variable	(Abdullah & Al-Mubarak, 2020)
Musharakah (Partnership)	A joint enterprise where partners share profits and losses	High	Yes	Proportional	(Usmani, 1999)
Mudarabah (Profit-Sharing)	One party provides capital, the other provides management	High	No	Pre-agreed Ratio	(Rahman, 2018)
Sukuk (Islamic Bonds)	Asset-backed securities representing ownership in tangible assets	Varies	Yes	Fixed/Variable	(Vishwanath & Azmi, 2009)
Takaful (Islamic Insurance)	Based on the concept of mutual assistance and shared risk	Pooled	No	Surplus-Sharing	(Khan, 2015)

principles, have enabled the IFSI to expand its reach and appeal across different market segments. This is evident in the industry's impressive growth trajectory, with the global IFSI reaching a valuation of USD 3.25 trillion in 2022, marking a 6.2% year-on-year increase. The versatility of Islamic financial instruments is reflected in the composition of the IFSI's assets (IFSB, 2023)[1].

2.3 Nexus Climate Finance vs. Islamic Finance

The integration of climate finance and Islamic finance has emerged as a significant area of research in recent years. This unique convergence of ethical financial principles and urgent environmental concerns presents potential solutions to the global challenge of climate change.

A foundational aspect of this research area is the exploration of conceptual alignment between Islamic finance principles and climate finance objectives. Scholars such as Dirie et al.,(2023) and Harahap et al.,(2023) have argued that the principles of Islamic finance naturally align with sustainable development goals, including climate action. This conceptual compatibility has provided a strong basis for further integration of these two financial domains.

This conceptual compatibility has provided a strong basis for further integration of these two financial domains. Expanding on this, Kamali (2012) and Nasir et al., (2022) highlights that the Islamic principles of stewardship (*khilafah*) and avoiding harm (*la darar wa la dirar*) are particularly relevant to environmental protection and climate change mitigation. These principles encourage responsible resource management and discourage activities that may harm the environment, aligning closely with climate finance objectives.

A significant body of literature focuses on the development of innovative financial instruments that combine Islamic finance principles with climate finance goals. Green sukuk, in particular, has received considerable attention. Saeed (2021) analyze the potential of green sukuk in financing renewable energy projects, noting its ability to attract both conventional and Islamic investors. Aassouli et al., (2018) explore the use of Islamic finance structures climate-resilient projects. They argue that these profit-sharing arrangements can effectively distribute risk and encourage long-term investment in climate adaptation.

Despite the potential synergies, researchers have identified several challenges in integrating climate finance and Islamic finance. Delle Foglie & Keshminder, (2022) highlight the lack of standardization in Shariah-compliant green financial products as a significant barrier to market growth. They argue for the development of clear guidelines and regulatory frameworks to facilitate the expansion of Islamic climate finance. Al-Roubaie & Sarea, (2019) points out the need for capacity building within Islamic financial institutions to effectively assess and manage climate-related risks. The study emphasizes the importance of developing expertise in both Islamic finance principles and climate science to create truly effective financial products.

The role of policy and regulatory frameworks in promoting the nexus between Islamic finance and climate finance is another key area of research. Jan et al., (2023) examine the policies of various Muslim-majority countries in promoting green Islamic finance. They find that

[1] For instance, the dominance of the Islamic banking segment, accounting for 69.3% of total IFSI assets, can be attributed to the successful implementation of instruments like Murabahah and Ijarah in retail and commercial banking services. Similarly, the significant share of Islamic capital markets (29.8% of IFSI assets) demonstrates the growing popularity of instruments such as Sukuk and Islamic funds. Even the smaller Takaful segment, representing 0.9% of assets, has shown strong growth, underlining the increasing adoption of Islamic insurance principles.

while some countries, such as Malaysia and Indonesia, have made significant progress, many others lack comprehensive strategies to leverage Islamic finance for climate action.

Several studies have attempted to quantify the market potential and economic impact of integrating Islamic finance and climate finance. According to projections from the UK Islamic Finance Council (UKIFC & UNDP, 2021), the market for green and sustainability sukuk is expected to experience significant growth. By 2025, these financial instruments could potentially mobilize between $30 billion and $50 billion in capital, which would be directed towards achieving the Sustainable Development Goals. This projection highlights the significant growth potential in this sector and suggests that such growth could have substantial positive impacts on both climate mitigation efforts and economic development in Muslim-majority countries.

The literature review reveals several areas that require further research, including the development of Shariah-compliant carbon trading mechanisms, the role of fintech in facilitating Islamic climate finance, the potential of Islamic social finance instruments (e.g., zakat and waqf) in supporting community-level climate resilience projects, and comparative studies on the effectiveness of Islamic vs. conventional climate finance instruments.

3 UAE's Climate Finance and Islamic Finance Landscape

The United Arab Emirates (UAE) has emerged as a global leader in both climate action and Islamic finance, demonstrating a commitment to sustainable development and ethical financial practices. The country's national climate strategies and the growth of its Islamic finance sector showcase its innovative approach. The UAE's initiatives in climate action, the evolution of its Islamic finance industry, and its efforts to combine sustainability with Shariah-compliant finance highlight how this country is positioning itself at the forefront of green economic transformation in the Middle East and beyond.

3.1 National Climate Action Strategies and Commitments

The United Arab Emirates has emerged as a regional pioneer in climate action, setting ambitious targets and implementing comprehensive strategies to address global environmental challenges. The country's approach encompasses a wide range of initiatives, from carbon emissions reduction to sustainable development across various economic sectors. These efforts not only position the UAE as a leader in the Middle East, but also contribute significantly to global climate action.

The UAE's climate strategies are characterized by their long-term vision, ambitious targets, and integration with broader economic and social development goals. These initiatives reflect a recognition of the urgent need to transition to a low-carbon economy while maintaining economic growth and improving quality of life for citizens.

These strategies and commitments (Table 2) demonstrate the UAE's holistic approach to climate action, integrating environmental concerns with economic development and social progress. By setting clear targets and implementing comprehensive plans, the UAE aims to transform its economy and society to meet the challenges of climate change while maintaining its position as a global energy leader.

3.2 Islamic Finance Sector in the UAE

The United Arab Emirates (UAE) has established itself as a global hub for Islamic finance, boasting a robust and rapidly expanding sector that adheres to Shariah principles. This journey began in the 1970s with the founding of Dubai Islamic Bank in 1975, widely acknowledged as the world's first commercial Islamic bank in the UAE (Iqbal & Mirakhor, 2011). This milestone marked the beginning of a transformative era in the UAE's financial landscape.

By 2023, the Islamic finance sector in the UAE had grown substantially, claiming a significant portion of the country's financial industry. The Islamic Financial Services Board (IFSB) reported that Islamic banking assets in the UAE represented approximately 20% of the total banking sector assets as of 2022 (IFSB, 2023). Key players in this sector include major Islamic banks such as Dubai Islamic Bank, Abu Dhabi Islamic Bank, and Emirates Islamic Bank.

The UAE's commitment to innovation has further solidified its position as a leader in Islamic finance. In recent years, the country has embraced fintech solutions to enhance Shariah-compliant financial services. For instance, the Abu Dhabi Global Market (ADGM) launched a regulatory framework for Islamic financial technology in 2022, fostering the development of digital Islamic banking products and services. This initiative has attracted numerous Islamic fintech startups to the UAE, creating a vibrant ecosystem that bridges traditional Islamic finance with cutting-edge technology (Gulrez, 2021).

Looking ahead, the UAE's Islamic finance sector is poised for continued growth, driven by factors such as increasing digitalization (Alblooshi, 2022), and the government's vision "We the UAE 2031", launched in 2022, which includes plans to further develop the Islamic economy, with a focus on creating a comprehensive ecosystem that supports Islamic financial institutions, halal industries, and Islamic lifestyle services.

Table 2. UAE's Climate Action Framework: Key Strategies and Commitments Source: by authors

Strategy/Commitment	Year Launched	Key Objectives
COP28 Presidency Agenda	2023	Focus on fast-tracking energy transition, fixing climate finance, and centering nature and people in climate action
National Hydrogen Strategy	2023	Establish UAE as a major hydrogen producer and exporter by 2031
Mangrove Initiative	2023	Plant 100 million mangroves by 2030 to support carbon sequestration
Net Zero by 2050 Strategic Initiative	2021	Achieve carbon neutrality by 2050; first such commitment in the Middle East
UAE Circular Economy Policy 2021–2031	2021	Promote efficient use of natural resources and sustainable economic growth
Sustainable Finance Framework	2021	Align UAE's financial system with sustainable development and climate change mitigation
Updated Nationally Determined Contribution (NDC)	2020	Reduce greenhouse gas emissions by 23.5% compared to business-as-usual projections for 2030
Abu Dhabi Global Market (ADGM) Sustainable Finance Agenda	2019	Develop a sustainable finance hub and ecosystem in ADGM
National Climate Change Plan 2017–2050	2017	Provide a comprehensive framework for climate change mitigation and adaptation across economic sectors
UAE Energy Strategy 2050	2017	Increase clean energy contribution to 50% of the energy mix by 2050
Dubai Islamic Economy Development Centre Strategy	2017	Promote Dubai as the global capital of Islamic economy, including green Sukuk
Dubai Green Fund	2016	Provide AED 100 billion in loans for clean energy projects at reduced interest rates
UAE Green Agenda 2015–2030	2015	Transition to a green economy and sustainable development
UAE Vision 2021	2010	Ensure sustainable development while preserving the environment

4 Integration of Climate Finance in UAE's Islamic Financial Institutions

The alignment between Islamic finance and climate-related initiatives has gained significant momentum in recent years, as evidenced by multiple authoritative reports. The Islamic Finance Council UK and Global Ethical Finance Initiative (UKIFC et al., 2021) highlighted how Islamic financial institutions are increasingly integrating climate-related goals through innovative products like green sukuk, while the Central Bank of the United Arab Emirates (2020)'s Financial Stability Report noted initial steps toward sustainable finance through the issuance of voluntary 'Guiding Principles on Sustainable Finance' to encourage UAE financial entities to incorporate sustainability practices in their risk management frameworks. This integration is further supported by S&P Global Ratings, (2022)'s Islamic Finance Outlook which emphasizes the Gulf Cooperation Council (GCC) region's, particularly the UAE's, leadership in green sukuk issuances, underlining the region's commitment to sustainable Islamic finance practices.

Additionally, many Islamic banks are incorporating environmental, social, and governance (ESG) criteria into their investment decisions and product offerings (Al Ansari & Alanzarouti, 2020; Muhamad et al., 2022). The Abu Dhabi Islamic Bank (ADIB), for example, has developed a comprehensive Sustainable Finance Framework that integrates ESG principles into their banking operations while maintaining Shariah compliance. As outlined in their 2023 framework (ADIB, 2023), ADIB aims to offer sustainable products, including green and social sukuks, covered cards, and financings, demonstrating the bank's commitment to combine Islamic banking principles with environmental and social responsibility. Furthermore, Islamic financial institutions are partnering with government entities and international organizations to develop innovative climate finance solutions (Kawabata, 2019; Kooli, 2020).

5 Challenges and Opportunities in Developing Shariah-Compliant Climate Finance Instruments in the UAE

Developing tailored Shariah-compliant instruments for climate finance in the UAE presents both significant challenges and opportunities. One of the primary challenges is the need to ensure that climate finance instruments align with both Shariah principles and international climate finance standards. This requires careful structuring of financial products to avoid interest-based transactions while still providing competitive returns for investors(Richardson, 2020). Another challenge is the lack of standardization in Shariah interpretation across different jurisdictions, which can create uncertainty for international investors and limit the scalability of Islamic climate finance instruments (Foglie & Keshminder, 2022). Additionally, there is a need for capacity building and education within the Islamic finance sector to enhance understanding of climate risks and opportunities (Kismawadi & Rafay, 2024).

On the opportunity side, the UAE's strong Islamic finance sector and its commitment to sustainability create a fertile ground for innovation in Shariah-compliant climate finance. The country's ambitious climate goals, including its Net Zero 2050 strategic initiative, provide a clear direction for financial institutions to develop new products. There is significant potential for green sukuk to finance large-scale renewable energy projects and sustainable infrastructure. The UAE's position as a global financial hub also offers opportunities to attract international capital providers for Islamic climate finance initiatives. Furthermore, the development of new technologies, such as blockchain, presents opportunities for creating more transparent and efficient Shariah-compliant climate finance instruments (Mohammed et al., 2022). The alignment between Islamic finance principles of social responsibility and environmental stewardship with the goals of climate finance also presents a unique opportunity to create authentic and impactful financial solutions.

6 UAE's Model for Integrating Islamic Finance and Climate Action: Implications for the Gulf Region and Muslim World

The UAE's approach to integrate Islamic finance and climate action has the potentials to serve as a model for other Gulf countries and the broader Muslim world due to several key factors. Firstly, the UAE has demonstrated strong leadership in both Islamic finance and climate action. The country has established itself as a hub for Islamic finance, with Dubai aspiring to be the global capital of the Islamic economy (Alkhan & Hassan, 2024). Simultaneously, the UAE has set ambitious climate goals and has been proactive in developing renewable energy projects, such as the Mohammed bin Rashid Al Maktoum Solar Park (MOCCAE, 2023). This dual focus positions the UAE uniquely to showcase how Islamic finance can be leveraged for climate action.

The UAE's strategic partnerships have been crucial in advancing this integration. For instance, the collaboration between the Dubai Islamic Economy Development Centre and the Climate Bonds Initiative to develop green sukuk standards can serve as a blueprint for other countries (IIFM, 2020). The UAE has also been active in international forums, such as the United Nations Climate Change Conferences, promoting the role of Islamic finance in addressing climate change (CBUAE, 2023). This international engagement can inspire other Muslim-majority countries to take similar initiatives.

In terms of capacity building, the UAE has invested in education and training programs to enhance knowledge of both Islamic finance and climate finance. The creation of specialized courses and research centers, such as the Dubai Centre for Islamic Banking and Finance at Hamdan Bin Mohammed Smart University, demonstrates a commitment to developing human capital in this field. Other countries can learn from this approach to build their own expertise.

The UAE's regulatory framework, which supports both Islamic finance and sustainable finance, can also serve as a model. The introduction of green sukuk guidelines by the Securities and Commodities Authority provides a template that other regulators in the region can adapt. Moreover, the UAE's efforts to align its Islamic finance sector with global sustainability standards, such as the United Nations Sustainable Development Goals, offer a roadmap for other countries seeking to integrate these two domains (CBUAE, 2023).

To illustrate how other Gulf countries can learn from the UAE's experiences, the following table outlines specific strategies for each country:

Table 3 provides a roadmap for each Gulf country to adapt the UAE's successful strategies in Islamic climate finance to their unique contexts. By learning from the UAE's experience and tailoring approaches to their specific needs and strengths, these countries can accelerate the development of Islamic climate finance in the region, contributing to both economic growth and environmental sustainability.

However, it is important to note that while the UAE's approach offers valuable lessons, other countries will need to adapt these strategies to their specific contexts, considering factors such as their level of economic development, regulatory environment, and specific climate challenges. Additionally, as the integration of Islamic finance and climate action is still evolving, the UAE's model will likely continue to develop, providing ongoing learning opportunities for other countries in the Muslim world.

While this study provides valuable insights, it is important to acknowledge that the rapidly evolving nature of both climate finance and Islamic finance means that some findings may require updating as new developments emerge. The UAE's unique economic position and regulatory environment may limit the direct applicability of some findings to other contexts. These limitations suggest several promising directions for future research, including comparative studies examining the adaptation of the UAE's model in different economic and regulatory contexts, and investigation of emerging technologies in Islamic climate finance. Furthermore, a continue analysis is needed on the roles of Islamic social finance instruments in supporting community-level climate resilience, development of standardized metrics for measuring the environmental impact of Islamic financial instruments, and exploration of potential synergies between Islamic fintech and climate finance innovations.

Table 3. Adapting UAE's Islamic Climate Finance Model: Strategies for Gulf Countries. Source: by authors.

Country	Key Learnings from UAE's Experience
Saudi Arabia	Leverage existing Islamic finance infrastructure to develop green sukuk markets; Align climate finance initiatives with Vision 2030 sustainability goals; Establish partnerships between religious authorities and financial regulators to ensure Shariah compliance in climate finance instruments; Develop capacity building programs focusing on both Islamic finance and climate science; Integrate climate risk assessments into Islamic banking practices
Qatar	Utilize natural gas wealth to fund transition to renewable energy through Islamic finance mechanisms; Develop a regulatory framework for green sukuk issuance, similar to the UAE's Securities and Commodities Authority guidelines; Leverage international sporting events to promote sustainable Islamic finance; Establish research centers focused on Islamic finance and sustainability; Enhance collaboration between Qatar Financial Centre and climate-focused organizations
Kuwait	Adapt the UAE's model of integrating ESG criteria into Islamic banking practices; Develop climate-focused takaful (Islamic insurance) products to manage climate risks; Establish green financing initiatives through the Kuwait Investment Authority, incorporating Islamic finance principles; Create educational programs to build expertise in Islamic climate finance; Implement policies to encourage private sector participation in green Islamic finance
Oman	Leverage Islamic finance to support Oman's renewable energy targets, particularly in solar and wind projects; Develop a green sukuk framework to finance sustainable infrastructure projects; Integrate climate considerations into Oman's existing Islamic banking regulations; Establish partnerships with international organizations to enhance capacity in Islamic climate finance; Promote fintech solutions that combine Islamic finance principles with sustainable investing
Bahrain	Utilize Bahrain's position as an Islamic finance hub to become a leader in Shariah-compliant climate finance; Adapt UAE's model of strategic partnerships between Islamic finance institutions and climate organizations; Develop specialized Islamic climate finance products tailored for small island nations facing climate risks; Enhance the role of Bahrain's central bank in promoting sustainable Islamic finance practices; Leverage Bahrain Fintech Bay to drive innovation in Islamic climate finance technologies

The UAE's approach in integrating Islamic finance and climate action represents a significant innovation in sustainable finance. As the world grapples with the urgent challenges of climate change, the combination of Islamic finance principles with climate finance objectives offers a promising pathway for mobilizing capital toward sustainable development. The UAE's experiences demonstrate that religious values and environmental sustainability can be effectively aligned through appropriate financial instruments and regulatory frameworks. The success of this integration suggests that Islamic finance could play a crucial role in addressing global climate challenges, while maintaining ethical and religious principles. As other countries seek to develop their own approaches to sustainable finance, the UAE's model offers valuable lessons and inspiration for combining traditional Islamic values with modern environmental imperatives.

7 Conclusion

This research provides a comprehensive analysis of the UAE's pioneering approach to integrate Islamic finance with climate action, offering valuable insights for both academic discourse and practical implementations. Through a careful examination of the UAE's strategies, challenges, and achievements, the study demonstrates the natural alignment between Islamic finance principles and climate finance objectives. The UAE's experiences show that Shariah-compliant financial instruments, particularly green sukuk, can effectively mobilize capital for climate mitigation and adaptation projects, such as sustainable development related projects, while adhering to Islamic principles. This alignment extends beyond mere financial compatibility to encompass shared values of environmental stewardship (*khilafah*) and prevention of harm (*la darar wa la dirar*).

The research reveals that successful integration of Islamic finance and climate action requires a comprehensive ecosystem approach. The UAE's model encompasses regulatory frameworks, capacity building initiatives, strategic partnerships, and innovative financial products. This holistic approach has enabled the country to overcome traditional barriers between conventional climate finance and Islamic finance, creating a unique framework that could be adapted by other nations. The findings have significant implications for policymakers, financial institutions, and regulators across the Muslim world, particularly in developing regulatory frameworks, creating innovative Shariah-compliant climate finance instruments, building institutional capacity, and fostering international partnerships to advance climate goals. The study's examination of the UAE's approach to integrating ESG criteria with Islamic banking practices offers valuable insights for financial institutions aiming to align religious compliance with environmental responsibility. It illustrates how traditional Islamic financial principles can be effectively adapted to address contemporary environmental challenges, all while preserving their ethical foundations. This adaptation process provides important lessons for other jurisdictions looking to develop their own sustainable Islamic finance frameworks.

This research contributes to both theoretical understanding and practical implementation of Islamic climate finance, while opening new avenues for future investigation. As the field continues to evolve, further research and practical experimentation will be essential to fully realize the potential of Islamic finance in addressing global climate challenges. The findings presented here serve as a foundation for future scholarly work and practical initiatives in the growing field of Islamic climate finance, offering both theoretical insights and actionable recommendations for practitioners and policymakers.

Competing Interests The authors have no conflicts of interest to declare that are relevant to the content of this paper.

Ethics Approval This study did not involve human participants or animals, therefore no ethics approval was required.

References

Aassouli, D., Asutay, M., Mohieldin, M., Nwokike, T. C.: Green Sukuk, Energy Poverty, and Climate Change: A Roadmap for Sub-Saharan Africa (2018). https://doi.org/10.1596/1813-9450-8680

Abdullah, A.B., Al-Mubarak, T.: Maqasid in risk management: an analysis of Ijarah contract with special reference to Malaysia. ICR J. (2020). https://icrjournal.org/index.php/icr/article/view/357

ADIB: Sustainable finance framework [ESG Finance framework/Ver 1/ Nov 2023] (2023). https://www.adib.ae/en/SiteAssets/sustainability/sustainable-finance-framework.pdf

Al Ansari, R., Alanzarouti, F.: ESG and Islamic finance: an ethical bridge built on shared values. J. Islamic Finan. Stud. **6**(1) (2020). https://journals.uob.edu.bh/handle/123456789/3859

Alblooshi, F.S.A.K.: FinTech in the United Arab Emirates: a general introduction to the main aspects of financial technology. In: Sindakis, S., Aggarwal, S. (eds.), Entrepreneurial Rise in the Middle East and North Africa: The Influence of Quadruple Helix on Technological Innovation, pp. 163–178. Emerald Publishing Limited (2022). https://doi.org/10.1108/978-1-80071-517-220221010

Alkhan, A.M., Hassan, M.K.: From revelation to modern Islamic finance: Islamic commercial law in the UAE as a case study. Int. J. Law Manage. (ahead-of-print) (2024). https://doi.org/10.1108/IJLMA-06-2024-0206

Al-Roubaie, A., Sarea, A.: Building capacity for green economy: the role of Islamic finance. TAFHIM: IKIM J. Islam Contemp. World **12**(2) (2019). https://doi.org/10.56389/tafhim.vol12no2.1

Ayub, M.: Understanding Islamic Finance. Wiley (2009). https://www.wiley.com/en-au/Understanding+Islamic+Finance-p-9780470030691

CBUAE: UAE Islamic finance report 2023: the year of sustainability. Central bank of UAE (2023). https://www.centralbank.ae/media/52fdsod3/cbuae-uae-islamic-finance-report.pdf

Central bank of the United Arab Emirates: Financial stability report 2020 (2020). https://www.centralbank.ae/media/2rgpcr2i/cbuae-fsr-report-2020-en-1.pdf

CPI: Global landscape of climate finance 2023. Climate policy initiative (2023). https://www.climatepolicyinitiative.org/publication/global-landscape-of-climate-finance-2023/

Foglie, A.D., Keshminder, J.S.: Challenges and opportunities of SRI Sukuk toward financial system sustainability: a bibliometric and systematic literature review. Int. J. Emerg. Mark. **19**(10), 3202–3225 (2022). https://doi.org/10.1108/IJOEM-04-2022-0601

Dirie, K.A., Alam, M.M., Maamor, S.: Islamic social finance for achieving sustainable development goals: a systematic literature review and future research agenda. Int. J. Ethics Syst. (2023). https://doi.org/10.1108/ijoes-12-2022-0317

Gulrez, T.: Strategic assessment of Islamic Fintech in GCC countries. In: Alam, N., Nazim Ali, S. (eds.), Fintech, Digital Currency and the Future of Islamic Finance: Strategic, Regulatory and Adoption Issues in the Gulf Cooperation Council, pp. 223–241. Springer International Publishing (2021). https://doi.org/10.1007/978-3-030-49248-9_12

Haites, E.: Climate change finance. Clim. Policy **11**(3), 963–969 (2011). https://doi.org/10.1080/14693062.2011.582292

Hanif, M.: Islamic Finance: Theory and Practice (2014). https://papers.ssrn.com/abstract=1731551

Harahap, B., Risfandy, T., Futri, I.N.: Islamic law, Islamic finance, and sustainable development goals: a systematic literature review. Sustainability **15**(8), 8 (2023). https://doi.org/10.3390/su15086626

Hong, H., Karolyi, G.A., Scheinkman, J.A.: Climate finance. Rev. Finan. Stud. **33**(3), 1011–1023 (2020). https://doi.org/10.1093/rfs/hhz146

IFSB: Islamic Financial Services Industry Stability Report, p. 133. Islamic Financial Services Board (2023). https://www.ifsb.org/wp-content/uploads/2023/10/Islamic-Financial-Services-Industry-Stability-Report-2023_En.pdf

IIFM IIFM Sukuk Report 2020: A Comprehensive Study of the Global Sukuk Market (9th edn.). International Islamic Financial Market (2020). https://www.sukuk.ir/wp-content/uploads/2024/08/IIFM-Sukuk-Report-9th-Edition.pdf

IPCC: Climate change 2023: Synthesis report. In: Contribution of Working Groups I, II and III to the Sixth Assessment Report of the Intergovernmental Panel on Climate Change, pp. 35–115. IPCC, Geneva, Switzerland (2023). https://doi.org/10.59327/IPCC/AR6-9789291691647

Iqbal, Z., Mirakhor, A.: An Introduction to Islamic Finance: Theory and Practice, vol. 687. Wiley (2011). https://onlinelibrary.wiley.com/doi/book/10.1002/9781118390474

Jan, A., et al.: Islamic corporate sustainability practices index aligned with SDGs towards better financial performance: evidence from the Malaysian and Indonesian Islamic banking industry. J. Clean. Prod. **405**, 136860 (2023). https://doi.org/10.1016/j.jclepro.2023.136860

Kamali, M.H.: Environmental care in Islam: a quranic perspective. Islam and civilisational renewal (2012). https://www.semanticscholar.org/paper/Environmental-Care-in-Islam%3A-A-Quranic-Perspective-Kamali/e38df64dc19d91df05d1c0ef0e5343048e1bd71b

Kawabata, T.: What are the determinants for financial institutions to mobilise climate finance? J. Sustain. Finan. Investment (2019). https://www.tandfonline.com/doi/abs/10.1080/20430795.2019.1611148

Khan, H.: Optimal incentives for takaful (Islamic insurance) operators. J. Econ. Behav. Organ. **109**, 135–144 (2015). https://doi.org/10.1016/j.jebo.2014.11.001

Kismawadi, E.R., Rafay, A.: Islamic finance: a catalyst for sustainable climate change solutions. In: Modern Concepts and Practices

of Climate Finance, pp. 205–228. IGI Global (2024). https://doi.org/10.4018/979-8-3693-2117-1.ch009
Kooli, C.: Islamic financing initiatives stimulating SMEs creation in muslim countries. Yönetim Ekonomi Edebiyat İslami ve Politik Bilimler Dergisi **31**, 266–279 (2020). https://doi.org/10.2139/ssrn.3726578
MOCCAE: Accelerating action towards a green, inclusive and resilient economy -Third update of second nationally determined contribution for the UAE, p. 69. Ministry of climate change and environment (2023). https://unfccc.int/sites/default/files/NDC/2023-07/Third%20Update%20of%20Second%20NDC%20for%20the%20UAE_v15.pdf
Mohammed, M., Khan, N., Ahmad, T.: Opportunity for Islamic finance in blockchain solutions for climate change: a case study of DCarbonX model. مجلة بيت المشورة, pp. 238–279 (2022). https://doi.org/10.33001/M0110202218/105
Moussaid, F.Z., Jerry, M., Qafas, A.: Examining the nexus water demand-economic growth in morocco: a tapio decoupling index analysis. Int. J. Account. Finan. Audit. Manag. Econ. **4**(5–2), 5–2 (2023). https://doi.org/10.5281/zenodo.8388151
Muhamad, S., Mohd Zain, F.A., Syafiqah, A., s, Rahman, A., & Yasoa, M.: Measuring sustainable performance of Islamic banks: integrating the principles of environmental, social and governance (ESG) and Maqasid shari'ah. IOP Conf. Ser. Earth Environ. Sci. **1102**, 012080 (2022). https://doi.org/10.1088/1755-1315/1102/1/012080
Nasir, N.M., Nair, M.S., Ahmed, P.K.: Environmental sustainability and contemporary Islamic society: a shariah perspective. Asian Acad. Manag. J. **27**(2), 2 (2022). https://doi.org/10.21315/aamj2022.27.2.10
Rahman, M.H.: Mudarabah and its applications in Islamic finance: an analysis. Asian J. Res. Bank. Finan. **8**(6), 33–46 (2018)
Richardson, E.: The UAE and responsible finance—Can responsible finance ṣukūk help the UAE in fulfilling its sustainability ambitions? (2020). https://doi.org/10.1163/15730255-BJA10013
Saeed, M.: Green Sukuk for financing renewable projects in Islamic social finance: Problems and solutions. In: Contemporary Issues in Islamic Social Finance. Routledge (2021)
S&P global ratings: Islamic finance outlook 2022 (2022). https://www.spglobal.com/ratings/en/research/pdf-articles/islamic-finance-outlook-2022-edition-28102021
UKIFC, ISRA, GEFI.: Islamic Finance and the Principles for Responsible Investment. Thought Leadership Series Part 2 March 2021, p. 40. Islamic Finance Council UK (UKIFC), International Shari'ah Research Academy for Islamic Finance (ISRA) and the Global Ethical Finance Initiative (GEFI) (2021)
UKIFC, UNDP: Innovation in Islamic finance: Green Sukuk for SDGs. Islamic finance council UK and united nations development programme Indonesia (2021). https://www.ukifc.com/wp-content/uploads/2021/09/UKIFC21_Green_Sukuk_v1-2_20210927.pdf
UNEP FI, IsDB, & CIBAFI: Mobilising Islamic banking for climate action. United nations environment programme finance initiative (2023)
UNEP, U.N.: Emissions gap report 2023: Broken record – Temperatures hit new highs, yet world fails to cut emissions (again). United nations environment programme (2023). https://doi.org/10.59117/20.500.11822/43922
UNFCCC: Standing committee on finance's fifth Biennial: assessment and overview of climate finance flows. United nations climate change secretariat (2022). https://unfccc.int/topics/climate-finance/resources/biennial-assessment-of-climate-finance
Usmani: The Concept of Musharakah and its Application as an Islamic Method of Financing (1999). https://doi.org/10.1163/026805599125826435
Vishwanath, S.R., Azmi, S.: An overview of Islamic sukuk bonds. J. Project Finan. **14**(4), 58 (2009)

Strategies for Enhancing Travel Choice Adaptation in Hot Weather Conditions: Insights from Stakeholder Consultation

Deepty Jain, Akanksha Bhardwaj, and Gyanada Panda

Abstract

In the past two decades, the rise in temperatures attributed to climate change has resulted in a 70% increase in extreme weather events, notably intense heat waves. These events pose significant risks to human health, exacerbating heat-related illnesses and vulnerabilities within affected communities while compounding urban challenges by increasing energy demands and straining infrastructure. The temperature increase in Indian cities has increased the population's vulnerability, leaving people exposed to extreme environmental conditions while travelling. Studies have shown that rising temperatures will significantly change travel patterns and preferred destinations, imposing infrastructural challenges and health risks. This paper aims to assess the acceptance and relevance of various adaptive and mitigative strategies at individual, community, and governmental levels to address vulnerabilities while commuting in extreme hot weather conditions. Input was collected from 13 stakeholders representing the government, community, and individuals during a workshop on "Gendered approaches for addressing adaptation capacity to hot weather" organised on 5th August 2023 in Delhi. Using exploratory data analysis, the consensus and conflicts for selected strategies are defined. The findings of the study indicate diverse stakeholder responses, highlighting the need to secure broad support and a comprehensive understanding of consensus and conflict, which can inform effective policymaking and enable governments to implement targeted strategies to address the vulnerabilities of commuters in hot weather conditions.

Keywords

Hot weather · travel choice · adaptation · stakeholder consultation · climate change

1 Introduction

1.1 Challenges Related to Heat Waves: Global and Indian Perspective

Globally, anthropogenic activities and climate change have influenced the local climatic conditions. One of the major harmful impacts observed is from the rising temperatures. According to the World Health Organisation (World Health Organisation, 2024a) 125 million people were additionally exposed to heat waves between 2006 and 2016. Heat waves can be defined as conditions of high temperatures, exposure to which can cause morbidity or mortality. However, the criterion for defining heatwave may vary by location. For example, in Indian conditions, a heatwave is to be declared when the actual maximum temperature remains 45 °C or more, irrespective of the normal maximum temperature (World Health Organisation, 2024b). Exposure to extremely high temperatures can lead to heat-related illnesses such as heat cramps, heat exhaustion, and heat stroke. Also, exposure to high ambient temperatures, above the long-term average, and extreme heat (heat wave) increases mortality and causes financial and life losses, especially in the summer months (Kiarsi et al., 2022).

Most megacities in tropical and subtropical climates are witnessing significant warming episodes, resulting in increased mortality and morbidity (Hess et al., 2018; Oudin Åström et al., 2013; Perkins-Kirkpatrick et al., 2012). Indian cities in the tropics have recently experienced an increased

D. Jain · A. Bhardwaj (✉) · G. Panda
TRIPC, Indian Institute of Technology, Delhi, India
e-mail: abhardwaj.iitd@gmail.com

B. He et al. (eds.), *Towards the Framework of Livable and Resilient Cities* Advances in Science, Technology & Innovation, https://doi.org/10.1007/978-3-031-97849-4_29

number of hot days during the summer, affecting the quality of life. The rising temperatures impact cities more because of the urban heat island effect, impervious surfaces, reduced green canopy cover, and intense urban activities. Further, poor urban planning and management coupled with climate change have resulted in deteriorated urban thermal conditions and micro-climates, which have a detrimental influence on people's health and quality of life (Li et al., 2022).

Disaster risk is an outcome of the interaction between hazard level, exposure to the hazard, and vulnerability of people to the hazard (UNISDR, 2015). In this equation, both the hazard level and exposures are easier to control than the population's vulnerability. This would require the adoption of both mitigation and adaptation strategies. People are exposed to ambient thermal conditions when participating in an outdoor activity, which includes travelling, participating in jobs pursued in outdoor settings, shopping, playing in open fields, picking/dropping off children, and others (Eslamirad et al., 2022; Taleghani et al., 2015). It is imperative to reduce their exposure to the hazards which would require improved thermal conditions and adjusting timings of the activities. Therefore, there is a need to improve the thermal conditions and reduce the exposure of commuters to high temperatures (Park et al., 2018; Yasumoto et al., 2019). This would require the adoption of both mitigation and adaptation strategies.

1.2 Types of Mitigation and Adaptation Strategies

Mitigation refers to actions taken to reduce temperatures and improve thermal comfort conditions in an urban environment. The mitigation strategies include, for example, upscaling green and blue infrastructure (Herath et al., 2018), landscaping strategies (Magotra, R., 2020; Park et al., 2018), provisioning shaded road networks (Dzyuban et al., 2022), changing infrastructure design(Fraser & Chester, 2017; Miao et al., 2019), and optimising the usage of appliances (Youssef et al., 2023; Yu et al., 2013). These strategies help reduce temperatures, thereby minimising health risks. Additionally, by implementing measures such as transitioning to renewable energy sources (such as solar and wind power), improving energy efficiency in buildings and industries, implementing sustainable land use practices (like afforestation and reforestation), and promoting sustainable transportation options (such as public transit and electric vehicles), helps in mitigating the extent of temperature increase and limit the severity of climate-related impacts over the long-term(IPCC, 2022). On the other hand, adaptation refers to adjusting to the impacts of climate change that are already occurring or are expected to occur in the near future (Ray Biswas & Rahman, 2023).These measures are designed to increase resilience and reduce exposure to climate-related risks. Adaptation strategies include enabling the change in the timing of outdoor activities, changing locations, keeping hydrated, changing modes, changing routes, and adopting technologies like air conditioners (NRDC, 2022; Rosenthal et al., 2022). Such strategies help in reducing exposure to high temperatures. These measures are essential for safeguarding human health, ensuring economic stability, and addressing the needs of vulnerable communities. Both mitigation and adaptation are essential components of climate action, necessary for building sustainable cities. Aligning to this, Sustainable Development Goal 11, target 11.5 aims to reduce the number of deaths and number of people affected by disasters with a focus on protecting people in vulnerable situations. It is important to note that heat wave and high temperature are recognised as disasters because it has severe impact on human health and infrastructure (World Health Organisation, 2024b).

1.3 Need for Understanding Stakeholders' Perception

Adopting mitigation and adaptation strategies requires multi-stakeholder involvement at all levels, including citizens, the community, service providers and operators, the private sector, and the government. Various strategies have been proposed for mitigating and adapting to the increasing temperature levels. However, stakeholders' roles and responsibilities shall vary for each strategy. Stakeholders' roles are determined by their knowledge and expertise, hierarchical position, and specific purpose or mandate (Dwivedi, 2021). For example, strategies like keeping hydrated will require strong acceptance by citizens where the community can take the role of enabler. Similarly, changing travel time would require establishments and communities to understand the diurnal temperature variation and the related health impacts. Their key role shall be to adopt an environment enabling the change in work timings. Individuals' acceptance will also be needed to enable the change. Here, the government can provide an appropriate landscape for individuals and the community to adopt the strategy.

Yeleliere (2022) studied the role of government and practitioners in climate change adaptation of agriculture sector in Ghana. As per the authors, the government, at regional, district, and local levels, assists in planning and implementing adaptation interventions despite shortfalls in skilled staff, financial capacity, and funding opportunities. Their prime role is to develop policies. The authors further discuss that the practitioner (here NGOs), on the other hand bring innovation, technical expertise, and financial resources therefore contributing effectively to policy implementation. Practitioners operate within the policy framework implemented by the respective government, utilising varied methods but aiming to achieve similar goals. Therefore, it is understood that if there is consensus on policies and strategies between government and practitioner, then the implementation is more effective. Collaborative partnerships among stakeholders foster knowledge sharing, resource mobilisation, and coordinated actions, strengthening resilience, and contributing to effective responses to the challenges imposed by rising temperatures (Yeleliere et al., 2022).

In their purview of duties, roles, responsibilities, and power, each stakeholder shall determine whether a strategy is relevant, acceptable, and adaptable. However, this may vary according to socio-demography, context, awareness, and local and regional governance mechanisms (Mehiriz, 2024; Mehiriz et al., 2018). Understanding the acceptance and relevance of potential mitigation and adaptation strategies can help design strategies to reduce both the hazard and exposure effectively (Wang et al., 2023). This would involve evaluating how well stakeholders, including individuals, communities, and government officials, embrace and support the potential measures. (Lawler et al., 2013). The assessment shall include factors such as stakeholder feedback on the effectiveness and feasibility of the strategies, their perceived relevance in addressing heat wave challenges, and the willingness to implement them (Klinsky et al., 2012; Watts et al., 2015). It also considers any concerns raised by stakeholders, as well as the overall level of support or resistance encountered during the planning and implementation phases of these strategies (Häberlein & Hövel, 2023).

Every stakeholder has their own understanding of what is acceptable and relevant (Guerrero-Villegas, 2019).This understanding not only varies across the stakeholders but factors such as age gender, context, awareness, and other variations like cultural background and socioeconomic status. These factors also play a significant role. Conflicts arise between stakeholders, impacting decision-making and the effectiveness of policy implementation (Lemke & Harris-Wai, 2015). For example, local communities prioritise immediate economic benefits over long-term environmental sustainability, while policymakers may focus on regulatory compliance. If stakeholders' conflict with certain decisions, their implementation can be significantly affected (Lemke & Harris-Wai, 2015). Therefore, it would need understanding which of potential adaptation and mitigation strategies are accepted and considered more relevant by different stakeholders. This will help us in pre-determining the potential conflict that may exist between the stakeholders for the same set of strategies (Pless et al., 2022). This would require an understanding of acceptance and relevance is essential for informing decision-making processes, determining policy interventions, and allocating resources effectively for ensuring successful implementation of strategies.

1.4 Aim and Contribution of the Paper

As discussed, people are more exposed to ambient temperatures in outdoor settings. Among these settings, travelling—which is typical for everyone, including those pursuing indoor or outdoor activities—can be identified exposed to ambient temperatures while travelling. Even those who are using public transport or personal vehicles for commute are exposed to extreme weather conditions during access and egress from public transport stops or parking lots. The study aims to map stakeholders' perceptions regarding potential strategies to address the impact of hot weather conditions, particularly for commuters in Delhi. The study is based on the findings from the workshop "Gendered approaches for addressing adaptation capacity to hot weather conditions" conducted on 5th August 2023 as part of a project funded by UKAID through the UK Foreign Commonwealth and Development Office under the High-Volume Transport Research Programme. The study addresses two key questions. The first question identifies which strategies are considered most relevant and acceptable by various types of stakeholders. The second question explores how different stakeholders position themselves in influencing government-led strategies. Government-led strategies would need acceptance from individuals, and communities, where the government provides an appropriate landscape for implementation. The study helps understand which of the pre-selected 35 strategies are more accepted by different stakeholders and what would be the level of conflicts during implementation. The findings can be used to define a potential set of actions that can be implemented with the least effort, providing maximum benefits and the strategies that would need methods to overcome implementation challenges and gain consensus from all stakeholders. The key contribution of the study is the presentation of diversity in stakeholder responses and a comprehensive understanding of the acceptance and relevance of the identified strategies.

This paper is structured in five sections. Following the introduction in the first section, the second section discusses the different methods for assessing stakeholders' perception. The third section present the methodology used for the study includes data collection procedures, approach for stakeholder assessment. In the fourth section, the methodology outlines the research design, including the selection of study areas, data collection techniques, and analytical methods used to evaluate the effectiveness of these adaptive and mitigative strategies in minimising the adverse impacts of extreme heat on urban commuters. The fifth and final section presents the findings and discusses policy implications, thereby concluding the study.

2 Literature Review

2.1 Data Collection Methods for Stakeholder Assessment

Various techniques and approaches are used to gather information, opinions, and feedback from individuals or groups with a vested interest or impact on a particular issue or project. These methods include structured surveys to collect information in a quantitative way and unstructured questionnaires to gather qualitative data on stakeholder perspectives, priorities, concerns, and preferences. Both methods can be administered through online platforms, mail, or in-person interviews.

Interviews, whether structured or unstructured, involve one-on-one reporting from stakeholders to gather quantitative or qualitative data, providing in-depth insights, opinions, and experiences (Mathers et al., 2000; Moser & Korstjens 2018). Focus group discussions are conducted with an assembly of small, representative groups of stakeholders in a moderated setting to discuss specific topics or issues, fostering interaction, brainstorming, and idea generation. (Then et al., 2014). Workshops and stakeholder consultations convene stakeholders in facilitated sessions to collaborate on key issues, strategies, or plans, promoting active participation and consensus-building (Nielsen et al., 2017). Participatory mapping and visualisation are developed using tools like geographic information system (GIS) to represent perspectives, relationships, and priorities (Cochrane & Corbett, 2018). Each of these methods has its strengths and limitations, and they can be combined as needed to achieve comprehensive stakeholder assessment based on the objectives, scope, and context.

Quantitative data can be in the formats of Likert scale rating systems, ranking, scaling, binary choices, or multiple choices. In the Likert scale rating system, stakeholders assign numerical values to a set of choices on a scale of 1 to 5 or 1 to 10 (Juhnke & Mühlbacher, 2013; Koo & Yang, 2025). Whereas a ranking system requires stakeholders to rank a set of alternatives in the order of preference, therefore allowing stakeholders to compare the alternatives (Chen & Cheng, 2010). Binary choices or multiple choices involve the selection of the most preferred or used alternative by the stakeholders amongst a given set of choices (Cinelli et al., 2022).

2.2 Methods for Assessing Stakeholders' Perception

For the successful implementation of policy initiatives or projects, understanding the perspectives of multiple stakeholders is essential. To achieve this, various stakeholder assessment methods are used, including statistical analysis, multicriteria decision analysis, stakeholder mapping, and qualitative techniques. Each method can help in systematically identifying, analysing, and assessing the roles, relationships, and perceptions of different stakeholders. These methods also help identify areas of consensus and conflict among stakeholders, ensuring their diverse perspectives are considered in the decision-making process.

Statistical methods in stakeholder assessment involve analysing quantitative data to identify patterns, relationships, and trends among stakeholders. The study by Karthikeyan et al. (2024) used descriptive analysis to evaluate variables such as power, position, and interest of stakeholders and ranked enablers and barriers using scores from 17 stakeholder representatives. Data was collected through a review of 45 articles and policy documents. Concordance and Discordance analysis is another statistical method used to measure the degree of agreement or disagreement among stakeholders' opinions or preferences. The method has been used to assess the agreement and disagreement between consumers and clinicians regarding the quality of end-of-life care using qualitative measures (Carlini et al., 2022; Kwiecien et al., 2011). The study involved interviews or surveys to collect data from both consumers and clinicians. Other methods like factor analysis are used to identify the underlying factors that influence stakeholder behaviour (Tavakol & Wetzel, 2020) and cluster analysis is used to identify the distinct groups within their stakeholder base (Migliore & Rossi-Lamastra, 2023). Additionally, the exploratory data analysis (EDA) method is a visual and statistical technique used to understand the data, identify patterns, detect anomalies, and test hypotheses (Abzalov, 2016).

Multicriteria decision analysis (MCDA) techniques offer a structured approach for evaluating and prioritising stakeholder interests by considering multiple criteria. One of the most common methods of MCDA is the Technique for order of preference by similarity to the ideal solution (TOPSIS) as described in the study by Madanchian & Taherdoost (2023). It involves determining criteria weights, constructing a decision matrix, and calculating TOPSIS scores to rank alternatives based on their closeness to the ideal solution. The analytic hierarchy process (AHP) structures a decision problem into a hierarchy and assigns weights to both criteria as well as alternatives. The method provides pairwise comparisons for both the criteria and alternatives that are used to determine priorities in the decision-making process. A study by Kwatra et al. (2021) involving key stakeholder groups (elected representatives, government, industry, NGOs/academia) used the AHP approach to assign weights to three themes (social, economic, environmental) and eighteen sub-themes (e.g., livelihood, waste management) for regional sustainability assessment. The ELECTRE is a family of multi-criteria decision analysis (MCDA) methods ranks alternatives through pairwise comparisons, considering both concordance and discordance indices in the decision-making process (Haider et al., 2015). Similar to ELECTRE, preference ranking organisation method for enrichment evaluation (PROMETHEE) helps rank alternatives by comparing the differences between each pair of alternatives, thereby, assessing the stakeholders' preferences (Uzun et al., 2021). Multi-Attribute utility theory (MAUT), assesses stakeholders' preferences based on utility functions for multiple attributes, further enriching the decision-making process. The method has been used by Taufik et al. (2021) to objectively select representatives for evacuation decisions. The study used variables like selection of foreign diplomats, criteria for suitability, and number of representatives and determined value-weighted criteria to balance conflicts on sample data points.

Stakeholder mapping techniques is used to map stakeholders' influence, interest, and relationships with the decision or decision-making process. This can be achieved by developing stakeholder 2 × 2 analysis matrix and can be used for evaluating diverse group of stakeholders (Brugha & Varvasovszky, 2000). Ginige et al. (2018) used grid-based analysis method to map stakeholders by power versus interest and to assess their involvement in addressing global societal challenges. The influence/impact Matrix used by Hernandez Villamizar et al. (2018) focuses on stakeholders' ability to influence decisions. Salience model determines stakeholders' position with regard to power, legitimacy, and urgency. It helps identify which stakeholders hold the most significant sway over the decision-making process (Adib & Zhang, 2019). Social network analysis (SNA) helps us analyse the relationships and flows between stakeholders (Saqr & Alamro, 2019).

Qualitative methods can also be used to understand the underlying motivations, perceptions, and attitudes of stakeholders through in-depth analysis of qualitative data. Within this, SWOT (strengths, weaknesses, opportunities, and threats) analysis is employed to assess an organisation or project's strengths, weaknesses, opportunities, and threats. This approach can also be modified to evaluate stakeholders by examining the influence of each stakeholder's attributes on the project (Christodoulou & Cullinane, 2019).Additionally, methods like PESTLE (political, economic, social, technological, legal, and environmental) analysis assess external macro-environmental factors that could impact stakeholders, providing insights into broader influences on decision outcomes (Christodoulou & Cullinane, 2019). Delphi method to assess stakeholders uses rounds of surveys to reach a consensus among a panel of stakeholders. Each round builds on the previous one, with feedback provided to participants to refine their responses, ultimately leading to a more accurate and agreed-upon outcome (Vatandoost et al., 2022). In this paper, we utilised EDA, a descriptive statistical method that facilitates understanding of the data through visualisations and summary statistics. This enables us to identify the divergent perception of the stakeholders with regard to the strategies. EDA is appropriate to understand the perception of 13 stakeholders across stakeholder groups and by type of strategies.

2.3 Summary

In existing literature, different data collection and analysis methods are applied to assess stakeholder perceptions and preferences. These data collection methods range from structured surveys to qualitative in-depth interviews. The data assessment method vary from using EDA approaches to MCDA techniques and grid-based mapping. The selection of data assessment method is based on the type of data collected and how it is collected. For instance, AHP requires stakeholders to pairwise compare both the criteria and alternatives, while the ELECTRE method employs concordance and discordance indices to rank alternatives and identify the best option. Statistical methods like descriptive analysis, factor analysis, and EDA are more useful for identifying patterns and relationships in stakeholder data. Specifically, EDA helps visualise and understand data, detect anomalies, and generate hypotheses. Each method, whether simple or complex, has its strengths and limitations with regard to the ease of data collection, the complexity of analysis, the depth of analysis, and the interpretability of the results.

3 Methodology

Various strategies can be adopted by cities, communities, and individuals to address hot weather conditions. Health institutions and respective authorities adopt various strategies, including issuing advisories, changing services operations, and provisioning amenities to help reduce vulnerabilities (IRADe, 2021). However, the implementation and effectiveness of the adopted strategy may depend on how different groups of stakeholders perceive the strategies as relevant or acceptable. Strategies that are widely accepted by all stakeholders may be more easily implemented and more effective in achieving the agenda. However, strategies that receive divergent views are likely to be questioned, debated, and contested, resulting in delays in implementation and, consequently, reduced effectiveness. Capturing stakeholder perceptions in the early design stages can help in better decision-making. Using appropriate methods, analysts and planners can identify if a certain strategy receives consensus 'yes' or consensus 'no' or conflict opinions. This can be used to prioritise strategies and devise methods for consensus building. For example, the strategies that receive consensus 'yes' may be identified as those strategies that can be easily implemented.

The strategies that receive consensus 'no' shall require gathering and sharing more evidence for the effectiveness of the strategy. However, the strategies that receive conflicting opinions shall require gathering evidence, awareness building, and addressing the concerns of the selected groups.

As discussed, most of the studies reviewed have investigated the degree of influence and role of stakeholders in the decision-making process. This study aims to understand the potential consensus and conflicts among stakeholders regarding a selected set of strategies that can be adopted to address the vulnerability of commuters during hot weather conditions in Delhi. To achieve this, data was collected in a structured format during the workshop titled "Gendered approaches for addressing adaptation capacity to hot weather conditions," conducted on August 5, 2023, in Delhi. The project is funded by UK-AID through the UK Foreign, Commonwealth, and Development Office as part of the High-Volume Transport Research Programme.

Drawing from the existing literature, we identified 35 strategies that can be used to reduce the exposure and vulnerabilities of commuters to hot weather conditions (Table 1). These strategies were bifurcated into three levels - individual-led (10), community-led (10), and government-led (15) strategies depending on who implements the strategy. For example, keeping self-hydrated is essentially individual led but can be facilitated by a community or government agency. Similarly, adopting climate-responsive urban design can be identified as a strategy that is primarily implemented by a government authority. While there can be overlaps between the roles of different stakeholders. For example, capacity building can be implemented by both the community and the government.

A total of 13 stakeholders participated in the workshop, which included three representatives from various government agencies in Delhi involved in implementation, decision-making, and enforcement. Additionally, seven practitioners from academia, private sector agencies, and NGOs working in the fields of transport and health were present, along with three community members, which included schoolteachers and parents. While in their capacity and roles, each stakeholder may switch from government to community and practitioners and vice-versa. For example, a practitioner may also be a parent or a community member. For ease of interpretation, we only considered each stakeholder's primary role based on the nature of their occupation.

During the workshop, the participants first discussed the potential impacts of hot weather conditions on commuters. They then presented the selected strategies and engaged in discussions organised into three groups: individual-led, community-led, and government-led strategies. A structured form was circulated after discussing the strategies in each group, listing the strategies discussed. The stakeholders were asked to report if they identified a strategy as relevant and acceptable using binary choice as a yes or no (Table 2). Here,

Table 1. Different strategies considered for evaluation bifurcated across different groups. (Source: By author)

Individual-led Strategies	Community-led Strategies	Government-led Strategies
I1: Avoid crowded spaces	C1: Provisioning cooling amenities like fans in local streets	G1: Shading streets by permanent and temporary structures
I2: Avoid travelling in peak heat hours	C2: Plant trees	G2: Low-energy ventilation systems for buses and metro
I3: Avoid travel	C3: Provisioning drinking water	G3: Heat-resistant paving materials
I4: Use AC bus/ Metro	C4: Sharing shaded spaces	G4: Restricting vehicle usage
I5: Change route – use shaded routes	C5: Shading spaces using temporary structures like awnings	G5: Water Misting stations at bus stops
I6: Change destination/location	C6: Providing umbrellas, caps etc	G6: Shaded vending zones
I7: Share ride	C7: Capacity building and awareness generation	G7: Changing bus routes to reduce access distance
I8: Keep hydrated	C8: Provisioning ride sharing	G8: Provisioning paratransit for last-mile connectivity
I9: Carry umbrellas	C10: Changing working hours	G9: Climate-sensitive urban design strategies
I10: Choose weather-appropriate clothing	C11: Community support for shared responsibilities	G10: Appropriately shaded transit shelters
		G11: Deployment of additional buses
		G12: Green scaping urban streets
		G13: Issue heat advisories
		G14: Changes in working hours, school hours
		G15: Capacity building of communities and awareness generation

Table 2. Method for data criteria for assessment for different types of stakeholders (Source: By author)

Types of strategies	Criteria for stakeholder analysis		
	Acceptance	Relevance	Degree of Influence (DOI)
Individual-led Strategies	Y/N	Y/N	
Community-led Strategies	Y/N	Y/N	
Government-led Strategies	Y/N		Likert scale 1 to 5

acceptance measures the willingness and ability of stakeholders to implement a particular strategy. High acceptance indicates that stakeholders are on board with the strategies and are likely to support their implementation. Relevance, on the other hand, assesses how pertinent and appropriate the strategy is in addressing vulnerability to hot weather conditions. High relevance indicates that stakeholders perceive the strategy as effective and beneficial. Assessing variation in relevance and acceptance between stakeholders can help in identifying whether there is a consensus or a conflict. Differences in levels of acceptance and relevance among stakeholders can significantly impact the project's progress and outcomes.

For government-led strategies, the stakeholders were also asked to report their degree of influence in decision-making or implementation of the strategy. Here the stakeholders were asked to rate the strategy on a Likert scale of 1 to 5, with 5 representing a high level of influence. This helped in determining the degree of role of each type of stakeholder in decision-making or its implementation. The structured approach facilitated the collection of quantitative data on stakeholder perspectives, enabling the identification of preferences and areas of consensus or divergence. The insights gleaned from this consultation process can be instrumental in formulating recommendations for prioritising and implementing strategies to address the exposure and vulnerabilities of commuters during hot weather conditions.

To assess the level of consensus or conflict between different groups of stakeholders. For this purpose, we use a benchmark of 75% in accordance with environmental impact assessment (EIA) policy guidelines (MoEFCC, 2021). If 75% or more of the stakeholders in all groups identify a strategy as relevant or acceptable, then it is identified that there is a consensus on the strategy. Since there are fewer respondents in each group, rounded-off values at 0 decimal places are used for the corresponding number of respondents. So, for example, within the community of the three stakeholders, a minimum of two stakeholders should give one response, and within the government and practitioners group, at least four stakeholders out of five should give one response. For three representatives from various government agencies based in Delhi or the community, at least 2 of the 3 members (67.7%) must identify the strategy as relevant or acceptable. Among seven practitioners, at least 5 of the 7 (71.4%) need to agree. At these thresholds, if a strategy is identified as irrelevant or unacceptable, then there is a consensus among stakeholders regarding its ineffectiveness. However, conflict exists if different shares of stakeholders in each group give different responses. Considering a hypothetical example, strategy A received a positive response from a minimum of 71% of the stakeholders in group 1, but less than 67.7% of the community representatives submitted a positive response. This situation represents conflict. Such strategies would require evidence generation, understanding the concerns of particular groups of stakeholders, incorporating changes to address concerns, generating awareness, and eventually building consensus.

4 Results

In this section, the findings from the stakeholder assessment using the EDA approach are presented. Two sets of results are discussed here. One refers to whether a strategy is identified as relevant and acceptable by different stakeholders. For this purpose, bar graphs present results for individual-led and community-led strategies. Here, the positive y-axis represents the percentage of stakeholders identifying a strategy as relevant, and the negative y-axis presents the percentage of stakeholders identifying the same strategy as acceptable. Four types of interpretations are drawn from here. 1) A long bar equally distributed on both sides of the x-axis represents that the strategy is relevant and acceptable. Such a strategy is easy to adopt and may not require awareness generation. 2) If a longer bar is on the positive axis only, then the strategy is relevant but unacceptable. This indicates that while stakeholders recognise the strategy as appropriate and beneficial, they may either identify difficulty in implementing it or are unwilling to support it. Implementing such strategies shall require interventions like changing infrastructure, enabling the environment and providing a landscape for improving acceptance. 3) If the overall length of the bar is shorter, then the strategy is irrelevant and unacceptable. 4) A strategy with low relevance but high acceptance needs to be tested for its effectiveness, and once tested, there would be a need to generate awareness regarding it. For government-led strategies, the relevance and degree of influence are discussed. The degree of influence is assessed for each stakeholder to understand which stakeholders can assume prominent roles as enablers, decision-makers, or executors.

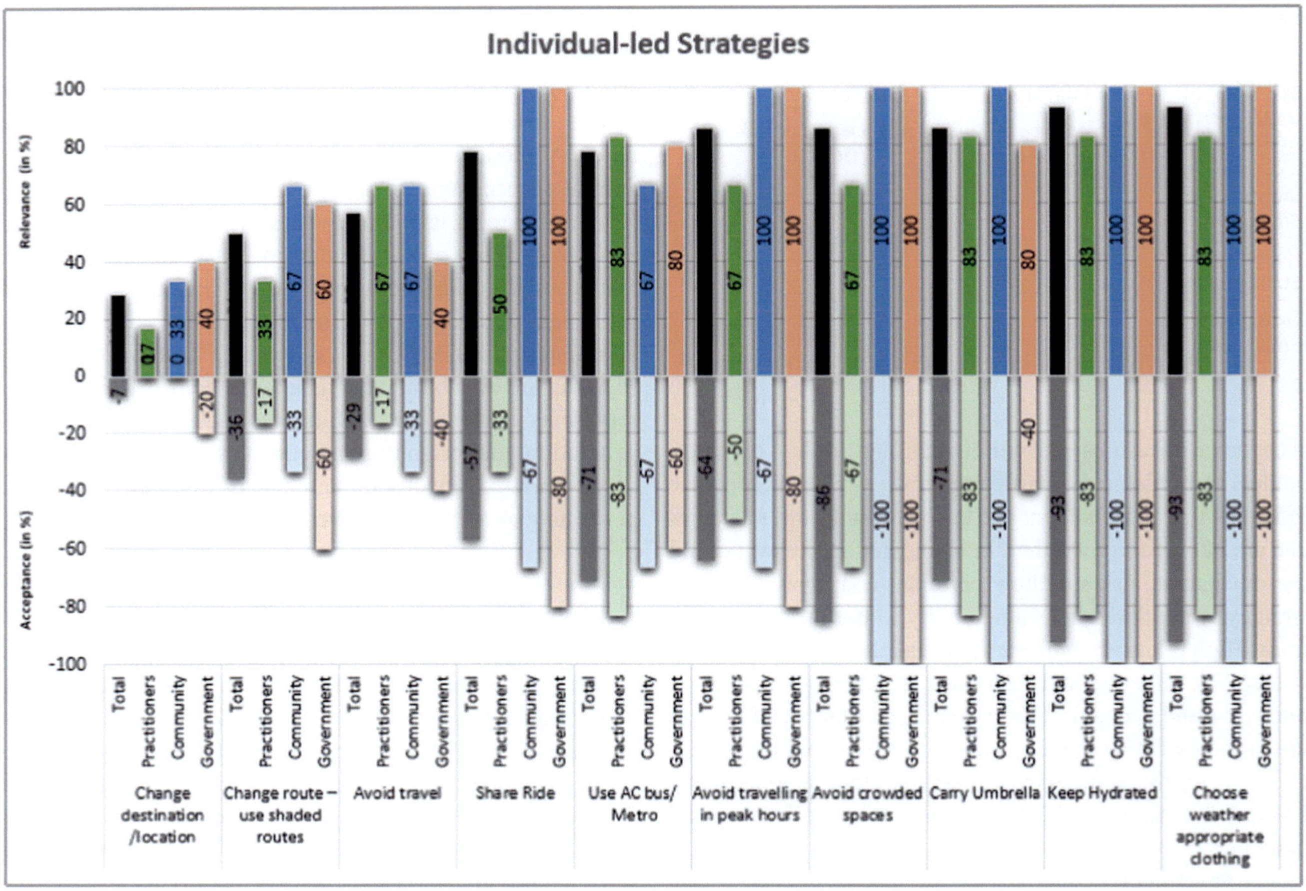

Fig. 1. Relevance and acceptance of individual-led strategies (Source: By author)

4.1 Findings

4.1.1 Individual-Led Strategies

In Fig. 1, the strategies are arranged in increasing order of relevance and acceptance by all the stakeholders. Of the ten selected strategies, wearing weather-appropriate clothing, and keeping yourself hydrated are considered relevant and acceptable by most stakeholders across all the groups. As per the Centre for Disease Control and Prevention (2022), these strategies help individuals better cope with extreme heat while commuting. Carrying an umbrella is identified as relevant strategy by majority of the stakeholders, across the groups however, a smaller share of government representatives identifies this as acceptable.

Over 75% of all stakeholders identify avoiding crowded spaces as acceptable and relevant. However, there is a potential conflict as only 67% of the practitioners identify it as both relevant and acceptable. This may be likely because of the inability of individuals to avoid certain public places, such as using a bus. Nuruzzaman (2015) found that crowded areas tend to have higher levels of heat retention due to the urban heat island effect, exacerbating the overall temperatures. Therefore, choosing less crowded routes for travel or modes of transportation can improve airflow and ventilation, enhancing thermal comfort for commuters (Rawat & Kumar, 2023).

Similarly, avoiding travel during the hottest part of the day, using AC bus, metro, or ridesharing, is identified as relevant by the majority of the stakeholders. However, less than 75% of the stakeholders, particularly only 50% of practitioners identify avoiding travel during peak hours as acceptable. The strategy is thus exposed to potential conflict regarding acceptability. Actions like providing the coercive environment, policy landscape and infrastructure interventions can improve the acceptability levels. Ridesharing, for example, is beneficial as it helps reduce the number of vehicles on the road, minimising emissions, and contributing heat dissipation. However, practitioners don't identify this strategy as both acceptable and relevant. To encourage ridesharing or avoid travel during certain hours of the day, the government may need to facilitate flexible work hours and telecommuting options and incentivise ride-sharing services. Of all the individual-led strategies, more than 75% of the stakeholders identified changing destinations as irrelevant and unacceptable. Strategies like changing routes are also deemed irrelevant and unacceptable. Here, the potential conflict exists where 33% of the stakeholders representing the practitioners feel that the strategy is relevant. Changing routes, altering destinations, and avoiding travel may result in decreased accessibility, increased inconvenience, disruption of routine, and potential cost implications. These strategies are therefore not recommended further for investigation or implementation. At the individual level, strategies that do not depend on external factors are preferred. These external factors may include service provision—such as changing routes, changing destinations, sharing rides and infrastructure improvements which would require community or government intervention.

4.1.2 Community-Led Strategies

Of all the community-led strategies, only the strategy of capacity building and awareness generation is identified as relevant and acceptable by more than 75% of the stakeholders in all the groups, as shown in Fig. 2. The strategies of community support for shared responsibilities and provisioning drinking water are also identified as relevant and acceptable by more than 75% of all the stakeholders. All other strategies within community-led initiatives are exposed to higher levels of conflicts with varying ranges of relevance and acceptance by different stakeholder groups. No one strategy receives consensus on being irrelevant. Changing working hours is perceived as unacceptable by more than 75% of all the stakeholders. Here, too, the conflict exists, where more than 75% of the practitioners identify it as unacceptable and a relatively higher share of government representatives identify it as acceptable. Only 33.3% of the community identify the strategy as relevant. Considering the observations made for avoiding travel in individual-led strategies, the similar level of interventions is needed to increase the acceptance by different stakeholders' groups. Furthermore, strategies such as community support, capacity building, and awareness, are deemed most relevant amongst all other strategies. In most cases, when a community faces a rise in certain incidents, it often responds by launching awareness programs to control the problem. However, studies have shown that these programs are frequently less effective than intended. Despite this, these strategies may be perceived as highly relevant because of their ease of implementation and seeing these strategies leading to early interventions, acceptance of other strategies and effective decision making. (Illahi, 2019).

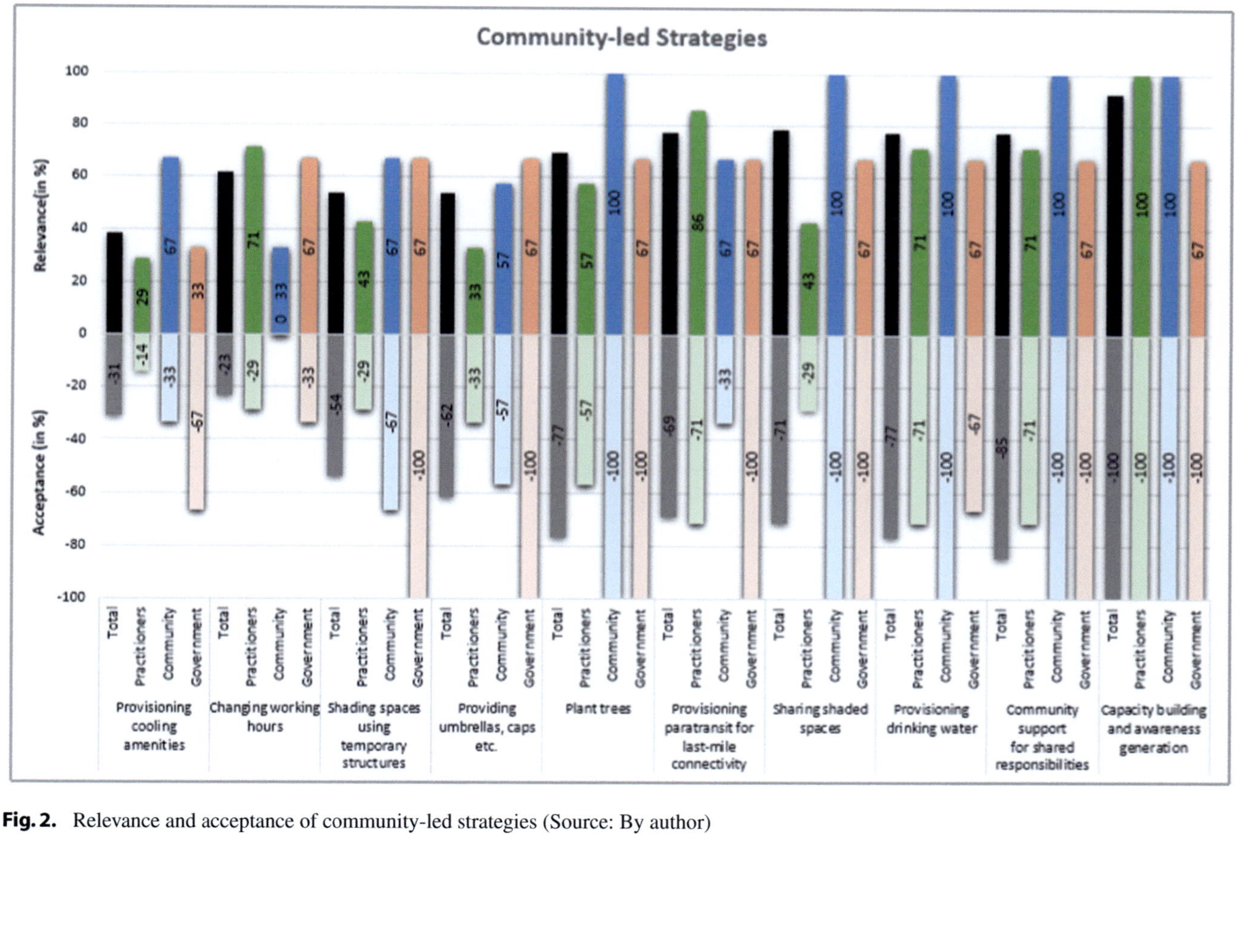

Fig. 2. Relevance and acceptance of community-led strategies (Source: By author)

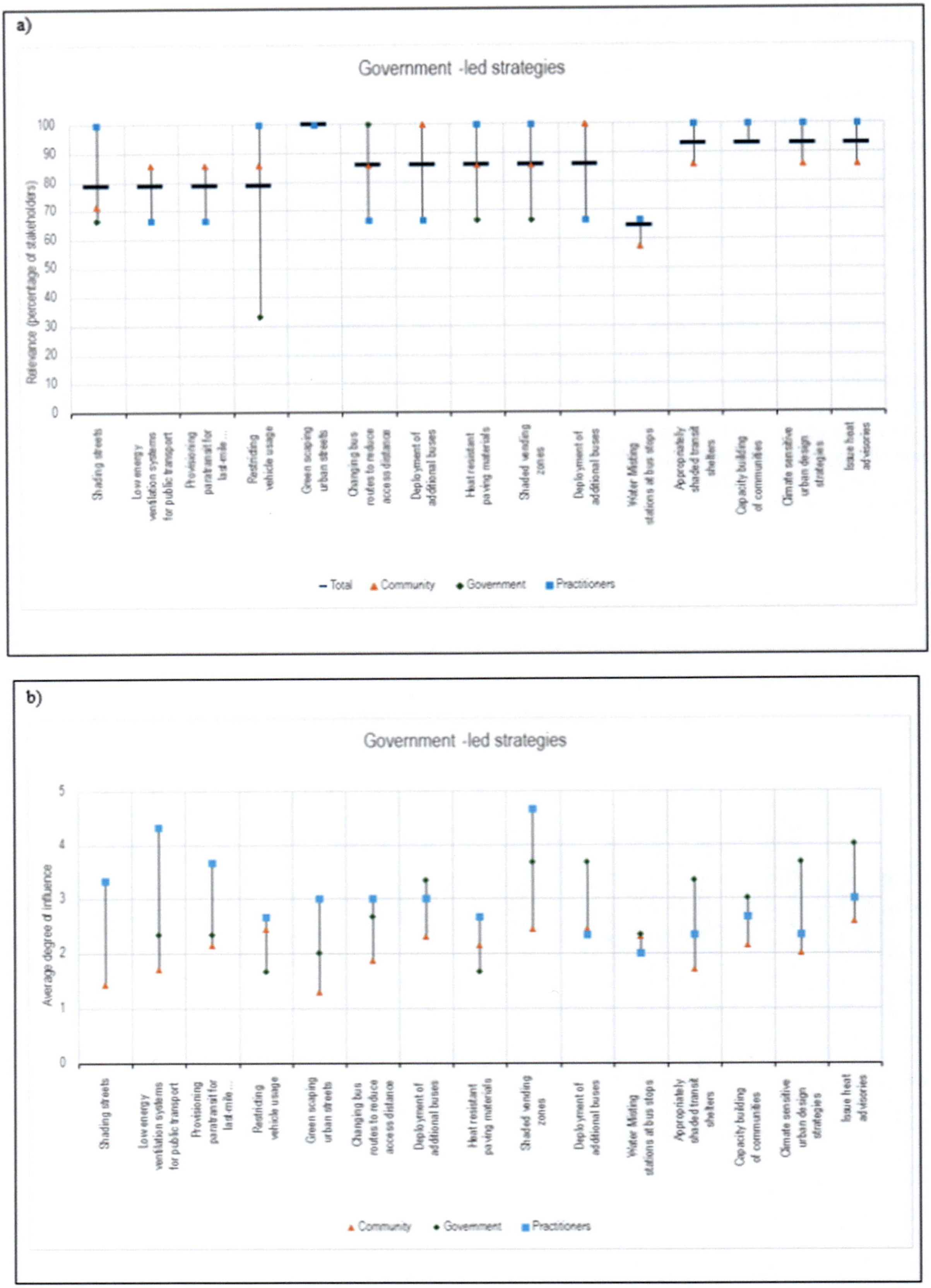

Fig. 3. a) Government-led strategies, a) level of relevance; b) degree of influence (Source: By author)

4.1.3 Government-Led Strategies

As compared and in contrast to the community-led strategies, most of the stakeholders in all groups identify the government-led strategies as relevant, as shown in Fig. 3. Two strategies – restricting vehicle usage and provisioning shaded streets are exposed to potential conflicts. Only 33%, that is one of the three community stakeholders identify it as relevant. This may be related to the lack of alternative choices, unable to avoid trips, especially the mandatory trips and being captive to the use of vehicles because of unsafe conditions. Sixty-seven percent of the government identify the strategy of provisioning shaded streets as relevant for addressing vulnerabilities. Additionally, Water misting stations at bus stops helps people adapt to heat by providing a cooling mist that reduces ambient temperature and improves comfort while waiting at these stations. However, less than 67% community finds the strategy as relevant. Implementing both strategies would require consensus building, which may be achieved by evidence generation and awareness generation. Restricting vehicle usage would require the provisioning of safe and shaded urban streets, which may also benefit public transport users. While most strategies are deemed relevant by most stakeholders, the degree of influence varies, particularly because these strategies may fall under the purview of roles and responsibilities of multiple agencies and not one. For example, any urban design related strategies which require infrastructure related improvisation, provision of shades, provision of greens may fall under different departments such as horticultural departments, road infrastructure department and development authority. Thus, no single agency can take full ownership of implementing a strategy, which leads to variations in the degree of influence across different strategies.

The degree of influence is read differently than the level of relevance. Higher variations lead to eliciting who among the three stakeholders can take a leading role, in terms of influencing decisions, decision-making or implementing strategy. The analysis shows that government can take a leading role in deployment of additional buses, provision of shaded transit shelters, climate sensitive urban design strategies, and issuing heat wave advisories. Both government and practitioners need to jointly build capacity of communities to address rising challenge of heat waves. Practitioners identify themselves as having more influential in shading streets, provisioning of paratransit services, or green scaping of streets.

4.2 Summary

Based on the workshop, the relevance and acceptance of strategies to address the exposure and vulnerability of commuters during hot weather conditions in Delhi were assessed by different stakeholders. The stakeholders represent the community (affected by decisions and can lead community-led initiatives), the government (implementing and decision-making agencies), and the individuals (affected by decisions and can change or adapt their lifestyle and behaviour). The results are discussed for individual-led, community-led, and government-led strategies. The results are useful to determine which set of strategies are identified as most relevant and acceptable and what is the level of consensus and conflict between the three stakeholder groups for each of the strategy. Further, regarding government-led strategies, the degree of influence as perceived by different stakeholders is determined.

As per the analysis, there is a consensus among the stakeholders on the relevance and acceptance of wearing weather-appropriate clothing, carrying an umbrella, and staying hydrated. However, strategies like avoiding travel and changing routes, though relevant, are exposed to conflicts regarding acceptance, necessitating infrastructure improvements and policy support. Amongst the ten community-led strategies, only capacity building and awareness generation are supported by all. There are potential conflicts for changing working hours as the strategy. Taking up such strategies would need pilot testing and awareness generation to resolve practitioner resistance. There was a strong consensus among stakeholders regarding the importance of public awareness campaigns at the individual and community level. These findings suggest the need for community engagement programs, public awareness campaigns, and educational initiatives to promote these strategies to enhance resilience and reduce health risks associated with extreme heat.

On the Government-led strategies, strategies such as developing green spaces in urban areas and implementing climate-sensitive urban designs have shown high acceptance. These strategies help reduce the urban heat island effect by providing cooling and shade, enhancing thermal comfort for commuters. Policy-level interventions to support these strategies include the integration of green space requirements into urban planning regulations, incentives for private developers to incorporate green designs, and investment in public parks and making streets walkable. Additionally, implementing zoning laws that mandate climate-sensitive designs for new developments and retrofitting existing infrastructure to include green roofs, walls, and other cooling features can further enhance the effectiveness of these strategies. Particularly restricting vehicle usage reveal potential conflicts due to concerns over alternative options and safety. Thus, a comprehensive approach integrating public education, infrastructure enhancement, and supportive policies is essential for improving the acceptability and implementation of these heat mitigation strategies.

Additionally, issuing heat advisories is an effective measure to guide individuals in taking appropriate actions to avoid exposure to extreme hot hours. It further highlights the need for government-led initiatives to provide early warnings and guidance to individuals, especially vulnerable groups. Changing school hours to minimise exposure to the hottest part of the day has also gathered considerable acceptance among stakeholders. These findings highlight the essential role of comprehensive and multi-faceted policies to effectively address heat mitigation. The analysis emphasises the need for collaborative efforts amongst the stakeholders to effectively implement the needed strategies to enable both mitigation and adaptation to hot weather conditions.

5 Conclusion

In this study, the acceptance levels of heat mitigation and adaptation strategies beneficial for commuters in Delhi are mapped across different stakeholder groups. In conclusion, the research reveals the levels of consensus and conflict among various stakeholder groups regarding the selected strategies. Furthermore, the study highlights the significance of collaborative efforts among government agencies, communities, and stakeholders in implementing targeted strategies to effectively address the challenges posed by hot weather during commuting. Individual-led strategies like wearing weather-appropriate clothing and staying hydrated are broadly accepted by all, while other strategies, such as changing travel routes, face conflicts and require infrastructural and policy support. Community-led strategies, particularly capacity building and awareness campaigns are widely supported, and strategies like changing working hours necessitate pilot testing and targeted awareness efforts. Government-led strategies, including developing green spaces and implementing climate-sensitive urban designs, garner high acceptance but encounter potential conflicts for restricted vehicle usage.

A comprehensive approach integrating public awareness, infrastructure enhancements, supportive policies, and early warning systems is essential to effectively mitigate heat-related risks during commuting. Collaborative efforts between government agencies, communities, and stakeholders are needed for the successful implementation of these targeted strategies. The level of consensus and conflicts for strategies among different stakeholder groups provide useful insights. Conflicts may stem from varying priorities, resource allocation, and perceived effectiveness. Addressing these conflicts through transparent dialogue and negotiation can further enhance the acceptability and implementation of the strategies. Understanding and managing stakeholder conflicts is essential in the policy formulation and implementation process. Yet, empirical validation and monitoring are essential to ascertain the impact of selected strategies on heat exposure reduction and mitigation. Incorporating scientific evaluation, data analysis, and feedback mechanisms into policy frameworks is imperative to translate identified strategies into tangible and impactful measures for mitigating hot weather conditions. This approach will ensure that strategies are validated and widely supported, enhancing their potential for success in reducing heat exposure and improving commuting conditions in Delhi.

Acknowledgement This study is based on the findings from the workshop titled "Gendered approaches for addressing adaptation capacity to hot weather conditions," conducted on August 5, 2023. The authors extend their gratitude to UKAID for funding this project through the UK Foreign, Commonwealth and Development Office under the High-Volume Transport Research Programme. They also acknowledge the contributions of all participants and stakeholders who provided valuable insights and expertise during the workshop, which were essential to the research.

References

Abzalov, M.: Exploratory data analysis. In: Abzalov, M. (ed.) Applied Mining Geology, pp. 207–219. Springer International Publishing (2016). https://doi.org/10.1007/978-3-319-39264-6_15

Adib, M., Zhang, X.: The risk-based management control system: a stakeholders' perspective to design management control systems. Int. J. Manag. Enterp. Dev. **18**, 20 (2019). https://doi.org/10.1504/IJMED.2019.10018989

Brugha, R., Varvasovszky, Z.: Stakeholder analysis: a review. Health Policy Plan. **15**(3), 239–246 (2000)

Carlini, J., et al.: Discordance and concordance on perception of quality care at end of life between older patients, caregivers and clinicians: a scoping review. Eur. Geriatr. Med. **13**(1), 87–99 (2022). https://doi.org/10.1007/s41999-021-00549-6

Centers for Disease Control and Prevention. The Use of Cooling Centers to Prevent Heat-Related Illness: Summary of Evidence and Strategies for Implementation (2022)

Chen, Y.-L., Cheng, L.-C.: An approach to group ranking decisions in a dynamic environment. Decis. Support. Syst. **48**(4), 622–634 (2010). https://doi.org/10.1016/j.dss.2009.12.003

Christodoulou, A., Cullinane, K.: Identifying the main opportunities and challenges from the implementation of a port energy management system: a SWOT/PESTLE analysis. Sustainability **11**(21) (2019). https://doi.org/10.3390/su11216046

Cinelli, M., Kadziński, M., Miebs, G., Gonzalez, M., Słowiński, R.: Recommending multiple criteria decision analysis methods with a new taxonomy-based decision support system. Eur. J. Oper. Res. **302**(2), 633–651. (2022) https://doi.org/10.1016/j.ejor.2022.01.011

Cochrane, L., Corbett, J.: Participatory Mapping, pp. 1–9 (2018). https://doi.org/10.1007/978-981-10-7035-8_6-1

Dwivedi, Ph. D., R.: Role of stakeholders in project success: theoretical background and approach. Int. J. Finance Insur. Risk Manag. **11** (2021). https://doi.org/10.35808/ijfirm/248

Dzyuban, Y., et al.: Evidence of alliesthesia during a neighborhood thermal walk in a hot and dry city. Sci. Total. Environ. **834**, 155294 (2022). https://doi.org/10.1016/j.scitotenv.2022.155294

Eslamirad, N., Sepúlveda, A., De Luca, F., Sakari Lylykangas, K.: Evaluating outdoor thermal comfort using a mixed-method to improve the environmental quality of a university campus. Energies **15**(4) (2022). https://doi.org/10.3390/en15041577

Fraser, A.M., Chester, M.V.: Transit system design and vulnerability of riders to heat. J. Transp. Health **4**, 216–225 (2017). https://doi.org/10.1016/j.jth.2016.07.005

Ginige, K., Amaratunga, D., Haigh, R.: Mapping stakeholders associated with societal challenges: a methodological framework. In: 7th International Conference on Building Resilience: Using Scientific

Knowledge to Inform Policy and Practice in Disaster Risk Reduction, vol. 212, pp. 1195–1202 (2018). https://doi.org/10.1016/j.proeng.2018.01.154

Guerrero-Villegas, J.: The Relevance of Stakeholders in an International Context, pp. 196–214 (2019). https://doi.org/10.4018/978-1-5225-5781-4.ch010

Häberlein, L., Hövel, P.: Importance and necessity of stakeholder engagement. In: González-Esteban, E., Feenstra, R.A., Camarinha-Matos, L.M., (eds.) Ethics and Responsible Research and Innovation in Practice: The ETHNA System Project, pp. 38–53. Springer Nature Switzerland (2023). https://doi.org/10.1007/978-3-031-33177-0_3

Haider, H., Sadiq, R., Tesfamariam, S.: Selecting performance indicators for small and medium sized water utilities: multi-criteria analysis using ELECTRE method. Urban Water J. **12**(4), 305–327 (2015). https://doi.org/10.1080/1573062X.2014.900089

Herath, H.M.P.I.K., Halwatura, R.U., Jayasinghe, G.Y.: Evaluation of green infrastructure effects on tropical Sri Lankan urban context as an urban heat island adaptation strategy. Wild Urban Ecosyst. Challenges Opportunities Urban Dev. **29**, 212–222 (2018). https://doi.org/10.1016/j.ufug.2017.11.013

Hernandez Villamizar, F.Y., Laguado Ramirez, R., Rodriguez, J.: Maturity analysis in project management in Colombian universities. J. Phys. Conf. Ser. **1126**, 012055 (2018). https://doi.org/10.1088/1742-6596/1126/1/012055

Hess, J.J., et al.: Building resilience to climate change: pilot evaluation of the impact of India's first heat action plan on all-cause mortality. J. Environ. Public Health **2018**, 7973519 (2018). https://doi.org/10.1155/2018/7973519

Illahi, U.: Transport planning and traffic safety: making cities, roads, and vehicles safer. Transp. Rev. **39**(6), 819–820 (2019). https://doi.org/10.1080/01441647.2018.1552632

IPCC. Climate change: A threat to human wellbeing and health of the planet. Taking action now can secure our future (2022). https://www.ipcc.ch/2022/02/28/pr-wgii-ar6/

IRADe. Review of Heat Action Plans (2021)

Juhnke, C., Mühlbacher, A.: Patient-centeredness in Integrated healthcare delivery systems—needs, expectations and priorities for organized healthcare systems. Int. J. Integr. Care **13**, e051 (2013). https://doi.org/10.5334/ijic.1114

Karthikeyan, V., Ghosh, N., Boro, P., Das, S., Chakraborty, S.: Stakeholder analysis for hypertension management in the context of the Indian public health system. Indian J. Public Health **67**, s27 (2024)

Kiarsi, M., et al.: Heat wave adaptation paradigm and adaptation strategies of community: a qualitative phenomenological study in Iran. J. Educ. Health Promot. **11**, 408 (2022). https://doi.org/10.4103/jehp.jehp_440_22

Klinsky, S., Dowlatabadi, H., Mcdaniels, T.: Comparing public rationales for justice trade-offs in mitigation and adaptation climate policy dilemmas. Glob. Environ. Chang. **22**(4), 862–876 (2012). https://doi.org/10.1016/j.gloenvcha.2012.05.008

Koo, M., Yang, S.-W.: Likert-type scale. Encyclopedia **5**(1), 18 (2025). https://doi.org/10.3390/encyclopedia5010018

Kwatra, S., Kumar, A., Sharma, S., Sharma, P.: Stakeholder participation in prioritizing sustainability issues at regional level using analytic hierarchy process (AHP) technique: a case study of Goa India. Environ. Sustain. Ind. **11**, 100116 (2021). https://doi.org/10.1016/j.indic.2021.100116

Kwiecien, R., Kopp-Schneider, A., Blettner, M.: Concordance analysis: part 16 of a series on evaluation of scientific publications. Deutsches Arzteblatt Int. **108**(30), 515–521 (2011). https://doi.org/10.3238/arztebl.2011.0515

Lawler, J., et al.: Mitigation and adaptation strategies to reduce climate vulnerabilities and maintain ecosystem services. In: Climate Vulnerability: Understanding and Addressing Threats to Essential Resources, vol. 4, pp. 315–335 (2013). https://doi.org/10.1016/B978-0-12-384703-4.00436-6

Lemke, A.A., Harris-Wai, J.N.: Stakeholder engagement in policy development: challenges and opportunities for human genomics. Gen. Med. Official J. Am. Coll. Med. Genet. **17**(12), 949–957 (2015). https://doi.org/10.1038/gim.2015.8

Li, J., Mao, Y., Ouyang, J., Zheng, S.: A review of urban microclimate research based on CiteSpace and VOSviewer analysis. Int. J. Environ. Res. Public Health **19**(8) (2022). https://doi.org/10.3390/ijerph19084741

Madanchian, M., Taherdoost, H.: A comprehensive guide to the TOPSIS method for multi-criteria decision making. Sustain. Soc. Dev. **1** (2023). https://doi.org/10.54517/ssd.v1i1.2220

Magotra, R.: Nature-based solutions for heat wave management in cities. Int. Conf. Sustain. Dev. (2020)

Mathers, N., Fox, N., Hunn, A.: Using Interviews in a Research Project, pp. 113–134 (2000)

Mehiriz, K.: The effects of attitudes, norms, and perceived control on the adaptation of elderly individuals and individuals with chronic health conditions to heatwaves. BMC Public Health **24**(1), 256 (2024). https://doi.org/10.1186/s12889-024-17712-w

Mehiriz, K., Gosselin, P., Tardif, I., Lemieux, M.-A.: The effect of an automated phone warning and health advisory system on adaptation to high heat episodes and health services use in vulnerable groups-evidence from a randomized controlled study. Int. J. Environ. Res. Public Health **15**(8) (2018). https://doi.org/10.3390/ijerph15081581

Miao, Q., Welch, E., Sriraj, P.S.: Extreme weather, public transport ridership and moderating effect of bus stop shelters. J. Transp. Geogr. **74**, 125–133 (2019). https://doi.org/10.1016/j.jtrangeo.2018.11.007

Migliore, A., Rossi-Lamastra, C.: Cluster analysis, pp. 95–107 (2023). https://doi.org/10.1201/9781003289845-7

MoEFCC. Environment Impact Assessment (EIA) (2021). https://www.nccr.gov.in/sites/default/files/EIA.PDF

Moser, A., Korstjens, I.: Practical guidance to qualitative research. Part 3: sampling, data collection and analysis. Eur. J. Gen. Pract. **24**(1), 9–18 (2018). https://doi.org/10.1080/13814788.2017.1375091

Nielsen, M., Bedsted, B., Bryndum, N.: Organising stakeholder workshops in research and innovation: between theory and practice. J. Public Deliberation **13** (2017). https://doi.org/10.16997/jdd.285

NRDC. Expanding heat resilience across India: Heat action plan highlights 2022 (2022). https://www.nrdc.org/sites/default/files/india-heat-resilience-20220406.pdf

Nuruzzaman, M.: Urban heat island: causes, effects and mitigation measures -a review. Int. J. Environ. Monit. Anal. **3**, 67–73 (2015). https://doi.org/10.11648/j.ijema.20150302.15

Oudin Åström, D., Forsberg, B., Ebi, K., Rocklöv, J.: Attributing mortality from extreme temperatures to climate change in Stockholm, Sweden. Nat. Clim. Chang. **3**, 1050–1054 (2013). https://doi.org/10.1038/nclimate2022

Park, C.Y., et al.: A multilayer mean radiant temperature model for pedestrians in a street canyon with trees. Build. Environ. **141** (2018). https://doi.org/10.1016/j.buildenv.2018.05.058

Perkins-Kirkpatrick, S., Alexander, L., Nairn, J.: Increasing frequency, intensity and duration of observed global heatwaves and warm spells. Geophys. Res. Lett. **39**, 20714 (2012). https://doi.org/10.1029/2012GL053361

Pless, N.M., Sengupta, A., Wheeler, M.A., Maak, T.: Responsible leadership and the reflective CEO: resolving stakeholder conflict by imagining what could be done. J. Bus. Ethics **180**(1), 313–337 (2022). https://doi.org/10.1007/s10551-021-04865-6

Rawat, N., Kumar, P.: Interventions for improving indoor and outdoor air quality in and around schools. Sci. Total. Environ. **858**, 159813 (2023). https://doi.org/10.1016/j.scitotenv.2022.159813

Ray Biswas, R., Rahman, A.: Adaptation to climate change: a study on regional climate change adaptation policy and practice framework. J. Environ. Manage. **336**, 117666 (2023). https://doi.org/10.1016/j.jenvman.2023.117666

Rosenthal, N., Chester, M., Fraser, A., Hondula, D.M., Eisenman, D.P.: Adaptive transit scheduling to reduce rider vulnerability during heatwaves. Sustain. Resilient Infrastruct. **7**(6), 744–755 (2022). Scopus. https://doi.org/10.1080/23789689.2022.2029324

Saqr, M., Alamro, A.: The role of social network analysis as a learning analytics tool in online problem based learning. BMC Med. Educ. **19**(1), 160 (2019). https://doi.org/10.1186/s12909-019-1599-6

Taleghani, M., Kleerekoper, L., Tenpierik, M., van den Dobbelsteen, A.: Outdoor thermal comfort within five different urban forms in the Netherlands. Spec. Issue Clim. Adapt. Cities **83**, 65–78 (2015). https://doi.org/10.1016/j.buildenv.2014.03.014

Taufik, I., Nurul Alam, C., Mustofa, Z., Rusdiana, A., Uriawan, W.: Implementation of multi-attribute utility theory (MAUT) method for selecting diplomats. IOP Conf. Ser. Mater. Sci. Eng. **1098**, 032055 (2021). https://doi.org/10.1088/1757-899X/1098/3/032055

Tavakol, M., Wetzel, A.: Factor analysis: a means for theory and instrument development in support of construct validity. Int. J. Med. Educ. **11**, 245–247 (2020). https://doi.org/10.5116/ijme.5f96.0f4a

Then, K.L., Rankin, J.A., Ali, E.: Focus group research: what is it and how can it be used? Can. J. Cardiovasc. Nurs. = Journal Canadien En Soins Infirmiers Cardio-Vasculaires **24**(1), 16–22 (2014)

UNISDR. UNISDR annual report 2015 (2015)

Uzun, B., Almasri, A., Uzun Ozsahin, D.: Preference ranking organization method for enrichment evaluation (PROMETHEE). In: Uzun Ozsahin, D., Gökçekuş, H., Uzun, B., LaMoreaux, J. (eds.) Application of Multi-Criteria Decision Analysis in Environmental and Civil Engineering, pp. 37–41. Springer International Publishing (2021). https://doi.org/10.1007/978-3-030-64765-0_6

Vatandoost, M., Ekhlassi, A., Yazdanfar, S.A.A.: Applying Delphi Method in decision-making: Selection of Architectural Optimization Problem case study: Extension and Renovation of Baharloo Hospital (2022)

Wang, F., et al.: Climate change: strategies for mitigation and adaptation. Innovation **1**, 100015 (2023). https://doi.org/10.59717/j.xinn-geo.2023.100015

Watts, N., et al.: Health and climate change: policy responses to protect public health. Lancet **386**(10006), 1861–1914 (2015). https://doi.org/10.1016/S0140-6736(15)60854-6

World Health Organisation. Heatwaves (2024a)

World Health Organisation. Heatwaves. World Health Organisation (2024b)

Yasumoto, S., et al.: Heat exposure assessment based on individual daily mobility patterns in Dhaka, Bangladesh. Comput. Environ. Urban Syst. **77** (2019). https://doi.org/10.1016/j.compenvurbsys.2019.101367

Yeleliere, E., Nyamekye, A.B., Antwi-Agyei, P., Boamah, E.F.: Strengthening climate adaptation in the northern region of Ghana: insights from a stakeholder analysis. Climate Policy **22**(9–10), 1169–1185 (2022). https://doi.org/10.1080/14693062.2022.2134085

Youssef, H., Kamel, S., Hassan, M.H., Nasrat, L.: Optimizing energy consumption patterns of smart home using a developed elite evolutionary strategy artificial ecosystem optimization algorithm. Energy, **278**, 127793 (2023)

Yu, T., Kim, D., Son, S.-Y.: Optimization of scheduling for home appliances in conjunction with renewable and energy storage resources. Int. J. Smart Home **7**, 261–272 (2013)

Mitigating Climate Change: A Literature Review Based on Individual and Corporate Carbon Reduction Behaviors

Lin Wang, Yujun Huang, and Simei Wu

Abstract

Global warming has become unstoppable, and extreme weather has become more frequent. How to deal with climate change is a current issue that must be addressed. Carbon emissions from human activities are an important influence on climate change. Carbon emissions resulting from human activities are a key influencing factor in climate change. In this review, we examine: a) literature related to individuals; and b) literature related to corporations. Individual influences factor focus on the following: personal norms, individual attitudes and perceptions, green commitment, green experience, social networks, green technology, cost, policy and education. The influences factor of corporations are centered on the following aspects: corporate financial performance, board diversity, corporate social responsibility, corporate technology, information technology, green finance, carbon prices, and policy environment. Through the analysis of the above, this paper proposes suggestions for improving carbon reduction behavior, mainly focusing on the following: 1) adopting the means of sharing data to improve the efficiency of public services; 2) adopting publicity and education, and knowledge sharing on carbon emissions; 3) adopting incentives to provide appropriate grants as well as subsidy policies; and 4) improving the carbon performance evaluation and supervision system.

L. Wang (✉) · Y. Huang · S. Wu
School of Management, Xi'an University of Architecture and Technology, Yanta Road 13, Xi'an 710055, China
e-mail: m17635997067@163.com

S. Wu
e-mail: simeiwu@xauat.edu.cn

Keywords

climate change · carbon emission · individual · corporate · improving carbon reduction

1 Introduction

Climate change has a significant impact on human production activities and daily life. It has become urgent to improve the harm caused by climate change, and personal carbon emissions and corporate carbon emissions account for a large part of carbon emission sources, so research focuses on personal carbon emissions and corporate carbon emissions. P. Syldon et al. (2024) point out that changes in the intensity and magnitude of floods due to climate change will have a devastating impact on agriculture [1]. N. Teasdale et al. (2023) found that climate change will affect human health, potentially leading to an increase in prevalence of cardiovascular respiratory diseases, neurological disorders, and lead to a detriment in the mental health [2]. B. Aylward et al. (2024) argue that climate change affects people psychologically in terms of helplessness, hopelessness or powerlessness [3]. At the same time, climate change will also affect corporations, not only in terms of their economic output, but also in terms of their work efficiency. C. Li et al. (2024) argue that climate change affects business output and increases business costs, thereby inhibiting sustainable business development [4]. V. Q. Do et al. (2021) argue that extreme weather events and climate change factors negatively affected business efficiency in the study area [5]. Carbon emissions are one of the major contributors to climate change. Carbon emissions are produced both in people's daily lives and in the production, transportation, and distribution processes of corporations. Carbon emissions from individual human and economic activities have reached

B. He et al. (eds.), *Towards the Framework of Livable and Resilient Cities* Advances in Science, Technology & Innovation, https://doi.org/10.1007/978-3-031-97849-4_30

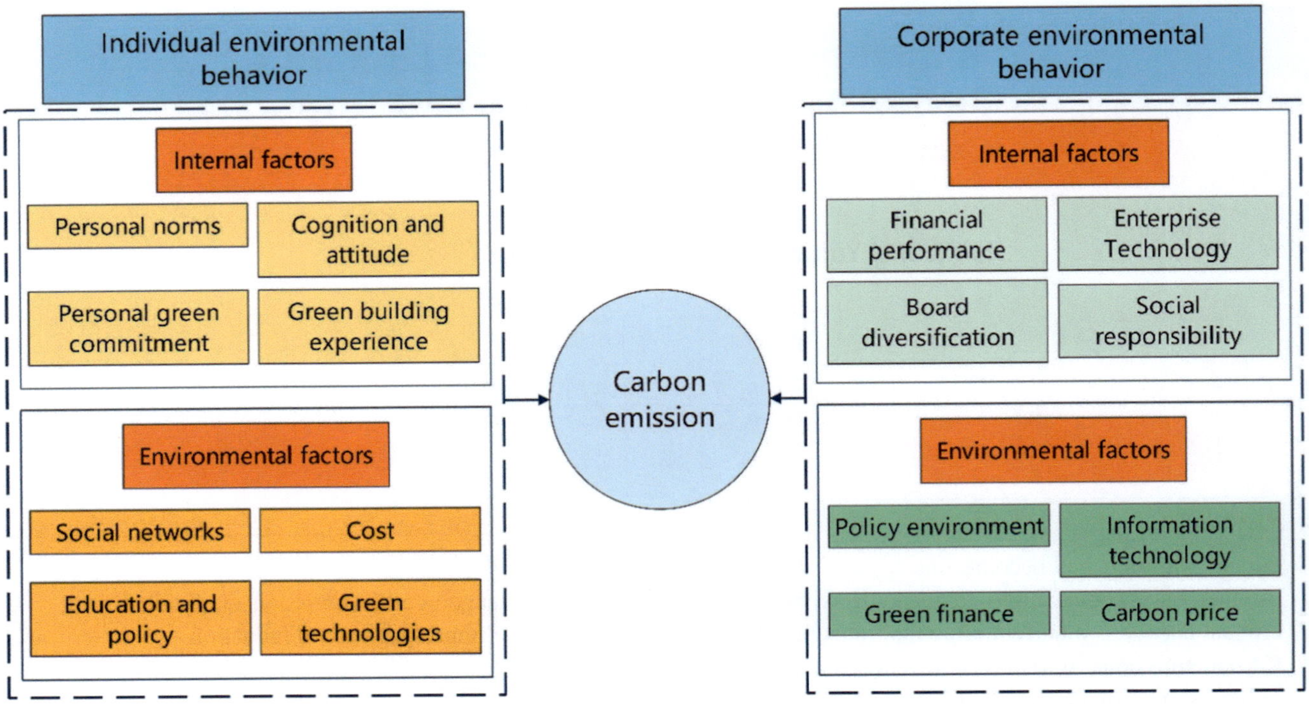

Fig. 1. Individual and corporate carbon reduction behaviors

36.2 Gt per year [6]. Taking the necessary measures to reduce carbon emissions has become a top priority for mitigating climate change. This paper focuses on the carbon reduction behavior from both individual and corporate perspectives. This paper sorts out the influencing factors of carbon reduction behavior of individuals and corporate, and puts forward suggestions for reducing carbon emissions based on existing research results. Figure 1 shows the influences of individuals and firms in carbon reduction behaviors.

2 Individual and Corporate Carbon Reduction Behaviors

2.1 Individual Carbon Reduction Behaviors

There are many types of individual carbon reduction behaviors, and the carbon emissions of energy use, travel and green building purchase are relatively large. In order to promote the sustainable development of the environment, this paper discusses the relevant research on individual carbon reduction behavior from three perspectives: energy use behavior, travel behavior and green building purchase behavior. In Abu Dhabi, U.A.E. E. Azar (2017) found that respondents' energy saving actions and need were significantly influenced by demographics, the level of control over building systems, and motivation drivers [7]. In Taiwan and Mozambique, S. P. Ho et al. (2024) indicate that economic capacity, environmental moral motivation, and psychosocial motivation have a significant impact on consumers' purchase intention of residential green buildings [8]. In addition, personal norms can significantly influence the carbon reduction behavior of individuals. N. Shanmugavel et al. (2023) argue that personal norms play an important role in developing positive behavioural intention towards e-vehicles [9]. In Kazakhstan and United States, K. Haldorai et al. (2024) point out that an individual's values indirectly influence their willingness to use electric vehicles, directly causing positive or negative effects on the environment [10]. In Pakistan, F. Baig et al. (2024) indicate that disseminating awareness of environmental threats and management applications will help people move to biogas-fueled public transportation systems [11]. In Taiwan, M.Y. Huang (2023) argue that consumers' perception and attitude toward green building housing products may affect their purchase intention [12]. In Bangladesh, I. Zahan et al. (2020) found that consumers' attitudes and perceived behaviors influence their willingness to purchase green homes [13]. In China, A. J. Shie et al. (2022) point out that personal green commitments positively affect their low-carbon travel motivation and intention [14]. In Hong Kong, C. K. Chau et al. (2010) indicate that residents with experience in green building were willing to pay (WTP) for improving the environmental performance of their green homes [15].

Some studies highlight the impact of social networks on individual carbon reduction behaviors.In Australia, A. Dowd et al. (2012) tested whether it could change the way individuals consume energy by encouraging people to go through a process in a social network, and found that the pre- and post-trial carbon calculator reported an average reduction of 20% in emissions [16]. In Rwanda, P. Kumar et al. (2024) point out that the use of social networks increased the probability of women using clean stoves [17]. In Europe, R. Curtale et al. (2023) found that a positive socio-environmental and social management system can increase the probability that consumers will use electric bike-sharing [18].

Green technology is an important factor influencing the willingness of individuals to adopt environmental behaviors. O. Awogbemi et al. (2024) indicate that the deployment of AI and digitalization will facilitate the use of clean energy in the energy sector. In China, A. Cao (2024) found that rural development e-commerce can greatly promote the use of clean energy in rural areas [19]. In Nepal, T. R. Pokare et al. (2021) discover that the growing trend of hydroelectricity production can shifting consumers' energy choices towards cleaner energy sources [20]. Not only are there technical reasons, but also cost influences people's willingness to take decarbonization behaviors. In Malaysia, M. Waqas et al. (2023) concluded that the ease of use and practicality of bike-sharing services can influences the continuance usage intention of bike-sharing service [21]. In Korea, J. Choi et al. (2023) found that cost-related factors are the most important factors influencing public acceptance of green buildings in Korea [22]. In Sardinia, E. Cristiano et al. (2023) argue that individuals prefer installing green roofs on public buildings due to the high installation costs [23].

Education and policy support are also thought to have an impact on individual environmental behavior. In Ghana, A. Darko et al. (2018) point out that increased media outreach, education and training programs related to green building technologies for key stakeholders can promote the adoption of green building technologies [24]. K. Ravindra et al. (2021) found that policies to provide free liquefied petroleum gas (LPG) to the poor households are key to driving the adoption clean cooking fuels [25]. In Israel, C. Cohen et al. (2019) indicate that government incentives can incentivize builders and consumers to adopt green buildings [26]. A summary of the factors influencing individual carbon reduction behaviour is shown in Table 1.

2.2 Corporate Carbon Reduction Behaviors

This review discusses the relevant research on corporate carbon reduction behavior from the perspectives of corporate architecture, environment and national level. In United States, M. S. Alam et al. (2022) indicate that carbon emissions are lower in corporations with higher corporate cash holdings [28]. In South Africa, F. Ganda et al. (2018) argue a negative correlation between corporate financial performance and carbon emissions [29]. T. Busch et al. (2018) concluded that good financial performance leads to good carbon performance [30]. A. Hyoung-Tae et al. (2020) found that there is a negative correlation between corporate performance and carbon emission intensity [31]. In United States, O. Ogunrinde et al. (2022) point out a U-shaped relationship exists between corporations' environmental performance outcome dimension and their financial performance that for corporations in the high-carbon intensive sector. At the same time, there is an inverted U relationship between corporations in low-carbon-intensive corporations [32].

The structure of investors and boards of directors can also significantly affect a corporate's carbon emissions. In Japan, T. Okimoto et al. (2024) concluded that investor stress has a significant decreasing effect on carbon emissions [33]. C. Gaganis et al. (2023) found that corporations with more capable managers have lower greenhouse gas (GHG) emissions [34]. E. Develay et al. (2024) discover that co-opted boards decrease GHG intensity in financially material industries, while there is no effect in non-financially material industries [35]. Not only does the structure of the board affect a corporate's carbon emissions, but the diversity of the board also affects a corporate's carbon emissions. In Nigeria, K. Aliani (2023) indicate that the diverse culture and skill diversity of the board of directors contributed significantly to the corporation's environmental performance and led to a reduction in CO_2 emissions [36]. In India, R. Konadu et al. (2022) found a significant negative correlation between board gender diversity and carbon emissions [37]. B. Oyewo (2023) indicate that board gender diversity is inversely correlated with carbon emission rates [38]. R. Barroso et al. (2024) argue that an increase in the number of female directors can reduce a corporate's carbon emissions [39]. In United States, R. Kizys et al. (2023) found that genetic diversity leads to improvements in environmental performance as well as carbon emissions [40]. Y. Shahab et al. (2022) point out that larger, gender-diverse boards can improve corporations' waste recycling practices and reduce corporate carbon emissions [41]. N. N. Goud et al. (2022) found that board size, board meetings, and board gender diversity were inversely correlated with carbon emission performance [42].

Corporate social responsibility (CSR) and technology are key factors influencing a corporate's carbon emissions. In South America, Córdova et al. (2018) indicate that the increase in size of the corporate and the presence of a CSR committee can lead to a decrease in the level of corporate carbon emissions [43]. In addition, the impact of corporate social responsibility on corporate carbon emissions was also verified in another article. In United States, K. Ahmad et al. (2023) indicate that there was a significant positive correlation between greenhouse gas emissions and corporate social responsibility [44].

Not only does a corporate's organizational structure have an impact on carbon emissions, but information technology

Table 1. Personal carbon reduction behavior

Types	Influencing factors	Country	Study results
Sociological factors of the population	Personal norms	Worldwide	Personal norms play an important role in cultivating positive behavioral intentions towards electric vehicles [9]
	Demographics, the level of control over building systems, and motivation drivers	U.A.E	Respondents' demographics, degree of control over building systems, and motivational drivers all had a positive impact on respondents' energy-saving actions and needs [7]
	Economic capacity, environmental moral motivation, and psychosocial motivation	Taiwan and Mozambique	Consumers' economic ability, environmental moral motivation and social psychological motivation have a significant impact on the purchase intention of green building housing [8]
	Individual's values	Kazakhstan and United States	An individual's values influence the willingness to use an electric vehicle [10]
	Awareness of environmental threats and management applications	Pakistan	Helping people switch to biogas-fueled public transport systems [11]
	Consumers' perception and attitude	Taiwan	Consumers' perceptions and attitudes towards green building residential products affect their willingness to purchase green homes [12]
	Consumers' attitudes and perceived behaviors	Bangladesh	Consumers' attitudes and perceived behavior control affect consumers' willingness to purchase green homes [13]
	Personal green commitments	China	Green commitments have a positive impact on the motivation and willingness of low-carbon travel, and a negative impact on the constraints of low-carbon travel [14]
	Experience in green building	Hong Kong	Residents with green building experience are willing to pay for the improvement of green homes [15]
Social networks	Social networks	Australia	The carbon calculator before and after the trial reported an average reduction of 20% in emissions [16]
	Social networks	Rwanda	Effective use of social networks can help women to use clean stoves [17]
Environmental factors	Socio-environmental and social management system	Europe	Environmental and social management systems can increase the outcomes of using electric bike-sharing trips [18]
Green technology	AI and digitalization	Worldwide	The deployment of AI and digitalization in the energy sector will facilitate the use of clean energy [27]
	E-commerce	China	The adoption of e-commerce can greatly facilitate the use of clean energy in rural areas [19]
	Hydroelectricity production	Nepal	The growing trend of hydropower is shifting consumers' energy choices towards cleaner energy sources [20]
	Bike-sharing services	Malaysia	The ease of use and practicality of bike-sharing services have an impact on consumers' willingness to continue to use them [21]
Cost	Cost	Korea	Cost-related factors are the most important factors influencing public acceptance of green buildings in Korea [22]
	Cost	Sardinia	Individuals prefer installing green roofs on public buildings due to the high installation costs [23]
Education and policy	Education and training	Ghana	Increased media outreach and education and training programs related to green building technologies for key stakeholders can promote the adoption of green building technologies [24]
	Clean Fuel Policy	Developing countries	Promote the adoption of clean cooking fuels [25]
	Reward system	Israel	To incentivize the adoption of green buildings, governments should reward their builders and consumers [26]

also has an impact on a corporate's carbon emissions. In Europe, M. A. Quttainah et al. (2024) examined how digital technologies affect corporate sustainability practices, and the results show that digital integration has a positive impact on corporate carbon reduction [45]. From a macro perspective, green finance can reduce carbon emissions by upgrading the industrial structure and technological progress [46]. While at the micro perspective, carbon emissions can be reduced through financing effects, innovation effects, and external regulatory effects [46].

The market environment has also had a non-negligible impact on corporate carbon emissions. In Europe, G. Adamolekun (2024) found that rising carbon prices reduce greenhouse gases for corporations through an analysis of 1591 corporations from 23 European countries [47]. X. Qian et al. (2024) discover that corporations reduce their carbon emissions that suffer losses due to extreme weather events [48]. National policies are also a significant factor affecting corporate carbon emissions. In United States, M. Safiullah et al. (2024) found that political risk negatively affects overall environmental performance and carbon emissions [49]. In China, A. Z. Han et al. (2024) point out that carbon trading policies will reduce the carbon emission intensity of corporations to varying degrees, There are significant differences in the carbon emission intensity of corporate under different policy types, marketization levels and digitalization levels [50]. In Europe, M. A. Qureshi et al. (2022) argue a significant negative correlation between the country-level eco-innovation index and the CO_2 emissions generated by European corporations [51]. A summary of the factors influencing corporate carbon reduction behaviour is shown in Table 2.

3 Discussion

The study finds that in the sociological factors of population, the economic ability of individuals affects individual carbon emissions, and when individuals have stronger economic ability, they are more inclined to choose green products. When individuals have more positive values, they have a higher sense of morality about the environment and are therefore more likely to adopt pro-environmental behaviors. Individuals' attitudes and perceptions towards the environment also affect their environmental behaviors, and individuals tend to adopt pro-environmental behaviors when they have a good attitude towards the environment. Among other factors, an individual's social network influences an individual's carbon emission behavior. When individuals discuss information about their environment in social networks, they reinforce pro-environmental behaviors. Green technology also affects the carbon emission behavior of individuals, and when green technology is used to improve the environment, it will strengthen the carbon reduction behavior of individuals. Cost is also one of the factors for individuals to choose cleaning products, and when the cost of green products is within the economic range of individuals, individuals will be willing to choose green products. The support of national policies will also promote individuals to adopt carbon emission reduction behaviors.

In terms of corporate carbon emissions, a company's financial performance affects its carbon emissions, and when a company has excellent financial performance, it reduces its carbon emissions. In addition, the size of a company's board of directors is also one of the factors that affect a company's emissions, and in general, when the board is gender-diverse and diverse, the company's carbon emissions are reduced. Green finance and digital technologies will affect a company's carbon emissions. Not only that, the political risk of the enterprise and the national policy will affect the carbon emissions of the enterprise, and when the political risk of the enterprise is large, it will have a negative impact on the carbon emissions of the enterprise. When the country's policy on carbon trading is better, the carbon emission intensity of enterprises will decrease.

4 Conclusion

In this article, we review research on individual carbon reduction behaviors and corporate carbon reduction behaviors.

From the perspective of individual behavior, carbon emissions can be mitigated in the following ways.

1. Shared data technologies should be promoted to improve the efficiency of public services. Government departments should strengthen information, logistics and energy infrastructure. The corporations has to modernize, upgrade and expand the equipment associated with these infrastructures. A data sharing mechanism between the government and operators should be established to improve the quality and efficiency of public services, attract people to adopt low-carbon lifestyles, and reduce carbon emissions.
2. Governments and corporations need to strengthen low-carbon publicity and education for professionals and the public to reduce carbon emissions. The government should direct academic and research institutions to cooperate with energy producing corporations. Governments should inculcate an awareness of sustainability in the renewable energy sector and promote the application of low-carbon technologies. Corporations need to instill the knowledge, corporate, and best practices they need to their workers. The government need to widely disseminates the need and awareness of low-carbon technologies among the population to promotes the adoption of low-carbon measures.

Table 2. Corporate carbon reduction behavior

Environmental behavior	Influencing factors	Country	Study results
Carbon performance	Financial performance	Worldwide	Good financial performance leads to good carbon performance [30]
Carbon emission intensity	Corporate performance	Worldwide	There is a negative correlation between corporate performance and carbon emission intensity [31]
Carbon emissions	Extreme weather exposure	Worldwide	Corporations reduce their carbon emissions that suffer losses due to extreme weather events [48]
Carbon emissions	Cash for corporate	United States	Corporations with higher corporate cash holdings have lower carbon emissions [28]
Carbon emissions	Financial performance	South Africa	There is a negative correlation between a corporate's financial performance and carbon emissions [29]
Corporate environmental performance	Financial performance	United States	A U-shaped relationship exists between corporations' environmental performance outcome dimension and their financial performance that for corporations in the high-carbon intensive sector. At the same time, there is an inverted U relationship between corporations in low-carbon-intensive corporations [32]
Carbon reduction	Carbon price	Europe	Rising carbon prices reduce greenhouse gases for corporations [47]
Carbon emissions	Corporate management ability	Worldwide	Corporations with more capable managers have lower greenhouse gas emissions [34]
Carbon emission intensity	Co-opted Board of Directors	Worldwide	Co-opting boards of directors reduced the intensity of greenhouse gas emissions in the financial materials sector, while the non-financial materials sector had no effect [35]
Carbon emissions	Board of directors	Worldwide	There is a statistically significant negative correlation between board gender diversity and carbon emissions [38]
Carbon emissions	Board of directors	Worldwide	An increase in the number of female directors can reduce a corporate's carbon footprint [39]
Carbon reduction	Board of directors	Worldwide	A larger, gender-diverse board of directors can improve a corporate's waste recycling practices and reduce carbon emissions [41]
Carbon emissions	Board of directors	Worldwide	Board size, board meetings, and board gender diversity were inversely correlated with carbon emission performance [42]
Carbon emissions	Investor pressure	Japan	Investor pressure has had a significant effect on carbon emissions [33]
Carbon emissions	Board diversity	Aliani	The diverse culture and skill diversity of the Board of Directors contributes positively and significantly to the outstanding environmental performance of the best citizen corporations and leads to lower CO_2 emissions [36]
Carbon performance	Board of directors	India	There is a significant negative correlation between board gender diversity and carbon emissions [37]
Environmental performance and carbon emissions	Board of directors	United States	The genetic diversity of the board leads to improvements in environmental performance as well as carbon emissions [40]
Carbon emissions	The size of the corporate, corporate social responsibility	South American	The size of the corporate and the existence of the CSR committee will lead to a decrease in the level of corporate carbon emissions [43]
Carbon emissions	Corporate Social Responsibility	United States	There is a significant positive correlation between greenhouse gas emissions and corporate social responsibility [44]
Carbon reduction	Digital technology	Europe	The digital integration of corporations has a positive impact on the carbon reduction of corporations [45]
Carbon emissions	Green Finance	China	Green finance can reduce carbon emissions by upgrading the industrial structure and technological progress [46]
Carbon emissions	Political risk for corporations	United States	Political risks have a negative impact on overall environmental performance and carbon emissions [49]
Carbon emission intensity	Carbon trading policy	China	Carbon trading policies will reduce the carbon emission intensity of corporations to varying degrees [50]
Carbon emissions	Ecological Innovation Index	Europe	There is a significant negative correlation between the eco-innovation index at the national level and the CO_2 emissions generated by European corporations [51]

3. The government should provides appropriate funding and subsidy policies. The government can provides incentives including research grants, scholarships, tax incentives, and soft loans to encourage young graduates to pursue research in the field of low-carbon technologies. The government need to increase disbursements to promoted subsidy programs. In this way, people are encouraged to adopt more low-carbon means of mobility, use cleaner energy and choose green buildings.

From the perspective of reducing corporate carbon emissions, the following measures can be taken:

1. Carbon performance assessment and monitoring policies should be improved to monitor corporations achieving carbon neutrality. The government should improve the statistics, verification and reporting system for corporate carbon emission data. The government should establish a well-organized and well-functioning trading platform and introduce third-party service providers. The government mobilizes the enthusiasm of corporate stakeholders by distributing green benefits and publicizing typical cases to help corporations conduct carbon performance assessments. Corporations should regularly assess their carbon footprint, raise awareness of information disclosure, and improve their carbon performance evaluation and monitoring systems to identify and mitigate impacts.
2. Knowledge sharing on digital technologies is being promoted to reduce corporate carbon emissions. Researchers should expand their research on the impact of digital technologies and sustainability, analyze the relationship between digital integration and corporate sustainability, and conduct cross-sector analyses to identify best practices across different industries. The government need to encourages industry associations to provide a platform for corporations to facilitate knowledge sharing and develop training programs to implement digital technologies to reduce corporate carbon emissions.
3. Incentives should be used to encourage corporations to reduce their carbon emissions. The government should improve the efficiency of environmental protection spending, stimulate the development of innovative technologies, and promote the efficient use of resources. Governments need to promote digital adoption through financial incentives and invest in the necessary digital infrastructure. Reduce corporate carbon emissions by supporting the integration of advanced technologies in traditional areas. At the same time, The government should reduce the carbon emissions of corporations by reducing the burden of corporations through welfare policies such as financial subsidies and improving the quality of green innovation of small non-state-owned corporations.

References

1. Syldon, P., et al.: Assessing the impact of climate change on flood inundation and agriculture in the Himalayan Mountainous Region of Bhutan. J. Hydrol. Reg. Stud. **52**, 101687 (2024)
2. Teasdale, N., Panegyres, P.K.: Climate change in Western Australia and its impact on human health. J. Clim. Change Health **12**, 100243 (2023)
3. Aylward, B., et al.: Examining the mental health impacts of climate change on young people in Canada: a national cross-sectional survey. Lancet Planet. Health **8**, S3 (2024)
4. Li, C., et al.: The impact of climate change on corporate ESG performance: the role of resource misallocation in enterprises. J. Clean. Prod. **445**, 141263 (2024)
5. Do, V.Q., et al.: The impact of extreme events and climate change on agricultural and fishery enterprises in central Vietnam. Sustainability **13**(13) (2021)
6. Yue, X.L., Gao, Q.X.: Contributions of natural systems and human activity to greenhouse gas emissions. Adv. Clim. Chang. Res. **9**(4), 243–252 (2018)
7. Azar, E., Al Ansari, H.: Framework to investigate energy conservation motivation and actions of building occupants: The case of a green campus in Abu Dhabi, UAE. Applied Energy, vol. 190, p. 563–573 (2017)
8. Ho, S.P., et al.: Raising the demand for residential green buildings: a general consumer behavior model, the evidence, and the strategies. Build. Environ. **252**, 111267 (2024)
9. Shanmugavel, N., Balakrishnan, J.: Influence of pro-environmental behaviour towards behavioural intention of electric vehicles. Technol. Forecast. Soc. Chang. **187**, 122206 (2023)
10. Haldorai, K., et al.: From personal convictions to collective action: predicting employees' pro-environmental behaviors in public and private spheres. Int. J. Hosp. Manag. **122**, 103864 (2024)
11. Baig, F., et al.: Willingness to Shift towards Biogas-fueled Bus Rapid Transit in Karachi Pakistan. KSCE J. Civil Eng. **28**(4), 1501–1512 (2024)
12. Huang, M.-Y.: Effects of consumer perception, attitude, and purchase intention on the willingness to pay for green building housing products. J. Housing Built Environ. **38**(3), 1559–1583 (2023)
13. Zahan, I., et al.: Green purchase behavior towards green housing: an investigation of Bangladeshi consumers. Environ. Sci. Pollut. Res. **27**(31), 38745–38757 (2020)
14. Shie, A.J., et al.: Diamond model of green commitment and low-carbon travel motivation, constraint, and intention. Int. J. Environ. Res. Public Health **19**(14) (2022)
15. Chau, C.K., Tse, M.S., Chung, K.Y.: A choice experiment to estimate the effect of green experience on preferences and willingness-to-pay for green building attributes. Build. Environ. **45**(11), 2553–2561 (2010)
16. Dowd, A., et al.: Energymark: empowering individual Australians to reduce their energy consumption. Energy Policy **51**, 264–276 (2012)
17. Kumar, P., Kalra, N., Shankar, A.: Role of personal network attributes in adoption of clean stoves among Congolese refugees in Rwanda. World Dev. **177**, 106539 (2024)
18. Curtale, R., Liao, F.X.: Travel preferences for electric sharing mobility services: results from stated preference experiments in four European countries. Transp. Res. Part C-Emerg. Technol. **155** (2023)
19. Cao, A., Su, M., Li, H.: Digitizing the green revolution: E-commerce as a catalyst for clean energy transition in rural China. Energy Econ. **137**, 107778 (2024)
20. Pokharel, T.R., Rijal, H.B.: Energy transition toward cleaner energy resources in Nepal. Sustainability **13**(8) (2021)
21. Waqas, M., Najmi, A.: Enhancing consumption of sustainable transportation: determinants and outcome of consumer engagement with bike-sharing service. Environ. Sci. Pollut. Res. **30**(18), 53411–53423 (2023)
22. Choi, J., et al.: Understanding heterogeneous consumer preferences for residential zero-energy buildings (ZEBs) in South Korea: a latent class approach. Sustain. Cities Soc. **97**, 104747 (2023)
23. Cristiano, E., Deidda, R., Viola, F.: Awareness and willingness to pay for green roofs in Mediterranean areas. J. Environ. Manage. **344**, 118419 (2023)

24. Darko, A., Chan, A.P.C.: Strategies to promote green building technologies adoption in developing countries: the case of Ghana. Build. Environ. **130**, 74–84 (2018)
25. Ravindra, K., Kaur-Sidhu, M., Mor, S.: Transition to clean household energy through an application of integrated model: ensuring sustainability for better health, climate and environment. Sci. Total. Environ. **775**, 145657 (2021)
26. Cohen, C., Pearlmutter, D., Schwartz, M.: Promoting green building in Israel: a game theory-based analysis. Build. Environ. **163**, 106227 (2019)
27. Awogbemi, O., Kallon, D.V.V., Kumar, K.S.: Contributions of artificial intelligence and digitization in achieving clean and affordable energy. Intell. Syst. Appl. **22**, 200389 (2024)
28. Alam, M.S., Safiullah, M., Islam, M.S.: Cash-rich firms and carbon emissions. Int. Rev. Financ. Anal. **81**, 102106 (2022)
29. Ganda, F., Milondzo, K.S.: The impact of carbon emissions on corporate financial performance: evidence from the South African firms. Sustainability **10**(7) (2018)
30. Busch, T., Lewandowski, S.: Corporate carbon and financial performance: a meta-analysis. J. Ind. Ecol. **22**(4), 745–759 (2018)
31. Hyoung-Tae, A., et al.: The correlation between climate change and corporate performance. J. Sci. Ind. Res. **79**(1), 38–43 (2020)
32. Ogunrinde, O., Shittu, E., Dhanda, K.K.: Distilling the interplay between corporate environmental management, financial, and emissions performance: evidence from US firms. IEEE Trans. Eng. Manage. **69**(6), 3407–3435 (2022)
33. Okimoto, T., Takaoka, S.: Credit default swaps and corporate carbon emissions in Japan. Energy Econ. **133**, 107504 (2024)
34. Gaganis, C., et al.: Managerial ability and corporate greenhouse gas emissions. J. Econ. Behav. Organ. **212**, 438–453 (2023)
35. Develay, E., Virk, N.S.: The effect of co-opted boards on corporate carbon performance: evidence from financially material industries. Econ. Lett. 111910 (2024)
36. Aliani, K.: Does board diversity improve carbon emissions score of best citizen companies? J. Cleaner Prod. **405** (2023)
37. Konadu, R., et al.: Board gender diversity, environmental innovation and corporate carbon emissions. Technol. Forecast. Soc. Chang. **174**, 121279 (2022)
38. Oyewo, B.: Corporate governance and carbon emissions performance: international evidence on curvilinear relationships. J. Environ. Manag. **334** (2023)
39. Barroso, R., et al.: Board gender diversity reform and corporate carbon emissions. J. Corp. Finan. **87**, 102616 (2024)
40. Kizys, R., Mamatzakis, E.C., Tzouvanas, P.: Does genetic diversity on corporate boards lead to improved environmental performance? J. Int. Finan. Markets. Inst. Money **84**, 101756 (2023)
41. Shahab, Y., et al.: Do corporate governance mechanisms curb the anti-environmental behavior of firms worldwide? An illustration through waste management. J. Environ. Manage. **310**, 114707 (2022)
42. Goud, N.N.: Corporate governance: does it matter management of carbon emission performance? An empirical analyses of Indian companies. J. Cleaner Prod. **379** (2022)
43. Córdova, C., Zorio-Grima, A., Merello, P.: Carbon emissions by South American companies: driving factors for reporting decisions and emissions reduction. Sustainability **10**(7) (2018)
44. Ahmad, K., et al.: Greenhouse gas emissions and corporate social responsibility in USA: a comprehensive study using dynamic panel model. Heliyon **9**(3) (2023)
45. Quttainah, M.A., Ayadi, I.: The impact of digital integration on corporate sustainability: emissions reduction, environmental innovation, and resource efficiency in the European. J. Innov. Knowl. **9**(3), 100525 (2024)
46. Zhao, X., et al.: The charm of green finance: can green finance reduce corporate carbon emissions? Energy Economics **134**, 107574 (2024)
47. Adamolekun, G.: Carbon price and firm greenhouse gas emissions. J. Environ. Manag. **349** (2024)
48. Qian, X., Qiu, S., Yang, X.: Extreme weather exposure and corporate carbon emissions management: evidence from forty countries. J. Multinatl. Financ. Manag. **75**, 100872 (2024)
49. Safiullah, M., Kabir, M.N.: Corporate political risk and environmental performance. Glob. Financ. J. **60**, 100939 (2024)
50. Han, A.Z., et al.: Study on the effect of carbon trading on the carbon emission intensity of enterprises-a mechanism test based on ESG performance. Front. Environ. Sci. **12** (2024)
51. Qureshi, M.A., Ahsan, T., Gull, A.A.: Does country-level eco-innovation help reduce corporate CO2 emissions? Evidence from Europe. J. Clean. Prod. **379**, 134732 (2022)

The Potential Civic Role of Universities in the Urban Regeneration of Local Communities: A Mutual Learning Process

Mariana Auad Proença
and Roberta Consentino Kronka Mülfarth

Abstract

The urgent need to identify leaders who can address effective alternatives to societal future changes related to socio-economics, environment, and cultural issues is becoming increasingly important. Universities have been playing a crucial role in this area by expanding their research strategies beyond the boundaries of the university campus and encompassing social and civic engagement within communities. By developing critical thinking about universities' civic role, this research was built based on the relevance of improving and changing behavior in educational practices, leading to environments for mutual learning that are effective in dealing with urban issues. Finally, it is intended to provide significant insights into the universities' civic role and how they deal with urban challenges approaching local communities of contemporary peripheries by presenting its concepts and the practical application through a case analysis of the Mapping San Siro (MSS) project.

Keywords

Universities · Communities · Civic engagement · Mutual learning · Urban regeneration

1 Introduction

This paper proposes an in-depth analysis of universities' civic role and their approach to urban challenges, a topic of significant interest to academic researchers, policymakers, and educators in urban studies. The focus is primarily on local communities of contemporary peripheries. The central idea of this research and its first insights were developed by considering the socio-economic development aspects of community life when approaching the framework of 'territorial fragilities[1] with a focus on urban regeneration. This research is crucial for understanding the evolving role of universities in addressing urban challenges.

In this regard, the main aim is to provide an investigative perspective to understand universities' civic and social role and how it impacts the direction of an innovative educational process and a cultural and environmental sustainability approach that effectively achieves urban regeneration.

Moving forward and regarding this context, peripheries were considered places geographically far from the central urban areas and related to weakness and cloudy concepts for different places. However, new peripheral conditions appear today, which go "beyond both the center-periphery debate and the meaning linked to physical distance from central areas: contemporary peripheries are simultaneously heterogeneous, different and specific. As in the past, they do not have a specific role, assuming it regarding another area" (Fontanella, 2019).

Moreover, urban challenges are related to the social, economic, environmental, and cultural issues that the world is facing; in other words, marginalized and fragile areas are facing urban challenges regarding housing, food, mobility, infrastructure, identity, quality of spaces (public spaces),

[1] The Politecnico di Milano is researching this theme within the Department of Architecture and Urban Studies (DAStU), which, during the 2018 to 2022 period, was awarded by the Italian Ministry of University and Research (MIUR) the recognition of "Excellent Department."

M. A. Proença (✉)
Urban Planning Design, Policy at Politecnico di Milano, Milan, Italy
e-mail: marianaproenca@usp.br

M. A. Proença · R. C. K. Mülfarth
Faculty of Architecture and Urbanism and Design, University of São Paulo FAUUSP, São Paulo, Brazil
e-mail: rkronka@usp.br

B. He et al. (eds.), *Towards the Framework of Livable and Resilient Cities* Advances in Science, Technology & Innovation, https://doi.org/10.1007/978-3-031-97849-4_31

access to education, safety, climate change, job opportunities, and so on. The concept and the idea to combine the urban regeneration approach in this work is a way to "solve the multifaceted problems of urban areas and to improve the economic, physical, social and environmental conditions of deprived areas" (Barosio et al. 2016, p. 367–380) and the local development of marginal and contemporary peripheries.

Understanding universities' civic and social roles is vital to approaching contemporary peripheries, urban challenges, and environmental sustainability themes and finding practical solutions. Moreover, and most importantly, innovating educational practices and highlighting universities' "Third Mission" (a movement of how universities connect with the public by bringing traditional education systems to a more realistic environment) is how we can change behavior and become critical thinkers and real-life problem solvers.

Therefore, this contribution is presented based on the reflection on the role of universities in addressing urban regeneration in contemporary peripheries, particularly in the context of community engagement and social impact. It also intends to provide a new way to see how different and innovative learning environments are vital to bring together students, professors, and local actors, who play a crucial role in a collaborative and participatory approach. Therefore, it also highlights how innovative learning environments are effective in understanding local fragile communities' needs and the positive impact of becoming more critical of urban challenges.

In this regard, and based on the research question methodology of this research, how are universities playing their civic role in dealing with local communities in the search for urban regeneration? The idea is first to present the design thinking of this investigation and the relevance of the universities' civic role in providing innovative educational processes that can effectively achieve results in seeking urban regeneration. The second is to present the main concepts regarding the university's civic approach and, finally, the leading case of the MSS project.

1.1 Methodology: Design Thinking

Throughout critical thinking of the possible role that the university can play in becoming a real active agent of urban change, the social and civic aspects were chosen because they deal with the aspirations and behavioral change of educational practices by training students to become better citizens capable of developing civic awareness and critical thinking about urban issues facing the world.

Universities are key players in urban regeneration. Their importance to communities and society, their increasing social responsibility and civic engagement within local communities, and their commitment to providing co-production of knowledge to their communities and society are all vital. It is equally important to understand how this role is being built and how the impact of their actions is being perceived. Therefore, the research was constructed based on five inputs to justify its importance.

Define: What is the problem? The problem is the urgent need for more recognition and importance of innovative educational processes among institutions. This recognition is crucial for addressing the evolving urban issues.

Understand: Why is it important? Universities are powerful organizations with the ability to use their teaching and research expertise to create actions that extend beyond the borders of the university campus, reaching and impacting communities.

Investigate: How do we identify the problem? We employ a comprehensive approach, using interdisciplinary academic areas combined with case analyses to contextualize the fragile local communities and the role of universities better.

Strategy: How do we proceed? By conceiving a specific methodology framework for this research, composed of the research question, theoretical concepts, literature review, and a case analysis.

Outcome: Does it work? The expected outcome of this research is to highlight the importance of developing new educational processes (mutual learning) that provide knowledge co-creation and co-design to achieve better urban regeneration alternatives facing the needs of fragile local communities.

In short, the main aim is to offer an original contribution by addressing the subject through critical thinking involving conceptual aspects and applications. Figure 1 is a diagram of the first insights that guided the development of this research's five inputs.

2 The Civic Role of Universities

Commencing this discussion, it's pertinent to highlight the diverse and multifaceted nature of the university's civic role. This role, which can be seen as a potential connector with society, is just one of the many hats the university wears, for example, as an institution, an economic and sustainable indicator, and a community in itself. Therefore, to unpack this role, it's crucial to mention, according to Kempton et al. (2021):

"But what are universities for? This simple question, which is as old as universities themselves, remains extremely topical and pertinent today. In fact, the rapid transformation experienced by contemporary societies implies the search for a better understanding of the role and purpose of universities. Due to the profound changes in the world economy, universities have been called upon to go beyond their traditional missions and therefore play a strategic role in the development process of their regions".

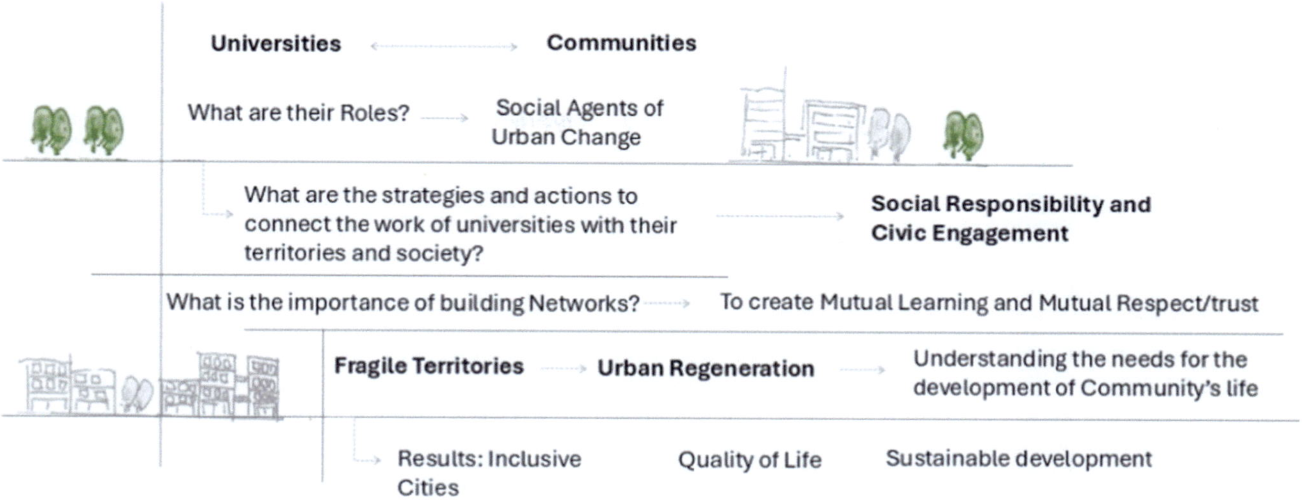

Fig. 1. Diagram of the first insights. Source: developed by the author, 2023.

In recent years, universities are playing an important role, "some would say a unique role, in their cities and regions in meeting a diverse set of needs, and, in doing so, contributing to the economic and social development of those cities and regions" (Charles et al. 2021. More than that, universities must confront their role to offer practical alternatives when dealing with their territories' urban issues. Hence, universities can become potential civic agents to achieve changes in society.

Approaching the social aspect in the context of engaged universities is crucial because it reflects on "universities as anchor institutions and new forms of the civic university. There is a proliferation of conceptual frameworks applied to the task of analyzing the engaged university at different levels of generality, some about the university sector as a whole, some to specific forms of institution, yet one common aspect is that all universities are to some degree unique" (Charles et al. 2021).

The idea of universities becoming a potential anchor for the benefit of their communities and the enhancement of policy-makers start to gain shape with "new universities since the 19th century has often been associated with the desire to support regional economies. This was especially the case in the US since the Land Grant colleges were established by the Morrill Act[2] of 1862" (Charles et al. 2021). Therefore, civic universities in the UK appear in provincial cities in England (19th Century), supporting local interests based on engaged education and research.

As a result, in the past decade, the third stream[3] or third mission, which refers to work with employers and community groups as part of the teaching and research, is considered as a movement of universities on the awareness of social responsibility that influences the level of study and research linking issues of local, national, and global concern. Therefore, universities are promoting co-production of knowledge and innovative educational practices that require the engagement of students and communities to build a better society.

2.1 Why Does the Social and Civic Engagement Approach Matter?

Although the social and civic role is necessary to change educational practices, defining them takes work. It faces challenges in being recognized and receiving financial support to enable projects and research development in various university areas. Therefore, the third mission, a burgeoning movement that champions the social and civic aspects of universities, underscores the pivotal role of these institutions in addressing social issues. It enhances the pedagogical and educational process, fostering critical thinking about urban and social problems and integrating teaching, research, and engagement. This approach highlights the potential of universities to be leaders in societal change.

---

[2] Sponsored by Senator Justin Morrill of Vermont, the "Act Donating public lands to the several States and [Territories] which may provide colleges for the benefit of agriculture and the Mechanic arts" marked the first federal aid to higher education. Source: https://www.archives.gov/milestone-documents/morrill-act (2023).

[3] The Third Stream is an international concept that "traces an emerging role of universities in innovation processes. Such processes require a more collaborative approach with other sectors and the gradual extension of this from working with the private sector to the consideration of collective actors working on civil society issues" (Millican and Bourner 2014).

From the conceptual perspective, the social and civic aspects, considering the participation of different actors, can help improve and provide new skills, co-production of knowledge, and innovative pedagogical and learning methods. Furthermore, mutual understanding, which results from merging scientific and shared knowledge, is essential to elaborate on when connecting the university with the public.

However, significant changes are imperative in academic environments for universities to effectively embrace the third mission. These changes, such as fostering an interdisciplinary approach and promoting innovative educational practices, are crucial for developing actions and projects that are responsive to society's evolving needs.

Finally, these concepts can add an interesting point of view: indeed, universities can be the leaders and the leading communicators in building networks for a better world; however, doubts remain about whether universities are prepared to assume this social role and whether students will be motivated to participate in this change in educational processes.

3 The Mapping San Siro MSS Project

The San Siro neighborhood, nestled in the western region near Milan's central area, was built between the 1930s and 1950s. It comprises nearly 6,110 dwellings with 11,000 residents and is considered Milan's most significant public housing neighborhood. Furthermore, it is managed by the Regional Agency for Public Housing of Lombardy (ALER). Despite its urban challenges and peripheral location, it is a unique urban space that thrives as a vibrant community where local initiatives and actions are driven by civil society and are constantly in motion to address their specific local needs, desires, and aspirations.

The MSS project, a significant initiative - under the coordination of Francesca Cognetti from the Department of Architecture and Urban Studies (DAStU) from the Politecnico di Milano and with the participation of Liliana Padovani from IUVA University of Venice and supported by the DAStU and the Polisocial Program - started as a workshop in 2013. This workshop was instrumental in identifying the local conditions and conducting a physical analysis of the San Siro neighborhood. The primary objective was to enhance the actions and interventions to foster social and urban transformation for the local community. In 2014, the project gained momentum by providing and utilizing a physical space in the neighborhood, which was made possible by the Regional Agency for Public Housing of Lombardy (ALER).

In April 2019, the Polisocial Programme - launched in 2012 in collaboration with the Politecnico di Milano and Fondazione Politecnico di Milano - facilitated the transfer of this laboratory to a new space, Off Campus (The Cantiere per le periferie) Via Gigante. This physical space was the first of four in the contemporary peripheries of the city of Milan, including the Nolo district (September 2020), the San Vittore detention center (October 2022), and the former farmhouse Cascina Nosedo (December 2022).

This physical space, now a 'living lab,' is a place to study and learn about complex issues related to fragile communities and housing policies. Still, it allows students to critically analyze and reinterpret the knowledge shared by community partners and residents. In this sense, it demonstrates to the neighborhood residents the possibility of change, the opportunity to build relationships and expertise, and, most importantly, work as a "window" to their needs and aims for the future. Being part of the Off-Campus space for almost three months was essential to understand that the main result of this project was not just a mere outcome but a significant achievement. It was to create a connection among the residents, to build networks with local communities, and to explain why the university was there. From a personal point of view, understanding effective alternatives for urban regeneration and environmental sustainability is how to approach and build a story with the daily life of a community and its residents. The importance and value of this project cannot be overstated.

More than that, when they decided to develop urban interventions in public spaces or when they decided to present social and cultural events using the Off Campus space, it was when behaviors started to change from both sides and to understand deeply that quality of life matters to everyone. The living lab approach, a valuable tool for understanding marginal neighborhood context, also empowers researchers and operators. According to Cognetti and Maranghi (2022)[4], the living lab approach induces reflections "on the one hand, on the skills that researchers and operators should develop involving the development of marginal contexts, on the other hand, on the conditions for producing a change durable and scalable in terms of local development and regeneration."

Living labs can be described as real environments in which they can experiment, develop, co-create, validate, and test services and systems with different stakeholders (Cogenetti 2023, p. 27). The So-called Urban Living Laboratories (ULLs) are capable of developing and strengthening people's capacity to reflect, imagine, aspire, and act for the transformation of their living context, becoming actual agents of change (Cognetti, Marangui, 2023: 74). Living labs are also urban because they refer to the political dimension of cities. From this perspective, they are environments that allow those involved to experiment and train "everyone's" civil capacity, thus challenging the function of local democracy. This implies a focus on residents, regional actors, and institutions. These

[4] Material provided by Prof. Francesca Cognetti: Cognetti, F. and Marangui, E. (2022). In QU3 – iQuaderni di U3. Attivare la rigenerazione locale: verso i Social Urban Living Lab. Activating local regeneration: towards the approach of Social Urban Living Labs.

MSS LIVING LAB ACTIVITIES

Themes	Approach	Tools	Goals	Learning Activities	Impact	Outcomes
• Housing and Living Conditions • Vacant and Empty Spaces • Public and Common Spaces	• Situating • Inquiry • Acting • (Network)	• Setting up a base; Local dialogues; networking; Local education • Participatory mapping; Map stories; San Siro 1:1; communication • Micro interventions; pilot projects; scenarios	• Understand complex dynamics and social practices in the neighborhood • Build up different relationships with institutions and social actors • Reshape the image of San Siro to improve public opinion • Provide tools for more effective initiatives and actual projects	• Micro interventions in public spaces; co-design activities • Networking with local partners; situated exhibitions; Caffè San Siro (local open lessons); learning activities • Interviews; observation; neighborhood walks; workshop with local partners; storytelling www.sansirostories.com • Working tables	• On the neighborhood: acquire more awareness and new knowledge • Tighten new alliances and networks • Micro transformations of space	• For students develop new sensitivities and awareness • Be able to understand urban and social dynamics in a critical way • Teamwork, communication abilities, cognitive and emotional empathy, problem-solving • Develop new capabilities and soft skills

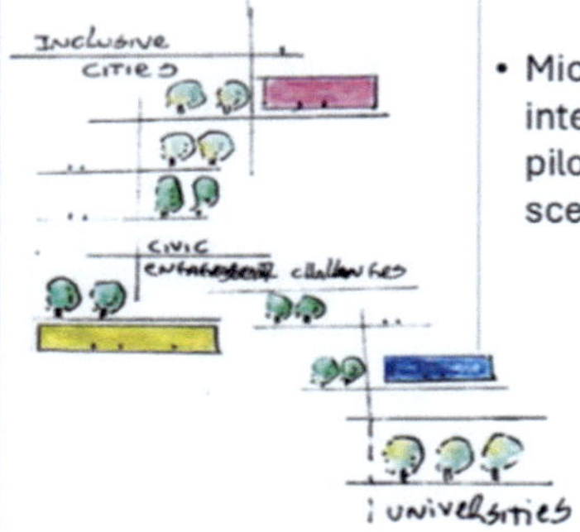

Fig. 2. Diagram of the MSS living lab activities. Source: developed by the author, 2023.

actors must be put in a position to learn from the local situation, introducing mechanisms that contribute to redefining citizenship habits and places and institutional routines and norms (Cognetti 2023, p. 28).

The Off Campus activities include research, education and culture, co-design, neighborhood archive, Sansheros's local network coordination, legal help desk, and legality education. It is important to note that the active network of Off Campus San Siro includes, among others, Bocconi University, Region Lombardy Department Social policies, housing and Disability and ALER Milano, Foundations and Companies, local network Sansheros, and the Municipality of Milan. Figure 2 shows activities developed regarding the living labs approach, such as themes, strategies, tools, goals, learning activities, impact, and outcomes.

The analysis of MSS was very relevant, independent of the social aspect; the Off Campus space can also be considered a living lab while being closer to the residents and understanding their needs and the dynamics of everyday life in the neighborhood. Therefore, as a living lab, it reinforces the importance of innovative educational processes for students to develop new skills and competencies and to model their critical thinking about a fragile local community in a marginal context. From the authors' point of view, the importance of this social approach also emphasizes that the university can indeed play an essential role as an agent of change.

Therefore, it should also be noted that the development of mutual learning between the university and the community is very relevant to building networks (a refreshing approach of MSS) and trust among local organizations, institutions, and residents. Moreover, in the fieldwork and interviews, the participation and engagement of researchers, professors, and students were paramount for the project. Furthermore, the Off-Campus space is a 'window' to the neighborhood. Figure 3 shows the Off Campus space layout.

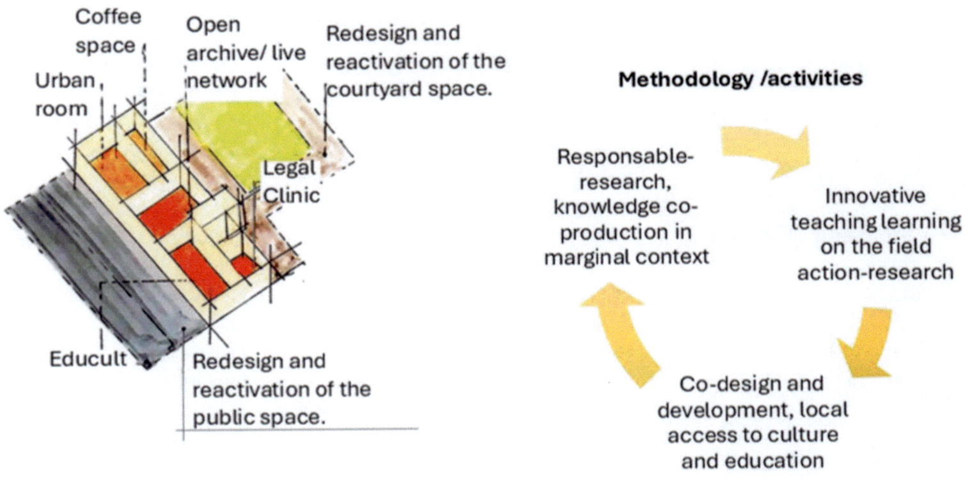

Fig. 3. Off Campus space layout and methodology. Source: image developed by the author from material collected at Off-Campus flyer, 2023.

4 Conclusions

This paper highlights the research's central idea: universities' civic role and how they deal with urban challenges by engaging with local communities of contemporary peripheries. In this regard, the active participation of local communities is essential and integral to the process. Universities have recently increased their engagement with local communities to prioritize learning environments for co-producing knowledge and innovative educational practices, aiming to build more resilient, effective, and sustainable societies.

Universities, as they face the myriad challenges of a transforming society, possess a transformative power that can serve as a beacon of hope. This potential strategy to deal with challenges involves serving the public through dynamic and mutual action. This transformative power 'presupposes the importance of greater involvement by the public in the design and implementation of collaborative activities and also demands that the faculty draw on their scholarship expertise for the benefit of the public as well as themselves' (Reich 2014, p. x).

It is essential to highlight that participation has become relevant once local communities play a vital role as active agents that can transform their environment and social and cultural needs, providing new ways to achieve urban regeneration (Proença and Balducci 2022). Therefore, critical reflection on the university's civic role can also mitigate the challenges of creating proximity and mutual learning within communities. In this sense, as a civic action, the living labs, which are collaborative spaces where students, faculty, and community members work together on real-world problems, can be considered a crucial step in creating a learning environment and a safe space to build networks and develop relationships with local organizations, institutions, and residents.

Other aspects can be identified when connecting research and community engagement, such as reflexive, developing individual and collective skills and competencies, project design, sharing and listening, and developing processes to support knowledge creation. It is essential to understand that these engagement activities involve, in addition to participatory co-design in urban interventions, cultural and artistic events, and the induction of networks as a tool to connect the community with local organizations, universities, and institutions.

It is crucial to reflect on the unique advantages and necessity of universities as civic agents. With their resources, expertise, and influence, universities are uniquely positioned to innovate their educational behavior. By bringing students and professors together, they can make the impact of their research more significant, underscoring the importance of their role.

Finally, the civic and social role of universities is not just essential, but crucial. More than training qualified personnel and civic citizens capable of developing critical thinking, universities can become responsible actors in increasing social cohesion. By fostering community among students, faculty, and staff and engaging with residents and organizations, they play a vital role in addressing the complex challenges that urban areas face today.

Therefore, it contributes to SDG Goal 11 - Make cities and human settlements inclusive, safe, resilient, and sustainable - by exploring the role of universities in urban regeneration, especially in fragile territories and vulnerable local communities. Integrating research, teaching, and social responsibility is not just an academic exercise. It is essential for tackling the contemporary challenges facing cities. It is directly aligned with the principles of SDG 11, validating the efforts of universities and offering practical solutions for sustainable urban development.

Acknowledgements The authors would like to thank you for contributing this paper to Prof. Alessandro Balducci and Politecnico di Milano, especially to the Department of Architecture and Urban Studies (DAStU) and FAUUSP for developing collaborative research. Thank you to Prof. Dr. Francesca Cognetti for supervising and providing consistent material and data for the Mapping San Siro (MSS) project.

References

Barosio, M., Eynard, E., Marietta, C., Marra, G., Melis, G., Tabasso, M.: From urban renewal to urban regeneration: Classification criteria for urban interventions. Turin 1995–2015: evolution of planning tools and approaches. J. Urban Regeneration Renewal **9**(4), 367–380 (2016)

Charles, D., Ahoba-Sam, R., Manrique, S.: Entrepreneurial Universities in Regional Innovation. A book from the Marie Skłodowska-Curie RUNIN Project-The Role of Universities in Innovation and Regional Development. Published by UK Book Publishing, pp. 1–312 (2021)

Cognetti, F.: Urban Living Lab for Local Regeneration. Beyond Participation in Large-scale Social Housing Estates. Chapter 2 Beyond a Buzzword: Situated Participation Through Socially Oriented Urban Living Labs. The Urban book series, Springer Nature, Switzerland, pp. 1–191 (2023)

Cognetti, F., Maranghi, E.: Urban Living Lab for Local Regeneration. Beyond Participation in Large-scale Social Housing Estates. Chapter 5 Adapting the Living Lab Methodology: The Prefix 'Co' as an Empowerment Tool for Urban Regeneration in LargeScale Social-Housing Estates. The Urban book series, Springer Nature, Switzerland, pp. 1–191 (2023)

Cognetti, F., Marangui, E.: In QU3 – iQuaderni di U3. Attivare la rigenerazione locale: verso i Social Urban Living Lab. Activating local regeneration: towards the approach of Social Urban Living Labs (2022)

Fontanella, E.: Peripheries. Temi e Progetti, Territorio 91, Politecnico di Milano, pp. 39–41 (2019)

Kempton, L., Rego, C.M., Alves, R.L., Vallance, P., Serra, A.M., Tewdwr-Jones, M.: Putting universities in their place. An evidence-based approach to understanding the contribution of higher education to local and regional development. Reg. Stud. Policy Impact Books **3**, i–86 (2021)

Millican, J., Bourner, J.: Learning to Make a Difference Student–community engagement and the higher education curriculum, The National Voice for Lifelong Learning, National Institute of Adult Continuing Education, pp.1–232 (2014)

Proença, M.A., Balducci, A.: Universities as potential social anchors in the development of mutual learning in local communities. The importance of proximity between the actors involved in collaborative research.' INFOLIO_39 Coesistere in Prossimità (2022). https://re.public.polimi.it/handle/11311/1233610

Reich, J.N.: Civic Engagement, Civic Development and Higher Education. Bringing Theory to Practice Monographs, The Civic Series, Washington, DC, viii-83 (2014)

The Impact of the Built Environment from the Local, Opportunities and Challenges: Affordable Regional Instrument for Healthy Housing in Queretaro, Mexico

José Granados Navarro, Edgar A. Rivas Araiza, and Edmundo A. Figueroa Viruega

Abstract

The adequate and efficient use of resources, as well as the knowledge of instruments that promote a healthy built environment in Latin America, has been limited; the conception that social technology currently allows the development of inclusive socio-productive processes that adapt quickly to small scales. Understanding these problems and limitations in the region, detected in the research process, the idea arose to create a practical tool for the general population to have healthy homes. This research develops an affordable regional instrument for the Metropolitan Area of Queretaro in the state of Queretaro, Mexico, which takes up some aspects of international and national experiences available, in addition to considering direct categories essential to have a healthy habitat, thus allowing the evaluation of the negative impact of buildings on the health of the inhabitant, and based on the results, prevention, adaptation, and mitigation strategies are implemented at a local level, through a series of recommendations based on the analysis of the parameters established accordingly to the reference values complying with the regulations and recommendations; Implementing everything from an easy-to-use and affordable mobile application prioritizing the social aspect.

J. G. Navarro (✉) · E. A. R. Araiza · E. A. F. Viruega
Faculty of Engineering, Autonomous University of Queretaro, Santiago de Querétaro, Mexico
e-mail: jose.granadosn@uaq.mx

E. A. R. Araiza
e-mail: erivas@uaq.mx

Keywords

Healthy building · social technology · environmental impact · healthy architecture · public health

1 Introduction

The impact of the built environment on health directly refers to its beginnings in the transformation of nature through the implementation of scientific and technological advances of each era. As the development of these processes advanced, various problems of adaptation to the environment were solved, leading to transformations, but also unconsciously, in many cases, threats and risks to the health of the inhabitants have arisen. (Mumford 1977, pp. 187–191). The greatest catastrophes experienced by humanity are linked to social processes of industrialization and modernization. However, the risk appears as a "historical, social construct" in the transition from the Late Middle Ages to the Early Modern Age, which is based on the determination that society considers each moment as usual and safe. Consequently, the risk is the "measure," the limited determination of chance according to the social perception of risk, emerging as a device for rationalization, quantification, metrization of chance, and reduction of indeterminacy (Beriain 1996, pp. 8–9), taking up Ulrich Beck's concepts about risk and its features.

In advanced modernity, the social production of wealth is systematically accompanied by the social production of risk. Therefore, the problems and conflicts of distribution of society of lack are replaced by those that arise from the production, definition, and distribution of risks generated in a scientific-technical way, with a clear tendency towards globalization

B. He et al. (eds.), *Towards the Framework of Livable and Resilient Cities* Advances in Science, Technology & Innovation, https://doi.org/10.1007/978-3-031-97849-4_32

(Beck 1998, p. 25). In this phase of modern society's development, new risks arise in economic conditions, living conditions, and the environment, which are outside the control of the institutions and protection of society (Ballesteros, 2014, p. 204). Consequently, modernization is conceived as an autonomous process of innovation, which must consider its deterioration. Giddens mentions the enormous potential for the destruction of modernity, producing a problematic coexistence, the expansion of options and the risks (Carreño 2015, p. 94), which, on the one hand, makes productivity grow, while on the other it causes diseases, the production of risks and their ignorance have their first foundation in a cyclopean vision of the economy on the part of techno-scientific rationality whose gaze is directed towards the productivity benefits.

At the end of the 20th century, the belief that rich countries had the financial and technological capacity to bring progress to developing countries constituted a political order that established differences, posed limits, and proposed actions aimed at the "eradication" of poverty. Through the implementation of development policies, although essentially what was sought was to establish control strategies over such nations and their resources (Carreño, 2015, p. 104), often moving risk industries, apparently not perceptible, under few or no protection and safety standards, without prevention policies existing to prevent the reality of their impacts and effects. The unilateral sense with which the intervention in the so-called third world countries was supported is notable, supported by the presumed existence of only one correct path, that of the so-called first world countries, whose experience was presented as exemplary and passable through implementation. Appropriate tools, such as science, technology, planning, and even help from international organizations. To offer the hidden source of social wealth through scientific-technical development, technology transfer became a type of practice inherent to development projects that promoted the production of scientific and technological knowledge, consolidating today a linear, reductivist, and deterministic development model.

2 Social Technology as an Inclusive Regional Scale Process

According to Fullat, technology is understood as the set of knowledge about the technical procedure of action in reality (Magallanes Udovicich & Zanotti 2018, p. 75), given that the technique satisfies basic needs and produces well-being, being constantly mutative and protean, which is critical in forming any culture. Technology is a problem or possibility demanding serious reflection due to its importance and the necessity to understand, value, and direct it. The social movement of Science, Technology, and Society "STS" was born in the United States in the sixties as a critical reaction against technology being out of control (Cutcliffe 2003, pp. 7–9) and the need for public regulation of scientific-technological change. Subsequently, the currents of Social Technology and Technology for Social Inclusion were made known in Latin America, which discuss the linear logic of technological development, seeking to address social problems from new models where technology changes society.

Serafim and Dagnino, in Brazil, defined *Social Technology* as the result of a collective of producers on a work process based on a socio-economic context that engenders collective ownership of the means of production and that, based on a social agreement that legitimizes productive associativism, generates self-managed and cooperative control, consequently allowing a modification to the product generated, according to the decision of the productive collective; However, it does not consider that sectors with lower economic capacity can acquire technological objects, but rather that the current Science and Technology model that promotes these innovations cannot respond to demands of another order. Technology is considered an independent and universal variable capable of determining the behavior of all other productive and social system variables. Reconstruction is intended to allow for the incorporation of diverse perspectives beyond the capitalist market and be compatible with the region's most relevant sectors of society (Magallanes Udovicich & Zanotti 2018, p. 81).

In Argentina, Hernán Thomas proposed, concerning Technology for Social Inclusion, the review of the different conceptions of technology with which problems linked to poverty in the region had been addressed, defining both the fields of production of scientific knowledge and technological as well as the organizations responsible for its execution as a way of designing, developing, implementing and managing technology aimed at solving social and environmental problems, through social and economic dynamics of social inclusion and sustainable development. Social technology is a complex plot between products, processes, and organizations that develop new sociotechnical solutions to local problems.

Identifying the appropriate technology and using it appropriately consists of developing technologies for people with low incomes and promoting the design and production of socio-technologically appropriate systems for the community. The main characteristics that technology must have to be socially and environmentally appropriate, according to the provisions of Sohn Alan, are the following:

- Do not cause foreseeable harm to people or other forms of life (animals and plants).
- Do not irrecoverably compromise the natural heritage of future generations.
- Improve the primary living conditions of all people, regardless of their purchasing power.

- Do not be coercive; respect the rights and possibilities of choice of its voluntary users and its involuntary subjects.
- Do not have generalized irreversible effects, even if they appear beneficial or neutral.
- Government investment in appropriate technologies must prioritize the satisfaction of human beings' basic human needs.

Technology is not independent of culture; it is integrated into an inseparable social and technical system and constantly changes according to the needs of the inhabitant and historical circumstances. It is contemporary with the human generation, which aims to respond to and transform the human environment to adapt it in the best possible way to the needs and desires of its inhabitants. (Gordón 2011, p. 132).

3 The Environmental Impact of the Construction Sector, Scope and Relationship with Healthy Building

The construction sector is responsible for 38% of energy-related carbon dioxide emissions; direct emissions from buildings must be halved by 2030, heading towards neutrality by 2050. (UNEP, Programa de las Naciones Unidas para el Medio Ambiente, 2020, p. web). Likewise, it represents one of the main reasons for the emission of greenhouse gases into the atmosphere, which directly harms the population's health. Since the 1990s, various certifications have been established (BREEAM, Building Research Establishment Environmental Assessment Methodology, 1990 in the United Kingdom; HQE, High-Quality Environmental Standard, 1996 in France; LEED, Leadership in Energy and Environmental Design 1998 in the United States; Green Globe, 2000 in Canada; CASBEE, Comprehensive Assessment System for Building Environmental Efficiency, 2001 in Japan; Green Star, 2002 in Australia; Passive House, 2003 in Germany among others) that have contributed to the mitigation and prevention involved in damaging agents in the built environment, resulting in essential improvements in the health of the inhabitant within its spaces.

In 1988, a conference was held in Stockholm, Sweden, to integrate technical and functional solutions to promote healthy buildings for their inhabitants. Among the documents collected, N. Carlson defined the term "healthy housing." as the impression that the occupants have of the combination of psychological, physical and social qualities (Berglund, Lindvall, & L., 1988, p. 313); later, in 2010, P. Bluyssen mentioned that a "Healthy Building is free of hazardous materials and is capable of promoting the health and comfort of occupants throughout their life cycle, supporting social needs and improving productivity." (Bluyssen 2010, p. 808).

Different disciplines and media have studied the health risks that the built environment represents for its inhabitants, developing tools and instruments that have led to the creation of certifications and regulations, primarily addressing issues related to the impact of buildings on the environment. However, the sustainable building market was equivalent to barely 1% of total construction worldwide in 2016, according to The Global Alliance for Building and Construction, published in its 2017 report, which indicates that the construction area reached approximately 235 billion m2 and in the next 40 years another 230 billion m2 will be reached, still having a minimal positive impact at a global level by implementing this type of instruments. (UNEP, Global Alliance for Building and Construction., 2017, p. web).

In addition to the above, as Jerry Yudelson comments in his book "Reinventing Green Building: Why Certification Systems aren't Working and What We Can Do About It," the central green building certification system in the United States, LEED, is growing, the sustainable building revolution is stagnant, and no easy solutions are in sight (Yudelson 2016, p. XXIX). In Mexico, the LEED certification is the one that has the most significant impact on the national market, having 2016 a total of 211 certified projects, while on the other hand, the Mexican Sustainable Building Standard NMX-AA-164-SCFI-2013 published in 2013 (NMX-AA-164-SCFI-2013, 2013). to date, it does not have any verified buildings in force according to the National Body for Standardization and Certification of Construction and Building, S.C. (ONNCCE) (ONNCCE 2020, p. web), a unique body endorsed in Mexico by the Mexican Accreditation Entity (EMA) to carry out the verification of this standard. The previous regulatory instruments, both private and public, clearly indicate the significant imbalance and limitation that exists in the country to carry out actions focused on improving the impact of the built environment on the health and well-being of the inhabitants.

4 Healthy Housing in Latin America: Challenges and Opportunities

In America, more than 60% of housing is self-built and in irregular settlements, with a precarious nature that makes the inhabitant's health more vulnerable. This type of housing progressively improves through continuous processes of cumulative actions (Barceló, 2018, p. 33); under this scheme, the possibility of developing healthy housing from its first conception is hardly achievable since housing is determined as a progressive, transitory, and incremental good. More critically, healthy housing results from various interventions for its improvement and consolidation.

In addition to the above, most of the improvements in housing have to do with the capabilities of its residents and their attitude toward adaptation. At this stage, the development process becomes reflective, with knowledge being determined and decisive for decision-making, avoiding risky situations, and the absence of experience and expertise leading to risky situations. Knowledge is directly linked to prevention, understood as preparation against future damage, destruction that has not yet occurred but is imminent. Prevention helps us reduce the magnitude or probability of the appearance of the risk. (Beck, 1998, p. 26).

Healthy housing results from multiple actions of improvement over time through the development of capabilities, taking advantage of existing knowledge and technology. (Barceló, 2018, p. 30). Unfortunately, inequality in the distribution of social goods, as well as that of scientific knowledge, compromises the health and well-being of the inhabitants, conceiving low-quality homes with inappropriate designs lacking ventilation and natural lighting, increasing thermal and acoustic discomfort among others, with limited access to services and a deteriorated environment, also contributing to the lack of guarantees that should cover the primary needs of human beings.

Scientific knowledge is not distributed socially equitably; modern capitalist society has been characterized by favoring practices where scientific knowledge predominates and has only privileged social groups that have access to it. The ecology of knowledge invites us to question the logic of the monoculture of knowledge and scientific rigor, identifying other knowledge and other criteria that operate assertively with social practices. The utopia of interknowledge consists of gaining new knowledge without leaving previous knowledge behind, overcoming the monoculture of scientific knowledge, and including other forms of knowledge that identify with the context. (Sousa 2015, pp. 113–115).

For this, it is desirable to consider social technology since *technological change* is defined as a system of actions that contemplates social, economic, and political interests and is not only seen as the result of a simple application of scientific knowledge that constitutes well-being to the population, this understood in the availability of goods and services based on technological innovation, derived from a local dynamic that involves specialized knowledge in inclusive and sustainable social development, resulting in policies appropriate to local conditions.

It establishes the basis of scientific-technological models from and for Latin America, where the ecology of knowledge is analyzed in terms of cognitive inclusion before thinking about social inclusion processes. That is, the mental dimension of inclusion that involves social technologies is intertwined with the decolonial perspective that deconstructs the colonial practices present in daily life, allowing us to expand the possibilities of inclusion thought only in terms of access to goods or services involving the conception of knowledge that is usually silenced.

5 Development of a Local Measurement Instrument of Healthy Housing in Queretaro from the Complementarity of Knowledge

In this stage of the research, an instrument is developed considering the concept of "Glocal Society" by U. Beck (Ballesteros 2014, p. 208), where supranational cooperation is essential as an instrument to reinforce local development and stability; the global is related to the regional based on their multiple interconnections and allows us to understand how globalization is localized. The development of the instrument was based on previous research, where the relationship between the various categories existing in the literature related to the impact of the built environment on the health of the inhabitant was established, coinciding with the analysis and emphasis of various authors for the review of seven essential direct categories (Table 1) considered as the minimum criteria to have a healthy built environment. (Granados 2024b, p. 748).

Consequently, the methodology established for developing the local instrument included the direct categories of the healthy built environment to be broader by considering the various factors that influence the well-being of the inhabitants and that are related to healthy housing. The bibliographic review carried out, as well as the analysis of similar research on the subject, have served as a reference for the development of this research that is applied in the specific area of Querétaro, Mexico, but considers a series of indicators that are essential to propose improvements in habitability, in addition to the fact that it was detected that the vast majority of the works only consider some of the seven variables (Indoor air quality, Hygrothermal range, Acoustic range, Humidity, Lighting range, Electromagnetic fields, Safety, and accessibility), so this work is adequately consolidated by integrating theory with technical development for the implementation of new technologies to improve habitability conditions, according to environmental, social and cultural requirements, providing a tool that helps the population to have a better quality of life.

5.1 Materials and Methods

The methodology used for this research has a quantitative approach, which is important because it aims to use

Table 1. Direct categories of the healthy built environment (Granados J., 2024a)

Categories	Description
Indoor air quality	It refers to chemical and biological contamination in a gaseous, liquid, or solid-state of the air inside the home, coming from both the environment and construction and furniture inputs (particulate matter PM and volatile organic compounds VOCs), as well as the natural environment through radiological dangers such as radon gas, to which we are exposed and potentially impact health. Natural ventilation is one of the main strategies for mitigating and maintaining adequate air quality from artificial ventilation
Hygrothermal range	It considers the mental condition that expresses satisfaction with the thermal environment and is evaluated subjectively. It is influenced by objective factors such as air temperature, radiant temperature, air velocity, and humidity, as well as personal characteristics such as metabolic activity and the degree of clothing. Meeting minimum thermal comfort standards for temperature and humidity is essential, as these can drastically affect health
Acoustic range	It refers to the design of spaces to maintain good hearing, incorporating, if necessary, the control and mitigation of sound sources that affect physical and mental health through noise exposure. Maintaining developmentally appropriate sound levels and quality for the different activities is required
Humidity	This addresses building water damage caused by inadequate execution or lack of maintenance. Excessive moisture accumulation creates conditions for mold growth and produces irritating substances responsible for respiratory conditions and musty odors
Lighting Range	Considering light's physical, physiological, and psychological aspects, natural light is one of the most abundant resources and essential in our physiological, metabolic, and behavioral regulation. Both natural and artificial lighting must be considered in terms of quality and quantity for the optimal development of various activities
Electromagnetic fields	It refers to low-and high-frequency electromagnetic pollution inside the home. Exposure to these fields is a health risk. To protect occupants, the health exposure limit values and the location of the generating sources must be considered
Security and accessibility	It considers accessibility, safety, and protection within the home against unintentional injuries known as accidents, which constitute a significant cause of mortality and morbidity. Different accessibility regulations and safety standards must be considered, such as fire protection systems, lighting, emergency exits, and adequate pavements

existing scientific and technological knowledge (Hernández, Fernández-Collado, & Baptista 2006, p. 755), that is, theoretical and practical knowledge, to develop new knowledge. It is intended to create an innovative local evaluation instrument that allows for an affordable diagnosis of the built environment's negative impact on the inhabitants' health. The type of research corresponds to the exploratory correlational, considering it at an application level that aims to resolve the problem posed, putting into practice the methodological instrument for its application to obtain knowledge that allows modifying reality, linking knowledge and the training, "Research to transform."

The research design is of an experimental quantitative type, having a first part of an exploratory type, a second correlational part, and finally one at an application level, with the general phases and goals of the research being the following:

For the first stage, we worked from the exploratory level, carrying out a bibliographic review of the different regulations from the global to the local, which made it possible to identify the criteria and indicators related to the seven direct categories of the healthy built environment, for this stage the techniques Data collection refers to documentary review. In the second stage, the type of correlational research used is information analysis, which takes advantage of the experience and knowledge of the existing regulations. This research's criteria and reference values are based on national and international regulations for evaluating sustainable and healthy buildings.

Finally, once the regulations that affect the essential direct categories were defined, they were linked to the development of the local instrument through applied research, which considered the discoveries and knowledge obtained in the study to achieve results exploitable as products, procedures, and tools, among others. It is about applying the knowledge acquired. The research is carried out in spaces separate from where the events occur, such as in universities and institutes that provide knowledge, determining the usefulness of knowledge in general terms and its application. (García C. Fernando., 2018, p. 63). The proposed intervention model for applying the instrument optimizes the innovation of existing technological tools for their development and local application.

5.2 Eligibility Criteria

Additionally, normative instruments, regulations, standards, guides, and recommendations at a global and local level considered by some countries and organizations with excellent representation and support in the scientific and technological fields were included in this research. First determining the consideration of the existing regulations at the national level (Mexico), which included official regulations published by the Official Gazette of the Federation that are directly related to the categories of the healthy built environment; once this first stage was established, it was They identified the categories that did not have applicable regulations that will limit the impacts at the local level, and the search was carried out at a global level; In this second stage, the Precautionary Principle was included (Escalante 2005, p. 134) considered as a risk management tool, as a resource in case of scientific uncertainty regarding suspected health risk.

Likewise, sources of information not found within official publications, both locally and globally, were excluded, and information from organizations or associations that did not have scientific support was included.

5.3 Search and Sources of Information

The search for regulations, standards, guides, and recommendations was carried out in various databases at a global and local level, initially considering Mexico, later Latin America and finally the rest of the world (Table 2), resorting to databases such as the Official Gazette. of the Federation, Official Gazettes of Latin American Countries, Legislation and international regulations about study, in addition to bibliographic databases such as Scopus, Google academic and Mendeley, without considering the language restriction.

The analysis of the various municipal, state, national, and even international legislations and regulations allows **us** to know the specific situation of the region by detecting the advances or omissions in the matter but also strengthens the methodological development because the gaps detected are supplied with some other legislation that marks the trends or work routes, with which a complete and complex tool is sought that has updated legal support, understanding that this aspect has been dramatically transformed in recent decades as a result of the global concern for environmental protection, but that it is precisely this interest that has achieved benefits for humanity and with which work for the development of new technologies as part of an ethical and social commitment.

5.4 Instrument Application Model

According to the International Telecommunications Union 2011, Latin America had an average of 103 telephone lines per 100 inhabitants, considerably increasing mobile telephony and information transfer speeds due to decreased costs and the evolution of mobile equipment technology. The availability of better network technology and mobile devices has allowed the development of endless benefits for people through applications, solving specific problems in various fields (business, commercial, academic, health, and social). (Gasca Mantilla & Camargo Ariza 2014, p. 21).

Starting from the previous premise, this research developed a computer application designed to be executed on mobile devices, taking advantage of the unique characteristics of mobile environments and agile methodologies for software development, allowing projects to be carried out quickly. The application enables the evaluation to be carried out through spatial analysis, taking as reference a series of parameters supported by the applicable criteria referred to in various national and international regulations and legislation, in addition to the precautionary principle, and consequently, generates recommendations based on the categories direct measures to achieve a healthy built environment, all of this, understood from a local context, creating a comprehensive and affordable solution that takes advantage of the use of technology and digital tools to innovate.

5.5 Analysis

For the analysis of requirements, the current regulations that allow the evaluation and determination of reference values (see Table 3) are considered from the direct categories of the healthy built environment, established by parameters that are first identified with correspondence within the local level, and subsequently at the global level, mainly for those categories that are not included in the first section. Once the evaluation regulations and the reference values for the optimal use of each category about the inhabitant's health have been identified, the process of each category is organized and outlined, as well as the standards used for their evaluation. Their reference values and recommendations are generated for aspects outside the allowable reference values, transferring the information to the design or analysis case.

5.6 Interface Design

As a result of the research and bibliographic review process, the next stage was taken, which consisted of the development of a mobile application that would take the detected variables and present them within a quantifiable scheme that is easy to use by any user, with the purpose that practically anyone would be able to take the required readings of the different spaces in their homes and, based on spatial analysis and comparison with the information preloaded and configured in the software for the series of desirable values and ranges for each of the seven indicators, deficiencies in the spatial configuration would be detected, and basic proposals would be made as recommendations so that the user knows how to improve their construction and therefore have a better quality of life.

In this phase, the application proposal and development design were conceived through the different stages, sequences,

Table 2. Review of local and international regulations (Granados J. 2024a)

Ambit	Location	Laws/Regulations/Norms/Standards/Codes
LOCAL	Queretaro, Mexico	Environmental Protection Law for Sustainable Development of the State of Queretaro
		Income Law for Fiscal Year 2014. Queretaro: Executive Branch of the State of Queretaro
		Construction Regulations of the State of Queretaro
		Regulations of the Housing Institute of the State of Queretaro
		Environmental Code of the State of Queretaro
		Urban Code of the State of Queretaro
	Mexico City	Mexico City Sustainable Buildings Certification Program
	Mexico	Regulations of the General Law of Civil Protection
		Regulations of the Federal Law of Environmental Responsibility
		Regulations of the General Law on Climate Change
		Regulations of the General Health Law
		Housing Law Regulations
		NOM-035-ECOL-1993, the Official Mexican Standard of the Ministry of Environment and Natural Resources, establishes the measurement methods for determining the concentration of total suspended particles in ambient air and the procedure for calibrating measurement equipment
		NOM-008-ENER-2001 Official Mexican Standard of the Ministry of Energy establishes energy efficiency in buildings**,** envelope of non-residential buildings
		NOM-018-ENER-2011 Official Mexican Standard of the Ministry of Energy that establishes thermal insulators for buildings. Features and **testing** methods
		NOM-020-ENER-2011, Official Mexican Standard of the Ministry of Energy, establishes building energy efficiency—building envelope for residential use
LOCAL	Mexico	NOM-081-SEMARNAT-1994, the Official Mexican Standard of the Ministry of Environment and Natural Resources, establishes the maximum permissible noise emission limits from fixed sources and its measurement method
		NOM-020-SSA1-2014, the Official Mexican Standard of the Ministry of Health, establishes the permissible limit value for the concentration of ozone (O3) in ambient air and the criteria for its evaluation
		NOM-021-SSA1-1993, the Official Mexican Standard of the Ministry of Health, establishes the criteria to evaluate ambient air quality concerning carbon monoxide (CO). Permissible value for the concentration of carbon monoxide (CO) in ambient air as a measure to protect the health of the population
		NOM-022-SSA1-2010, the Official Mexican Standard of the Ministry of Health, establishes the criteria for evaluating the quality of ambient air with respect to sulfur dioxide (SO2)
		NOM-023-SSA1–1993, the Official Mexican Standard of the Ministry of Health, establishes the criteria to evaluate ambient air quality concerning nitrogen dioxide (NO2). Regulated values for the concentration of nitrogen dioxide (NO2) in ambient air are used to protect the population's health
		NOM-025-SSA1-2014, the Official Mexican Standard of the Ministry of Health, establishes permissible limit values for the concentration of suspended particles PM10 and PM2.5 in ambient air and criteria for their evaluation
		NOM-026-SSA1-1993, the Official Mexican Standard of the Ministry of Health, establishes the criteria for evaluating ambient air quality with respect to lead (Pb)—the regulated value for the lead concentration (Pb) in ambient air to protect the population's health
		NOM-001-STPS-2008, Official Mexican Standard of the Ministry of Labor and Social Welfare, establishes safety conditions in buildings, premises, facilities, and areas in work centers
		NOM-010-STPS-2014, Official Mexican Standard of the Ministry of Labor and Social Welfare regarding chemical polluting agents in the work environment, recognition, evaluation and control
LOCAL	Mexico	NOM-011-STPS-2001, the Official Mexican Standard of the Ministry of Labor and Social Welfare, establishes safety and hygiene conditions in workplaces where noise is generated
		NOM-013-STPS-1993 Official Mexican Standard of the Ministry of Labor and Social Welfare regarding workplace safety and hygiene conditions where non-ionizing electromagnetic radiation is generated
		NOM-025-STPS-2008, Official Mexican Standard of the Ministry of Labor and Social Welfare, establishes workplace lighting conditions

(continued)

Table 2. (continued)

Ambit	Location	Laws/Regulations/Norms/Standards/Codes
		NMX-C-460-ONNCCE-2009, Mexican Standard for the Construction Industry - Thermal insulation - "R" value for housing envelopes by thermal zone for the Mexican Republic - Specifications and verification
		NMX-AA-164-SCFI-2013 is the Mexican Standard of the Ministry of Commerce and Industrial Development, which establishes the minimum environmental criteria and requirements for Sustainable Buildings
		Sustainable Housing Standard NAMA (Nationally Appropriate Mitigation Measures)
		Housing Building Code, National Housing Commission
		Green Housing Assessment System
		Guide for adopting normative technical guidelines for sustainability in construction and urban development regulations
		Program to Strengthen the Capacity of Local Governments for the Construction of Sustainable Housing, National Housing Commission
		Adequate Housing Criteria, National Housing Commission
GLOBAL	Venezuela	Social Interest Housing Evaluation Instrument
	Chile	Sustainable Construction Standards for Chilean Housing
	Germany	Technical Measurement Standard SBM-2015
		ICNIRP guidelines for limiting exposure to electromagnetic fields (100 kHz a 300 GHz) International Commission on Non-Ionizing Radiation Protection
	United States of America	Healthy Building Certificate
	Austria	TBQ Assessment, Austrian Society for Sustainable Building Practices

Table 3. Reference regulations (Granados J., 2024a)

Ambit	Category	Location	Laws/Regulations/Norms/Standards/Codes
LOCAL	Hygrothermal comfort	Mexico	Adequate Housing Criteria, National Housing Commission
	Acoustic comfort		NMX-AA-164-SCFI-2013, Mexican Standard of the Secretary of Commerce and Industrial Development, establishes the minimum environmental criteria and requirements for Sustainable Building
	Lighting comfort		Housing Building Code, National Housing Commission
	Security and accessibility		Housing Building Code, National Housing Commission
GLOBAL	Humidity	United States of America	Healthy Building Certificate
	Electromagnetic fields	Germany	Technical Measurement Standard SBM-2015

and interactions, depending on the user's decision-making and the selection of patterns according to the service scenario. The interface is the external presentation the user sees in their mobile application. It allows communication through interaction within a digital environment, calling said application HabitaSano© about the purpose of this digital tool, which is to create healthy built spaces that will enable users' enjoyment, considering the required specificities of optimal spatiality and habitability for human beings. Below are some examples of the interface design of the developed application.

The development of a mobile application emerges as a viable and accessible response so that the general population has a measurement mechanism to detect problems in their buildings and even look for options that help correct them, understanding the suitability parameters, taking as a reference the national and international legal frameworks, with which a general vision is formed; based on this, as well as on the existing and easily accessible operation and technology, that is, a cell phone, this becomes the fundamental tool of the analysis.

By the limitations and general characteristics of the population of Mexico and specifically of Querétaro State, in addition to the problem of access to technology, the application implements basic elements that practically the majority of today's devices have: developing the application in the first instance for models with Android system (the most used in Mexico), and making programming based on Java language version 23, which through the configuration of a code implements the use of the light sensor, access to the microphone, magnetometer sensor of the mobile device, in addition to using the iTexy library version 7 to create a report in PDF format.

In addition to the above, the minimum requirements for the mobile device are the Android 9 operating system, 400 MHz processor, 3 GB RAM, the use of applications in the background, and the installation of a PDF reader on the mobile

Fig. 1. Captures of different screens of the HabitaSano2024a © application, showing the operation and the measurements generated from the information captured and recorded through the mobile device's sensors (Granados,)

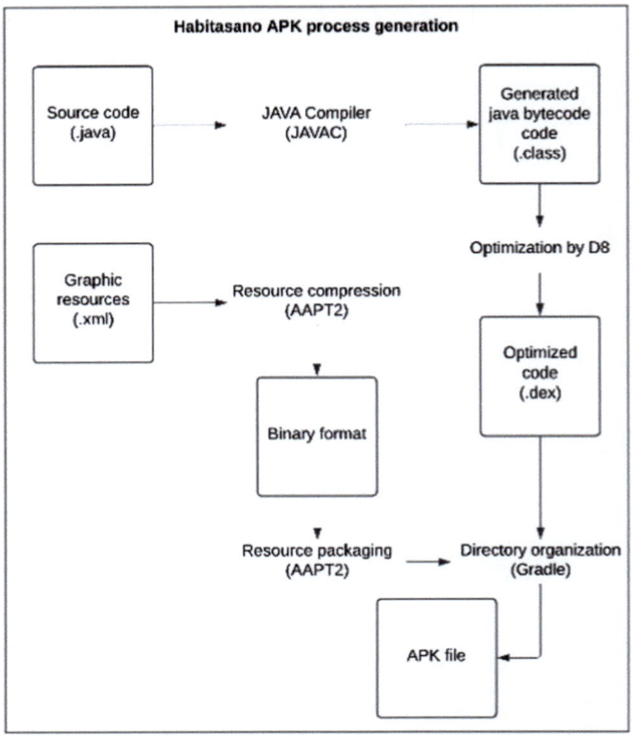

Fig. 2. Process generation source code diagram of the HabitaSano © application. Source: By author.

device. The download and installation are practical and intuitive since they are generated within current technological environments. It is important to note that the application has completed the testing phases, so it is not yet available to the public. However, this will be done in the coming weeks. However, it already has a Copyright Registration (No.: 03-2023-070313212200-01) with the participation and endorsement of the Autonomous University of Querétaro.

The operation and functioning of the application are very graphic and intuitive, allowing the user to operate it efficiently without the need for previous training or preparation; each of the screens allows the user to enter specific data (see Fig. 1), and in case of any omission or error, the system detects and points it out, marking the appropriate ranges or parameters, the above accompanied by the taking of readings that the device automatically does with the measurement tools that are configured from the code to avoid or reduce the margins of error in filling. Although the application was developed in such a way that allows the user to understand its operation, there is a User Manual that is downloaded with the application, which will enable you to consult and review the operation step by step or clarify any doubts.

Although the purpose of this article is not to delve into the development of the application but rather to see how it is a tool to detect the specific conditions of the constructions, taking as reference a series of variables and

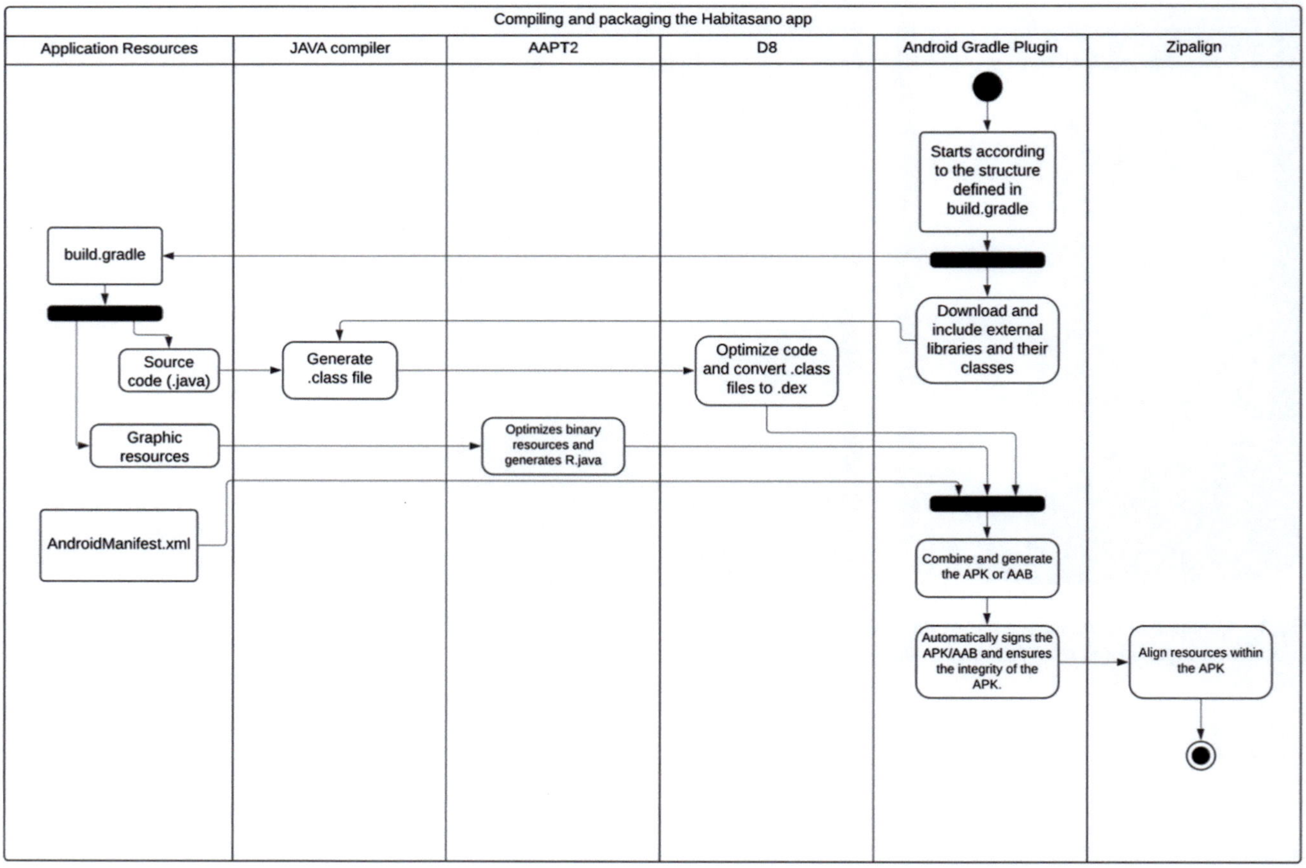

Fig. 3. Scheme of the source code of the HabitaSano © application. Source: By author.

general and particular regulatory aspects, the structure of the source code of Habitasano app (Figs. 2 & 3) consists of a series of processes that are linked and articulated, going through various stages, such as Android Gradle Plugin, D8, AAPT2, Java Compiler, Application Resources and Zipalign, which culminate in the configuration of the APK. The app's resources are compiled and packaged using AAPT (Android Asset Packaging Tool), using an Android Interface Definition Language (AIDL), compiling it within Java, complementing the interface with libraries and other sources; the command-line tool used to sign and verify Java Archive (JAR) files using jarsigner; finally, zipalign is used to optimize the application files.

5.7 Software Structure Design

This phase translates, through the analysis of various algorithms, the different processes implemented for modular development, taking as a basis the habitability requirements and comparing them with what exists, generating the most objective diagrams according to what is referred to through the readings and the registration of the information provided by the user as base requirements, obtaining a diagram that objectively describes the service implemented in the HabitaSano© application.

Figure 4 identifies the main development phases and their requirements. The HabitaSano© application begins with the property registration module, where the building data is entered with the climatic zone and entity. Subsequently, the following module is used to record the various spaces of the property and their characteristics, which provide the data for evaluating the seven direct categories of the healthy built environment. These data are analyzed, and finally, in the final module, a report is generated that indicates the reference status of the construction concerning the minimum criteria evaluated, as well as recommendations in cases where some requirements are outside the regulations.

In Fig. 5, the tools that make up the application are seen; an essential part of these are acquired through **mobile** equipment since the measurement is carried out using tools such as microphone sensors, compass, camera, and measurements. The development was carried out using SDK (Software Development Kit, level 29) for versions of Android 10 onwards using Java programming language and including several libraries, compiling each of the defined parts to run and implement the HabitaSano© application.

Figure 6 shows the relationship between the requirements analysis and the tools for application development. Each box (Orientation, dimensions, height, windows and doors, sound level, light level, EMF, and floor obstructions) displays a menu that is linked to its respective sensors and library developed in the source code for data collection, which is subsequently compiled and analyzed to generate a report on the status of the property along with its recommendations.

5.8 Source Code Generation

The application development was designed to generate responses agilely, so the JAVA programming language was chosen. The benefits of interoperability, scalability, and adaptability were also understood, and the capacity for updating and massification was provided according to the different characteristics and qualities that would be integrated in the future, giving a tool with more significant potential and certainty. Therefore, each defined part was coded according to the diagrams made as part of the design and interface. The objective of this phase was to implement the design in a software product that would allow the use of the specific characteristics of the seven categories analyzed.

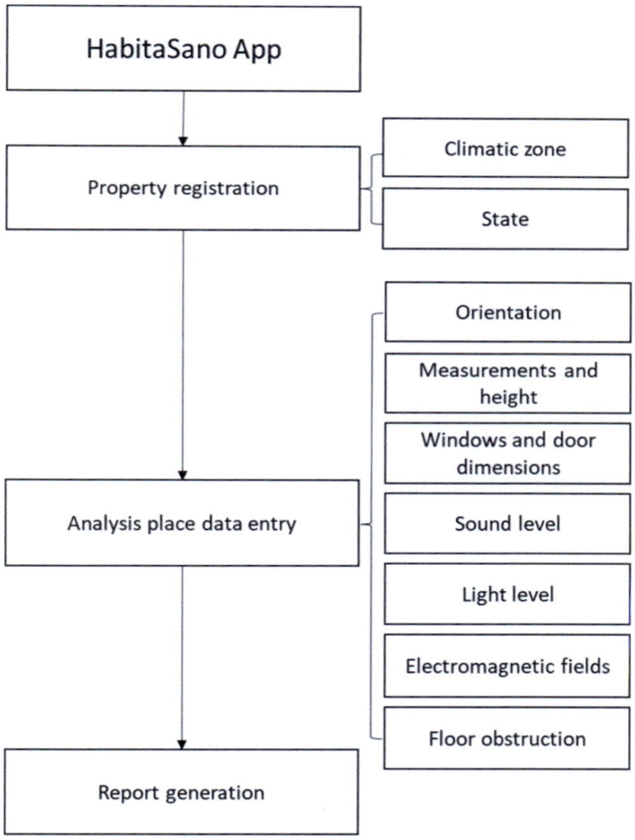

Fig. 4. Design of the structure of the HabitaSano2024a © application, considering the seven direct categories (Granados J.,)

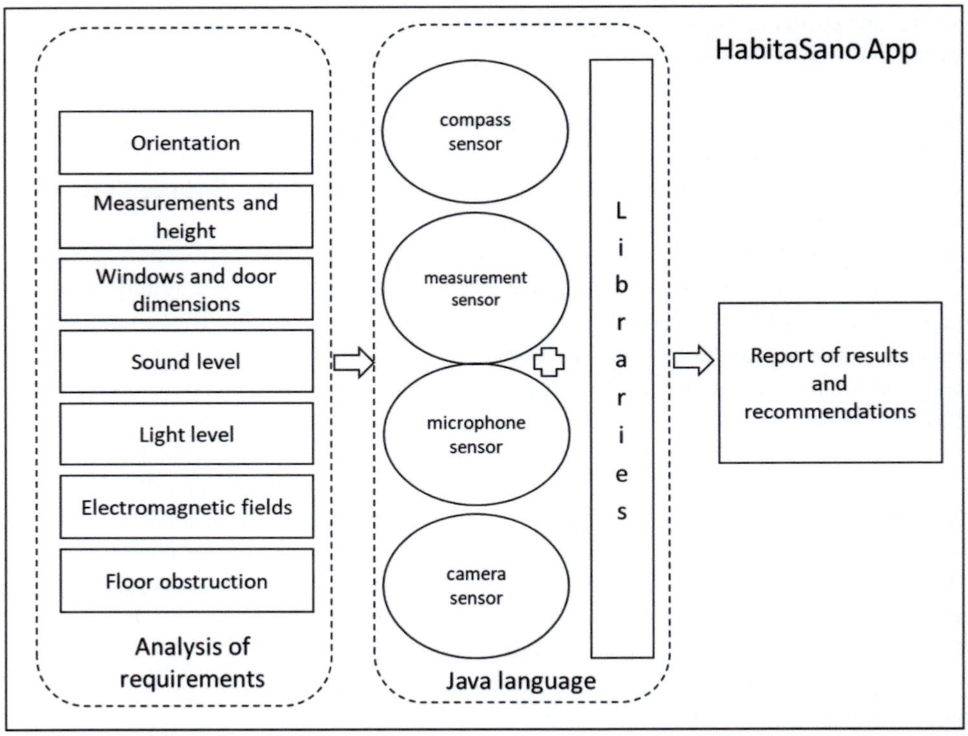

Fig. 6. Scheme of the relationship between spatial requirements and programming of the HabitaSano2024a © application (Granados J.,)

Fig. 5. Use mobile sensors to operate the HabitaSano2024a © application as a more accurate and reliable measurement tool (Granados J.,)

5.9 User Testing

This phase is relevant, demonstrating the operability, efficiency, and ease of understanding of the interface between users. The primary purpose is to verify the operation of the HabitatSano© application, correctly checking the operation of the elements developed individually and subsequently in all the elements, checking the interrelation between them. In this stage, the results obtained are executed and observed, verifying the correct functioning of the application on real devices, considering different scenarios and conditions with which it is possible to measure the performance and performance, exploring all the utilities, checking if there is any failure within of the algorithms and, where appropriate, make the pertinent adjustments or complements.

5.10 Implementation

Once the testing phase has been completed, the executable is implemented and delivered, in addition to the source code, documentation, and manuals that have the objective of training, such as the system manual, which indicates the installation process, the possible failures, and the minimum required technical specifications; there is also the user manual that aims to train users for the correct use of the HabitaSano© application, to finally begin the distribution phase of the application in the sectors of interest, seeking permeability in the construction sector and specialists in the field of sustainable buildings but also in the general public, with the firm purpose of bringing knowledge closer to society to create quality built spaces, reversing the deficiencies, omissions or problems that that buildings bring with them and that are detrimental to the physical, mental and psychological health of users. (Gasca Mantilla & Camargo Ariza 2014, p. 28).

6 Results

Currently, the development of the Habitat Sano© application is in the design phase of the software structure, work that is carried out jointly with Dr Edgar Rivas, so, based on the subtopic 5.7, the activities are mentioned by way of example. The remaining time for development is two months since the application processes are relatively short, allowing them to finish within the proposed period.

7 Conclusions

Although the instrument is still being developed for its improvement, it is essential to highlight the approximations of the development phase, where the original intention of the development of the instrument was linked to all the previous theoretical research that serves as documentary support and gives arguments and certainty to what is referred to, considering the seven direct categories, which are essential to consider to achieve a healthy built environment; The main objective being to shorten the gap between the dispersed knowledge existing in the various media and disciplines, to comprehensively consolidate the minimum requirements of a healthy space in an affordable way for all inhabitants at the local level, in this case in Queretaro, Mexico.

Derived from this research, the minimum criteria to be considered to generate a healthy built environment were taken advantage of, transferring said knowledge to a platform initially of local scope, starting from the concept of social technology, an instrument aligned with the existing glocal regulations, a simple and accessible evaluation of the spaces we inhabit, obtaining at the end feedback in the form of recommendations that arise from the analysis and disclosure of the parameters where the reference values indicate whether the required level of agreement is met. to what is established in the regulations, also considering these recommendations accessible to the local context.

The possibility of having an instrument such as the HabitatSano© application, which considers and integrates these characteristics within a handy and innovative mobile application, allowing knowledge to reach various sectors of the population, promoting interest in this type of dynamics that brings with them an essential social benefit, improving the health and well-being of the inhabitants, demonstrating that built spaces have a series of factors that must be considered and attended to achieve the enjoyment of the space with everything that this entails from the neuropathic aspects to the psychosomatics; prioritizing health in the various built environments.

The development of the application is quite accessible in every sense, given that it has been developed in the short term and with exciting alternatives that are reflected in improvements to what was initially proposed. The social acceptance of this first stage with the various actors in the environment has been well received and has generated positive expectations for its implementation from different scenarios. During the testing periods and the implementation of the application as a beta version, the results have been satisfactory, complying with the aspects stipulated within the research and strengthening the creation of healthy homes. Consolidating the instrument from this moment as an essential innovative technological proposal to prevent the risks that are triggered in a habitat built without adequate conditions, leaving users vulnerable to this type of phenomena, which has not been given the degree of necessary importance, nor has they been observed by interest groups at the local level.

References

Ballesteros, T.: Reflexión sobre la teoría de la Sociedad de riesgo. Temas Sociales (35) (2014)

Barceló, P.C.: Vivienda saludable, medioambiente y salud. Cuba. Ruht casa editorial (2018)

Beck, U.: La sociedad del riesgo, hacia una nueva modernidad. Paidós , Barcelona ediciones (1998)

Berglund, B., Lindvall, T., L., M.: Abstract Guide: Healthy Buildings´88. Estocolmo, Suecia: Swedish Council for Building Research (1988)

Beriain, J.: Las consecuencias perversas de la modernidad (compilación). Anthropos editorial, Barcelona (1996)

Bluyssen, P.: Towards new methods and ways to create healthy and comfortable buildings. Build. Environ. **45**(4), 808–818 (2010). https://doi.org/10.1016/j.buildenv.2009.08.020

Carreño, A.A.: Los postulados fundamentales de la teoría de la modernidad reflexiba de Anthony Giddens. Acta Sociológica **67**, 87–110 (2015). https://doi.org/10.1016/j.acso.2015.04.004

Cutcliffe, H.S.: Ideas, máquinas y valores, los estudios de ciencia, tecnología y sociedad, vol. 14. Anthropos, Barcelona (2003)

Escalante, J.M.: Principio de precaución y medio ambiente. **79**, 133–144 (2005)

Fernando, G.C.: La Investigación Tecnológica. Limusa Noriega Editores, México (2018)

Gasca Mantilla, M.C., Camargo Ariza, L.L.: Metodología para el desarrollo de aplicaciones móviles. Tecnura **18**(40), 20–35 (2014)

Gordón, F.A.: Reflexiones filosóficas sobre la tecnología y sus nuevos escenarios. Sophía, Bulgaria (2011)

Granados, J.: Desarrollo del modelo “Ambiente Construido Saludable” mediante herramienta metodológica innovadora para su prevención y mitigación en el estado de Querétaro. Tesis doctoral. Universidad Autónoma de Querétaro, Querétaro (2024). https://ri-ng.uaq.mx/handle/123456789/10686

Granados, J.R.: Criterios mínimos del ambiente construido saludable. Revista Internacional de Contaminación Ambiental (40), 741–758 (2024). https://doi.org/10.20937/RICA.55149

Hernández, S., Fernández-Collado, C., Baptista, L.: Metodología de la Investigación. McGraw-Hill, México (2006)

Magallanes Udovicich, M.L., Zanotti, A.: Diálogos en ciencia, tecnología y sociedad. Conocimiento, producción colaborativa, innovación. In: B. M., Innovación e Inclusión, pp. 74–99. Brujas editorial (2018)

Mumford, L.: Técnica y civilización. Alianza editorial, España (1977)

NMX-AA-164-SCFI-2013. Edificación sustentable: criterios y requerimientos mínimos. Diario Oficial de la Federación (2013)

ONNCCE. (2020). *Organismo Nacional de Normalización y Certificación de la Construcción y Edificación.* Retrieved from https://onncce.org.mx/es/certificados/certificados-emitidos?view=category

Sousa, S.: Una epistemología del sur. Clacso editorial, Argentina (2015)

UNEP. Global Alliance for Building and Construction (2017). Global Status Report.: https://www.unep.org/news-and-stories/story/buildings-and-construction-sector-grows-time-running-out-cut-energy-use-and. Accessed 10 diciembre 2024

UNEP. Programa de las Naciones Unidas para el Medio Ambiente (2020). Emisiones del sector de la edificación.: https://www.unep.org/es/noticias-y-reportajes/comunicado-de-prensa/emisiones-del-sector-de-los-edificios-alcanzaron-nivel. Accessed 10 diciembre 2024

Yudelson, J.: Reinventing Green Building: Why Certification Systems aren't Working and What We Can do About it. New Society Publishers, Cánada (2016)

Building Typologies and the Effects of Spatial Configurations on the Quality of Living Case Study: The City of Sarajevo

Mladen Burazor, Senaida Halilović-Terzić, and Amar Zuka

Abstract

Urban areas present a variety of urban structures with diverse forms and dimensions. Upon analyzing office buildings, one may observe certain similarities, especially in their architectural design, which is shaped by the widespread implementation of the "desk system". Upon analyzing office buildings, one may observe certain parallels, especially in their architectural design, which is shaped by the predominant utilization of the "desk system". However, when analyzing residential buildings, variations in spatial arrangements become more evident as a result of distinct floor layouts and whether the housing is individual or collective. Each housing unit is dependent on its urban surroundings, which subsequently influences the quality of living in such places, with focus on health and urban pollution. So, it is reasonable to suppose that certain buildings, based on their designs, have a more significant influence on the quality living in urban areas compared to others. To determine the veracity of this claim, one can conduct an investigation by gathering data on the air quality in the vicinity of specific structures, the quantity of sunshine present, and the volume of noise generated at ground level. Aim of this research is to determine does and how layout of urban residential forms influence on quality of living. This study focused on a designated region inside the municipality of Novo Sarajevo in the city of Sarajevo. This region contains multiple types of residential structures with diverse spatial configurations. The objective of this selection is to compare data pertaining to various housing typologies. The scope area is defined by the streets that surround the block, measuring approximately 400 by 400 m. The analysis of the collected data has validated the theory, and the conclusions can be used to future developments.

Keywords

housing typology · quality of living · urban setting · spatial configurations · urban environment indicators

1 Introduction

The annual ranking of cities worldwide is determined by several sources and approaches, such as the "Quality of Life Index", "City quality of life" or "Quality of living". Several factors must be considered in order to choose the optimal place to live, however, three of the most important categories are: economic, social, and environmental impact. (Milivojević, Cvetić, Kokić Arsić, Nikolić & Petronijević, 2017). In this paper, environmental impact in the city of Sarajevo will be emphasised for a several reasons. Firstly, the capital of Bosnia and Herzegovina frequently holds the title of the most polluted capital city in the world during winter, as determined by the air quality index (AQI) (AG, 2023). Secondly, the link between air pollution and heating season can be easily established and since the residential sector accounts for the majority of the energy needs in Sarajevo, it is important to target that sector. Thirdly, there are many initiatives that target energy efficiency measures in the building sector (Arnautović-Aksić et al., 2016) but they don't address spatial configurations.

In this paper, urban quality living is mainly considered from a perspective of environmental impact on the health of

M. Burazor · S. Halilović-Terzić (✉) · A. Zuka
Faculty of Sarajevo, University of Sarajevo, Sarajevo, Bosnia and Herzegovina
e-mail: senaida.halilovic@af.unsa.ba

M. Burazor
e-mail: mladenb@af.unsa.ba

A. Zuka
e-mail: amar.zuka@af.unsa.ba

B. He et al. (eds.), *Towards the Framework of Livable and Resilient Cities* Advances in Science, Technology & Innovation, https://doi.org/10.1007/978-3-031-97849-4_33

inhabitants and how the state of the natural environment could be influenced by the built environment. Health is regarded as one of the six primary categories in the index of life measurement (Yonk, Smith & Wardle, 2017). As new residential developments must adhere to existing regulations, the variable remaining is the spatial configuration of buildings (shapes and placements) and its impact on quality of life, which requires additional research. In this paper the intention is to establish some causal relationships between spatial configuration and urban surroundings.

1.1 Urban Setting

Over the centuries, the city of Sarajevo has gone through several phases of development. When it was founded, the country was ruled by the Ottoman Empire and the ways of living reflected the ways cities were formed. The main division was between the working area of the city where the wokplaces and retail stores were located along with administrative, educational and main religious buildings, and the living areas located around the city center. The dwelling units were scattered on the nearby hills and urban plans did not exist as such. The urban sprawl was "organic", where roads were aligned with terrain and houses were placed along those roads (Grabrijan & Neidhardt, 1953).

Once the political and social system underwent changes and the Austro-Hungarian empire replaced the Ottomans as the country's government power, the image of the city underwent a dramatic urban transformation over a period of 40 years. The city introduced a new building typology: multi-family houses, serving as a form of collective housing. Furthermore, the ground floor was now reserved for the workshops and stores which was opposite to the previous division of zones in terms of land use (Kurto, 1998). This change alone, along with the regulated urban matrix had most significant impact on the city image. Attached apartment buildings of four stories in height, where now forming urban blocks very much like it could be found in the rest of the Europe. As early as at the beginning of the 20th century, the centre of the city of Sarajevo had this "European appeal" but with the visible "Ottoman legacy". The effects of the changes can also be seen through the scope of demographical changes*:"* During the four decades of Austro-Hungarian rule, the number of inhabitants in Bosnia and Herzegovina doubled (20,944 people lived in Sarajevo in 1876, and about 59,000 in 1918), which was largely influenced by mechanical growth, i.e. migration. By 1910, the number of immigrants in Sarajevo from territories outside the borders of Bosnia and Herzegovina exceeded a third of the city's total population, and the influence of foreigners is dominant in the economy and administration*."* (Kurto, 1997, p. 49).

The industrialization process that followed World War II shaped the city's third phase of development. The prospects of a better life drew people from the rural parts of the country to the cities. Many manufacturing plants were opened and a work force was needed (Islambegović, 2022). Again, if we compare the census data on the number of inhabitants in the city of Sarajevo, we can see the extents of these changes: "According to the first post-war census carried out in 1948, 111,087 inhabitants lived in the innercity area of Sarajevo, which then consisted of the municipalities of Centar, Novo Sarajevo and Ilidža. The opening of new industrial facilities and other economic capacities and the opening of various educational, scientific and cultural institutions lead to a constant increase in the population of Sarajevo and its spatial expansion far beyond the old city limits. Thus, in 1971, the city already had 359,448 inhabitants, and ten years later 448,519 inhabitants." (Grad Sarajevo Historija, 2023). However, in that period of time, the so-called "socialist urbanisation" could not keep up with the demands of housing units for those in need (Džananović, 2016, p. 133). This led to many individual developments where people built their houses illegally on the outskirts of the city wherever they could find free land (Islambegović, 2022). This problem of illegal settlements has not been resolved to this day and many of those individual fireplaces (where coal, fuel oil or wood is used) are the major contributors to the air pollution in the city. This conclusion is easily drawn based on the fact that the air quality drops significantly in the winter (during the heating period) compared to the rest of the year, yet the number of motor vehicles remains in the same range and it can be assumed that the influence of traffic is constant.

1.2 Spatial Configurations

In urban planning, the terms "typology" and "morphology" are very common and are used to describe landscape features and the nature of buildings. Concept of understanding architecture through typologies and types is fundamentally based on the possibility of grouping objects by certain inherent structural similarities (Moneo, 1978). Upon analyzing the configurations of building floor plans, they are generally observed to be either rectangular or square. Other shapes, such as triangles or circles, may also be encountered, but they are less prevalent than ones founded on right angles. This is rational given the characteristics of our living environments and the furnishings and apparatus we utilize in our daily routines. Moreover, the design of many cities such as Madrid, Barcelona, and New York follows an orthogonal grid, reflecting the shape of the blocks and the building itself. Individual houses most commonly derive their shapes from the square, and different layouts typically correspond to non-typical building plots that emerge when streets intersect at varying angles.

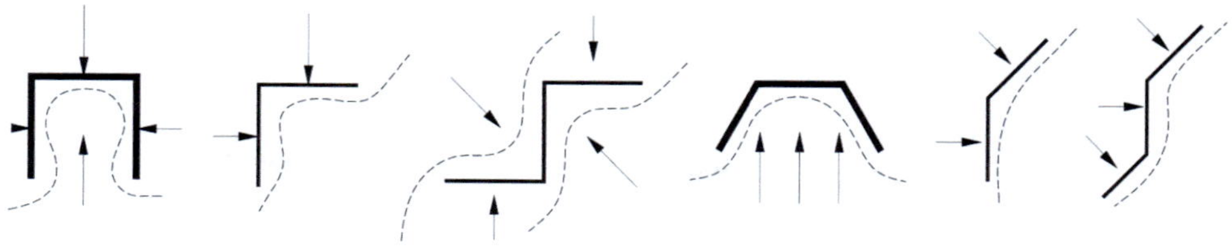

Fig. 1. Common layouts of building linear shapes (Source: authors)

Examining apartment buildings reveals their linear shapes, see Fig. 1, showing some of the most common layouts. Depending on the depth of the residential building, the most common is a corridor system with flats on each side of the building. This is related to new fire and safety requirements where additional fire escape staircases have to be built for all buildings with 4 floors and above (Federalna uprava civilne zaštite, 2009). In the socialist era, flats were designed according to the modernist principles and flats had a two-way orientation to maximize the effects of natural lighting and ventilation. Previously conducted studies clearly show connection between spatial configuration and different aspects of quality of life (Sapena, Wurm, Taubenbock, Tuia & Ruiz, 2021).

1.3 The Site

There were a number of reasons why this part of the city in the municipality of Novo Sarajevo was chosen. Firstly, historically speaking it was mostly developed in the 1960–1970's and is based on (at the time) contemporary urban planning strategies. However, at the north-east part of the site, there is still small urban block from the previous ear which makes it ideal for comparison between two approaches in the urbanism. Secondly, on this site we can find many different building typologies such as: administrative buildings, a school, kindergarten, commercial buildings and residential buildings. When we further examine housing units we can find: single-family houses, multi-family houses, apartment blocks and high-rise buildings as seen in Fig. 2.

The third reason for choosing this site was the variety of spatial configurations we can observe, particularly when we examine residential buildings. Single-family houses and high-rise buildings primarily have a square base, while the rest of the buildings use a rectangular base. Finally, we can compare the data and effects of linear configuration versus concentration and an open block system on a given site.

This site exhibits a clear hierarchy in the applied road layout, making it suitable for investigation in terms of traffic frequency and noise production effects. The site's south side borders a 6-lane road with three lanes in each direction, a 4-lane road with two lanes in each direction on the east and west, a 2-lane road with one lane in each direction on the north side, and internal roads. For this reason alone, the collected data should provide enough information to draw a conclusion about the noise impact of the surrounding traffic.

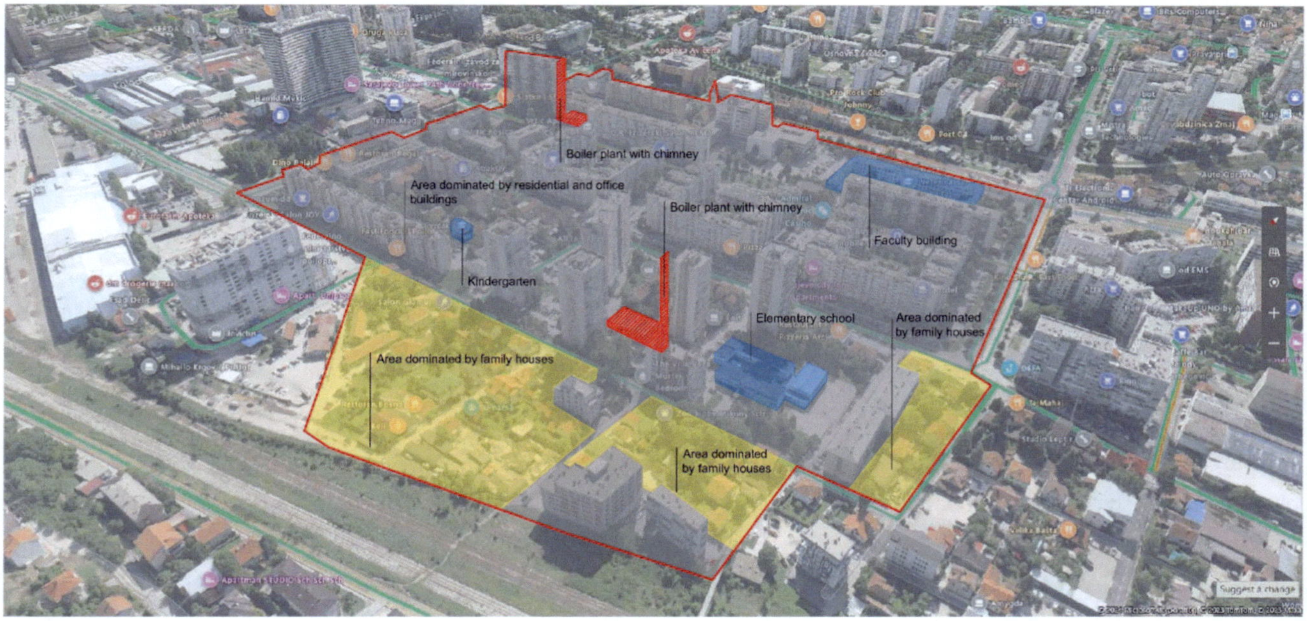

Fig. 2. The chosen site source: (authors over Windows maps)

Table 1. Specific energy needs of buildings in Bosnia and Herzegovina (Arnautović-Aksić, et al., 2016).

	SINGLE-FAMILY HOUSING		COLLECTIVE HOUSING			
	SINGLE FAMILY HOUSES (SFH)	TERRACED HOUSES (TH)	MULTI-FAMILY HOUSE (MH)	APARTMENT BLOCKS (AB 1)	APARTMENT BLOCKS (AB 2)	HIGH-RISE BUILDINGS (H)
A < 1945	639/452	256/183	309/231	248/176	–	–
B 1946–1960	602/474	449/321	279/216	224/159	249/177	–
C 1961–1970	582/465	254/196	265/188	214/153	229/170	273/193
D 1971–1980	485/382	251/199	192/147	–	162/130	176/126
E 1981–1990	176/136	294/219	259/189	132/94	147/111	–
F 1991–2014	150/128	–	87/65	86/68	67/55	–

Natural light conditions ensure that buildings do not have a predominant orientation, and in collective housing, we can observe linear configurations arranged in both north-south and east-west directions. Buildings constructed after 2000 typically design their flats with a single orientation, either east or west. Flats in buildings from earlier eras typically benefit from two or more sides that receive direct sunlight. The shadows cast from one building to another, due to their varying heights, significantly impact the urban setting and quality of living. Therefore, it is interesting to learn of the effects present at the site which can be used for comparison with other parts of the city.

1.4 Urban Environmental Setting

There is a constant energy need in industry, transport but also in building and according to IEA: “Energy use in buildings increased from 115 EJ in 2010 to almost 135 EJ in 2021, representing 30% of global final energy consumption.” (IEA, 2022). When we investigate residential sector, according to the same source it accounts for 16, 6% of energy needs. Furthermore, according to estimates for UNECE region, on “average over 80 per cent of household energy consumption is spent on heating and hot water supply.” (UN, 2013). This is the reason why so many countries have devised national plans to deal with energy efficiency especially since many buildings that are in use are due to be renovated.

Scaling down from global to local, when we consider Bosnia and Herzegovina, we learn that more than 85% of residential buildings have extremely poor energy characteristics, where the end use of energy is extremely inefficient with an estimated 30% of energy losses (Hood & Ćosić, 2012). Average annual energy consumption of 180 kWh/m2 places Bosnian residential buildings in the lowest energy class according to the European classification (EMAS) (FZOFBiH, 2014).

A research project aiming at the prospects of increasing energy efficiency of the residential building stock in B&H was conducted in year 2015 and 2016 and in a process 13.044 buildings were surveyed (Arnautović-Aksić, et al., 2016). When we considered the results, we could see what kind of impact each type of residential building has in terms of improvement measures and the overall impact that is related to the number of dwelling units. The ratio of single-family and collective housing by total number of dwelling units is 66, 47% vs. 33, 53% and clearly, single-family housing should be a priority in incorporating EE measures simply based on the number of those units. Additionally, if we take into consideration the specific energy needs of different typologies, we learn that single-family housing has by far the largest energy needs, see Table 1 (Arnautović-Aksić et al., 2016).

At the site, we can see both extremes in terms of energy efficiency in the residential sector. The high-rise buildings that we can find at our chosen site have been one of the most energy-efficient housing typologies over the course of 20 years since they were massively introduced in BiH. When compared to single-family houses from the same period of time (1971–1980), the specific energy need for intermitted heating in high-rise buildings is three times smaller (126 vs. 382 kWh/m2/year). We expect these results because high-rise buildings benefit from their compact form and minimal ground contact. Additionally, due to the relatively small roof surface and the height of 8+ stories, many flats benefit from such layouts in terms of heat loss. However, high-rise buildings have negative effects in urban settings. They cast long shadows, but more importantly, they can be obstacles to natural ventilation in cities. Their shape, which is usually rectangular with more than 15 stories in height, often prevents the flow of air if not positioned well: “In terms of the layout and shape of the buildings, Sarajevo has several examples of very good construction practices. Regarding the positioning of the buildings in the Hrasno settlement, there are several skyscrapers that they represent an example of a good arrangement of buildings in terms of urban ventilation … Four skyscrapers that are arranged in a grid of two columns and two rows, leaving a large space between each of them.” (Hadžiabdić, Midžić-Kurtagić, Arnaut, Begić, & Ćorović, 2019, p. 34).

Majority of the collective housing units at the site are connected to the district heating system in particular to one

Fig. 3. Power plants and their chimneys (Source: authors)

of the two heating plant that use natural gas. The first is located at the south-eastern part of the site next to the tallest building and the second one is placed in the northern part of the site alongside another high-rise building and the primary school, see Fig. 3. Majority of the single-family houses use individual means of heating and use different energy sources.

In this research there are two assumptions that form a working hypothesis:

1. the quality of air in winter time, around individual houses, will be poorer compared to other parts of the site regardless of the fact that traffic load around collective housing units is larger, and
2. air quality within urban blocks (inner courtyards) will be either higher or lower compared to the outside perimeter of residential buildings.

The first assumption is based on the fact that district power plants use natural gas, which means that emissions from chimneys next to the tallest buildings won't negatively affect ground level as much as emissions from individual homes using crude or liquid fuel.

The second assumption pertains to air speed, and it is reasonable to anticipate that air pollutants will persist in inner courtyards for a longer period than outside, where air currents and wind enhance ventilation. Alternatively, the concentration of air pollutants will be lower within the courtyards since buildings serve as physical barriers to influences from the outside, where the roads are.

2 Methodology

Methodological framework is based on several methods used to verify given assumptions. Primarily, various measurements can be made at the site at the number of places in order to gain diverse, accurate and comparable data. Secondly, the same measurements will be taken in three different parts of the same day, so that any changes in natural conditions will affect each measuring point equally. Collected data will be evaluated using the comparative method as a part of the process of analysis and reaching conclusions. Observation method is used to collect relevant information about building typologies, height of buildings, placement of the traffic lights and district heating powerplants.

In order to collect data, 39 measurement points were designated across the entire site. Those points were placed on pedestrian zones: alongside the road (1–22 points) and within building blocks (A-M), see Fig. 4. Points were placed around specific building typologies and in public spaces which are accessible. Through many measurement locations within a limited area, where environmental effect is relatively consistent, we aim to establish a correlation between urban design and quality of life mainly concerning health issues. The primary objective is to investigate a potential correlation between urban design and air quality.

These mesaurement could provide starting point for guidelines on building layout regarding the further development of the city in corelation to quality of living. We are attempting to ascertain whether urban structure and layout might influence environmental impact.

Measurements on site were taken by using Environmental Meter with sensors for measuring sound, light temperature and relative humidity (LUTRON LM8102) and Particle counter (VPC300) Extech Instruments. All measurements were later transferred into single spreadsheet for comparison. Two persons were in charge of gathering data and altogether 6 persons were involved in a process of gathering readings. Readings were taken at different times from 8:30, 13:00 and 16:30 measuring the: lighting level [lx], noise level [dB], airflow speed [m/s], air temperature [^{0}C], amount of 2,5 μm (fine) particles [$\mu g/m^3$], the amount of 10 μm [$\mu g/m^3$] particles and air humidity [%]. Different times were selected in order to observe different dynamics of the city (people going to work and coming back home, etc.). Due to the nature of measuring devices (handheld devices), readings were taken at the height of 1.5 m from ground level. The limitations were the number of available devices – only two and the factory accuracy of the devices.

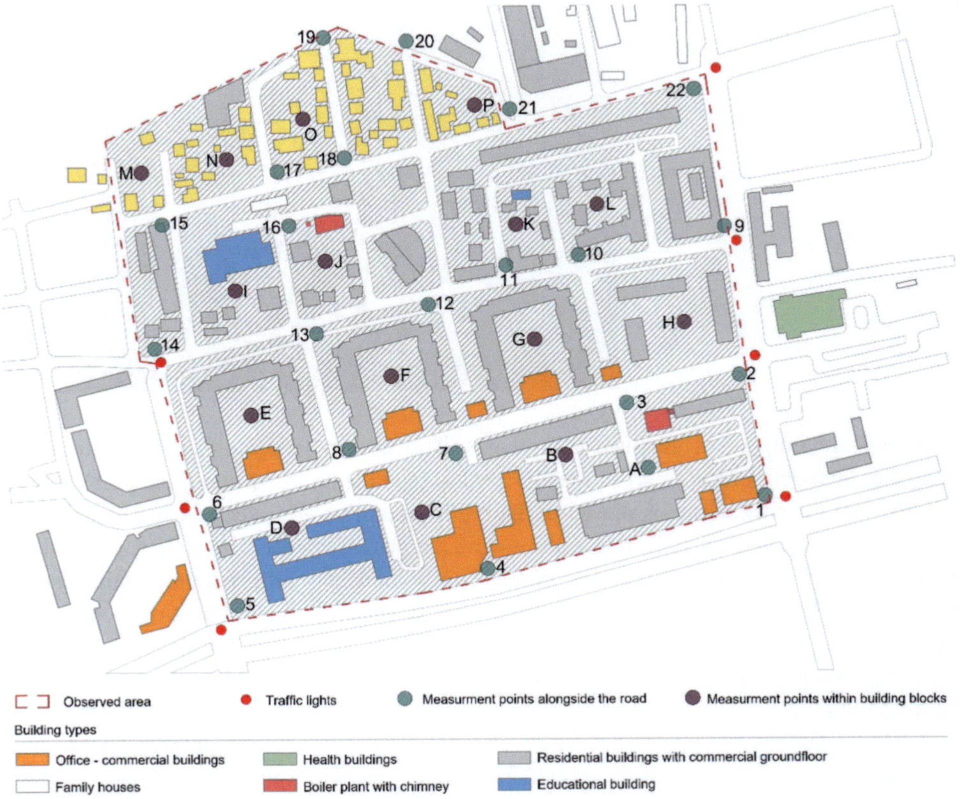

Fig. 4. Measurement points (Source: authors)

3 Results

Measurements for the given categories, were taken on the 3rd of February 2023. Weather conditions were stable thought the day and for that reason temperature readings, air humidity and air flow measurements did not show extreme deviations at given points and were consistent thought the site. For that reason, they are not discussed into detail and in this paper only average values are presented. Average humidity at 8:30, 13:00 and 16:30 was 60, 41%, 38, 00% and 41, 68% respectively. Average temperature at 8:30, 13:00 and 16:30 was 5, 09 ^{0}C, 18, 61 ^{0}C and 12, 31^{0}C respectively and the average air speed at 8:30, 13:00 and 16:30 was 0, 36 m/s, 0, 31 m/s and 0, 32 m/s respectively.

In terms of evaluating air quality conditions at the site, detailed readings are provided for the fine particles (2, 5 μm [μg/m^3]) and large particle (10 μm [μg/m^3]), see Table 2.

Lighting level and noise level are presented in the form of charts relative to the selected points, see Fig. 5 and 6.

4 Discusion

To get an overall assessment of the selected conditions on the site, an integral map was made showing the points with largest values relevant to the quality of living. This map shows conditions at the specific time for air, noise and lighting and 4 largest values are presented, see Fig. 7. Those values (4 highest out of 38) give sufficient insight of distribution of negative effects thought the site.

Furthermore, largest values for given period of the day are also graphically represented so that results could be interpreted relative to the way how people travel and use buildings.

4.1 Air Quality

The site's northwest part has the highest recorded values for air quality. As the map illustrates, individual housing is the predominant building typology in this area. Also at point "16", we can find the central heating power plant which certainly contributes to the air pollution. However, because the power-plant's chimneys are located in high-rise buildings (residential skyscrapers), the air pollution at ground level is heavily dependent on atmospheric pressure. Since there is another power plant at point "A" close to the most frequent road and the air pollution levels are lower compared to point "16" it can be deduced that the individual housing, where different energy sources are used (mainly coal and wood) is primarily the cause of poor air quality. Observations beyond the site's boundaries further support this conclusion. Individual housing continues towards the northwest and north parts of the city, across the railway lines. Therefore, the influence of those houses spread over the observed area, see Fig. 8.

Table 2. Values for air particles (2, 5 μm and 10 μm) the at the given points (Source: Authors)

	2,5 μm particles			10 μm particles		
POINT	8:30	13:00	16:30	8:30	13:00	16:30
1	822.00	565.00	1042.00	25.00	19.00	35.00
2	1035.00	882.00	789.00	22.00	28.00	38.00
3	1154.00	849.00	852.00	33.00	36.00	43.00
4	794.00	842.00	1326.00	20.00	32.00	45.00
5	1309.00	562.00	1334.00	37.00	23.00	46.00
6	1989.00	562.00	1255.00	57.00	23.00	66.00
7	1224.00	697.00	651.00	29.00	22.00	27.00
8	1328.00	664.00	783.00	31.00	19.00	35.00
9	915.00	909.00	398.00	23.00	30.00	13.00
10	1120.00	901.00	422.00	23.00	33.00	26.00
11	1196.00	1538.00	667.00	33.00	74.00	40.00
12	1213.00	897.00	792.00	35.00	32.00	36.00
13	1142.00	906.00	505.00	27.00	38.00	23.00
14	1552.00	1028.00	1498.00	78.00	57.00	49.00
15	1221.00	1122.00	1201.00	29.00	43.00	38.00
16	1636.00	651.00	804.00	52.00	39.00	24.00
17	1328.00	800.00	1346.00	42.00	28.00	47.00
18	1360.00	895.00	664.00	36.00	47.00	36.00
19	1198.00	835.00	806.00	32.00	39.00	32.00
20	1186.00	873.00	730.00	28.00	36.00	18.00
21	1120.00	792.00	648.00	25.00	36.00	28.00
22	1059.00	1093.00	829.00	34.00	69.00	34.00

	2,5 μm particles			10 μm particles		
POINT	8:30	13:00	16:30	8:30	13:00	16:30
A	1077.00	612.00	1147.00	25.00	30.00	67.00
B	1048.00	633.00	806.00	30.00	22.00	31.00
C	1064.00	534.00	596.00	16.00	13.00	28.00
D	1304.00	609.00	188.00	29.00	18.00	13.00
E	1168.00	586.00	646.00	27.00	22.00	34.00
F	1286.00	592.00	352.00	34.00	24.00	12.00
G	1207.00	634.00	459.00	28.00	23.00	26.00
H	1038.00	529.00	405.00	25.00	14.00	14.00
I	1105.00	865.00	999.00	24.00	39.00	41.00
J	1347.00	778.00	518.00	47.00	41.00	16.00
K	1465.00	698.00	528.00	39.00	24.00	13.00
L	1465.00	772.00	470.00	39.00	21.00	25.00
M	1359.00	1194.00	1399.00	47.00	47.00	34.00
N	2169.00	1090.00	1195.00	50.00	25.00	32.00
O	1654.00	878.00	914.00	47.00	39.00	22.00
P	1063.00	813.00	714.00	27.00	42.00	21.00

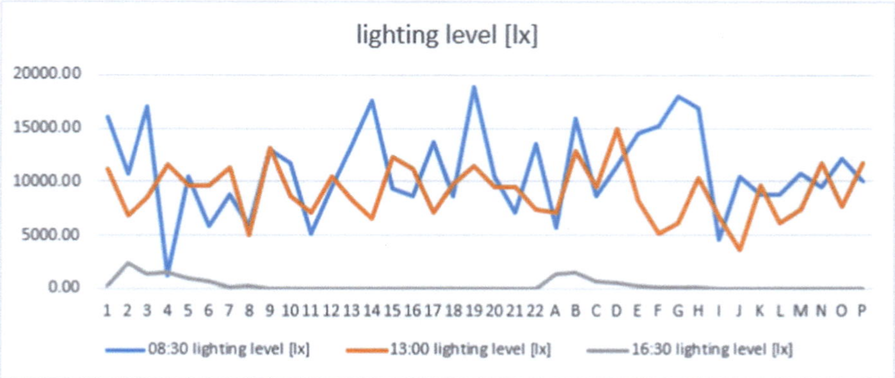

Fig. 5. Lighting conditions on the site (Source: authors)

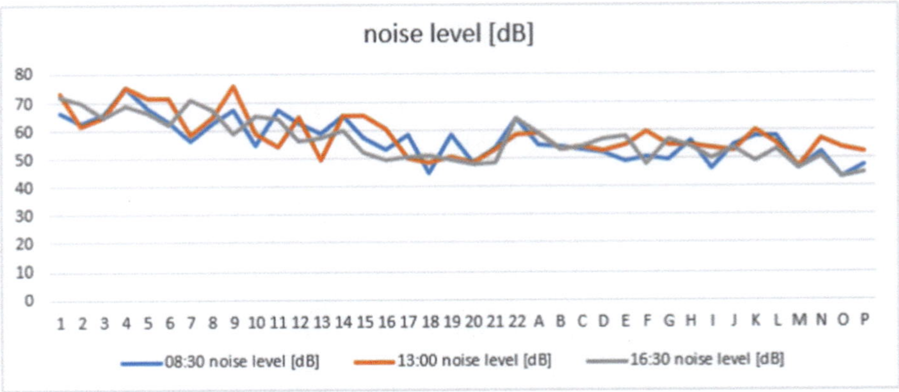

Fig. 6. Noise conditions on the site (Source: authors)

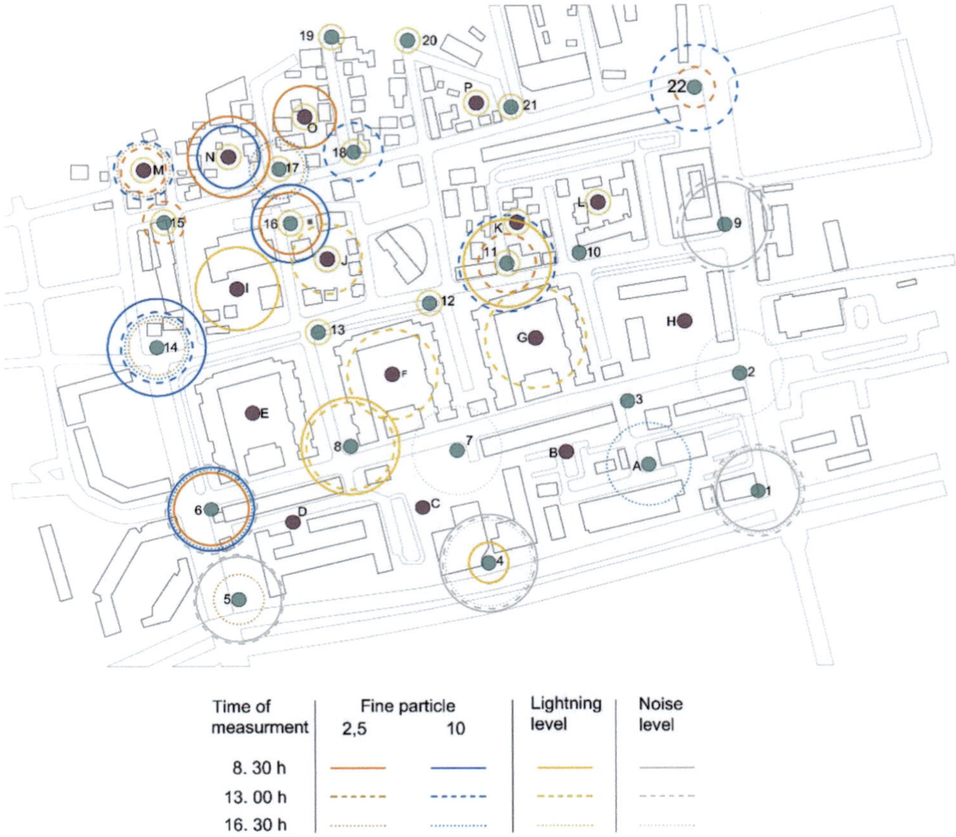

Fig. 7. Integral map of air, noise and lighting effects (source: authors)

In terms of building configurations, air quality data shows the air quality within the courtyards (points "E, F, G, H") is on average better compared to the readings on the outside perimeter, see Fig. 9.

Contrary to the first expectation, the concentration of dust particles is not higher inside compared to the outside perimeter, despite the lack of ventilation. Under calm weather conditions and without any wind, structures provided protection for inner courtyards, resulting in lower than usual concentrations. This demonstrates that introverted building layouts have a beneficial impact on air quality when compared to linear layouts.

4.2 Lightning Conditions

The highest recorded lighting conditions are located in the further north part of the site, among individual housing areas. At such a micro-location, other buildings are not blocking the views or casting shadows, so naturally, the lighting conditions

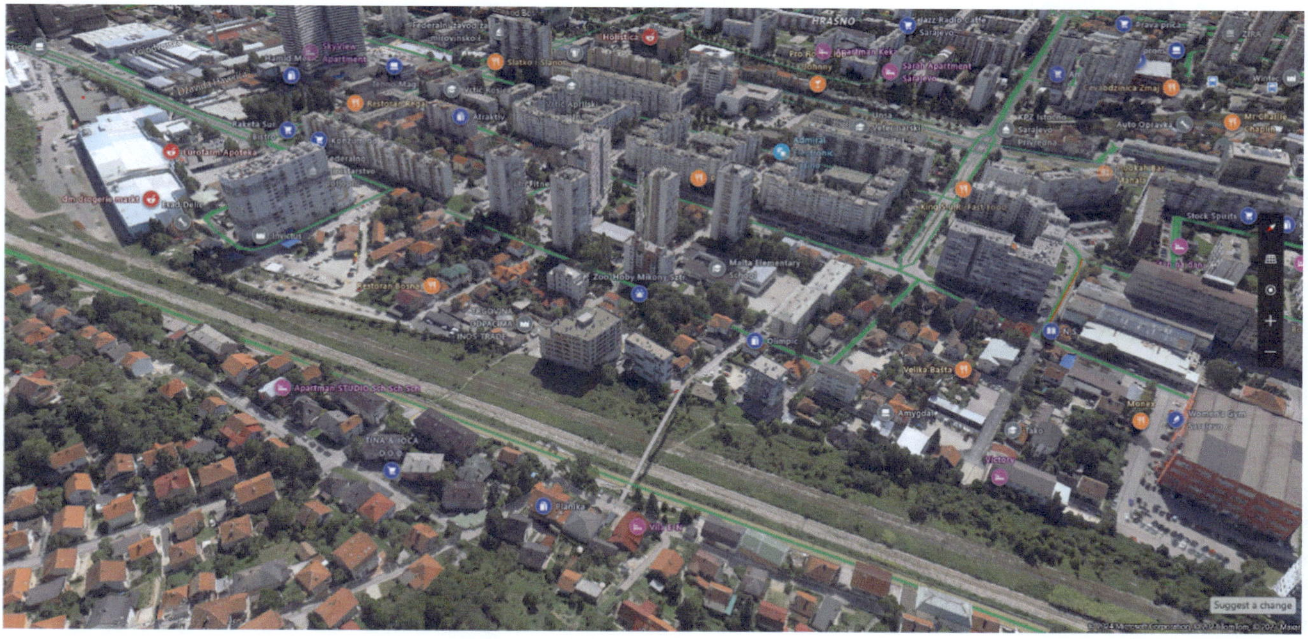

Fig. 8. Surrounding area (Source: Windows Maps)

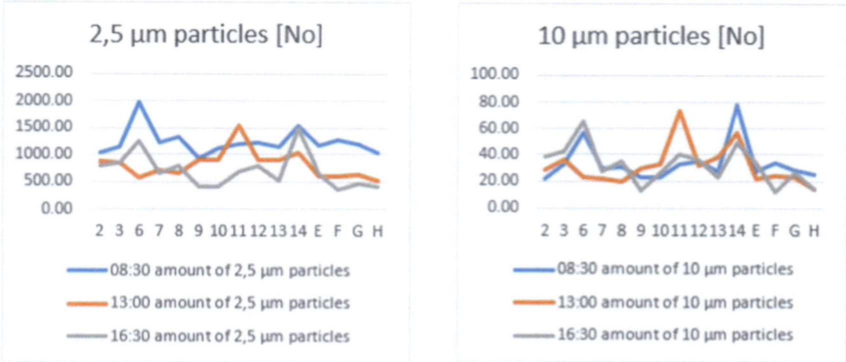

Fig. 9. Air quality within courtyards compared to the outside points (Source: authors)

there are the best. Within the residential blocks, more specifically at the point "F", the lighting conditions are the lowest. Within that courtyard, views are obstructed on all four sides and furthermore, eight-story buildings at the south which cast shadows on nearby buildings is degrading the light conditions. Readings taken in the afternoon session (from 16:30) were not considered since lighting conditions changed rapidly while taking measurements (at points "1–8" and "A, B, C and D") due to the sunset, which occurred at 16:59. For that reason, the data is not comparable.

4.3 Noise Conditions

Based on the collected data, points 1, 4, and 9 exhibit the highest levels of noise. The presence of cars on the site directly affects the noise level at points 1 and 9, which are located next to the traffic signals. Despite point 1 being situated at the main junction, the values at point 9 are significantly greater, which may be unexpected. The variation in sound levels can be ascribed to the phenomenon of sound reflection caused by the close buildings, known as the tunneling effect, at point 9. In contrast, at point 1, the buildings are significantly farther distant from one other. The maximum value is seen at point 4, situated adjacent to the busiest road and equidistant from the traffic lights. Consequently, the majority of vehicles are in constant motion when passing this site, resulting in increased noise levels. Furthermore, due to the proximity of tram traffic, the noise levels on this side of the site are expected to be highest in the south-eastern area.

5 Conclusions

Despite the higher traffic load surrounding collective housing units, the measurements of air quality and noise that were taken in the Municipality of Novo Sarajevo confirm that the wintertime air quality around individual residences is worse than the air quality in other parts of the site. The influence of traffic on the buildings in the surrounding area during the winter is not nearly as substantial as the effect of the actual means of heating for the dwelling units, despite the fact that traffic is a contributor to air pollution (due to the higher frequency of cars). In addition, we are able to draw the conclusion from the data that was collected that, in the absence of strong air flow, the air quality within urban blocks (inner courtyards) is lower than the air quality located outside the perimeter. In such a setting, closed building layouts have a more favorable impact on the air quality in comparison to linear building layouts, which are open to the environment. For further research, scenarios with significant air flow should be the primary emphasis, and data from sites that are comparable should be compared. In addition, additional research should be conducted to determine the extent to which the amount of traffic does, in fact, have an effect on the air quality; hence, the measurements should be repeated once more during the summer months.

We must consider several methods to enhance the quality of this research. Initially, it is important to obtain supplementary measurement equipment to enable the simultaneous recording of data at all locations, which would be the optimal situation. Ideally, there should be a minimum of ten measuring devices provided, allowing each individual to document four local locations within a four-minute timeframe using the equipment. The outcome of this will yield more precise data that is impacted by meteorological conditions such as air velocity, illumination, and pollution levels. A second aspect to take into account is the possibility of conducting measurements at various elevations. The presence of pedestrians and moving cars impacts the conditions at ground level. However, a more accurate spatial map can be created by using a drone to take measurements at different heights of buildings. The collection of these measurements may span multiple days, contingent upon the prevailing weather circumstances, such as variations in air pressure (either low or high) and the presence or absence of wind. Furthermore, it would be beneficial to conduct a more extensive analysis of vehicle counts over an extended duration and observe the fluctuations in frequency.

References

AG, I. (Ed.). Air quality in Sarajevo. Retrieved from Air quality index (AQI) and PM2.5 air pollution in Sarajevo (2023). https://www.iqair.com/bosnia-herzegovina/federation-of-b-h/sarajevo

Arnautović-Aksić, D., et al.: Typology of residential buildings in Bosnia and Herzegovina. (D. Arnautović-Aksić, & N. Zagora, Eds.) Sarajevo, Bosnia and Herzegovina: Faculty of Architecture, University of Sarajevo (2016)

Džananović, M.: Phases of urban development in Zenica during socialist Yugoslavia. (1945. - 1992. year). ResearchGate, 133–153 (2016). Retrieved from https://www.researchgate.net/publication/328175135_Faze_urbanistickog_razvoja_Zenice_za_vrijeme_socijalisticke_Jugoslavije_1945_-_1992_godine

Federal Civil Protection Administration. Law on Fire Protection and Firefighting. Official Gazette of the Federation of B&H, 64(09), 30–58 (2009). Retrieved from http://www.fucz.gov.ba/zakon-o-zastiti-od-pozara-i-vatrogastvu/

EPFFB&H. Environmental Protection Fund of the Federation of B&H. Retrieved from Energy efficiency (2014). http://www.fzofbih.org.ba/v2/news.php?id=3

Grabrijan, D., Neidhardt, J.: Architecture of Bosnia and the Way Towards Modernity. Sarajevo (1953)

City Sarajevo History. Retrieved from Sarajevo in the period 1945–1991 (2023). https://www.sarajevo.ba/bs/article/1222/sarajevo-u-periodu-1945-1991-godina

Hadžiabdić, M., Midžić-Kurtagić, S., Arnaut, S., Begić, T., Ćorović, F.: Green action plan of Sarajevo Canton. Sarajevo: European Bank for Reconstruction and Development (2019). Retrieved from https://mpz.ks.gov.ba/sites/mpz.ks.gov.ba/files/studija_o_urbanim_ventilacionim_koridorima_i_uticaju_visokih_zgrada.pdf

Sapena, M., Wurm, M., Taubenböck H., Tuia, T., Ruiz, L.A.: Estimating quality of life dimensions from urban spatial pattern metrics, Computers, Environment and Urban Systems, vol. 85 (2021). ISSN 0198-9715, https://doi.org/10.1016/j.compenvurbsys.2020.101549

Hood, R., Ćosić, D.: Bosnia and Herzegovina: Challenges and recommendations for reforms. The World Bank (2012)

IEA. Buildings (2022). Retrieved from https://www.iea.org/energy-system/buildings

Islambegović, V.: Beyond Context: Three Architectural Case Studies From Bosnia and Herzegovina. Sarajevo: Faculty of Architecture, University of Sarajevo (2022). Retrieved from https://af.unsa.ba/pdf/publikacije/V.Islambegovic.Beyond.Context_Three.Architectural.Case.Studies.from.Bosnia.and.Herzegovina.pdf

Kurto, N.: Sarajevo MCDLXII-MCMXCII. Sarajevo: OKO (1997)

Kurto, N.: Architecture of Bosnia and Herzegovina The development of the Bosnian style. (M. Mažar-Numankadić, & I. Krzović, Eds.) Sarajevo: Sarajevo publishing International Peace Center (1998)

Milivojević, J., Cvetić, T., Kokić Arsić, A., Nikolić, N., Petronijević, M.: QUALITY OF LIFE IN CITIES – RESEARCH ASPECT. 2nd International Conference on Quality of Life, (pp. 101–111). Kragujevac (2017). Retrieved from https://www.researchgate.net/publication/321886201_QUALITY_OF_LIFE_IN_CITIES_-_RESEARCH_ASPECT

Moneo, R.: On Typology, Oppositions, A Journal for Ideas and Criticism in Architecture, The MIT Press (1978)

UN. Good practices for energy-efficient housing in the UNECE region, United nations economic commission for Europe - United Nations Human Settlements Programme, New York and Geneva (2013). https://unece.org/DAM/hlm/documents/Publications/good.practices.ee.housing.pdf

Yonk, R.M., Smith, J.T., Wardle A.R.: Building a Quality of Life Index in Quality of Life and Quality of Working Life ed. Boas, A.A.V., IntechOpen (2017). ISBN 978-953-51-3446-6, https://doi.org/10.5772/6519

Author Index

Abbosova, Nozima. *See* 207
Abdelfatah-Aldayyat, Ebithal. *See* 151
Aliev, Rayimjon. *See* 207
Alsayed, Shaimaa S.. *See* 15
Álvarez Domínguez, Servando. *See* 247
Anand, Rishikesh. *See* 91
Araiza, Edgar A. Rivas. *See* 359
Attia, Shady. *See* 3

B. He et al. (eds.), *Towards the Framework of Livable and Resilient Cities* Advances in Science, Technology
& Innovation, https://doi.org/10.1007/978-3-031-97849-4

FSC
www.fsc.org
MIX
Papier aus verantwortungsvollen Quellen
Paper from responsible sources
FSC® C105338

If you have any concerns about our products,
you can contact us on
ProductSafety@springernature.com

In case Publisher is established outside the EU,
the EU authorized representative is:
Springer Nature Customer Service Center GmbH
Europaplatz 3, 69115 Heidelberg, Germany

Printed by Libri Plureos GmbH
in Hamburg, Germany